Tropical Forage Legumes

Harnessing the Potential of *Desmanthus* and Other Genera for Heavy Clay Soils

---

FSC
www.fsc.org
MIX
Paper from responsible sources
FSC® C013604

Tropical Forage Legumes

Harnessing the Potential of *Desmanthus* and Other Genera for Heavy Clay Soils

Edited by

John R. Lazier

Formerly International Livestock Centre for Africa

Nazeer Ahmad†

Formerly University of the West Indies, Trinidad and Tobago

CABI

CABI is a trading name of CAB International

CABI
Nosworthy Way
Wallingford
Oxfordshire OX10 8DE
UK

Tel: +44 (0)1491 832111
Fax: +44 (0)1491 833508
E-mail: info@cabi.org
Website: www.cabi.org

CABI
745 Atlantic Avenue
8th Floor
Boston, MA 02111
USA

Tel: +1 617 682 9015
E-mail: cabi-nao@cabi.org

A catalogue record for this book is available from the British Library, London, UK.

Library of Congress Cataloging-in-Publication Data

Names: Lazier, John R., 1941- editor. | Ahmad, N. (Nazeer), 1932- editor.
Title: Tropical forage legumes : harnessing the potential of Desmanthus and other genera for heavy clay soils / edited by John R. Lazier, Nazeer Ahmad.
Description: Boston, MA : CABI, [2016] | Includes bibliographical references and index.
Identifiers: LCCN 2016002563 (print) | LCCN 2016005979 (ebook) | ISBN 9781780646282 (hbk) | ISBN 9781780646299 (ePDF) | ISBN 9781780648774 (ePub)
Subjects: LCSH: Legumes. | Forage plants.
Classification: LCC SB203 .T88 2016 (print) | LCC SB203 (ebook) | DDC 633.3--dc23
LC record available at http://lccn.loc.gov/2016002563

ISBN-13: 978 1 78064 628 2

Commissioning editor: Rachael Russell
Editorial assistant: Emma McCann
Production editor: James Bishop

Typeset by SPi, Pondicherry, India
Printed and bound in the UK by CPI Group (UK) Ltd, Croydon, CR0 4YY

Contents

Five *Desmanthus* species

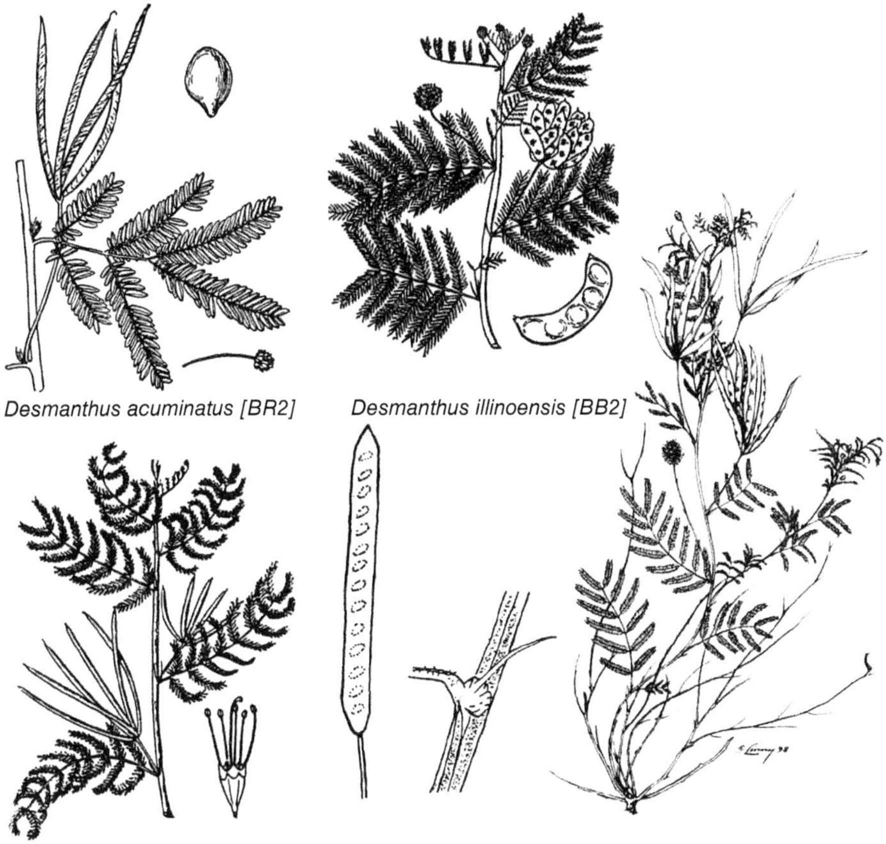

Desmanthus acuminatus [BR2] *Desmanthus illinoensis [BB2]*

Desmanthus leptolobus [BB2] *Desmanthus virgatus [F&4]* *Desmanthus velutinus [HEA]*

Source: Jardin Botanico Viera y Clavijo, Cabildo de Gran Canaria. Reproduced with permission.

This book is dedicated to Robert L. Burt and Nazeer Ahmad who were personal friends and valued colleagues. Bob's untimely death shortly after stimulating the commencement of work on this volume has denied us his rich experience and deep knowledge in its preparation. Nazeer Ahmad provided guidance on its preparation and was arranging publication when, after a sudden and brief illness, he too passed away.

Right to left: Nazeer Ahmad, Bob Burt, John Keoghan in Belize, at Mountain Pine Ridge Rio On strip trial, Site 12 (see Chapter 15, this volume). In September 1978, *Pueraria phaseoloides* covered the ground and was climbing the trees because of the fertility added by the trial, greatly increasing the biomass and nutritive value of the fodder available.

Cover photo Steers grazing *Desmanthus* 'Progardes' in Queensland, Australia. Photo by Chris Gardiner

John R. Lazier
15 Lawson Road
Courtice, Ontario
Canada, L1E 2K3
Tel: 1-905 438 8176
john.lazier@sympatico.ca

List of Line Drawings of Species with Forage Potential

Foreword

The research reported here is mainly the result of collaboration between the Crops and Pastures Division of the Commonwealth Science and Industry Research Organization of Australia (CSIRO) and the Soil Science Department of the University of the West Indies in Trinidad. It involved germplasm collection in centres of legume origin in the Caribbean and South, Central and North America, with initial characterization and screening of the acquired germplasm for utility as fodders at multiple heavy clay and other sites in the Caribbean, Central America and Australia. Since then the germplasm has had wide exposure in Africa and South America. From the collection and research undertaken both during the period reported here and thereafter, numerous species have proven useful and have been widely disseminated. Of the native plants, *Acacia angustissima*, *Desmodium cinerium* and species of *Desmanthus* and *Centrosema*, among others, and *Codariocalyx gyroides* (an exotic line) have all had their utility demonstrated and niches for their use identified globally.

These chapters were prepared on the initiative of Bob Burt. He had been very aware of the lack of adapted leguminous forage species for heavy clay soils and for many years was active in the process of collection, evaluation and domestication of promising material for these environments. In recent years, the area of research in plant introduction, which had large and vibrant development programmes in a number of countries, has fallen out of fashion worldwide and the number of researchers involved and the level of activity in such work have been greatly reduced. In addition, because much of the research had been published before the advent of widespread computerization, most is unavailable on the internet and so is not easily accessed by the few scientists who are still working in the field.

After some years of retirement Bob decided to make available the results of the work he had already undertaken but that had not been published, so that it would not be lost. I had not had direct contact with him for 22 years when he phoned me in September 2007 and began persuading me to work with him in publishing the results of the research based on the collections undertaken under the IDRC-funded University of the West Indies Forage Legume Programme. At that point I had not been directly engaged in forage research for 18 years and was reluctant to comply. He was very concerned that the scientific community be aware of the importance and potential benefits of further research and development of new species and genotypes, and that the research completed clearly pointed the way to further work, and promised considerable benefits to productivity on these soils. Sadly, relatively early in the preparation of this book, his health suddenly declined and he passed away on 11 February 2009. His chapters were completed based mainly on his handwritten drafts and his original field notes.

Prof. Nazeer Ahmad was Head of the Department of Soil Science at the University of the West Indies for many years and a world authority on heavy clay soils. In the 1970s he and Bob Burt were instrumental in developing a collaborative forage research programme with CSIRO, Australia. The resulting project was located in Trinidad, Antigua and Belize, and funded by IDRC Canada. As coordinator of the project he provided leadership and technical support to the three researchers involved, one of whom was me. After my departure from Belize in 1977, the project continued until 1986 and was successful in expanding the range of germplasm with forage potential for Belizean pastures and similar environments globally. A number of the species and genotypes collected have since proven their value in livestock operations both in the Caribbean and in other tropical regions. After retirement, Nazeer returned to Guyana, but then took up an Emeritus position at the University of the West Indies. A fuller account of his accomplishments is given at http://www.niherst.gov.tt/icons/profiles/nazeer-ahmad.htm

It was Bob Burt's wish that all of those involved in the research in the Caribbean and Australia be involved in the production of this book. However, in the end only he, Nazeer and I have been able to contribute. During the writing and shaping of this volume, Nazeer provided encouragement and advice. He also sought support in the University of the West Indies for them to publish it electronically. We had a final meeting in April 2013 when he was briefly in Canada, and he provided advice on the final publication details. Unfortunately at the end of his visit he suddenly experienced symptoms of the illness that resulted in his death on 6 June 2013, shortly after his return to Trinidad.

Kendrick Cox and Chris Gardiner have made a major contribution to this volume by summarizing the more recent Australian environment for research on tropical forage legumes and the results achieved, particularly with regard to *Desmanthus*, in Queensland, a major centre of research and development in tropical livestock production.

John Lazier

The authors

N. Ahmad† Formerly Head, Department of Soil Science, University of the West Indies, St Augustine, Trinidad and Tobago

R.L. Burt† Formerly CSIRO, Davies Laboratory, Townsville, Queensland; James Cook University, Townsville, Queensland, Australia

K.G. Cox Queensland Department of Agriculture and Fisheries, PO Box 1054, Mareeba, QLD 4880, Australia. E-mail: Kendrick.Cox@daf.qld.gov.au

C.P. Gardiner Livestock Production Systems, College of Public Health, Medical & Vet Sciences, James Cook University, Townsville, Queensland, Australia. E-mail: Christopher.Gardiner@jcu.edu.au

J.R. Lazier Formerly University of the West Indies/IDRC Forage Legume Project Belize; ICARDA, Syria; ILCA Nigeria and Ethiopia. Currently: 15 Lawson Road, Courtice, Ontario, Canada L1E 2K3. E-mail: john.lazier@sympatico.ca

Contributors to the research

W.M. Beattie Formerly CSIRO Division of Tropical Crops and Pastures, Davies Laboratory, Townsville, Queensland, Australia. Now: Box 212, Freshwater, Cairns, Australia 4872. E-mail: wmbeattie@hotmail.com

J. Carter Formerly Queensland Department of Primary Industries, Australia

J.M. Keoghan Formerly with the University of West Indies/IDRC project in Antigua, West Indies. Now Rolleston, Canterbury Plains, New Zealand. E-mail: enquiries@gardenviewbandbnz.com

M. Long Formerly Head, Central Farm Analytic Laboratory, Central Farm, Belize

A.C. Schlink Formerly CSIRO Livestock Industries, Wembley, Western Australia, Australia. Now: 29 Carrick St., Woodlands, Western Australia, Australia

W.T. Williams† Formerly CSIRO, Davies Laboratory, Townsville Queensland; University of Queensland, Townsville, Australia

Acknowledgements

Contributions to this book

The editors acknowledge the provision of information on the success of *Desmanthus* lines in a number of countries by Alan Robertson, and advice and comments on parts of the initial manuscript provided by one of the authors of a chapter in this volume, Chris Gardiner of James Cooke University. Belize reports have been reviewed for accuracy by the Belizean forage experts Mr Rodney Neal and Mr Harold Parham.

Australian/international studies (Bob Burtt)

Acknowledgements for contributions to the research reported herein include: W.T. Williams in particular for assistance with the analysis and interpretation of the data for Chapters 3, 5, 6, 7, 8 and 9 as well as J.M. Keoghan (Chapters 3 and 7), W.M. Beattie (Chapters 5 and 8) and J. Carter and A.C. Schlink (Chapter 9).

Some of the work detailed, and many of the philosophies embraced, were developed over a long period of time. In this time it has been possible to acknowledge the inputs of many colleagues and technical staff, because, in dealing with 'new plants', many skills are needed and great powers of observation are required. But it has not been possible to adequately acknowledge similar inputs by the owners and users of land in those areas from which the plants were collected and to thank organizations and staff of the various institutes in these countries. I am particularly grateful to SUDENE, Adelberto, Prof. Geraldo Pinto and the car driver – an unenviable task on a plant-collecting mission – and the people of Brazil. And to Bill Bryant in Venezuela, the IDRC teams (including Dan) in Trinidad, Belize and Antigua and Prof. N. Ahmad of the University of West Indies. In Australia I was given long-lasting support and encouragement by the Australian graziers and their organization, the AMRC. Graziers such as 'Bunty' Allingham, Harry Clark and Jack Brownson taught me a very great deal about the nature of the environment and the vegetation. And my two 'god professors', A.H. Bunting and W.T. Williams, showed great patience and provided encouragement and support as required. Finally, I gratefully acknowledge the total support afforded by family, past and present throughout. My thanks to all.

The Belize studies (John Lazier)

Numerous organizations and individuals contributed to the success of the project and to the results reported here.

IDRC provided the financial resources and enthusiasm for initiating and supporting the project for the period reported here and for a further period until 1985, a total of some 13 years. Staff involved in providing critical financial, technical and monitoring support included Dr Barry Nestel, Dr Jose valle Riestra and Mr Joe Hulse. The financial support to me extended to 9 months of analysing data at the Davies' Research Laboratory in Townsville Australia and time in the UK mainly to identify the accessions at the Royal Botanic Gardens Kew with the support of Dr Roger Pohill, Dr Bernard Verdcourt and Dr Steve Renvoize.

The University of the West Indies in Trinidad was the executing agency of the three project locations and provided an employment niche, administrative back-up, and the Coordinator, Prof. Nazeer Ahmad, Head of the Soils Department, who provided technical guidance and much other advice and support. The Central Analytic Laboratory at the University of the West Indies campus in Trinidad undertook the analysis of about 5000 plant and soil samples. Access to the extensive Australian experience in tropical pastures, which was fundamental to the success of the project, was provided with great humour and sensitivity by the IDRC technical monitor Dr Bob Burt of the CSIRO Davies Laboratory, Townsville.

Within Belize, tremendous assistance was received from the Government of Belize, which contributed financially to its support. The Chief Agriculture Office Mr Eric King, the Pasture Officer, Mr Rodney Neal, and many other members of the administrative, research and technical staff, too numerous to mention here, contributed either directly or indirectly. Mr Neal joined in one of the collection missions to the Yucatan, as did Dr John Keoghan who managed the Antigua portion of the project. Mr Mark Long and the Central Farm Analytic Laboratory assisted through the analysis of the fodder samples reported in Chapters 13 and 18. The project staff all had specific responsibilities and carried these out conscientiously with minimum guidance. Without them and their cheerful natures and hard work, there would be no data and the project would not have been successful. They contributed not only to the mechanics of the project but also to shaping it. They did some collections as well. For example, Mr Dan Bellini collected *Desmodium cinerium*, which has potential as a cut and carry fodder.

For 9 months in 1977–1978 I was fortunate to be sponsored by IDRC to analyse the Belize data at the CSIRO Davies Laboratory in Townsville where all of the staff were very open and generous in providing valuable advice, and assistance with analysis of the data. Besides Bob Burt, major contributors included Dr W.T. Williams, Mr John Clay, Dr R. Jones and Mr A. Watkinson. The computer section undertook manual double-digitalizing of vast amounts of data. Further IDRC support through subsequent years enabled screening of the project's germplasm in Nigeria and Ethiopia.

The Queensland studies: Chapter 11, the Progardes story (Chris Gardiner)

The assistance and collaboration of Agrimix Pty Ltd, JCU School of Veterinary and Biomedical Sciences and Dr Ross Coventry (Soils Horizons Pty Ltd.) are kindly acknowledged.

1 Providing Pasture and Ley Legumes for Use on Clay Soils in Tropical and Subtropical Environments

R.L Burt†, J.R. Lazier*[1] and N. Ahmad†
*Formerly International Livestock Centre for Africa

Almost 50 years ago, the FAO published a very informative review on the dark clay soils of the tropics and subtropics (Dudal and Bramao, 1965). It showed that the tropical and subtropical worlds contain very large areas of alkaline and clay soils; 83 million ha, one-third of the clay soils in sub-Saharan Africa, are vertisols; the Republic of the Sudan has 40 million ha of 'dark clay soils', India has 60 million ha and Australia 70 million ha. The soils are generally characterized by higher montmorillonitic clay content and are often situated under wetter conditions. They occupy very large areas and are of major importance for cropping and pastures because of generally higher fertility and their ability to 'hold water' and thus extend the growing season. Indeed, plants that form effective rhizobial relationships with bacteria from such soils are deemed to have 'special rhizobial relationships'.

It has long been known that such soils are of huge agronomic significance for all types of farming (e.g. Dudal and Bramao, 1965). Eponymously termed black cotton soils, these soils are justly world famous, however in India smallholder cotton farmers have been faced with major challenges, including rising costs of inputs, erratic rainfall and an inefficient manufacturing industry. It may thus be necessary for such farmers to diversify their production in order to ensure income security (Ramesh *et al.*, 2004). Yet sources of N for pasture growth and animal nutrition are currently largely limited to chemical fertilizers, which are uneconomic for many pasture-based livestock operations. Dudal and Bramao point to the need to utilize productive legumes to assist in improving productivity on these soils through enhanced soil fertility, organic matter content and greater water retention, and as a source of high-quality fodder. While there is already considerable interest in greater use of legumes in 'no-till agriculture' and 'organic farming', many livestock production systems are currently largely based on legumes. Their use can increase animal production levels and allow other land to be conserved. However, there is a general lack of pasture and ley legumes that are productive and persistent for such edaphic environments, and the need is worldwide not just restricted to India (Clem and Hall, 1994; Pengelly and Conway, 2000).

The need to collect, conserve and utilize the plant genetic resources of the world has long been recognized: one of the earliest recorded plant-collecting missions is that organized by Queen Hatshepsut of Egypt in 1481 BC (e.g. http://www.plantlives.com/docs/A/Ammi_majus.pdf). Today the importance of plant biodiversity is widely acknowledged and there is both national and international attention being paid to the collection and maintenance of genetic variability throughout the world.

[1] E-mail: john.lazier@sympatico.ca

In the 1960s the lack of adapted legumes for tropical and subtropical pasture conditions generally (e.g. Burt and Williams, 1975) resulted in the development of worldwide forage germplasm collection and evaluation programmes. In Australia, programmes initiated in the late 1960s had, by the early to mid-1970s, resulted in the commercial release of previously undomesticated species, mainly of the genus *Stylosanthes*. The International Centre for Tropical Agriculture in Colombia (CIAT), which commenced research in 1969, also identified and released domesticated lines of legumes suitable for livestock fodder, mainly additional *Stylosanthes* species. The bulk of the research was focused on adaptation to acid, sandy-textured soils, which are of major importance in the higher rainfall areas of Australia and South America. Today, *Stylosanthes* is undoubtedly the genus of major utility for tropical pastures throughout the tropics and the genus most widely tested on clay soils (e.g. Ramesh *et al.*, 2004 for India). Because little emphasis has been placed of the provision of legumes adapted to dark clay soils, the genotypes of productive legumes identified and disseminated are in general not very productive or persistent on them (Cook *et al.*, 2005).

As there are only a limited number of pasture cultivars currently available and these do not provide varieties that are consistently productive over the range of farming systems and environments requiring them, there is a need for new cultivars to expand the range of environments in which legumes can increase the efficiency of cropping and livestock production. These cultivars may be found in species, or within species' forms already in use, and could provide such characteristics as improved disease resistance or slightly different agronomic characteristics (e.g. Jones and Clem, 1997). However, other situations, such as the hugely important tropical clay soil areas of Sudan, India and Australia (Dudal and Bramao, 1965), require the development of 'new genera' and 'new species' (Burt, 1993a; Clem and Hall, 1994; Chapter 6, this volume). There is a similar need for new legumes for use in ley cropping systems (Pengelly and Conway, 2000).

Enthusiasm for the development of leguminous species for pastures grew from the 1960s to the 1980s, after which interest faded for a number of reasons: new areas of research became popular and attracted funds, it was thought that the genetic variation had been well sampled, a lack of interest in agricultural development in general and livestock and complex pasture and ley agricultural systems in particular. National and international funding has dried up, and scientific research establishments drastically reduced or eliminated.

This need for adapted forage legumes is exacerbated by the rapidly rising price of chemical fertilizers because of the high cost of energy, the increasing awareness of the pollutant effects of these fertilizers, the pressure for greater productivity on the limited global arable land resources and the deteriorating condition of much of the world's cultivated soils. The identification of productive and persistent species and genotypes for heavy clay soils is increasingly threatened by expanding land use and global climate change, which result in the loss of potentially valuable native plant germplasm. These facts emphasize the urgent need for the collection and conservation of genetic resources with agricultural potential.

The requirement for new adapted and productive legume germplasm is likely to increase with global warming and climatic change. Even slight rises in temperature or changes in patterns of rainfall are already causing reductions in the livestock carrying capacity of many traditional grazing areas, resulting in overgrazing, plant loss and soil erosion (Anon, 2006; DeMeo, 2006). Re-establishing native pasture species, many of which were poorly adapted in the first instance (Moore, 1967) – and as shown herein – may not always be possible. Diversification of the native sclerophyll vegetation to meet such climatic changes, such as that which occurred during earlier long-term climate changes in Western Australia (White, 1986, p. 191), may well be too slow to cope with the relatively rapid changes of climate likely to occur (e.g. Bättig, 2007). Such changes are likely to be most rapid and most severe in tropical South America, the region most likely to be the source of new pasture legumes of value (Burt and Reid, 1976).

A review of available leguminous fodder species (Cook *et al.*, 2005) has shown that, as had been suggested earlier by Dudal and Bramao (1965), there are few leguminous species available for heavy clay soils: 'Legumes have not yet come to occupy the place they deserve for soil improvement, pasturage and fodder production.

Research and extension in this field are still needed for tropical and subtropical countries' and, as will be shown subsequently in this book (Chapter 6, this volume), development of legume use in agricultural production in the tropics lags far behind the temperate areas. As an example, the genus *Medicago*, which has provided many temperate legume cultivars for use on clay soils, has a USDA genetic resource collection of over 7000 accessions, while *Desmanthus*, its tropical equivalent (see later chapters in this volume), has 300 accessions in the largest collections.

Desmanthus is a leguminous genus whose species have particular potential for clay soils and has had little research attention. What information has been available has indicated that they have important potential as long-term pasture legumes (e.g. Clem and Hall, 1994), as ley legumes in crop/pasture systems (Cook *et al.*, 2005), as cut and carry fodder in fodder banks, and for revegetation programmes (e.g. Muir *et al.*, 2005 for use of both native and introduced *Desmanthus* species to regenerate and renew blackland prairie in Texas). A further approach to utilizing legumes and their genetic variation is the development of pulse/forage crops (e.g. Glover *et al.*, 2007); species such as *Desmanthus illinoensis* can be used as a perennial pulse crop and soil cultivation avoided.

The research reported herein is international in scope; it involves two Australian programmes, one designed to provide pasture legumes for use on tropical clay soils and one undertaken to provide tropical legumes in general (Chapter 6, this volume). It also involves an IDRC-sponsored programme at the University of the West Indies (UWI) based in Belize, Antigua and Trinidad (Spurgeon, 1977), which was developed based on the need for such plants in the West Indies and Central America. The Caribbean has areas of economically important heavy clays, although these are too small to appear on the world soil map used by Dudal and Bramao (Fig. 1.1). These are of particular potential for global pasture development as their unique and diverse leguminous flora have provided adapted material of potential for similar soils in tropical and subtropical environments worldwide. The UWI research was loosely coordinated with that undertaken in Australia. Many of the results obtained have remained unpublished or, if published, not in a format that allows their ready appreciation. This volume seeks to address this situation.

The research also updates information provided by Dudal and Bramao, particularly that which relates to the distribution and major characteristics of the soils involved. It is now possible to give some specific examples of areas with heavy clay soils, problems involved in utilization, and relevant programmes undertaken to overcome these. Mention is made, for instance, of mineral nutrition and rhizobial studies, as the result of which legume growth can be made possible or stimulated (e.g. P and trace elements; Chapter 17, this volume), thereby promoting N fixation and alleviating the need for frequent application of costly chemical N fertilizer or scarce animal manure. Such strategic use of fertilizers to improve 'soil health' can be a major factor in promoting a 'Green Revolution' for smallholder farmers and commercial farms in Africa, for example. Appendices to chapters provide more detailed information about the clay soils that have yielded the majority of the important legume accessions currently in genebanks.

A further aim of this volume is to highlight the fact that considerable knowledge exists on numerous leguminous genera with forage potential for tropical and subtropical conditions, but is not readily accessible to today's researcher having been acquired before digitalization of reports. As a result, research may be repeated, and some views may be currently held that are not in accordance with previously established results. The approaches utilized in these studies to identify, collect and screen genera, species and genotypes for particular environments are presented as illustrative of efficient and effective approaches to identification of adapted forage germplasm.

A global perspective was utilized in undertaking the research. It involved the worldwide identification of areas of legume genetic diversity, heavy clay soils and climatic conditions similar to those of target environments. Collection has been undertaken on a broad scale internationally with the collection of legume germplasm from selected previously unexplored environments in South and Central America, and the Caribbean representative of important livestock areas elsewhere in the tropics and subtropics. These were then tested for adaptation and persistence in a range of target environments along with promising materials from earlier global collections.

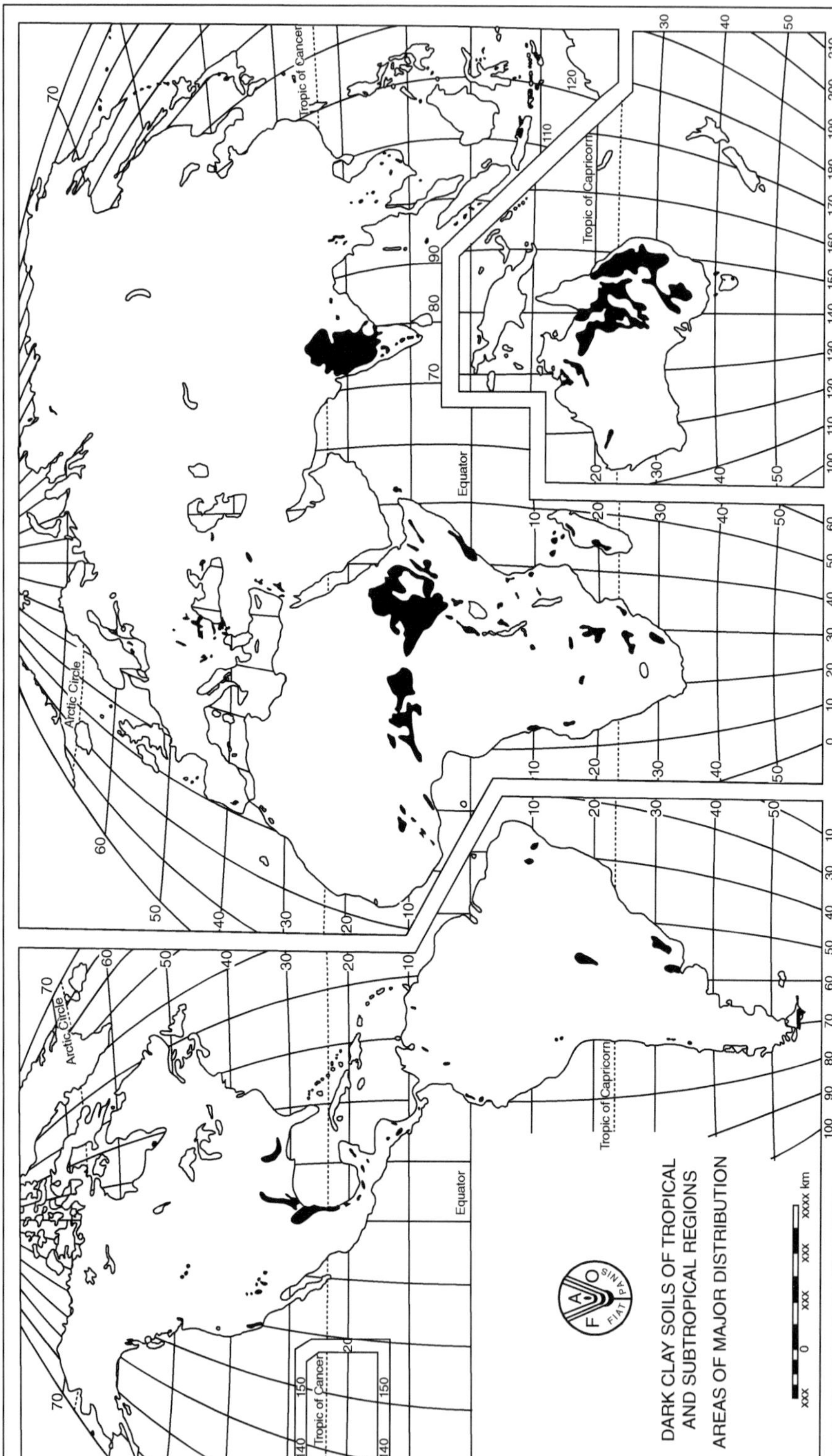

Fig. 1.1. Dark clay soils of tropical and subtropical regions: Areas of major distribution. (Source: Food and Agriculture Organization of the United Nations (Dudal and Bramao, 1965). Reproduced with permission.)

The sites in which the research was undertaken have contrasting rainfall regimes. Central Belize has higher rainfall, 1460–1860 mm, with a variable but relatively brief dry season; Antigua is drier, with 890–1280 mm and a 4–5-month dry season; and western Queensland is driest, with erratic rainfall of 760–890 mm and long dry seasons. As would be expected, accessions of a wide range of genera survived and grew under higher rainfall conditions in Belize, but in the harsh environment of western Queensland only *Desmanthus* accessions persisted. The information gained not only assists in predicting what accessions are of potential value but has also been utilized to forecast what additional genetic material of promise could be collected, where it can be found and in what environments it is most likely to succeed. The authors emphasize that considerable valuable genetic diversity has yet to be exploited.

The organization of the volume

The results of the collaborative research presented herein commences with a report on the collection of a broad range of genetic material of fodder potential in Belize, the Yucatan and Guatemala, the initial evaluation of some of this material in Belize and a brief review of the impact of the collection on forage development over the past 30 years (Chapter 2); in fact this germplasm collection stimulated much of the work reported in this volume. An analysis of the value and potential of *Stylosanthes* genetic material adapted to heavy clay soils follows. This study mainly focuses on *S. hamata* and its relatives, and is based in part on collections made concurrently in the Caribbean islands and South America as part of the collaborative programme (Chapter 3). The Belize collections of *Desmanthus* species stimulated interest in the genus and based on those collections and others made later by other collectors, studies were undertaken on the potential of the genus generally, and specifically for its potential for development for heavy clay soils (Chapters 4 and 5). Chapters 6 through 9 report on the steps undertaken to select and test germplasm for adaptation for the heavy clay soils of the dry Australian tropics and subtropics where *Desmanthus* species were the most successful of those tested. The variation within the genus as related to its distribution and its adaptation to Australian environments is discussed. Chapter 10 provides recent advances in the development of forages for heavy clay soils in Australia. The final series of chapters (11 through 19) describes the Belizean environments for which forage germplasm was being selected and the testing of the regionally collected legume accessions for their adaptation to, and suitability for, pastures on clay soils in a range of environments compared to exotic material. The results reported in this volume have been either part of or based on the collaborative CSIRO–University of the West Indies research activities. Numerous other results have been achieved by these programmes that have been published elsewhere and these are referred to in the text.

2 The Collection and Initial Evaluation of a Wide Range of Pasture Legumes From Mexico, Belize and Guatemala: Implications for Genetic Resource Development for Tropical Countries

J.R. Lazier*[1] and R.L. Burt†

**Formerly International Livestock Centre for Africa*

Abstract

There is a dearth of pasture and ley legume cultivars for use on alkaline and clay soils in tropical areas. Recognizing this, the collaborative Australia–UWI Forage Research Programme, with field studies in Belize and Antigua, took advantage of the proximity of these soils, the rich floras with high endemism and researchers' sustained residency in the area, and assembled a genetic resource collection of potentially useful legumes. This chapter provides details of an initial assessment for local needs of part of this collection undertaken in Belize.

The evaluation of the accessions began at the point of collection, with the acquisition of information on the climates and soils. It was continued with a simple field experiment designed to provide specimens for identification, for the taxonomic status of many was uncertain, and to broadly categorize a selection of the accessions in terms of perenniality, persistence and vigour of growth in the wet and dry seasons under 1460–1860 mm of rainfall. Evaluation was carried out on a heavy clay soil with a neutral pH in a wet tropical environment that contrasts with the much lighter acid soils that most previous forage research had been carried out on.

It was found that some species that had been successful and provided cultivars for wetter conditions elsewhere were also successful here, notably *Centrosema pubescens* and *Macroptilium atropurpureum*, both of which demonstrated much potentially useful variation. *Desmodium incanum*, a species that has shown promise elsewhere, also showed potential. However, *Stylosanthes guianensis* and the widely used dryland species *S. scabra* both failed to survive. 'New species' of particular interest were found in *Galactia* and *Desmanthus*, with the latter genus affording a new form of pasture legume in between the tree legume *Leucaena leucocephala* and suffruticose perennial herbs such as *Desmodium incanum*. The potential of *Desmanthus* as fodder is discussed.

This retrospective analysis, when viewed in the light of more recent findings, illustrates the potential value of a genetic resource component in aid projects. It has resulted, for instance, in the first domestication of the species *Desmanthus pubescens* and has demonstrated that novel genetic variation was captured in this and other species. The results focus attention on potentially useful genotypes both in species already in domestication and others that remain undeveloped. Brief mention is made on the value of germplasm collections generally and their utility in the face of global climate change and the loss of plant biodiversity.

[1] E-mail: john.lazier@sympatico.ca

2.1 Introduction

Much diversity in the family *Leguminosae* is centred on South and Central America, and so this region was the initial focus for plant collection, particularly in South America (e.g. Burt and Williams, 1979a; Edye and Maass, 1997), and most commercialized tropical pasture legumes have originated there. Because South America has limited areas of alkaline soils, few collections have been made from them. However, the Yucatan Peninsula in Central America does have substantial areas of heavy clay soils that are primarily derived from limestone. The climates of the area range from dry tropical, deciduous cactus-thorn vegetation, to tropical semi-deciduous forest. Floristically it is very rich, containing numerous legume species with potential for fodder, many of which are endemic (Good, 1947; Estrada-Loera, 1991). Many of the native genera and species such as *Leucaena leucocephala* and *Macroptilium atropurpureum* have long been recognized as valuable pasture plants. However, the potential of the flora remains incompletely explored; for example, recent evidence shows that 'new' variation may be found in Central America in native species of *Stylosanthes* (see information quoted by Maass and Sawkins, 2004).

The IDRC-funded University of the West Indies project's field operations were undertaken in Belize and Antigua between 1973 and 1977 to develop pasture plants for alkaline, heavy soils in Belize and Antigua, the Caribbean generally and elsewhere in the tropics and subtropics. Collections of native legumes were made in Belize, the Yucatan, Guatemala and the Caribbean islands. The total collection held at Central Farm (CF) in Belize to July 1977 numbered some 670 accessions, and with further IDRC funding by January 1981 the collection had expanded to 1245 accessions (see Appendix 2.3). This very important genetic resource is now housed, at least in part, in more general national and international germplasm collections.

It has been suggested that evaluation should begin at the point of collection, where, in order to help predict environments to which the plants will be adapted, information on climate, soil, associated species and local anecdotal information have commonly been recorded (Burt *et al.*, 1979). Knowledge of soil characteristics has been used to predict likely edaphic adaptation (cf. Burt *et al.*, 1979; Cook *et al.*, 2005), as has associated species. In the latter instance, for example, *Stylosanthes scabra* has been shown to be associated with specific grass species, thus assisting in predicting its environments of adaptation (Burt *et al.*, 1983). Locally gained anecdotal information about such relevant characteristics as animal acceptability is also useful in predicting plant utility elsewhere, a feature that can be readily incorporated into new 'search engines' using Bayesian statistics.

This chapter presents work that was undertaken in the first stage in the process of germplasm evaluation, research on the collection sites of part of the Belize Central Farm germplasm collection. This was followed by a simple, short-term field experiment designed to show the main performance characteristics of a range of these accessions, often 'new' to agriculture, in wet tropical conditions on a heavier soil in order to identify species and genotypes for later, more rigorous agronomic testing (as shown by other chapters in this volume and later references). The initial experiment also provided the opportunity to obtain seed for multiplication, and pressed specimens for taxonomic classification[2].

These initial results demonstrated that some of the new native material had considerable potential for development for use in farming systems in the environment in which they were tested, and some commonly used species and cultivars were unsuitable. The results are clearly of importance, but have not been previously published. With the benefit of research carried out more recently, as reviewed, for instance, by Cook *et al.* (2005), it is possible to review the value of the studies reported herein and to suggest where other studies might be of value. This is done in section 2.4 'General discussion', in this chapter.

2.2 Materials and methods

2.2.1 Collection sites

Belize and the adjacent areas of Mexico and Guatemala are climatically diverse and their herbaceous legume floras were poorly known, so there

[2] Kindly undertaken by Drs Bernard Verdcourt and Roger Polhill at Royal Botanic Gardens, Kew, Richmond, Surrey TW9 3AB, UK.

was interest in collecting throughout the area. A transect was thus collected from the very north of the Yucatan Peninsula to southern Guatemala near the Pacific Coast. The climates are driest at the north-west tip of the Yucatan Peninsula in the area of Progreso, become somewhat wetter toward the north-east, and then substantially wetter to the south (Fig. 2.1). An area of particular interest in the north is located between Ticul and Pisté (Fig. 2.1), where there is a wetter climate, similar to those found in northern Belize and neighbouring Guatemala, and reportedly contains dark clay soils (Dudal and Bramao, 1965). The climates of southern Belize and neighbouring areas of Guatemala are moist to wet-tropical and support rainforests. Further south in Guatemala, collections were obtained in a transect from the Caribbean lowlands to the highlands in the Guatemala City area, a 'terra templada' type of climate, and from a warmer, wet-tropical lowland climate further south, near the Pacific Ocean.

The initial collection relied heavily on road transects and roadside collections because in many areas intensive agriculture, overgrazing or dense bush made other forms of collection difficult. More intensive collection was then carried out in areas of particular interest as determined by considerations of the climates, the occurrence of interesting plant species or the soils.

Intensive collection was also carried out in Belize because the sub-project was based there; legumes were abundant and it was possible to revisit sites and collect plants when they were seeding. Useful agronomic observations were also made; for example, a native *Desmanthus* was found growing in a highly competitive Pangola grass sward (*Digitaria eriantha*) in a grazing experiment. This was significant because persistent companion legumes for this grass are very limited. On another occasion when small amounts of fertilizer were applied to a very difficult, infertile, waterlogged soil, carrying an exceptionally restricted range of unproductive, low-nutrient content grass species, native legumes appeared.

The soil types associated with the various collections were recorded (Table 2.1; Appendix 2.2). Most of the soils on the Yucatan Peninsula are derived from limestone and are alkaline. The most common soils of the northern portion of the Peninsula are rendzinas and these are said to be atypical in that they have above average clay content (Gama Castro, 2006) with primary and secondary carbonates low or absent. To the uninitiated plant collector, these may simply be referred to as 'alkaline red clays', if noted at all, and without analytical data this renders comparison of results from different collectors difficult. However, where such analytic data are available, much can be learned about the various species and their likely edaphic adaptation elsewhere (cf. results in Burt *et al.*, 1979; Cook *et al.*, 2005; and for this project in Antigua, Chapter 7, this volume).

2.2.2 Climatic comparisons

It is well known that plant distribution can be affected by many environmental variables: climatic, edaphic, management system, and for legumes, the availability of a suitable rhizobial flora. Experience has also shown that some genotypes and species can be much more tolerant of a wide range of environmental conditions than others. It would greatly simplify germplasm evaluation and testing if it were possible to predict or hypothesize the climates and soils to which plant material is likely to be adapted early in the assessment process.

Such predictions are likely to be most accurate when, as with many crop plants, the climatic adaptation and limitations of the plants concerned are well known. Papadakis (1966), by combining the requirements for suites of crop plants, was able to produce a classification using climatic data as the primary grouping tool. This approach has proven to be useful on a world scale. With less well-known genetic material, as is the case here, it may be possible to display climatic types using multivariate analysis (e.g. Burt *et al.*, 1976; Burt, 1983; Sawkins *et al.*, 1999). By utilizing the climatic parameters derived from the sites of collection, some indication of genotypic adaptation can then be hypothesized. The same procedure has been proven to be useful for grouping genetic material using edaphic information (Burt *et al.*, 1979).

The analysis of climates included data from 580 of the 670 legume accessions acquired over the period of the UWI–Belize Forage Programme Phase I (1972–1977). The accessions utilized originated from Belize (189), the Mexican Yucatan Peninsula (304), Guatemala (30) and the West Indies (26). The 31 remaining were cultivars or accessions that had performed well elsewhere and could also be regarded as 'standards'.

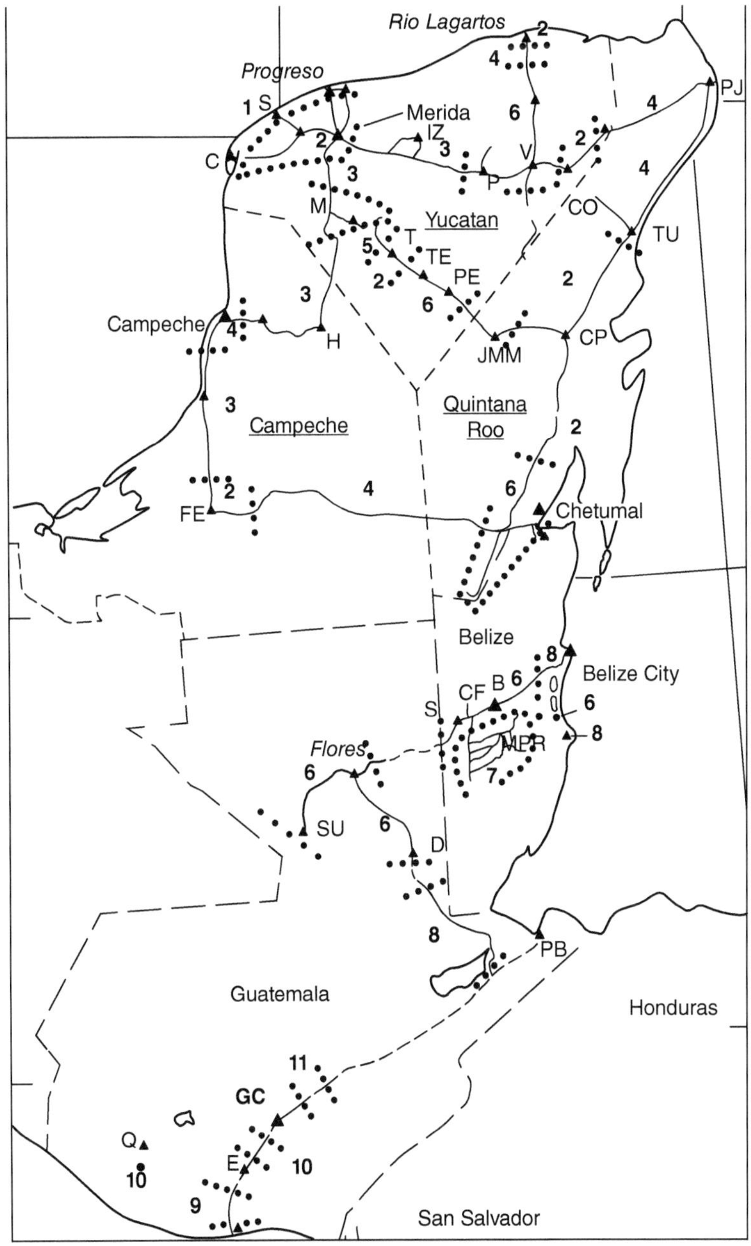

Underlined names: States of Mexico,
Solid triangles: towns:

Mexico: C - Celestun, CO - Coba, CP - Felipe Carillo Puerto, FE - Francisco Escarga, H - Hopelchen IZ - Izamal, JMM - Jose Maria Morelos, M - Muna, Pe - Peto, P - Pisté, PJ - Puerto Juarez, Progreso, Rio Lagartos, S - Sisal, T - Ticul, TE - Tekax, Tu - Tulum, V - Valladolid,

Belize: B - Belmopan, CF - Central Farm, MPR - Mountain Pine Ridge, S - San Ignacio

Guatemala: D - Dolores, E - Escuintla, GC - Guatemala City, PB - Puerto Barrios, SU - El Subin, Q - Quetzaltenango

Fig. 2.1. Collecting missions undertaken by the Belize-based project from 1972 to 1977. Routes traversed are shown by single lines (solid = collections; dashed = no collections). Dots indicate the boundaries of climates on the collection routes, and in the case of Mountain Pine Ridge in Belize, a more intensively collected area. Numbers 1 to 11: climates as defined in the text.

The areas from which plants were collected in Belize, the Mexican Yucatan Peninsula and Guatemala are shown in Fig. 2.1.

The strategy used in the analyses of the data generated was determined in part by the very large size of the data set and by the environmental information available. With such a large data set, graphical presentation of the results, and their interpretation, is usually very difficult. In many instances the individual vectors are concerned with the relationships of attributes that are of little interest or are difficult to interpret; some may be concerned with only a small part of the collection (see Chapter 7, this volume).

The second factor affecting the analytic strategy was the type of environmental information available. Many of the accessions were collected from roadsides, which are subjected to run-on and run-off of water, very disturbed or artificial soil conditions and atypical management strategies. As such, they are quite different from the environments of the areas that surround them and for which climatic data are available. Thus only a broad, general classification was warranted. The data set chosen for use depended on the availability of relevant data for the entire area in question and was drawn from a variety of sources, that of Walker (1973) and UNAM (1970) being of particular importance for Belize and Mexico respectively (Table 2.2).

An initial simplification of the data set was necessary and for this purpose the rapid computer programme POLYDIV was employed using only three attributes: number of months receiving 85–95% of the total annual rainfall, average total annual rainfall and average minimum temperature (Table 2.2). These attributes had previously been shown to be most important in similar climatic classifications. The associated programme GROUPER was then used to define the group characteristics at the 175 group level of the hierarchy. The mean values of the 175 groups were then analysed using the MULCLAS programme, a system that has been shown heuristically to produce meaningful results for the present purpose. Finally Cramer values were

Table 2.1. Tentative classification of edaphic environments of collections made in Belize, Guatemala and the Mexican Yucatan Peninsula.

Parent material	Likely FAO classification[a]
Fluvial deposits	Fluvisols
Coarse calcareous gravel with sand (road fill)	–
Marls	Vertisols
Corals	Vertisols
Limestone	Rendzinas
Granite	Lithosols
Slates	Aerisols
Recent volcanic deposits	Andosols
Ancient beach deposits	Planosols
Sand deposits	Arenosols
Various, well weathered	Ferralsols

[a]FAO/UNESCO (1973).

Table 2.2. Climatic attributes used in pattern analysis.

a. Number of months to receive 85–95% of total annual rainfall
b. Average annual rainfall
c. Average minimum monthly temperature (°C)
d. Average monthly temperature (°C)
e. Lowest minimum monthly temperature during wet period as defined in (a) (°C)
f. Average maximum monthly temperature (°C)
g. Altitude (m)
h. Latitude (degrees), no distinction made as to north or south

calculated to explore the contributions of the eight attributes listed in Table 2.2 to the definition of the 16 groups nominated. Where detailed examination of the groups revealed too much heterogeneity they were split.

2.2.3 Edaphic conditions at the collecting sites

General purpose soil classifications may have limited use for specific purposes because some species may be found over a wide range of soil types while others are adapted to very specific conditions. *Stylosanthes capitata*, for example, is only found on and adapted to very acid conditions, even on a 'yellow earth' (Burt *et al.*, 1979; Cook *et al.*, 2005). Edaphic conditions on roadsides may be substantially altered and difficult to sensibly categorize. Only a broad, general classification was therefore attempted using the FAO classificatory system (FAO/UNESCO, 1973) and complementing this with information about the parent materials (Table 2.1). A summary of soil types for those accessions from the Mexican Yucatan Peninsula is presented in Appendix 2.2 and for the entire Belize collection in Appendix 2.3.

2.2.4 Agronomic testing

Plant material

At the time at which the agronomic evaluation was undertaken in Belize, the Central Farm (CF) legume germplasm collection numbered 520 accessions. The origins of these were Belize 194, the Mexican Yucatan Peninsula 187, Guatemala 32 and the West Indian islands 51. The remaining 56 accessions were composed of cultivars and standard lines and accessions that had performed well with CSIRO at Townsville Australia, at CIAT (Centro Internacional de Agricultura Tropical, Cali, Colombia) or at IRFL (Indian River Field Laboratory, Florida, USA) and were deemed most likely to be adapted in Belize.

Little or nothing was known about most of the accessions, particularly their performance in the environment in question – the wet tropics on a relatively heavy soil. For many, even their taxonomic classification was uncertain and, as is the case of *Desmanthus*, the situation was worsened by the lack of a relatively recent taxonomic revision of the genus. Because some of the facilities were relatively limited, such as those needed for drying of plant samples, it was necessary to limit the size of the experiment to a selected sample of the collection and to keep experimental treatments and numbers of replicates to a minimum. This was reasonable as only the broad agronomic characteristics of the plants were required, not detailed comparisons.

The plants selected for testing (Table 2.3) consisted of 182 accessions comprising 22 genera and 50 species. The accessions were from the Mexican Yucatan Peninsula (80), Belize (34), Guatemala (7) and the islands of the West Indies (42). To these were added relevant 'standards' from other collections (6 cultivars and 13 standards and promising accessions). The bulk of the West Indian accessions were from the Antigua-based programme of the IDRC/UWI project. For further details on the species and accessions involved, see Appendix 2.1.

The site

The experiment was located on Central Farm, the research farm of the Belize Department of Agriculture in Cayo District. This is situated on a high alluvial terrace of the Belize River, on a soil described as a silty clay of the Esperanza Series (Jenkin *et al.*, 1976) or as a mixed isohyperthermic mollic Tropudalf (McKinzie, 1977). The A horizon is neutral (pH 7.0) and the soil is of moderate fertility but low in phosphorous (2 to 8 ppm). Situated on a moderately well-drained native grass pasture, the site represented the basic, black clay soils that dominate the higher terraces of the Belize River and adjacent areas in central Belize. Soil analysis results are presented in Table 14.1 (Chapter 14, this volume).

Coastal and sub-coastal Belize is said to have a humid, semi-hot tropical climate, too cool for truly equatorial crops such as rubber or palm oil, but suited to such crops as banana and sugarcane (Papadakis, 1966). The mean annual rainfall is 1500 mm with a dry season from March to May. The dry season is preceded by a relatively cool, cloudy period in December and January, which limits the growth of some species.

Over the experimental period from mid-June 1976 until October 1977, above average rainfall was experienced in the cool season, but rainfall in February, before the usual onset of the dry season, was below average (Fig. 2.2).

Table 2.3. Species, origins and numbers of the 182 accessions in the agronomic trial.

Species	Origin and number of accessions				
	Mexico	Belize	Guatemala	Caribbean Islands	Promising lines and standard cultivars
Aeschynomene americana var. *glandulosa*			1		
Aeschynomene elegans		1			
Aeschynomene vaginalis		1			
Calopogonium mucunoides			1		
Cassia aeschynomene		1			
Cassia sericea	1				
Centrosema plumieri		3		1	
Centrosema brasilianum × *C. virginianum*					CIAT 17-33-F7
Centrosema pubescens	1			4	IRFL 980, 1238,1239, 'Centro'
Centrosema schottii	6				
Centrosema virginianum	1			6	
Chaetocalyx scandens	1				
Clitoria rubignosa		1			
Clitoria ternatea	1	1		3	
Cracca greenmanii	3				
Cracca spp.	1				
Crotalaria pumila	1	1			
Crotalaria sagittalis var. *fruticosa*		1			
Desmanthus virgatus	2	2			
Desmodium barbatum		2			
Desmodium canum	7				CSIRO unknown
Desmodium distortum					Q8378
Desmodium heterophyllum					cv. Johnstone
Desmodium intortum		1			
Desmodium tortuosum	7	1	1		
Galactia striata	12				
Indigofera mucronata	1				
Leucaena leucocephala	6			5	cv. Cunningham, Peru
Macroptilium atropurpureum	8			2	cv. Siratro
Macroptilium lathyroides	1			2	
Macrotyloma axillare					IRFL 1805
Pachyrrhizus erosus		1		1	

Continued

Table 2.3. Continued.

Species	Origin and number of accessions				
	Mexico	Belize	Guatemala	Caribbean Islands	Promising lines and standard cultivars
Pachyrrhizus palmatus	1				
Rhynchosia longeracemosa	3	2			
Rhynchosia minima	4	1		1	
Rhynchosia yucatanensis	1				
Stylosanthes calcicola	7				
Stylosanthes hamata		1		11	CIAT 122
Stylosanthes guianensis		7	1		cv. Libertad, Endeavour, Schofield (2 selections)
Stylosanthes mexicana			2		
Stylosanthes scabra					CPI 40205
Stylosanthes viscosa		1			
Tephrosia littoralis	1				
Tephrosia noctiflora		1			
Teramnus labialis		1		6	
Teramnus uncinatus		1			IRFL 1550
Vigna luteola		1			
Vigna speciosa	3	1			
Totals	80	34	7	42	19

CIAT, Centro Internacional de Agricultura Tropical; CPI, Commonwealth Plant Introduction number (Australia); IRFL, Indian River Field Laboratory, Florida, USA.

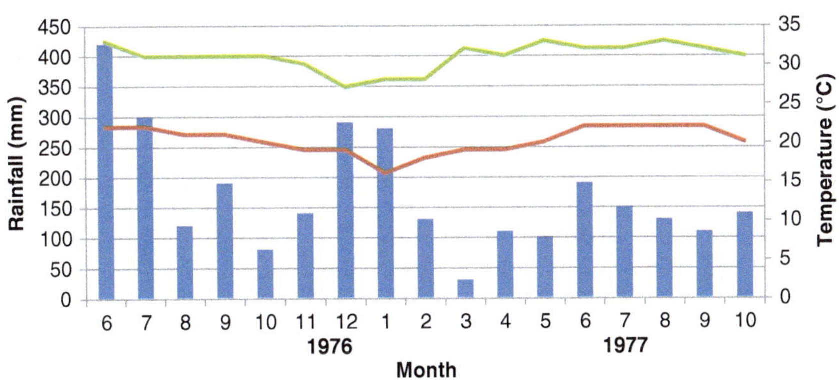

Fig. 2.2. Monthly rainfall (bars) and average maximum and minimum temperatures over the period of the trial.

Experimental design and plant culture

Each of the 182 accessions was represented by three plants, one of each being located at random within each of three blocks. The plants were raised in small pots and transplanted into the field on the square at 1.5 m intervals in mid-June 1976. Plants that died before mid-August were replaced (Fig. 2.3)

Fig. 2.3. Initial agronomic evaluation of 182 accessions as single-spaced plants in three replicated blocks (middle trial). See also Fig. 14.1 for a closer view of plant spacing and management. Foreground is 12 legume upper river terrace trial (Chapter 17, this volume).

Before planting, the soil had been cultivated to a fine tilth and fertilized with triple superphosphate and muriate of potash applied at the rate of 12 and 20 kg/ha of elemental P and K respectively. This same rate of fertilizer was applied annually, as split applications after alternate yield harvests.

Soon after transplanting it became apparent that weed control between transplants would be difficult and that damage to transplants might result. A small area around each plant was therefore hand weeded throughout the experimental period and interplant spaces were mowed as required (Figs 2.3 and 14.1).

Yield harvests commenced in mid-September 1976 and continued at six-weekly intervals thereafter until October 1977. At each harvest, with the sole exception of the tree legume *Leucaena leucocephala*, plants were trimmed to a cylindrical shape with a radius of 15 cm around the point of insertion of the main stem into the soil, and 15 cm high. For *L. leucocephala* only the leaf was harvested. Harvested material was dried and weighed and the number of surviving plants in each accession recorded.

2.3 Results and preliminary discussion

2.3.1 Climatic analysis

The second hierarchical classification of the climatic data was truncated at the 16 group and is presented in Table 2.4 in the order in which the groups appeared on the dendrogram. The three attributes used in the classification (number of rainy months, rainfall and average minimum temperature) have, as expected, high Cramer values. They account for much of the discrimination between the groups. The remaining attributes, particularly the geographical features altitude and latitude, show considerably more variation within the groups. The variation for latitude is not unexpected because, for instance, tropical latitudes may have subtropical or even temperate environments and suitably adapted plants at high altitudes.

The location of the sites of collection of the 182 accessions is presented in Tables 2.5 and 2.6 along with other areas of interest relevant to

Table 2.4. Characteristics of the 16 climatic groups.

	Climatic groups																
Attributes[a]	I	II	III	IV	V	VI	VII[b]	VIII	IX	X	XI	XII	XIII	XIV	XV	XVI	Cramer values
Rainy months[c]	5	5.8	6	6	6	6	6	7	7	7.5	8	8	8	8.4	8.7	9	0.97
Rainfall[c] (cm)	167	119	75	94	128	300	9620	116	133	362	70	144	160	255	107	95	0.97
Average minimum temperature[c] (°C)	26.3	22.5	16.2	27.3	15.2	26.1	20.6	22.8	20.6	19.2	22.2	22.6	20.5	23.8	25.8	14.9	0.96
Average temperature (°C)	27.8	26.1	17.9	26.9	21.5	27.0	23.7	26.1	25.1	23.1	25.5	25.6	23.7	25.8	27.3	21.0	0.90
Lowest wet season minimum temperature (°C)	27.4	24.9	17.4	26.2	23.0	27.8	23.3	25.0	22.6	19.4	23.0	23.2	20.9	23.3	26.0	17.1	0.85
Average maximum temperature (°C)	29.1	29.0	19.5	28.5	27.2	28.5	27.4	28.4	26.8	26.4	26.7	27.6	26.7	27.1	28.6	26.5	0.78
Altitude (m)	58	150	1400	50	375	450	921	46	66	630	48	68	408	319	14	52	0.57
Latitude (°)	13.7	18.7	14.7	14.7	19.1	14.4	13.0	19.4	20.4	10.8	19.2	18.1	23.5	11.3	15.6	26.9	0.27

[a]Attributes use in hierarchical analysis. [b]Group ignored due to error in rainfall data, 962 cm is the correct figure. [c]As shown on Table 2.6.

Table 2.5. Locations of research stations and areas with potential for livestock used in pattern analysis.

Research stations		
Australia	South Queensland:	Gayandah, Grafton, MacDonnell, Samford, Rodds Bay
	North Queensland:	Atherton, Kangaroo Hills, Lansdown, South Johnstone
	Northern Territory:	Darwin, Katherine
	West Australia:	Kununurra
Malaysia		West Kluang, Serdang
Costa Rica		Turrialba
Colombia		Carimagua
Venezuela, Guarico state		Calabozo
Areas with potential for livestock		
Vietnam		Hanoi, Nha Trang
India		Bangalore
Ghana		Accra
Central African Republic		Berberati

this study – tropical pasture research stations and areas of heavy clays where the plant material being tested may be adapted.

Inspection of the results revealed that one group, VII, contained only a single entry (Table 2.6); the climatic information for this entry was incorrectly entered and the group has been discarded. Another group, V, combined low minimum temperatures with a relatively short wet season, but the annual rainfalls for the three collection sites within the group varied considerably. The group is thus of little interest and has been discarded. In similar situations, but where the sites concerned are of interest, II, IV, VIII, XIII, XIV and XVI, the groups have been split to provide more homogeneous groupings (Table 2.6).

Groupings and sites of no interest for this study were excluded and the remaining groups and split groups have been renumbered from 1 to 15 in a manner that reflects their location in the collection area and labelled climate types (Tables 2.6 and 2.7). The numbering begins with the climate types in the northern portion of the Yucatan Peninsula, passes into Belize and neighbouring Guatemala and concludes with southern Guatemala. Information for relevant testing sites or the origins of cultivars of interest (the standards) is presented in the groupings as they are sequentially listed or, where necessary, appended at the end.

The climates of the Yucatan Peninsula vary considerably. Using the agro-climatic classification of Papadakis (1966), the northerly part is described as 'dry, semi-hot, tropical'. It is generally suitable for maize and cassava, but too dry for oil palm, for instance. Within this region Papadakis separates three sub-regions: that near Progreso (Fig. 2.1, a type 1 climate) which has 6 or more dry months, that near Merida (climate types 2 and 3) with 4 or 5 dry months and that near Valladolid (type 3) with 0 to 3 dry months. The distinctions are biologically meaningful because Merida can only produce one rain-fed crop of maize annually while Valladolid can produce two.

While type 4 climate is the wettest in the study in the Yucatan, Belize and neighbouring parts of Guatemala are wetter, and have the longest wet seasons (types 6, 7, 8). These are the 'humid, semi-hot climates' of Papadakis that are suited to such crops as banana and sugar cane.

Finally there are the 'terra templada' climates of southern Guatemala, types 10 and 11, and the low-altitude climate (type 9) nearer the coast. Terra templada climates such as those near Guatemala City (Fig. 2.1) are suitable for Arabica coffees, those at the lower elevation for sugar cane.

The climatic classification produced herein clearly is in accord with that of Papadakis, and knowledge of the areas concerned validates the predictions of their suitability for crops. The classification is thus biologically meaningful and readily useable by those seeking new material of potential from the resources of germplasm banks, such as agronomists seeking material for use in a given environment.

Four climatic types, ranging from very dry to relatively wet, provided the bulk of the collection. The observed distribution of some of the better-collected species within these climates in the Mexican Yucatan Peninsula is presented in Table 2.8. Although it is tempting to attribute

Table 2.6. Constitution of the groups: geographic locations and the ranges in values of the three main discriminating climatic attributes.

Classificatory group		Climate type[a]	Sites[b]	Location/ origin	No. months	Annual rainfall (mm)	Minimum monthly temperature (°C)
I		9	Darwin, NT	Australia	5	1490	25.3
			Guatemala		5	1500–2000	26.4–27.0
			S. guianensis cv. Endeavour	Guatemala	5	2160	–
II	a	1	Kununurra, WA	Australia	5	740	21.1
			Katherine, NT	Australia	5	950	21.6
			MacDonnell, Qld	Australia	5	1680	23.2
	b	3	Lansdown, Qld	Australia	6	910	21.7
			Yucatan, Mex.	Mexico	6	1000–1300	22.5–23.8
			Turrialba	Costa Rica	6	1770	21.1
III		11	Guatemala		6	750	16.2
IV	a	15	*S. hamata* cv. Verano	Venezuela	6	390	27.3
	b		Calabozo	Venezuela	6	1320	26.6
V			Kangaroo Hills, Qld	Australia	6	660	13.3
			Atherton, Qld	Australia	6	1370	15.8
			Hanoi	Vietnam	6	1810	16.6
VI		10	Guatemala		6	2800–3200	22.7–27.0
VII			Bangalore	India	6	9620[c]	20.6
VIII	a	2	*S. scabra* cv. Seca	Brazil	7	693	–
	b	2	Yucatan	Mexico	7	750–1400	22.1–23.7
			Nha Trang	Vietnam	7	1100	24.0
			Trincomalee	Sri Lanka	7	1650	25.0
IX		4	*S. viscosa* CPI 34904[d]	Brazil	7	1280	22.0
			Yucatan	Mexico	7	1000–1500	20.3–20.8
X		12	South Johnstone, Qld	Australia	7	3620	19.2
XI		1	*S. scabra* cv. Fitzroy	Brazil	9	520	21.6
			Yucatan	Mexico	8	470–670	2.7–22.9
XII		6	Belize		8	1460–1860	22.3–23.5
			Berberati	Central African Republic	8	1520	23.7
			Guatemala		8	1750–2000	22.0–23.0
			Yucatan	Mexico	8	1050–1500	22.0–23.0
XIII	a	7	Belize		8	1530–2930	20.0–21.2
	b		Miami	USA	8	1520	19.4
XIV	a		*S. guianensis* cv. Cook	Colombia	11	3970	–
	b	8	*C. pubescens* cv. Belalto	Guatemala	–	3000	–
			Belize		8	2120–2400	22.2–23.2
			Carimagua	Colombia	8	2560	25.0
			Kluang	Malaysia	9	2790	25.0
			Serdang	Malaysia	9	2840	25.0
XV		5	Accra	Ghana	8	790	25.0
			Antigua		9	890–1270	26.1
			Barbados		8	1270	25.2
			Barbuda		–	680	–
			Trinidad		8	1730	24.4
			Yucatan	Mexico	8	1200	24.3
XVI	a		*S. guianensis* cv. Oxley	Argentina	10	1380	15.7
	b	13	Rodds Bay	Australia	8	890	16.6
	c	14	Gayndah	Australia	9	760	13.9
			Grafton	Australia	9	880	14.2
			Samford	Australia	9	860	14.0

[a]Derived from the Classificatory Groups. [b]Sites used in analysis and locations of accessions in groups. [c]Incorrect figure used in analysis, actual is 962. [d]CPI, Commonwealth Plant Introduction number.

Table 2.7. The principal climatic features of the various collecting zones, sites of origin of standard cultivars and relevant evaluation sites.

Country	Climate type[a]	Site	No. months	Annual rainfall (mm)	Minimum monthly temperature (°C)
Yucatan, Antigua	1	Yucatan, Mexico	8	470–670	22.7–22.9
		S. scabra cv. Fitzroy (Brazil)	9	520	21.6
	2	Yucatan, Mexico	7	750–1400	22.1–23.7
		S. scabra cv. Seca (Brazil)	7	693	–
	3	Yucatan, Mexico	6	1000–1300	22.5–23.8
		Landsown, Australia	6	910	21.7
	4	Yucatan, Mexico	7	1000–1500	20.3–20.5
		S. viscosa CPI 34904[b] (Brazil)	7	1280	22.0
	5	Yucatan, Mexico	8	1200	24.3
		Antigua	9	890–1280	26.1
		Barbados	9	1270	25.2
		Yucatan, Mexico	8	1050–1500	22.0–23.0
Belize, Guatemala	6	Belize	8	1460–1860	22.2–23.5
		Central Farm, Belize	8	1490	19.3
		Guatemala	8	1750–2000	22.0–23.0
	7	Belize	8	1530–1930	20.0–21.2
	8	Belize	8	2120–2400	22.2–23.2
		Carimagua, Colombia	8	2560	25.0
		Kluang, Malaysia	9	2790	25.0
Guatemala	9	Guatemala	5	1500–2000	26.4–27.0
		S. guianensis cv. Endeavour (Guatemala)	5	2160	–
		Darwin, Australia	5	1490	25.3
	10	Guatemala	6	2800–3200	22.7–27.0
	11	Guatemala	6	750	16.2
Australia	12	South Johnstone, Australia	7	3620	19.2
	13	Rodds Bay, Australia	8	890	16.6
	14	Gayndah Australia	9	760	13.9
		Grafton, Australia	9	880	14.2
		Samford, Australia	9	860	14.0
	15	*S. hamata* cv. Verano (Venezuela)	6	390	27.3

[a]As renumbered, Table 2.6. [b]CPI, Commonwealth Plant Introduction number.

the differences in species distribution to climatic effects, it must be remembered that other factors, such as soil type, land use and history are also of importance. Hypotheses should be generated about likely plant adaptation that may be explored in later evaluation work, laboratory studies or more detailed ecogeographical surveys. This is a particularly valuable objective in situations such as this with poorly known species and limited resources.

2.3.2 Environmental effects on the distribution of the various legume species

Many of the legume collections made were in the Mexican Yucatan Peninsula. It is those areas that are of particular interest here because almost all of their soils are alkaline, there are many endemic species and there is a range of climates of wide general interest, but little information is available for the region. An analysis of the observed distribution of selected legume species at collection sites across climatic types is presented in Table 2.8.

Large differences were apparent between the climatic types, with the wettest (type 6), located in the central Yucatan, providing most accessions and the greatest number of accessions per collecting site. It is richest in *Vigna speciosa* and *Macroptilium atropurpureum*, and *Desmodium incanum* is also well represented. It is poorest in *Desmodium tortuosum*. The driest area (type 1) is highest in *Centrosema schottii* and *Rhynchosia minima* and it was relatively rich floristically, lacking only two of the legume

Table 2.8. Observed distribution of selected legume species at collection sites across climatic types in the Mexican Yucatan Peninsula.

	No. sites observed				
	Climate type				
Genus/species	1	2	4	6	Total
Aeschynomene spp.	11	4	0	0	15
Centrosema schottii	30	11	0	12	53
Centrosema virginianum	11	19	0	12	42
Centrosema spp.	7	13	5	19	44
Chaetocalyx scandens	3	4	14	6	27
Desmanthus spp.	0	10	0	6	16
Desmodium incanum	11	13	33	25	82
Desmodium tortuosum	22	6	14	0	42
Galactia striata	26	38	29	25	118
Leucaena leucocephala	7	17	5	6	35
Macroptilium atropurpureum	7	4	0	31	42
Phaseolus spp.	0	4	0	12	16
Rhynchosia minima	30	19	0	19	68
Rhynchosia longeracemosa	6	11	0	12	29
Vigna speciosa	2	25	0	33	60
Number of sites inspected	36	58	41	27	162
Number of species observations	173	198	100	218	689
Average number of observations per site	4.8	3.4	2.4	8.1	

species and providing several accessions at each collecting site.

The type 4 climate, which is relatively favoured, has a much poorer flora than even the most arid (type 1). This probably reflects the lack of agricultural development and associated disturbance in these regions. Most of the area was still forested and the limited undergrowth was lacking in legumes.

Differences between species are also evident. *Desmodium incanum* and *Galactia striata* are environmentally tolerant. They are found in large numbers across all environmental types. *Macroptilium atropurpureum* dominates only in the wettest region (6) and is totally absent from the type 4 climate.

Most of the soils encountered in the Mexican Yucatan were shallow red clays of the rendzinas or road fills. Vertisols, which are of particular interest, were only found in a small part of the Mexican Yucatan from the northern Belize border to around Lake Balcalar. Wright (1967) notes that only small patches are found elsewhere in the Mexican portion of the Peninsula, again mainly in the extreme south where the limestone has been more weathered. This patchy distribution, together with the limitations inherent in general purpose soil classifications (see section 2.2.1), renders sensible investigation into plant/soil relationships difficult. But the paucity of such information in tropical areas means that even generalized relationships may be useful for those seeking to efficiently utilize the genetic resource collections available. Additional interest is provided in this study by the somewhat different climates of the three geographical locations where plant collecting was carried out most intensively: the Mexican Yucatan, on the Belize/Mexican border around Corozal and Chetumal, and around Central Farm in central Belize (see Fig. 2.1).

The Mexican Yucatan dark clay soils, although of relatively limited area, were of particular interest because of the need for adapted legumes for similar soils elsewhere (e.g. Ethiopia, India). The species found in the south-central and south-west portion of the peninsula were *C. pubescens*, *D. incanum*, *Galactia striata*, *Leucaena leucocephala*, *Rhynchosia longeracemosa* and *Vigna speciosa*. In the Corozal–Chetumal area (the extreme south-east of the Yucatan and across the border in Belize) the same species occurred, plus *Desmanthus pubescens*, *Desmodium tortuosum*, *Centrosema plumieri*, *C. macrocarpum* and *Rhynchosia minima*. In central Belize, in the area of Central Farm, most of the species mentioned above occurred as well as numerous other species, notably *D. cinerium*, *Pachyrrhizus erosus* and *Tephrosia noctiflora*. *Macroptilium atropurpureum* from clay soils have provided cultivars for worldwide use as has *Chamaecrista rotundifolia*. *Desmodium incanum*, another plant noted here, has not yielded cultivars but is a widely recognized pasture legume in these situations. Some of these could be 'new species' of potential utility in similar clay soil situations elsewhere. Reference to Cook *et al.* (2005) lends weight to this belief.

Legume species observed on the soils of Mexico are dominated by *Aeschynomene*, *Macroptilium* (especially *M. atropurpureum*) and *Desmanthus* (especially *D. virgatus*). The clay soil areas are poor in the genera associated with similar soils in the old world tropics, such as *Vigna* (Chapter 6, this volume). The Mexican collections make a very significant contribution to this particular germplasm bank, with well-known species such as *Macroptilium atropurpureum* and 'new' genera such as *Desmanthus*. The Central Farm collection in its entirety adds species not previously recorded, such as *Desmanthus pubescens*, *Acacia angustissima* and *Centrosema macrocarpum* as well as accessions of other species, such as *Leucaena leucocephala* and *Centrosema plumieri* not previously collected from vertisols.

2.3.3 Environmental tolerance of cultivars

Most tropical legume cultivars are derived directly from collections made in the wild. *Stylosanthes hamata* cv. Verano was from an acid, sandy earth soil and a semi-arid type 15 climate (Table 2.6). *S. scabra* cv. Seca was from a type 2 climate. Both are used on acid, sandy infertile soils over a much larger range of climates. In Australia, for example, they are the most widely used cultivars in climates 1, 2 and 3, and extend into the subtropical climate 13. They are now often sown as a mixture of cultivars and can become naturalized. Although a composite of cultivars has much to recommend it, native genotypes of legumes may have advantages over cultivars of the same species, as the cultivars in many cases were developed to suit a particular environment elsewhere. Native Kenya material of *Glycine wightii* (syn. *Neonotonia wightii*) for instance yielded 'Kenya white glycine' and 'Kenya violet glycine', which were better performing than imported cultivars.

The screening results showed that the *Stylosanthes* cultivars tested failed at the Central Farm Belize trial site. While it could be argued that the environment was too wet, this does not hold true for *S. guianensis*, whose cultivars encompass climate type 9, which yielded the cultivar 'Endeavour'. This cultivar is a South American type of the species that thrives in South Johnstone, climate type 12, which is wetter that that at Central Farm. Because *S. guianensis* occurs naturally on Mountain Pine Ridge near the trial site on acid soils (Appendix 2.3), almost certainly this is an edaphic rather than a climatic limitation.

2.3.4 Agronomic evaluation

Performance patterns

Many of the species to be evaluated were very poorly known, and the performance characteristics of all – even their persistence – on this particular combination of soil and climate conditions were uncertain. While major known differences in habit were evident, from long-lasting tree legumes such as *Leucaena leucocephala* to the smaller shorter lived herbs such as *Stylosanthes hamata*, differences within species also became apparent as some accessions were high yielding, some were unusually productive in the dry season and some behaved as annuals.

Ten performance pattern groups based on yield were delimited and examples are presented in Table 2.9. Groups 1–4 consist of the most successful plants. Those in groups 5 and 6 had yields that increased with time for each of the first three harvest periods. Plants in group 5 had relatively constant yields thereafter, while those in group 6 had a high initial dry season yield followed by very poor growth in the second season. Groups 7–10 characterized plants that failed in the second wet season and were either poorly adapted perennials such as *Stylosanthes guianensis* cv. 'La Libertad' or annuals such as *Centrosema schottii* CF 329. The performances of the best yielding species are summarized in Table 2.10.

Legumes with known ability to grow productively in heavy clay soils are those listed in Table 2.10 and they generally performed well in the experiment. However, others that are often recommended for such soils did not produce sufficient biomass to be included in the list of productive species. These included *Macroptilium*

Table 2.9. Agronomic types in the Central Farm germplasm collection based on seasonal growth patterns.

Growth type		Seasons			Example species	CF no.	Agronomic characteristics				
							No. living plants Oct 1977	Seasonal[a] average yields (g/plant/week)[b]			Total yields[b] 60 weeks g/plant
		Wet 1976	Cold/dry 1977	Wet 1977				Wet	Cold/dry	Wet	
Increasing	1	Continuous increase			*Leucaena leucocephala*	334	3	2.83	5.79	10.44	381[d]
	2	high	lower	high	*Desmanthus pubescens*	453	3	4.67	4.5	9.28	339
	3	high	low	high	*Centrosema plumieri*	28	2[c]	8.28	5.33	6.72	426
	4	low	low	high	*Vigna speciosa*	522	3	2.72	4.39	7.28	285
	5	Relatively constant			*Desmodium incanum*	417	3	2.02	2.63	2.44	144
Decreasing	6	low	high	low	*Stylosanthes guianensis*	509	2	2.56	4.25	1.56	176
	7	high	low	medium	*Desmodium tortuosum*	470	3	7.06	1.83	3.17	228
	8	high	low	very low	*Stylosanthes guianensis*	cv. La Libertad	1	14.28	6.17	2.17	444
	9	high	low	zero in final harvest	*Centrosema schottii*	329	0	9.72	3.38	2.00	292
	10	low	low	zero in final harvest	*Desmodium incanum*	424	1	0.44	0.42	0.39	25

[a]Wet season 1976 harvests 1, 2, 3 as 18 weeks to December 1976; cold/dry season harvests 4, 5, 6, 7 to May 1977; wet season 1977 harvests 8, 9 and 10 to October 1977. [b]Yield is the mean of three plots, including dead plants. [c]Accessions present in two replicates at beginning of trial. [d]Least significant difference of $P = 0.05$ for full data set (as shown in Appendix 2.1) is 142.

Table 2.10. General performance of the more productive species in the trial. For more detail, see Appendix 2.1.

		No. accessions					
				Yield (g/plant)			
				Total			Cool/dry season
Species	Habit	Planted	Yield pattern[a]	250–349	350–449	>450	>95
A. Adapted and productive species of known value							
Perennial species							
Leucaena leucocephala	Tree/shrub	13	5	5	5		8
Desmodium incanum	Erect/ sub-erect	4	5				4
Desmodium tortuosum[b]		9	1	1			
Centrosema pubescens	Climbers/ ramblers	10	6	2			1
Macroptilium tortuosum[c]		9	1	1			
Annual/weakly perennial species							
Desmodium distortum	Erect/ sub-erect	1	6			1	1
Stylosanthes guianensis (cv. Libertad only)		12	1		1		5
B. 'New species'							
Perennial species							
Desmanthus pubescens	Tree/shrub	1	1	1			1
Desmanthus leptophyllus		2	6	1			
Centrosema plumieri	Climbers/ ramblers	4	2	2	1	1	4
Teramnus uncinatus		2	1	1			1
Annual/weakly perennial species							
Centrosema schottii	Climbers ramblers	6	6, 10	2	2	1	2
Rhynchosia minima		6	10	1			
Vigna speciosa		4	4	1	1		3

[a]Seasonal growth pattern as shown in Table 2.9. [b]An annual/weak perennial, relatively low yielding, which is difficult to categorize in this type of experiment. [c]A very variable species sometimes stated to be an annual.

lathyroides, *Clitoria ternatea*, *Desmodium heterophyllum* and *Galactia striata*. *Desmodium incanum* also is known to thrive on heavy clays and did so here. Though its seasonal growth patterns generally indicated good persistence, its small plant size resulted in low per plant biomass yields.

WOODY LEGUMES. *Leucaena leucocephala*: All accessions of this well-known tree legume planted were persistent and high yielding with only one exception, CF 207. The cultivar 'Cunningham' yielded slightly, but not significantly, more than CF 469. The latter was collected from a vertisol and could possibly be of more value on such soils.

Desmanthus species: When the Belize project began, *Desmanthus* was not a genus of particular interest compared with, for example, *Stylosanthes*, *Desmodium* or *Centrosema*. The one accession that was acquired by the project early in its activities did not survive on the heavy clay soils of the testing site. A more concerted effort was made to collect the genus only after a number of accessions collected in the Yucatan showed good promise in introduction plots for use in pastures and for cut and carry feeding. All accessions at the time were considered to be *D. virgatus* due to lack of a revision of the genus, and finer differences between the accessions were not noted.

Since then some of those accessions that were duplicated in Australia have been identified, and those new names available through the AUSPGRIS website are used here.

Of the eight *Desmanthus* accessions collected by the Belize programme to the date of establishment of this trial, six were also planted in unreplicated introduction plots on a heavy clay soil on Central Farm in mid-1976 and were assessed visually a year later against the 543 accessions then growing in the introduction plots at the same site, some of which had been grown much longer than the *Desmanthus* lines. Three *Desmanthus* lines, *D. pubescens* CF 453, *D. leptophyllus* CF 406 and *Desmanthus* sp. CF 491, were judged to be among the 73 most promising accessions. A Mulclas Grouper Minspan analysis of 244 of these plots planted on the same date was undertaken, utilizing observational ratings for plot cover, vigour, height and leafiness taken at five critical times of the year. It classified five of the six *Desmanthus* lines among the 64 most promising lines. These five included the three mentioned above plus *D. virgatus* CF 317, and *Desmanthus* sp. CF 448 (Lazier, unpublished data).

Desmanthus pubescens is an erect woody shrub up to 1.5 m in height. It is much branched from the base, the stems arising from a woody lignotuber, which probably contributes to its grazing resistance and ability to be shaped into 'hedge lucerne'. Although only represented by a single accession in the trial it was very persistent, behaving in much the same way as *Leucaena leucocephala*; however, unlike the latter, it does not contain any anti-nutritional factors.

ERECT/SUB-ERECT PLANTS. *Desmodium incanum* (syn. *D. canum*) was found widely throughout the Yucatan and Belize in a wide range of climates and soils. It exhibited a wide range of performance patterns and persistency but never yielded highly on an individual plant basis. In a concurrent trial this legume proved to be persistent under cutting and grazing in mixed sward with Para grass (*Brachiaria mutica*, syn. *Urochloa mutica*) (Lazier, 1981a,b). An accidental introduction in Fiji, Hawaii and various Pacific Islands, it has become valued there as a pasture species. In Central America, where it is known as 'pega' and is a common component of pastures, its contribution is largely ignored and emphasis is given to improving grass yields. However its ubiquity, persistence and moderate yield indicate that it may be an important component of pasture flora and animal nutrition.

Macroptilium lathyroides is a high-yielding annual from which a cultivar 'Murray' has been selected. It has long been used as a ley legume. However, generally, on a per plant basis, its productivity was low in this trial.

CLIMBERS/RAMBLERS. Several of these that were tested in the trial (*Centrosema pubescens*, *Desmodium heterophyllum*, *Macroptilium atropurpureum* and *Clitoria ternatea*) have already yielded well-known, characterized and documented cultivars (e.g. Cook *et al.*, 2005). A number of cultivars were included in trial as 'standards': *Centrosema pubescens* 'Centro', *Macroptilium atropurpureum* cv. 'Siratro' and *Desmodium heterophyllum* cv. Johnstone (an exotic, and the only accession of this species). While *D. heterophyllum* did not perform well, Centro and Siratro were the highest yielders among the genotypes of their species tested, but only marginally, and not significantly. Most species showed some agronomic variation. *M. atropurpureum* CF 623 showed more winter growth than cv. 'Siratro' and such variation could be very useful. *Desmodium intortum*, 'Greenleaf Desmodium', which is often considered only for use in subtropical or higher altitude situations, was found in a range of tropical conditions and on several soil types. Persistence and yield patterns also varied considerably with one accession, CF 470, showing high persistence and consistent growth over the seasons.

Galactia striata is a well-known species having the local names 'Galactic' in Brazil, 'Frij-olillo' in Panama and 'Florida milkpea' in the USA. It is quite variable because it grows well on the very acid soils of the Cerrado in Brazil, where it has out-yielded such legumes as Siratro and Stylo, and occurs in the area surveyed in this study, where it has been found on alkaline rendzinas and lithosols. It occurred plentifully across all climatic types in the Mexican Yucatan Peninsula. This wide variation was mirrored in the wide range of performance patterns described herein (Appendix 2.1). Some accessions, such as CF 416 and CF 374, were persistent with relatively high yields that increased with time.

Two of the 'new' species described in this study, *Centrosema plumieri* (a perennial) and *Centrosema schottii* (an annual), were both capable of producing very high yields. Some accessions, for instance *C. plumieri* CF 28 and *C. schottii* CF

443, provided considerable growth in the cooler drier season. Both species demonstrated much potentially useful agronomic variation; one persistent and productive accession of *C. plumieri* had been collected on a vertisol. Another (CF 16-1) had also proven to be persistent and productive in a concurrent mixed sward trial under cutting and grazing on a heavy clay (Lazier, 1981a,b). Other climbing species persisting on vertisols include *Vigna speciosa* and *Rhynchosia longeracemosa*.

With the possible exception of the diploid form of *Stylosanthes hamata*, all species of *Stylosanthes*, probably the most successful genus of tropical pasture plants to date, failed. Even the species adapted to wetter climates, and successful cultivars thereof, died out in the first wet season. *S. guianensis* cv. Libertad was the only accession to do well in the one growing season, acting like a weak perennial. The poor performance of the *Stylosanthes* species has been supported by other trials and plantings undertaken by the same programme (Lazier, unpublished data). Half of the genotypes tested gave only moderate yields initially, which can be attributed to the adaptation of these species to lighter infertile acid soils. The diploid form of *S. hamata* (Stehle, 1956; Keoghan, 1980) is an exception. *Calopogonium mucunoides* ('Calopo') and *Teramnus labialis*, which are both known to grow on heavier soils elsewhere, also failed.

2.3.5 Predicting the likely utility of the Central Farm germplasm collection

The environmental factors that govern the ability of a tropical legume to thrive are numerous and complex, and for poorly known plants they remain obscure. Day length, moisture and temperature requirements, resistance to frost and response to grazing are all interrelated, and an understanding of the impact of these variables is very expensive and time-consuming to achieve. Further critical variables such as palatability and the farming system utilized still remain to be studied. Increasingly in recent years, the effects of such issues are not explored. *Stylosanthes scabra*, *Desmanthus pubescens* and many others all lack this basic information (e.g. Cook *et al.*, 2005). Thus provenance records from the point of collection and anecdotal information are now being progressively more formally used in the evaluation process, and 'new' search engine techniques and probability mathematics, such as Bayesian statistics, are being used to aid this process. Such approaches have long been used intuitively. Knowledge that one genotype of *Desmanthus* was found growing in a Central Farm grazing experiment with the highly competitive, often exclusive, Pangola grass (*Digitaria eriantha* subsp. *pentzii*) would ensure its inclusion in many evaluation programmes. Similarly, when seeking plants for a particular soil type, such as clay soils, it is usual to list accessions found in those particular edaphic environments.

The Central Farm collection is quite possibly unique in that most of it was collected from neutral and alkaline soils. Most of the accessions are probably 'calcicoles', having what is often deemed to be a 'selective' requirement for rhizobial strains adapted to alkaline conditions. *Desmanthus* accessions are often such plants and in Australia these accept *Bradyrhizobia* from *Neptunia*, a plant native to alkaline clays. Such an adaptation is a requirement for long-term pasture yields, for appropriate rhizobia must be able to regenerate. While *Bradyrhizobia* may not survive in the soil in the long term, they may in the short term, and thus may be useful in leys and cropping systems, despite the additional cost and logistical support needed for inoculation.

The data presented here at the species level are too sparse, and the experimental screening was too short term, for sensible comparisons between accessions to be drawn to justify further detailed analysis. However, some useful hypotheses can be developed even at this early stage that can be tested for validity in later stages of evaluation, or in dedicated experiments. This approach has been proven to be useful when resources are limited and the initial objectives are difficult to select. For the present purposes the question is posed: 'Do plants collected on vertisols perform better than others, as evaluated in the Belizean environment'?

Data presented in Table 2.11 allow such a hypothesis to be drawn, although statistical analysis is not warranted. In all but two species, *Desmodium tortuosum* and *Leucaena leucocephala*, plants from the vertisols were more persistent than plants from other soils. The short-term persistence of *Leucaena leucocephala* was high, irrespective of the climate or soils from which the accessions were collected. This may be due to its large seeds, which allow early seedling growth, its strong rooting, and its tree habit. For *Desmodium tortuosum* persistence was low, with only one exception.

Table 2.11. Persistence of legumes from vertisols and non-vertisols.

	Collected from			
	Vertisol		Non-vertisol	
Species	No. accessions[a]	Mean persistence/ accession[b]	No. accessions[a]	Mean persistence/ accession[b]
Centrosema plumieri	1	3	1	0
Desmanthus pubescens	1	3	–	–
Desmanthus leptophyllus	–	–	1	2.0[c]
Desmanthus virgatus	–	–	1	3.0[c]
Desmodium tortuosum	1	0	7	1.4
Leucaena leucocephala	1	3	5	3.0
Pachyrrhizus erosus	1	3	1	0.0[d]
Rhynchosia longeracemosa	1	3	4	2.0
Rhynchosia minima	1	1	4	0.0
Vigna speciosa	1	3	1	4.0
Total no. accessions	8	–	25	–
Mean persistence/accession	–	2.4	–	1.71

[a]Included in the agronomic evaluation, more collected. [b]Measured 0 to 3 in the agronomic evaluation. [c]Included for comparative purposes. [d]Included for comparative purposes, soil of origin not known.

Similar information relating persistence of performance to the various climatic regions is not presented. Comparisons between individual plants are based on only a few accessions, and some species, such as *Leucaena leucocephala*, were only found in the wetter environments. Unlike soil groupings, climatic environments tend to 'run into' each other, thus an accession from a 'run on' niche may have in fact been collected from a wetter environment than would appear from the data available.

In the Mexican Yucatan the species *Desmodium tortuosum* and *Macroptilium atropurpureum* were widely found in semi-hot tropical conditions, but in Australia, where these plants were first domesticated, they have been found to be most useful in higher altitude tropics or in more subtropical regions (e.g. Cook *et al.*, 2005). Possibly the Yucatan material is adapted to more tropical conditions than the Australian cultivars, a hypothesis that could be tested in evaluation studies. In addition, *Neonotonia wightii* (syn. *Glycine javanica*), an African species also developed for use in subtropical conditions in Australia, proved to be well adapted to more tropical conditions when grown on calcium-rich vertisols in Barbados (Ahmad, 1980) and various other Caribbean Islands (Keoghan, 1980). All three species seem to prefer more fertile soil situations. An FAO summary refers to *N. wightii* as performing best on 'deep freely-draining latosolic soils derived from basic igneous rock or self-mulching black soils and fertile alluvials ... with pH > 6.5' and forming effective nodules with rhizobia from '*Glycine tormentela* and *Glycine tabucani*', both native plant species that thrive on heavy dark clay soils in Australia. Although an African species, *N. wightii*, like the instance mentioned above with *Desmanthus*, is pre-adapted to the more fertile heavier soils, to which it has been introduced. Edaphic rather than simply climatic adaptation may be of prime importance.

2.4 General discussion

Much of the discussion about the Central Farm germplasm collection, its assembly and initial evaluation to facilitate its immediate use has been covered in the previous section and in other related publications (e.g. Lazier, 1981a,b). Its further evaluation for use in Central America and the Caribbean has also been documented (Ahmad, 1980). It only remains to illustrate its long-term value as part of a global germplasm resource and to illustrate how this resource should be developed for future use and for the preservation of biodiversity. It is now over 35 years since the initial collection was assembled. A retrospective view provides some relevant examples for the present purpose.

2.4.1 Examples from the genus *Desmanthus*: agronomic development

At the time this work was undertaken there was a slight interest in the genus *Desmanthus* (see information in Burt, 1993), but it was poorly known. A major international review of work on legumes (Summerfield and Bunting, 1980) mentions *Desmanthus* only twice, and both references are somewhat equivocal (Chapter 4, this volume). In the same publication, the genus *Medicago*, its temperate equivalent, requires 65 page references and the up-and-coming *Stylosanthes*, then recommended for sandy and acidic tropical soils has 50 page references. *Desmanthus pubescens* received no mention at all. A taxonomic revision of this genus, essential for sensible development and utilization of a germplasm collection, was not available until 1993 (Luckow, 1993).

Part of the Central Farm collection was incorporated into what is now known as the Australian Tropical Forages Genetic Resource Collection. It was used in early 'agronomic/morphological' studies at the CSIRO Davies Research Laboratory, Townsville and evaluated in various field tests, which were later expanded to clay soil areas in western Queensland (Chapters 8 and 9, this volume). A representative range of accessions was selected to represent the genus on the basis of the earlier part of these studies and incorporated in field trials throughout the state of Queensland. In 1991 after further testing, often in subtropical areas, the first domesticant from the genus was made commercially available as the cultivars 'Uman' (*D. pubescens*), 'Marc' (*D. virgatus*) and 'Bayamo' (*D. leptophyllus*). 'Uman' was one of the Central Farm collections from near the Belize–Mexico border. Its performance since that time has been summarized by Pengelly and Conway (2000) and a fact file on the species has been provided by Cook *et al.* (2005). The latter authors note that 'Uman' often solely represents the species, and that it tolerates alkaline and clay soils, withstands heavy grazing, shows poor regeneration from seed in subtropical environments and requires 'specific rhizobia' collected from plants native on alkaline clay soils. These characteristics are very much in accordance with intuitive assumptions gleaned from provenance data from the Central Farm collection of the species (Table 2.12). Note also that the species *D. virgatus* and *D. leptophyllus* were found with other genera and species noted for their occurrence on clays (Chapters 6 and 9, this volume), including the tree legume *L. leucocephala*. Recent advances in the development of cultivars from *Desmanthus* are presented below (Chapters 10 and 11, this volume).

Cook *et al.* (2005) also suggest that *D. pubescens* has potential as a 'phase legume' in cropping systems, a type of plant very much in demand in tropical agriculture, especially such systems as the Yucatan slash-and-burn 'milpa' system. *D. pernambucanus*, introduced into India in the 1800s, has been pressed into service for this use on clay soils in that country. It acts as a protein bank and is commonly called 'hedge lucerne'. It is somewhat difficult to maintain and must be trimmed with care. *D. pubescens*, however, has many more basal branches than *D. pernambucanus* (e.g. Luckow, 1993) and forms a dense 'crown' when cut (Burt, 1986). When mixed with grass and fed to sheep as hay it greatly stimulates wool production (Rangel *et al.*, 1996). *Desmanthus* seed, often produced in large amounts, is nutritionally rich for stock (Schlink and Burt, 1993). This study has indicated that the species withstands regular cutting and produces yields equivalent to those of *Leucaena leucocephala*, a widely touted 'phase legume'. Unlike *L. leucocephala*, however, it has no potential toxicity problems. These attributes strongly support the views of Cook *et al.* (2005) on the potential of the species. Indeed, had the information from this early screening been used in a search engine to predict likely outcomes rather than intuitive assertions, the suggested use of *D. pubescens* and probably *D. virgatus* and *D. leptophyllus* may have emerged at a much earlier date.

2.4.2 Contribution of the UWI/IDRC collections to global germplasm collections

Desmanthus and Stylosanthes as examples

The UWI/IDRC project had two major field components, one in Antigua and one in Belize. Both sites placed considerable emphasis on the provision of pasture legumes for clay soils and the germplasm so obtained was shared for evaluation in these two contrasting environments. To avoid undue repetition elsewhere, the examples chosen to illustrate the value of this approach

Table 2.12. Provenance data for some early *Desmanthus* accessions in the Central Farm collection.

Species/CF no.	Latitude (°, ′N)	Altitude (m)	MAR (mm)[a]	Climate[b]	Soil	Associated species/comments
D. pubescens						
453 Sta. Elena	18° 29′	0–10	1300–1500	6	Deep black alkaline clay, coral gravels	Dry tropical deciduous forest, rich in lime-loving species
491	18° 26′	10–15	1200–1300	6	Red clay from limestone	Road edge, a few patches
543	18° 29′	0–10	1300–1500	6	Black clay soil area	Scrubby regrowth along roadside. *Rhynchosia minima*, *Desmodium incanum*, *Phaseolus elegans* and *Indigofera* in nearby guinea grass (*Panicum maximum*) pasture
551 near Chetumal (Fig. 2.1)	18° 26′	10–15	1300–1500	6	Road margin	In scrub along roadside. *Desmodium tortuosum*, *Phaseolus* spp., *Galactia striata*
D. virgatus						
317	21° 01′	0–10	940	1, 3	On road edge, heavy clay on limestone	*Desmanthus* in patches on road edge with *Rhynchosia minima* and *Indigofera littoralis* (see Fig. 2.4)
448 near Kabah	20° 11′	100–150	1000–1200	3	In low dip, brown powdery soil	In low dip, *Leucaena leucocephala*, thorny scrub (see Fig. 2.4)
649 near Tucacab	20° 30′	40	1200	3	Deep clay, gravelly and sandy	Occasional on road edge and margin, scrub with some buffel grass (*Cenchrus ciliaris*). *Centrosema schottii*, *Rhynchosia minima*, *Indigofera* spp. Nearby and in pasture *Leucaena leucocephala*, *Cracca greenmanii*
617	20° 25′	100	1125	3	Thin red soil among rocks	Common on road edge with *Stylosanthes calcicola*, Jaragua grass (*Hypharrenia rufa*), *Indigofera littoralis* (common), *Centrosema schottii* (common)
626	20° 30′	50	1100	3	Thin red soil among rocks in road cut	Jaragua grass (*H. rufa*) backed by scrub. On roadside and margin. *Centrosema virginianum* (common), *Stylosanthes calcicola* (very common), *Desmodium incanum*, *Desmodium tortuosum*
D. leptophyllus						
343 near Progreso (Fig. 2.1)	21° 5′	0–10	800–1000	1, 3	Very shallow on cracking soils	Plants on road edge near maize and bean patch. *Leucaena leucocephala* frequent in bare areas
					Powdery, dark brown soil on cracked limestone pavement and rubble	*Galactia striata*, *Centrosema virginianum*
406 near Valladolid (Fig. 2.1)	20° 18′	10–50	1300–1400	6	Road margin	Dry scrub: frequent in patches on roadside. *Rhynchosia reticulata*, *Galactia striata*, *Desmodium incanum*, *Centrosema elegans*

[a]Rainfall data would not accurately reflect the moisture available to road edge collections. [b]Climate as described in the text, and shown in Tables 2.4 and 2.6, and Fig. 2.2. MAR, Mean annual rainfall.

Fig. 2.4. Yucatan collection sites of *Desmanthus virgatus*: (a) CF 448 near Kabah; (b) CF 317 on the Hunucma–Sisal road.

are *Desmanthus* and *Stylosanthes hamata*, the first being shown to be of value in the Belizean component of the work and the latter in the Antiguan section.

DESMANTHUS. In their genetically based study of the *Desmanthus* accessions in the Australian Tropical Forage Resource Collection, Pengelly and Liu (2001) draw attention to the contributions of the Central Farm collection. *D. pubescens* is noted to be 'restricted in its natural distribution to ... the Yucatan Peninsula and surrounding regions' and 'A large number of accessions in the collection were from Belize'. A close examination of the information presented in that study indicates that some *D. pubescens* accessions were genetically similar, and must be placed in a single group (the Pengelly and Liu group 4). However, other accessions were sufficiently genetically different as to be placed in four other groupings (G3, G7, G8 and G12), all located in a different branch of the hierarchical classification. In these groups the accessions concerned are deemed to be more similar to other species – *D. tatuhyensis, D. glandulosus, D. aciminatus, D. papalaceous, D. leptophyllus* and *D. virgatus* – than to other members of the main *D. pubescens* group. The accession deemed to be the most divergent, CF 551 from the Yucatan, was placed with *D. glandulosus*, found only in Texas, USA and Northern Mexico.

The bulk of the accessions of *D. virgatus* are placed in the Pengelly and Liu group 1. However, three of these accessions are noted to be very different. Two are from the Caribbean islands – Cuba (CPI 79653) and Antigua (CPI 73446) collected by the UWI/IDRC project. The third is CF 649, collected from the Yucatan but grouped genetically with *D. leptophyllus*.

Cook *et al.* (2005) commented on the lack of collections of *Desmanthus* from the Caribbean islands. In view of the increasing pressure on land, and the effects of global warming and the potential utility of *D. virgatus* and *D. leptophyllus*, further collection of these species is urgently needed. As described in more detail elsewhere (Chapters 4, 5 and 9, this volume), there are at least two other *Desmanthus* species where collection to preserve biodiversity is warranted due to the rarity of their occurrence and pressures on their environments, *D. oligospermus*, which is unique in having non-dehiscent seed, and *D. interior*, which is found in high-altitude grasslands and is palatable to stock.

STYLOSANTHES HAMATA SENSO STRICTO. It has long been known that this species is widely distributed throughout the Caribbean islands, where it is a valued grazing plant (Stehle, 1956; Keoghan, 1980). Although present in global germplasm collections, such accessions were few in number; in 1969 the ATFGU collection contained only eight accessions of this highly variable species (Chapters 7 and 9, this volume). Work by the project clearly demonstrated the value of *S. hamata senso stricto* (Keoghan, 1980) and it has proved to be well adapted to clay soils in India. Indeed, in a widespread evaluation programme, the sole representative of this species was deemed to be 'the most consistent across all sites' (Ramesh *et al.*, 2004). The Australian Tropical Forage Genetic Resource Collection contains just over 200 accessions of *S. hamata sensu stricto*, of which more than 100 are from the Caribbean islands, and 92 were provided by the project. These are quite variable (Chapter 3, this volume), making a very valuable addition to the germplasm bank for the species. Because many were collected as part of an ecological appraisal, it has been possible to gain an insight into their climatic and edaphic adaptation (Chapters 3 and 9, this volume) and to examine this further in evaluation studies.

Other species that warrant further study

The examples quoted above have shown that information recorded while the collection was assembled has been proved useful in predicting the potential utility of the plants involved in the preliminary evaluation work at Central Farm. Can these same ideas now be applied to other material in the collection? One general principle applies to most of the collections from the Yucatan. They are adapted to alkaline soils and most are unlikely to require pelleting with rhizobial inoculation when used on similar soils elsewhere.

SPECIES WITH EXISTING CULTIVARS. The genus *Desmanthus* contains three species with cultivars, *Desmanthus leptophyllus*, *D. pubescens* and *D. virgatus*. Despite these species having shown great potential, little has been done to capitalize on the range of variation within the genus (e.g. Chapters 4 and 9, this volume). Pengelly and Liu (2001) state that for efficient use of existing germplasm collections, 'careful selection of core representatives needs to be undertaken before full evaluation of these taxa can take place'. Because there is a total existing collection of some 400 *Desmanthus* accessions (some 50 of which are duplicates), it should be possible to explore the agronomic variation within one species with particular potential, for instance the 30 or so accessions of *D. pubescens*. Although most of these are deemed to be genetically uniform, they may in fact have considerable agronomic variation, as is the case with *D. virgatus*.

Desmodium tortuosum: Once highly regarded as a pasture species in subtropical environments, this species was found over a range of soils and climates throughout the Yucatan. As noted elsewhere, the species is very variable in terms of persistence and yield. High yielding/low persistence forms might offer possibilities as 'green manure' or 'cover crops' in cropping systems. There is deemed to be a requirement for such plants in 'milpa' farming in the Yucatan (Flores, 1995; Parsons, 1996) and familiar indigenous plants may be more readily accepted by farmers than exotic introductions.

Leucaena leucocephala: In this study some locally available accessions of *L. leucocephala* compared favourably with the two cultivars used as standards and one (CF 469) has the added advantage that it is known to grow on the local vertisol. The species had also performed well when planted with grasses under grazing in a trial run concurrently (Lazier, 1978).

Macroptilium atropurpureum: One accession, CF 223, showed better growth than the cultivar 'Siratro' in the cool and dry periods. It is clearly adapted to a range of tropical conditions, due to the contrast between the environment in which it was collected and that of Central Farm, where it was tested. Siratro showed good productivity in mixed sward with grasses under cutting and grazing in a trial run concurrently (Lazier, 1978).

SPECIES WITH NO EXISTING CULTIVARS. *Centrosema schottii*: A very high-yielding annual, collected over a range of environments from rendzina soils, this species showed considerable variation. It might well be of value in cropping systems and has shown promise in Australia (Pengelly and Conway, 2000). It is found in disjunct regions across a wide latitudinal range (Clements and Williams, 1980) and may show considerable genetic variation. This may account for the large number of synonyms for this species. Further collection and evaluation for use in clay areas may be warranted.

Desmodium incanum: Long known to have use in agroforestry and as a persistent component of long-term pastures, most attention has been placed on this species in the Pacific Islands, where it has been introduced. In Central America, as is shown here, it is adapted to a wide range of soils and climates. However, as 'familiarity breeds contempt', it has sometimes been manually removed from grass pastures. Native to the Americas, from southern USA to Argentina, it is a very variable species growing over a great range of soils and climates. Some form of agronomic/morphological classification is warranted (see comments on the *Desmanthus* species).

Galactia striata: Like *Desmodium incanum*, this well-known species was found across a wide range of soils and climates in the Mexican Yucatan Peninsula and the Central Farm collection and shows a great deal of agronomic and morphological variation. It is a slender, much branched, climbing perennial producing good yields, outperforming many 'standard' cultivars such as Siratro and Stylo on sandier, very acid 'cerrado' soils in Brazil and also occurring plentifully in the area surveyed in this study on alkaline rendzinas and lithosols. This wide variation was mirrored in the wide range of performance patterns described herein (Appendix 2.1). While production of many of the accessions in this trial was not high, several accessions tested had yield patterns indicating persistence and improving productivity over time. Accessions CF 416 and CF 374 had reasonable dry matter yields. Further studies, possibly based on an ecogeographic survey (Guarino *et al.*, 1999), are warranted.

Pachyrrhizus erosus: A very low-yielding perennial, but capable of growth on a vertisol, this is known as 'Mexican potato' or 'Yam Bean' and has an edible tuber. A medicinal herb, it contains some potentially anti-nutritional factors and is said to have weed potential in Taiwan.

Vigna speciosa. Collected from a vertisol, this plant showed high persistence and yield in the screening trial. Known as 'Wandering Cowpea' or 'Prairie Vetch', it is a very vigorous vine that can be invasive if underutilized.

2.4.3 The significance of the UWI/IDRC collection

Although small by international standards, the project described has had some major global impacts on the search for and development of tropical pasture legumes, and the development of a bank of relevant germplasm resources. It has, for instance, drawn attention to a number of previously unknown or understudied species, including *Acacia angustissima*, *Centrosema macrocarpum*, *Codariocalyx gyroides*, *Desmanthus pubescens* and *Desmodium cinerium*, most of which are now being promoted by fodder development programmes for appropriate environments and farming systems. In the case of *Desmanthus pubescens*, a previously undomesticated species, it provided virtually all of the accessions to be found in germplasm banks, including the sole cultivar, cv. 'Uman'. It has provided accessions with agronomic characteristics different from those in established cultivars, drawn attention to other species of interest and shown that some of the species and cultivars deemed to be suitable for tropical climates are simply not adapted to heavy non-acid soils in wet tropical conditions.

Agronomists have long used provenance data, and the results of simple agronomic observations, to predict the likely utility of pasture legumes (e.g. Burt and Ebersohn, 1969). When applied to the present collection, this heuristic approach proved to fairly accurately predict results obtained at a much later date, after considerable investment in time and facilities. It seems plausible to believe that modern 'search engines' using a Bayesian probabilistic approach could sensibly be applied to such data. However, this requires the efforts of a suitably qualified team in a dedicated research centre.

This project, with the two field components in Belize and Antigua, was successful because it was strategically sited in two legume-rich floristic zones, each with a flora with many endemics. It had access to relevant, contrasting climates and soils for evaluation. At this stage of the collection and evaluation of genetic resource material, where even the taxonomic status of the plants is uncertain, simple tests on the soils of interest were sufficient to show major differences between genera and species and to illustrate major features of interest. Once these have been determined, and material selected for testing, then more detailed analysis of yield and other attributes may be needed. This requires the facilities found in major research establishments, references for taxonomic identification, and relevant specialists with appropriate facilities including large-scale drying and analytical laboratories, ecogeographic analysis for species with wide distributional patterns etc.

In the past, projects such as the one described here have provided information and germplasm collections for some of the major research institutes. The latter, in turn, have been able to demonstrate, for instance, the genetic variation to be found within and between species. This information can be used with other phytogeographic information to show the need for further plant collection in other regions. Cook *et al.* (2005) mention the need for further collections of *Desmanthus virgatus* and *D. leptophyllus* from the Caribbean archipelago and specify the lack of collections of this genus from the coast of Ecuador, Peru and the Galapagos.

Previous studies (Burt and Reid, 1976) noted that Central America and the Caribbean is a very distinct phytogeographic zone, with very different legume floras (Chapters 7 and 9, this volume). Unlike most of South America, the region in which tropical forage legumes are most commonly sought, they have clay soils on which *Desmanthus* species – many not featuring in germplasm collections – are to be found (Luckow, 1993).

There is a need for legumes adapted to clay soils in the high-altitude tropics, for Ethiopia alone has some 10 million ha of such soils (Dudal and Bramao, 1965). The planting of adapted species, possibly *Desmanthus interior*, might reduce fertilizer use in such environments. It is found in scrub grassland at 1850 m to 2000 m from 17°N to 25°N in Mexico, where it is 'intensively grazed, which probably contributes to the rarity of this species' (Luckow, 1993). The collection and conservation of this biodiversity is important. However, because it is difficult to time visits so as to find plants in seed, small, on-site research projects undertaking collection and initial evaluation for specific environments within the target environment (as was the case with the UWI/IDRC project) may offer the best chance of success. Such projects have the opportunity to access local anecdotal information about the species concerned, which can be of considerable value.

Appendix 2.1

Table 2.13. Provenance and agronomic characters of accessions planted in trial (best results shaded).

Accession		CF no.[a] or cv.	Provenance		Agronomic			Yield (g/plant)	
			Climate[b]	Soil	Surviving plants[c]	Month of last yield	Yield pattern[d]	Total[e]	Cool/dry season
Aeschy-nomene	*americana* var. *glandulosa*	76	6	fluvisol[f]	0	12	10	77	0
	elegans	279	7	oxisol	0	11	10	22	0
Alysicarpus	*vaginalis*	202	6	fluvisol[g]	1	–	7	28	4
Calopogonium	*mucunoides*	75	9	fluvisol[f]	0	1	10	100	6
Cassia	*aeschynomene*	292	7	arenosol	0	1	10	115	1
	sericea	437	2	road fill	0[j]	1	10	78	10
Centrosema	*plumieri*	28	6	–	2[j]	–	2	442	128
		16–2	5	–	3	–	1	364	107
		16–1	6	fluvisol[f]	1	–	6	286	98
		458	6	vertisol[h]	3	–	5	276	106
	pubescens	Centro	–	–	3	–	4	304	108
		234	–	–	3	–	4	285	75
		461	6	–	2	–	6	233	77
		233	–	–	2	–	5	221	59
		IRFL 980	–	–	1	–	4	173	62
		231	–	–	2	–	4	173	58
		IRFL 1239	–	–	2	–	4	168	93
		236	5	–	2	–	4	127	39
		6–3	5	fluvisol[f]	1	–	6	81	36
		218	–	–	0	12	10	19	0

	brasilianum × *virginianum*	CIAT 17-33-F7	–	–	2	–	6	198	90
	schottii	443	1	rendzina[i]	2	–	6	515	106
		363	1	road fill	1	–	6	366	69
		430	2	rendzina[h]	1	–	6	353	112
		329	2	rendzina[h]	0	7	10	293	81
		388	6	rendzina[h]	0	1	10	235	22
		452	1	rendzina[h]	0[j]	3	10	110	11
	virginianum	256	5	–	1	–	6	60	17
		220	–	–	1	–	6	49	7
		352	5	–	1[j]	–	7	28	0
		219	5	rendzina[h]	0[j]	3	7	22	5
		255	5	–	0	11	7	17	0
		221	–	–	0	12	7	10	
		222	–	–	0	11	7	6	
Chaetocalyx	*scandens*	351	1	lithosol	1	–	7	7	
Clitoria	*rubignosa*	204	6	planosol	0	3	7	10	3
	ternatea	212	5	–	2	–	4	90	17
		213	–	–	2	–	4	70	21
		211	–	–	2	–	7	46	11
		51	8 b	arenosol	0	7	7	41	3
		78	10	andosol	0	1	7	23	1
Cracca	*greenmanii*	339	1	lithosol	1	–	4	27	0
		319	2	rendzina[h]	0	4	7	8	4
		376	2	fluvisol[f h]	0	1	7	4	2
	spp.	379	4	rendzina[h]	2	–	7	15	5

Continued

Table 2.13. Continued.

Accession		CF no.[a] or cv.	Provenance		Agronomic			Yield (g/plant)	
			Climate[b]	Soil	Surviving plants[c]	Month of last yield	Yield pattern[d]	Total[e]	Cool/dry season
Crotalaria	*pumila*	368	6	road fill	1	–	6	108	24
		142	7	ferralsol	1	–	6	52	4
	sagittalis var. *fruticosa*	278	7	ferralsol	0	12	7	4	0
Desmanthus	*pubescens*	453	6	vertisol	3	–	1	340	108
	leptophyllus	343	2	rendzina[h]	2	–	6	310	80
		406	2	rendzina[h]	2	–	6	215	31
	virginianum	448	1	road fill	3	–	5	116	40
Desmodium	*barbatum*	146	6	ferralsol	1	–	7	39	6
	barbatum	143	7	ferralsol	0	5	7	30	1
Desmodium	*incanum*	417	4	rendzina[i]	3	–	3	144	63
		364	1	road fill	3	–	5	132	67
		420	4	road fill	3	–	6	99	53
		440	2	road fill	1[j]	–	6	73	29
		399	6	road fill	2	–	6	72	35
		313	–	–	1	–	6	68	30
		64	7	ferralsol	1	–	1	48	13
		424	4	rendzina[h]	1	–	7	25	10
	distortum	Q 8378	–	–	1	–	6	485	155
	heterophyllum	Johnstone	–	–	3	–	2	185	53
	intortum	139	7	ferralsol	0	3	7	18	9

	tortuosum	476	4	rendzina	1	–	5	284	17
		470	4	lithosol	3	–	5	228	44
		387	6	lithosol	2	–	5	181	50
		481	4	rendzina	2	–	5	138	6
		455	6	vertisol[h]	0	3	10	131	19
		79	11	andosol	1	–	6	122	19
		366	6	road fill	0	5	10	114	22
		332	2	rendzina	0	3	10	82	8
		485	4	road fill	1	–	4	54	30
Galactia	*striata*	416	4	rendzina[h]	3	–	2	212	74
		73	9	fluvisol[f]	2	–	5	190	37
		374	2	lithosol	3	–	4	143	33
		378	4	rendzina[h]	2	–	2	128	48
		330	2	rendzina[h]	3	–	5	97	38
		482	2	lithosol	2	–	2	91	31
		347	2	lithosol	1	–	1	80	30
		356	1	lithosol	2	–	1	78	20
		402	2	rendzina[h]	1	–	6	69	20
		484	4	rendzina[h]	1	–	7	39	16
		346	2	lithosol	0	5	7	10	10
		471	4	rendzina[h]	0	11	7	2	0
Indigofera	*mucronata*	489	2	road fill	1	–	6	42	11

Continued

Table 2.13. Continued.

Accession		CF no.[a] or cv.	Provenance		Agronomic			Yield (g/plant)	
			Climate[b]	Soil	Surviving plants[c]	Month of last yield	Yield pattern[d]	Total[e]	Cool/dry season
Leucaena	*leucocephala*	Cunning-ham	–	–	3	–	1	464	143
		496	4	lithosol	3	–	9	438	146
		469	6	vertisol[h]	3	–	9	401	144
		325	6	lithosol	3	–	1	396	172
		334	2	lithosol	3	–	9	382	139
		208	–	–	2	–	9	365	140
		Peru	–	–	3	–	1	340	103
		209	–	–	3	–	1	314	113
		447	1	rendzina[i]	3	–	1	303	108
		206	5	–	3	–	9	293	105
		205	5	–	2	–	9	281	101
		395	6	lithosol	3	–	9	228	84
		207	5	–	1	–	9	92	36
Macroptilium	*atropurpureum*	Siratro	–	–	3	–	1	362	72
		223	–	–	3	–	1	356	171
		360	1	lithosol	3	–	2	340	92
		386	6	rendzina[h]	3	–	1	320	73
		371	6	lithosol	3	–	1	319	91
		409	2	rendzina[h]	3	–	2	310	86
		217	2	–	3	–	1	252	62
		411	2	rendzina[h]	2	–	1	248	62

		396	6	lithosol	3	–	2	245	69
		439	2	road fill	1	–	6	235	62
		442	2	road fill	1	–	7	83	33
	lathyroides	373	2	fluvisol[f]	0	8	10	90	31
		240	–	–	0	3	10	33	7
		239	5	–	0	7	7	10	2
Macrotyloma	*axillare*	178	–	–	0[j]	8	5	194	72
Pachyrrhizus	*erosus*	501	6	vertisol[i]	3	–	4	42	10
		238	–	–	0	12	7	13	0
	palmatus	408	2	rendzina[h]	0[j]	1	7	7	1
Rhynchosia	*longeracemosa*	457	6	vertisol3	3	–	5	228	93
		367	6	rendzina[h]	2	–	2	169	59
		419	4	road fill	3	–	2	139	58
		422	4	rendzina[h]	2	–	6	138	59
		493	2	road fill	1	–	6	116	41
	minima	429	2	rendzina[h]	0	4	10	279	11
		460	6	vertisol[h]	1	–	6	202	41
		321	1	road fill	0	3	10	166	3
		487	2	rendzina[h]	0	7	10	140	44
		385	4	rendzina[i]	0	7	10	129	30
		215	5	–	1	–	5	75	9
	yucatanensis	413	4	rendzina[h]	2[j]	–	4	124	54
Stylosanthes	*calcicola*	451	1	road fill	2	–	7	41	7
		435	2	road fill	0	3	10	34	1
		365	6	road fill	1	–	7	29	2

Continued

Table 2.13. Continued.

Accession		CF no.[a] or cv.	Provenance		Agronomic			Yield (g/plant)	
			Climate[b]	Soil	Surviving plants[c]	Month of last yield	Yield pattern[d]	Total[e]	Cool/dry season
		441	2	road fill	1	–	7	29	7
		449	1	road fill	1	–	7	18	3
		423	4	road fill	0	3	7	16	5
		393	6	road fill	1	–	7	15	3
	hamata	251	5	–	1	–	6	160	20
		25	5	–	1	–	6	148	15
		253	5	–	2	–	5	103	17
		246	5	–	0	8	10	100	26
		245	–	–	2	–	5	98	13
		244	–	–	2	–	6	91	22
		CIAT 122	–	–	0	3	10	86	19
		249	5	–	1	–	6	84	12
		243	–	–	1	–	6	82	27
		247	5	–	2	–	6	64	7
		250	5	–	1	–	6	61	12
		12	5	–	0	3	10	44	7
		151	6	planosol	0	5	7	24	9

	guianensis	Libertad	–	–	1	–	6	444	148
		81	11	lithosol	0	7	10	285	80
		507	6	planosol	2	–	5	239	106
		504	7	ferralsol	3	–	5	232	106
		119	7	vertisol[i]	0	8	5	205	118
		509	7	ferralsol	2	–	8	176	102
		165	–	–	1	–	8	106	85
		Endeavour	–	–	1	–	4	97	30
		163	–	–	1	–	6	92	48
		520	6	road fill	1	–	6	78	23
		505	7	ferralsol	1	–	6	78	42
		109	7	lithosol	0	1	7	27	3
	mexicana	82	11	lithosol	1	–	10	128	1
		72	10	andosol	0	3	10	99	4
	scabra	CPI 40205	1	–	1	–	8	114	65
	viscosa	108	7	lithosol	0	7	6	57	25
Teramnus	*labialis*	259	5	–	0	4	10	194	81
		230	–	–	1	–	6	151	70
		258	5	–	0[j]	5	10	137	60
		226	5	–	2	–	6	120	53
		224	5	–	2	–	6	115	52
		260	5	–	0	8	7	48	23
		293	6	arenosol	0	1	10	38	2
	uncinatus	IRFL 1550	1	–	1	–	2	291	99
		25	6	–	1	–	10	89	20

Continued

Table 2.13. Continued.

Accession		CF no.[a] or cv.	Provenance		Agronomic			Yield (g/plant)	
			Climate[b]	Soil	Surviving plants[c]	Month of last yield	Yield pattern[d]	Total[e]	Cool/dry season
Tephrosia	*littoralis*	336	1	arenosol	1	–	6	57	8
	noctiflora	158	–	vertisol[i]	0	8	8	164	103
Vigna	*luteola*	27	6	–	1	–	6	225	96
	speciosa	466	6	vertisol[h]	3	–	1	360	105
		522	7	–	3	–	4	286	105
		427	2	rendzina[h]	1	–	4	139	58
	speciosa (?)	490	2	rendzina[h]	3	–	2	109	25

[a]Accession number (Central Farm); [b]climate of origin as defined in text; [c]of 3 original plants; [d]as shown in Table 2.10; [e]total yield least significant difference (P=0.05) 14; [f]recent deposits; [g]older deposits; [h]less than 15 cm deep, 4–15 cm or deeper: 5 of 3 original plants; [i]15 cm or deeper; [j]lines present in 2 replicates at the beginning of the trial.

Appendix 2.2

Table 2.14. Number of collection sites for climates and soils in the Mexican Yucatan Peninsula. All sites also included areas with road fill substrate.

	Number of sites/climate						
Soils	1	2	3	4	5	6	15
Skeletal	4	4	1	5		2	
Arenosol	2		3				
Rendzina ≤15 cm	2	47	27	21	1	16	2
Rendzina >15 cm		4	4	4		1	
Vertisol ≤15 cm		4		10		7	
Vertisol >15 cm			1			1	
Swamp	2			1			
Total sites examined	10	59	36	41	1	27	2

Table 2.15. Climate type[a] and frequency of occurrence[b] of legume species collected on acid and basic soils of entire Belize germplasm collection to July 1977 from Belize, the Yucatan Peninsula, Mexico and the Peten, Guatemala.

Accession		Soil type											
		Basic							Acid				
					Rendzina		Vertisol						
		Skeletal	Road fill	Alluvium	≤15 cm	>15 cm	≤15	>15	Skeletal	Arenosol	Ultisol	Oxisol	Metamorphic parent material
Aeschynomene	*americana* var. *americana*											7[c]	
	americana var. *glandulosa*			6[c]									
	elegans											7[c]	
	fascicularis				1 (50)								
	spp.				1 (11) 2 (4)		6 (17)						
Alysicarpus	*vaginalis*			6[b]									
Calliandra	*haustoniana*										6[b]	7[b]	
Calopogonium	*caeruleum*			6[b]									
	galactoides			6[c]									
	mucunoides			6[b]	3 (7)		15 (50)	3 (100)					7[c]
Canavalia	*mexicana*	4 (20)											
	villosa			6[b]									
	spp.			6[c]	4 (5)								
Cassia	*aeschynomene*			6[c]								7[c]	
	bartletii										6[b]		
	diphylla										6[b]		
	mimosoides			6[c]									
	sericea	2 (25)											

	tagera										6[a]		
	tora			6[c]	6[c]								
	cf. *tetraphylla*											7[c]	
	spp.										6[b]	7[b]	
Centrosema	*plumieri*			6[b]	2 (4) 5 (100) 6[b]		6 (17)	12[c]					
	pubescens			6[a]	6[b]			12[b]			6[c]		
Centrosema	*schottii*		3 (11) 5 (100) 6 (4)		2 (11) 3 (30) 6 (12)	2 (25) 3 (100)							
	virginianum	3 (100)	1 (10) 2 (2) 4 (2) 5 (100)	6[c]	2 (19) 3 (11) 6 (12)	3 (25) 4 (25)						7[c]	
	spp.	2 (25)	4 (2) 15 (50)		2 (13) 3 (7) 4 (5) 6 (19)	3 (25)	6 (50)	12[c]					
Chaetocalyx	*scandens*	3 (100) 4 (20)	3 (2) 2 (5) 4 (2) 6 (7)		2 (4) 3 (3) 4 (14) 5 (100) 6 (6)	2 (25)							
Clitoria	*guianensis*										12[b]	7[b]	
	rubignosa										12[c]		
	ternatea		2 (2)				2 (30)		8[c]				
Cracca	*greenmanii*	6[c]	2 (2)		2 (2) 1 (50)	4 (25)							
	spp.				4 (5)	4 (25)							

Continued

Table 2.15. Continued.

Accession		Soil type											
		Basic							Acid				
					Rendzina		Vertisol						
		Skeletal	Road fill	Alluvium	≤15 cm	>15 cm	≤15	>15	Skeletal	Arenosol	Ultisol	Oxisol	Metamorphic parent material
Crotalaria	*pumila*		3 (3)		3 (3) 6 (6)	13[b]	4 (10)					7[b]	
	retusa	6[b]											
	sagittalis var. *fruticosa*										6[c]	7[c]	
	verrucosa		8[c]										
	vitellina		6[c]	6[c]								7[c]	
	spp.											7[c]	
Desmanthus[c]	*leptophyllus*		2										
	pubescens		6, 4			6							
	virgatus		2, 3		6			6					
	spp.		1 (10) 2 (8) 3 (11) 4 (5)	6[b]	2 (10) 1 (50) 6 (6)	3 (25)	4 (20) 6 (50)						
Desmodium	*axillare*			6[c]								7[c]	
	barbatum										6[a]	7[a]	
	incanum	4 (40)	2 (19) 4 (4) 5 (100) 6 (26)	6[a] 8[c]	2 (13) 3 (11) 4 (33) 6 (6)	4 (25) 6[a]	2 (25) 4 (20) 6 (50) 6[b]	3 (100) 6[a]				7[c]	
	cinerium			6[c]									
	intortum			6[c]								7[c]	
	scorpiurus			6[a] 8[a]	2 (4)								

Galactia	*anomala*											7[b]	
	striata	4 (80)	2 (15) 3 (14) 4 (15) 6 (15)	6[c]	2 (38) 3 (26) 4 (29) 6 (25) 7[a]	3 (50) 4 (75) 6 (100) 7[a]	2 (38) 4 (20) 6 (17)	3 (100)		3 (33)		7[b]	
Indigofera	*lespedeziodes*											7[c]	
	mucronata			6[b]	2 (2) 3 (7) 6 (6)		6 (33)	3 (1?) 12[b]					
Leucaena	*leucocephala*	3 (100) 2 (50)	2 (3) 3 (14) 4 (5) 15(50)		1 (50) 3 (7) 4 (5) 6 (6)	2 (25) 4 (50)	2 (25) 6 (17)			3 (33)			
Macroptilium	*atropurpureum*		3 (11) 2 (3) 6 (15)		2 (4) 3 (7) 6 (31)	2 (25)	15 (50)	3 (100)		3 (100) 7[c]		7[c]	
	lathyroides				2 (2)	4 (25)	6[a]						
	spp.		4 (2)									7[c]	
Mucuna	*andreana*			6[c]									
	pruriens	3 (100)		6[b]	2 (2) 3 (3) 4 (5) 6 (6)		6 (33)						
Pachyrrhizus	*erosus*			12[c]									
	palmatus				2 (2)						6[b]	7[b]	
Phaseolus	*linearis*											7[b]	
	lunatus	3 (100)		6[b]	2 (2) 4 (5)		4 (10) 6 (17)						

Continued

Table 2.15. Continued.

Accession		Soil type											
		Basic							Acid				
					Rendzina		Vertisol						
		Skeletal	Road fill	Alluvium	≤15 cm	>15 cm	≤15	>15	Skeletal	Arenosol	Ultisol	Oxisol	Metamorphic parent material
Rhynchosia	*longeracemosa*		3 (3) 4 (2)	6^{b}	2 (2) 4 (5)		2 (25) 6 (17)						
	minima	4 (20)	3 (6) 4 (5) 10 (30) 6 (4) 2 (12)	6^{a}	2 (19) 3 (30) 6 (19)	3 (50) 4 (50) 6 (100)	2 (25) 6 (50) 15 (50)						
	reticulata		4 (12) 3 (3)		2 (4) 3 (3) 4 (5)								
	yucatanensis		3 (3)		4 (5)								
	spp.	4 (20)	2 (3) 4 (2)		2 (11) 3 (6) 6 (12)	4 (25)							
Stylosanthes	*calcicola*		3 (3) 4 (2) 6 (7)		2 (4) 3 (3)								
	calcicola?	2 (25)	2 (20) 3 (8) 15 (50)		3 (3) 6 (6)					3 (33)			
	guianensis								7^{b}	7^{c}	6^{b}	7^{b}	7^{b}
	ingrata								7^{c}	7^{b}		7^{b}	
	viscosa									7^{b}		7^{b}	
	spp.				2 (2)				7^{c}		6^{c}	7^{b}	7^{c}

Tephrosia	*littoralis*		2 (12) 3(38) 6 (4) 15 (100)		3 (25)					3 (33) 1 (50)			
	nitens											7[a]	
	noctiflora												
	spp.				2 (2)								
Teramnus	*labialis*									1[c]			
	uncinatus			6[b]	6 (7)								
Vigna	*luteola*									7[c]			
	speciosa	3 (100)	2 (2) 3 (3) 6 (4)		2 (15) 4 (10) 6 (12)	4 (50)	2 (25) 6 (33) 6[b]						
	vexillata			6[b]	6[c]								
Zornia	*reticulata*											7[a]	
	spp.										6[c]		

[a]Climatic type: Initial number in cells in table as defined in Table 2.6. [b]Frequency: Mexican accessions given in brackets as percentage of sites examined on that soil (Appendix 2.2); other accessions (Belize, Peten): (a) ubiquitous, (b) common, (c) rare. [c]As *Desmanthus* species were not defined at the time of collection, climates of collection are given for the individual species, but not percentage of collection sites. The remainder of the data for the *Desmanthus* collections are presented under *Desmanthus* sp.

3 Developing and Utilizing Genetic Resource Collections: An Example From the Tropical Leguminous Genus *Stylosanthes*

R.L. Burt†

Abstract

A experiment was undertaken to explore the patterns of phenotypic variation within the Australian collection of *S. hamata* and accessions of the related species *S. calcicola*, *S. humilis*, *S. scabra*, *S. subsericea* and *S. sympodialis*. Hierarchical and minimum spanning tree classifications were compared, with the latter found more useful. The results provide a synoptic overview of the collection and thus a convenient framework that can include relevant information from other disciplines and sources and can be used to aid the efficient use of genetic resources. Examples are provided showing how the approach assists in choosing a range of representative material for testing from a large collection. Included is a general discussion on the collection, evaluation and utilization of genetic resource collections, primarily within *Stylosanthes*, but also drawing comparisons with relevant work with other tropical leguminous genera.

3.1 Introduction

The bulk of the tropics is non-humid (e.g. Papadakis, 1966), and *Stylosanthes* is by far the most important source of pasture legume cultivars for such regions (Burt and Williams, 1975; Edye and Maass, 1997; Cook *et al.*, 2005). Although some *Stylosanthes* species have long been known to be of value by local graziers – *S. scabra* was known locally as 'alfalfa do nordeste' in Brazil – the first domesticants of the genus for use in the dry tropics were not made available until the late 1960s. These were of the accidentally introduced and adventive *S. humilis*, 'Townsville Stylo', a species that out-performed any other cultivars available at that time. This was quickly followed by the commercial release of cultivars of two other species new to domestication, *S. hamata* (cv. 'Verano') and *S. scabra* (cv. 'Fitzroy' and cv. 'Seca'). These two species are the most important globally for sandy-textured tropical soils of medium acidity (Cook *et al.*, 2005). Other 'new species', *S. capitata* and *S. macrocephala*, were developed subsequently for much more acid soils in South America.

There are, however, major areas of heavier, often alkaline clay soils throughout the world that are of particular agronomic and economic significance because of their ability to 'hold water into the dry season' and their generally higher soil fertilities (Dudal and Bramao, 1965); many are used for cropping or ley farming systems. It has long been known that some legumes thrive in these situations, for instance *Stylosanthes hamata sensu stricto* (Stehle, 1956; Burt and Miller, 1975; Spurgeon, 1977) and *Desmanthus* (see information quoted in Burt, 1993a), but few cultivars have been made commercially available from these sources. Those that are available are the *S. humilis* cv. 'Khon Kaen', in Thailand, one promising accession of *S. hamata* in Java and India (Ramesh *et al.*, 2004) and other local types in the Caribbean (Spurgeon, 1977). Cook *et al.* (2005) noted that these, together with *S. seabrana* cv. 'Primar' and cv. 'Unica' (for use in ley systems in more subtropical areas), are the only *Stylosanthes* plants available for these soils. Their use is not problem-free because their productivity can be reduced by the fungal disease anthracnose (Chakraborty, 2004).

Research interest in *S. hamata* commenced at roughly the same time as the start of the UWI forage research programme reported in this volume. While only one accession was made by the Belize-based programme, on a sandy coastal soil, in common with all of the *Stylosanthes* species and genotypes tested, it did not persist in programme plantings. However, there was particular interest in the numerous collections of lines made in the West Indian islands by the Antigua-based UWI programme.

The rapid development of the 'new species' *S. hamata* and *S. scabra* was made possible by provision of what were termed 'morphological–agronomic' descriptions (Burt *et al.*, 1971; Edye *et al.*, 1974): these greatly aided communication between would-be users of the available genetic resources and allowed a coordinated approach to be developed (Burt, 1983). Genetic resource collections have grown markedly since that initial work was undertaken. At the time there was, for instance, only one accession of the tetraploid type of *S. hamata*, and this became cv. 'Verano'; by 1997 there were over 170 accessions (Edye and Maass, 1997) and there was additionally over 130 accessions of a diploid type termed *S. hamata sensu stricto*. The methods of data analysis employed with the earlier collection (less than 300 accessions and 17 named species) were found to be inappropriate for dealing with the continuous variation of characteristics encountered within a single species (Burt and Williams, 1979b) or within closely related plant forms (e.g. Jansen and Edye, 1996). There have been no morphological–agronomic descriptions (MA descriptions or MA groups) published for the expanded collections of *S. hamata* and the relationships between the various forms of *S. hamata* (as listed in Cook *et al.*, 2005) with each other and with *S.* sp. aff. *hamata*, *S.* sp. aff. *scabra* and *S. seabrana* (Burt, 1984a; Edye and Maass, 1997; Cook *et al.*, 2005) are difficult to visualize. This lack of published information on the species could help to explain why it has not figured more in *Stylosanthes* evaluation. In India, for instance, although there was only one representative of *S. hamata sensu stricto* among a collection of 27 accessions grown over a wide range of climatic and agro-ecological conditions, it was the most consistent performer (Ramesh *et al.*, 2004).

In the work that follows, we begin by describing an experiment that was primarily designed to explore the patterns of variation within the Australian collection of *S. hamata* as then assembled. To this were added accessions of related species. The aim is to provide a synoptic overview of the collection. This overview provides a convenient framework on which to 'hang' relevant information from other disciplines and sources, a process necessary to validate the meaningfulness of the framework provided. By doing so, we propose to develop what Harlan (1983) has described as a 'body of theory and general principles' that can be used to aid the efficient use of the genetic resource. We conclude with a general discussion concerning the collection, evaluation and utilization of genetic resource collections, primarily within *Stylosanthes*, but also drawing comparisons with relevant work with other tropical leguminous genera. Recent advances in the development of the potential of *Stylosanthes*, *S. seabrana* in particular, can be found (Chapter 10, this volume).

3.2 Experimental work

3.2.1 Aims

The Australian collection of *Stylosanthes*, originally compiled by Mr R.J. Williams at the CSIRO Laboratory in Brisbane, initially contained few accessions of those types of *S. hamata* and *S. scabra* that were later found to be so useful. There was a single accession of the tetraploid type of *S. hamata* that became cv. 'Verano', one of the type of *S. scabra* that became cv. 'Fitzroy' and two of the type that yielded cv. 'Seca'. Holdings of *S. hamata* increased rapidly after a subsequent collection mission to Brazil and Venezuela (Burt *et al.*, 1979), and liaison with an IDRC–University of West Indies project in Antigua (Spurgeon, 1977; Appendix 3.4, herein). Collection in north-east Brazil yielded a wide range of the promising *S. scabra* types and parts of a complex that includes

S. scabra, *S.* sp. aff. *hamata*, *S.* sp. aff. *scabra* and *S. seabrana* (Burt and Williams, 1979b; Edye and Maass, 1997). It is from this collection, then housed at the CSIRO Davies Laboratory in tropical Queensland, that the *Stylosanthes hamata*, *humilis* and *scabra* cultivars were domesticated and the cv. 'Khon Kaen' and an *S. hamata* accession deemed to be promising in Java were selected. The collection is now held in the Australian Tropical Forages Genetic Resource Centre.

Many of the accessions were being found to be composed of more than one species. For instance, in earlier work Edye *et al.* (1974) found one accession Commonwealth Plant Introduction number (CPI) 40264 contained two species, *S. viscosa* and *S. hamata*. In addition, *S. fruticosa* (syn. *mucronata*, syn. *bojeri*) produced many 'off types' and was probably cross pollinating. Particularly within *S. humilis*, one accession could contain several variants of the species, probably because at that time the seeds for one accession were collected from several plants. Many of the newer accessions also showed signs of admixture and this would later cause serious problems in evaluation work and cultivar release. One of the main aims of the experiment to be described was to separate the components of such admixtures and to re-label the constituents. The other aim was to explore the patterns of variation to be found within *S. hamata* and to show the relationships between the various types – as listed for instance by Cook *et al.* (2005) and Edye and Maass (1997) – and the closely related species of *S. scabra* and *S. humilis* (Burt and Williams, 1979b; Curtis *et al.*, 1995; Liu and Musial, 1997; Edye and Maass, 1997).

3.2.2 Choice of material: background information

In their publication, Cook *et al.* (2005) accept four types of *S. hamata*. The first, the diploid *S. hamata sensu stricto*, has been variously known as the Caribbean 'Island' or 'alkaline' type and it is found primarily on these islands and in adjoining areas on the coasts of Colombia and Venezuela. The second, the allotetraploid *S. hamata*, is found naturally only in the coastal areas mentioned and it is almost certainly a hybrid of *S. humilis* and the diploid *S. hamata* (Burt and Williams, 1979b; Curtis *et al.*, 1995). It has yielded the cultivars 'Verano' and 'Amiga'.

The third type is referred to as being a woody perennial, quite different from the diploid and allotetraploids detailed above, which are herbaceous annuals/biennials. It has been found only in sub-coastal areas in north-east Brazil. Although initially identified as *S. hamata* (Burt *et al.*, 1979), this type resembles *S. scabra* with which it is sympatric and to avoid confusion was labelled *S.* sp. aff. *hamata* and *S.* sp. aff. *scabra* (Burt *et al.*, 1980b), a name subsequently employed by Maass (as quoted in Edye and Maass, 1997 and Date *et al.*, 1996). Part, possibly all (Maass and Sawkins, 2004), of the latter was raised to the species level as *S. seabrana* (Maass and t'Mannetje, 2002) and two cultivars of this species, cv. 'Unica' and cv. 'Primar', released for use in more subtropical areas and heavier soils as a ley plant. It is diploid and seen as a putative parent, with *S. viscosa*, of *S. scabra* (Liu and Musial, 1997).

The final listing is for a 'short-beaked' plant from Florida; this resembles the tetraploid and allotetraploid *S. hamata* in overall appearance and agronomic performance (Chapter 8, this volume), but has seed and chemotaxonomic characteristics (Robinson *et al.*, 1980) that link it with *S. scabra*, a species imported for evaluation in Florida where *S. hamata* is native or adventive. It was listed by the finders as an 'off-type' found in the progeny of a local tetraploid.

All of the accessions of *S. hamata* then available were included in the experiment: the last two types were labelled *S.* sp. aff. *hamata* W (woody) and *S.* sp. aff. *hamata* SB (short beaked) respectively. The same epithets were applied to a woody perennial imported from Honduras as *S. hamata*, but of uncertain provenance, and two accessions from Guatemala noted by one of the authors (RLB) to be atypical of the species: the former was labelled as *S.* sp. aff. *hamata* H and the latter *S.* sp. aff. *hamata* G respectively (see Appendix 3.1).

S. humilis is of importance in that it is one of the parents of the allotetraploid *S. hamata* and because it has yielded some cultivars in its own right. The species has two distinct morphological–agronomic forms (Burt *et al.*, 1971; Edye *et al.*, 1974), the larger leaved form being South American and the smaller leaved type from Venezuela and Central America. The larger leaved form, adventive to the Australian tropics, yielded the early cultivars for sandy soils of medium acidity, while an accession that proved to be adapted to heavier, more alkaline soils was collected initially from the Paraguana Peninsula in Venezuela. 'Khon Kaen' stylo, *S. humilis*, is adapted to similar soils and comes from the same area. This peninsula is of interest floristically (Andrade de Lima, 1975, personal communication) and is very atypical for the region in that it has primarily alkaline soils, quite different from those nominated as the source of *S. hamata* in the region by Edye and Maass (1997). However, the location is close to, for instance, the island of Curaçao, on which the species is endemic and adjoins the Maracaibo region in which the allotetraploid *S. hamata* was first collected. It is the most arid region in which these two species have been collected and this could be the reason why Maass and Sawkins (2004) showed it to be unsuitable for *S. humilis*. An arid area in Brazil inland from Salvador is also deemed to be unsuitable for *S. humilis*, but it was found growing there by one of the authors. This accession, together with those from the Paraguana Peninsula, have been added to representatives of the other forms and included in the experiment.

A representative range of *S. scabra*, thought to be an allotetraploid derived from *S. seabrana* and *S. viscosa*, was also included. Preference was given to accessions that were found growing in the same areas as *S.* sp. aff. *hamata* W. and *S. seabrana*, accessions of the latter not then being available (the other parent of *S. scabra*, *S. viscosa* had been included in previous studies (Burt *et al.*, 1983)). A range of *S. subsericea*, a Central American species generally similar to *S. hamata* – and linked to it genetically (Maass and Sawkins, 2004) – was added. Before it proved to be very susceptible to anthracnose, this species was considered to have considerable agronomic potential (Burt *et al.*, 1970). *S. sympodiales*, a plant also generally similar to *S. hamata*, and capable of growth on heavy clay soils in dry tropical regions, was also added. It had not previously been described in morphological/agronomic terms and is of considerable botanical interest in that it evolved to the east of the Andes, away from other *Stylosanthes* species. An artificially produced hybrid of *S. hamata* × *S. humilis* (courtesy Dr D.F. Cameron) and two accessions of *S. calcicola* were included because all three had shown the ability to grow on alkaline soils. The final additions, one of a variant of *S. hamata* cv. 'Verano' and three of the cv. 'Seca', courtesy of Dr J.M. Hopkinson, completed the collection of 178 plants. These are listed in Appendix 3.1.

3.3 Experimental methods and observations taken

Unlike previous studies with this genus (e.g. Burt *et al.*, 1971), which contained many different species, we are here primarily concerned with one species, *S. hamata*, and other generally similar species such as *S. humilis*, *S. subsericea* and *S. sympodialis*; and within *S. hamata* we are trying to locate differences within accessions. Many of the differences of interest between plants will thus be relatively small and the resultant data sets likely to exhibit continuous variation (Burt and Williams, 1979b). To facilitate comparisons between plants, the experiment was therefore located in a shade house, rather than in the field, where plants could be temporarily placed next to each other at a convenient height for detailed observation and data recording. Controlled environment studies of this type have been shown to be suited for such a purpose (Burt and Williams, 1979a; Pengelly and Eagles, 1995).

The experiment was carried out in pot culture in a shade house at the CSIRO Davies Laboratory in Townsville, Queensland, Australia. In November 1975 pre-germinated seed was sown in 18-cm pots, each pot containing a weighed amount of a loamy sand/peat mixture that had previously been found to be suitable for growth of the species. There were four pots for each of the 150 accessions and each pot was planted with four seeds. As the *Bradyrhizobium* requirements of many of the newer accessions were not known, the plants were watered with an appropriate mixture of strains in the seedling stage and received a liquid nutrient solution at 2-weekly intervals thereafter. The resultant seedlings were examined for uniformity and subsequently reduced to one/pot. There were four blocks, each with a single pot of each accession randomly located within.

Many of the measurements and observations taken were those that had proved to be useful in previous studies, for instance an examination of the root nodules to assess their rhizobial effectiveness. Some observations, such as persistence, were deleted as being inappropriate in pot studies. Additional observations were taken of performance characteristics deemed to be of likely significance: from earlier observations in the quarantine glasshouse, the presence or absence of a dark 'blotch' in the centre of the flower; from chemotaxonomic studies (Robinson *et al.*, 1980) the presence or absence of a 'shoulder' on the seed; and, from unpublished studies (Burt, in 1972), the presence or absence of seed polymorphism. This latter character involved seed colour, because light- and dark-coloured seeds have different responses in germination – a factor likely to increase environmental plasticity in a given genotype (*sensu* Baker, 1974). A list of the characters measured is presented in Table 3.1.

3.3.1 Data analysis

The most informative form of analysis for the type of data collected here is, as far as we can ascertain, the multiple nearest graph theoretic technique. When used with a collection of *S. viscosa*, *S. scabra* and *S.* sp. aff. *S. scabra* or *S.* sp. aff. *hamata* W it delimited forms of *S. scabra* that equated to 'ecotypes' and that proved to be differentially adapted in field evaluation studies in Australia (Burt *et al.*, 1983). But, in its present form, it can only be used in relatively small collections, and its validation, an essential process with pattern-seeking methodologies (Burt *et al.*, 1983; Burt, 1983) (and one often now omitted), is onerous. A minimum spanning tree (MST) analysis is less informative but can deal with bigger collections and patterns of continuous variation (Burt and Williams, 1979b): it was used in the present instance and the size of the collection to be analysed set heuristically at 150 accessions.

It is necessary to complement such an analysis with data produced from other forms of classification, such as ordinations, principal components analysis or hierarchical classifications (now often referred to as 'cluster analysis') (Burt and Williams, 1979b). The latter is often used in genetic resource studies (Burt *et al.*, 1971; Pengelly and Liu, 2001; Maass (1989, as cited in Jansen and Edye, 1996; Edye and Maass, 1997; Liu, 1997) and is able to handle non-numerical data: it has therefore been used in the present studies.

Details of the methods used have been described and compared previously (e.g. by Burt *et al.*, 1983). We simply note that of the 'clustering' strategies employed, the incremental sums of squares (Burr, 1970) has proved to be the most generally useful for the present purpose. But when used in this manner it is intensely clustering and thus it can produce some misclassifications. As a result, the meaning and utility of any groupings produced requires checking and validation (in the sense of Heywood, 1973; Burt *et al.*, 1983).

3.4 Results: the hierarchical classification

3.4.1 Definition of the groups

The classification was arbitrarily truncated at the 20 group level and the attributes that had contributed to their delineation examined with the aid of the programme GROUPER. The first 'outside information' (information that had not been used in the analysis) to be applied for validation was the taxonomic identification of the species contained in the various groups.

The first dichotomy, which separates groups 1 to 11 (A) from 12 to 20 (D) (Fig. 3.1), is well founded. Groups 12 to 20, for instance, contain plants with rounded leaf tips, have almost prostrate main stems and pale seed: they contain all but two of the diploid *S. hamata sensu lato* accessions and are admixed with only a single accession of *S. subsericea* and two accessions of *S.* sp. aff. *hamata* G (from Guatemala).

These groups then divide into D1 and D2 (Fig. 3.1) and both are well defined: 98% of the D1 accessions have a dark blotch in the centre of the flower and red coloration in the stem, D2 accessions do not. Subsequent dichotomies are, however, less secure: D3 groups, for instance, usually have stems that are red all over while those of D4 do not – but this sole consistent difference could change with age or mineral content. After further examination, detailed in the next section, several of the initial groups were fused. In section A, which contains more species and types of species, few of the groups merged and further subdivision was called for in several instances.

The first dichotomy places all but two perennial accessions, of *S. calcicola*, into group B whose members have shorter beaks on the pods, lower beak/pod ratios (a characteristic much used in classical taxonomic methods, for instance by

Table 3.1. Data recorded.

Character	Observation
Morphological	
1. Floral characters	
Dark centre to flower (Q5)	(0) absent, (1) present
Mature raceme length (mm) (N10)	
Length inflorescence (mm) (N17)	
Breadth inflorescence (mm) (N18)	
Flower colour (OM6)	(1) white, (2) cream, (3) yellow
Flower size (OM7)	(1) small, (2) medium, (3) large
Length axis rudiment (mm) (N23)	
2. Seed and pod characters	
Polymorphic seed (Q1)[a]	(0) absent, (1) present
Shoulder on seed (Q3)[b]	(0) absent, (1) present
Pointed embryonic end (Q4)	(0) absent, (1) present
Seed colour (OM4)	(1) light, (2) medium, (3) dark, (4) very dark, (5) black
Seed shape (DM1)	(1) pointed, (2) normal, (3) square
Basal articulation (Q1)	(0) absent, (1) present
Length of beak on pod (mm) (N1)	
Length of pod + beak (mm) (N2)	
Length of pod (mm) (N3)	
Ratio of N2/N3 (N5)	
Pod beak curl (OM1)	(1) straight, (2) slightly curled, (3) curled, (4) very curled
Pod colour (OM2)	(1) light, (2) medium, (3) dark

Character	Observation
Morphological	
Hairiness of pod (OM3)	(1) glabrous, (2) sparse, (3) many, (4) very hairy
Weight of 10 seeds (mg) (N4)	
3. Leaf characters	
Hairs on upper leaf surface (Q9)	(0) absent, (1) present
Bristles on upper leaf surface (Q10)	(0) absent, (1) present
Hairs on lower leaf surface (Q11)	(0) absent, (1) present
Bristles on lower leaf surface (Q12)	(0) absent, (1) present
Hairs on leaf margins (Q13)	(0) absent, (1) present
Bristles on leaf margins (Q14)	(0) absent, (1) present
Hair type on upper leaf surface (DM4)	(1) tomentose, (2) villous, (3) velutinous, (4) puberulent, (5) woolly, (6) sericeous
Hair type on lower leaf surface (DM5)	As above
Distribution of hair on upper leaf surface (DM8)	(1) all over, (2) only on veins, (3) only near margin
Distribution of bristles on upper leaf surface (DM9)	As above
Distribution of bristles on upper leaf surface (DM10)	As above
Distribution of bristles on lower leaf surface (DM11)	As above
Apex of terminal leaflet (OM5)	(1) pointed to (9) round
Length/breadth ratio of last expanded leaf (N6)	

Table 3.1. Continued.

Character	Observation
Morphological	
Stipule length (mm) (N20)	
Length of stipule arm (mm) (N21)	
Ratio stipule length/awn length (N20/N21) (N22)	
4. Stem characters	
Hairs on stem (Q7)	(0) absent, (1) present
Bristles on stem (Q8)	(0) absent, (1) present
Hair type, if present (DM3)	(1) tomentose, (2) villous, (3) velutinous, (4) puberulent, (5) woolly, (6) sericeous
Distribution of hairs, if present (DM6)	(1) all over, (2) only on veins, (3) only near margin
Distribution of bristles, if present (DM7)	As above
Colour (DM2)	(1) all green, (2) lower nodes slightly red, (3) lower nodes red, (4) all nodes red
Length of main stem (cm) (N11)	
Angle of main stem from vertical (°) (N12)	

Character	Observation
Morphological	
Thickness of old stem (mm) (N14)	
Length of longest lateral branch (cm) (N15)	
No. axillary branches on main stem (N16)	
Performance	
Nodulation (Q6)[c]	(0) not effective, (1) effective
First flowering, days of year (Julian calendar) (N7)	
First ripe seed, days of year (Julian calendar) (N10)	
Plant dry matter yield, (g/plant) (N8)	
Crown branches below 5 cm (N13)	

Seed Types

A B

DM, disordered multistate attributes numbered 1 to 11; N, numerical attributes numbered 1 to 23; OM, ordered multistate attributes numbered 1 to 7; Q, qualitative attributes numbered 1 to 14.
[a]Polymorphic seed as evinced by having ripe seed of different colours.
[b]Shoulder on seed (A) as opposed to no shoulder (B) (see diagrams).
[c]Nodulation deemed to be effective when root nodules were red in cross section.

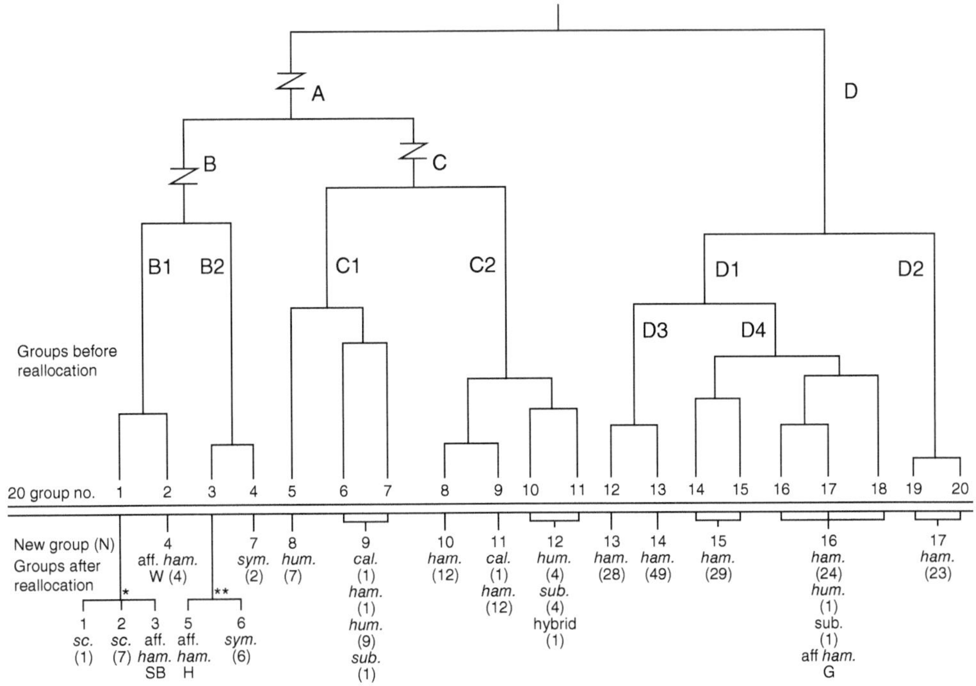

Fig. 3.1. ISS hierarchy before 20 group and after 17 group reallocation. For meaning of letters A to D_4 see section 3.4.1. Species are denoted by the first letters of their names. * One accession of *S. hamata*; ** one variant of cv. Verano excluded.

Maass and t'Mannetje, 2002), hairs distributed all over the stems (as opposed to being located on one side only), longer and more erect main stems, and so forth. This group contains *S. scabra*, the woody *S.* sp. aff. *hamata* W and *S.* sp. aff. *hamata* H and the more herbaceous *S. sympodialis*. Included also (Table 3.2) are two accessions of *S. hamata*, one, 95, appears to be an accession of *S. hamata sensu stricto* (and from a region in which that plant is very common), and one a locally occurring variant of the cv. 'Verano'. They appear to have been misclassified and their cases will be considered in the next section. B then separated into B1 and B2 on the basis of seed shape, square in B1 and normal in B2, and raceme length, which is longer in the former.

For the sake of brevity, information about the dichotomies in section C are not presented. We simply note that the first of these separates most of the accessions of *S. humilis* from those of the allotetraploid *S. hamata*: but most of the groups are multi-specific with much admixture of *S. hamata*, *S. humilis* and *S. subsericea*. The allotetraploid *S. hamata* is placed in two groups, 10 and 11. These two groups have considerable overlap in the attributes that have been used to separate them (e.g. hair distribution and hair type on the lower surface of the leaf and the presence or absence of polymorphic seeds). This suggests that they exhibit continuous variation.

3.4.2 Adjustment of the groups

In the previous section it was shown that, at the 20 group stopping level, the meaningfulness of many of the groups was open to question: many contained mixtures of species, some had been based on poorly defined differences and some contained misclassifications. This is not unusual (e.g. Jansen and Edye, 1996 where accessions of *S. scabra* are widely separated and associated with a range of other species). It is necessary to inquire into the meaningfulness of the groups and this is done herein using two sets of information, the taxonomic identification of the accessions and the data on their geographic provenance (Table 3.3): for misclassifications, reference is also

Table 3.2. Constitution of groups before and after adjustment.

Group and action[+]	Plant no.[++]	Species
1. split[+++]	95[h]	*S. hamata*
	136, 137[k], 138, 139[g], 212[j], 213[j], 214[j] 215, 148[q], 149[t]	*S. scabra* *S.* sp. aff. *hamata* SB
2.	146[m], 147[m], 225[m], 226[m]	*S.* sp. aff. *hamata* W
3. split	199, 143	*S. hamata*, *S.* sp. aff. *hamata* H
	141,142, 221, 222, 223, 224	*S. sympodialis*
4.	140, 220	*S. sympodialis*
5.	128, 135[d], 205, 206, 208, 209, 210[e]	*S. humilis*
6. bulked	34 127, 131, 133[f], 200.	*S. hamata* *S. humilis*
7.	1 56 202, 203, 204	*S. calcicola* *S. hamata* *S. humilis*
8.	8[a], 18, 20[b], 21, 22, 23, 24, 25, 42, 43, 44, 152	*S. hamata*
9.	2 17, 19, 26, 27, 32, 36, 49, 50, 51, 52, 53, 54	*S. calcicola* *S. hamata*
10. bulked	129 217 150	*S. humilis* *S. subsericea* *S. hamata/S. humilis* hybrid
11	130, 134, 201, 207, 216, 218.	*S. humilis* *S. subsericea*
12	3, 33, 70, 74, 85, 87, 97, 99, 100, 101, 106, 110[s], 111, 116, 119, 120, 124, 125, 157, 170, 171, 172, 177, 180, 186, 191, 192, 195	*S. hamata*
13	14, 29, 31, 48, 57, 58, 59, 60, 61, 62[p], 63, 66, 68, 72, 73, 91, 94, 96, 98, 103, 104, 105, 108, 109, 112, 113, 115, 117, 121, 123, 126, 151, 156, 158, 159, 160, 161, 162, 163, 166, 175, 176, 179, 182, 184, 188, 193, 197, 198	*S. hamata*
14 bulked	10, 11, 12, 13, 39, 40, 41, 45, 46, 64, 75, 76, 79, 83[r], 89, 93, 112, 167	*S. hamata*
15	15, 38, 190	*S. hamata*
16	28, 30, 35, 37, 55, 65, 69, 71, 102, 144, 145, 164, 165, 168, 169, 185, 189, 194	*S. hamata*
17 bulked	118,154,183,187 132	*S. hamata* *S. humilis*
18	47, 153, 155, 196 219	*S. hamata* *S. subsericea*
19 bulked	9, 67, 77, 78, 114, 173	
20	4, 5, 6, 7, 16, 80, 81, 82, 84, 86, 88, 90, 92, 107, 174, 178, 181	*S. hamata*

[+]At the 20 group level; [++]as listed in Tables 3.7 and Tables 3.8; [+++]re-adjustment.
Cultivars: *S. hamata*: [a]cv. 'Verano'; [b]cv. 'Amiga'. *S. humilis*: [d]cv. 'Paterson'; [e]cv. 'Gordon'; [f]cv. 'Khon Kaen'. *S. scabra*: [g]cv. 'Seca'.
Naturally occurring variants: [h]cv. 'Verano' (of *S. hamata*); [j]cv. 'Seca' (of *S. scabra*); [k]From previous studies accession similar to *S. scabra* cv. 'Fitzroy' (Burt *et al.*, 1983).
Promising accessions (as listed in Cook *et al.*, 2005; Chapter 9, this volume): *S. hamata* [m]ex. Brazil, perennial types; [n]CPI 61670; [p]CPI 70564; [r]CPI 72852; [s]CPI 73507; under group 14 *S.* sp. aff. *hamata* (short-beaked Florida tetraploids) [q]CPI 70522; [t]CPI 70529.

made to the raw data sets. The groups are further examined in light of the results from the MST.

The remaining grouping, C (Fig. 3.2), joins an accession of *S. scabra* from subtropical, southerly Brazil, with what appears to be an accession of *S. hamata sensu stricto* from Antigua, an island in which such plants are very common. It is a heterogeneous grouping in which the *S. scabra* accession differs morphologically from the more tropical accessions in N2, Fig. 3.1, and is again linked to *S. hamata*, an association that has been shown chemotaxonomically (Robinson *et al.*, 1980). For our present purposes we have retained the *S. scabra* accession as a separate, single-membered group and – after examination of the raw data – will explore the positioning of the *S. hamata* accession on the MST.

Group 1 (as shown in Fig. 3.1) contains three species. The next division (Fig. 3.2a) splits off the two accessions of *S.* sp. aff. *hamata* SB, the tetraploid from Florida, USA, on the basis of their longer, more curled pod beaks, shorter more prostrate main stems, heavier branching pattern and so on. They are designated as a separate group (N3, Fig. 3.1). Groupings C and D (Fig. 3.2) are then separated. Group D consists entirely of *S. scabra*, the first three plants (213, 212 and 139) being naturally occurring variants of cv. Seca and the cultivar itself. The remainder, like cv. Seca, are from north-east Brazil. The variants differ from cv. Seca in agronomic characteristics, such as the length of the main stem and its uprightness, length of the longest lateral and flower colour; morphological characterizations remained stable. This *S. scabra* association was nominated as a separate group, N2 (Fig. 3.1).

Group 2 contains the accession listed by Cook *et al.* (2005) as being a strongly perennial, woody type of *S. hamata* from Brazil: it is listed herein as *S.* sp. aff. *hamata*. This type is quite different from *S. scabra*, although sympatric with it, but is itself quite variable: details of this variation are presented elsewhere (Burt *et al.*, 1983) and the grouping is left unchanged.

Group 3 (Fig. 3.1) is composed of two species. Branch B (Fig. 3.2b) consists of six accessions of *S. sympodialis* characterized by their much shorter pod beaks, lower beak/pod length ratios and paler and less hairy pods (they differ from the accessions of the same species placed in the neighbouring group N7 (Fig. 3.1) in that they are more vigorous and have a shorter awn on the stipule). There are two accessions in section A. The first, the sole accession of *S.* sp. aff. *hamata* H, is a woody perennial, unique in its morphological/agronomic characteristics (Edye *et al.*, 1974) and, unusually, from Honduras. It is of doubtful taxonomic provenance (Williams *et al.*, 1984) and has been designated as a single-membered group, N5 (Fig. 3.1). The other constituent was a naturally occurring variant of cv. 'Verano', found growing on a particular heavy clay soil (by RLB) where the original cultivar had failed to persist. It was placed in this group, away from the other accessions of the allotetraploid *S. hamata*, because it typically possessed bristles on the stem and had a lower beak/pod ratio. Examination of the raw data showed that all four plants of this accession showed marked differences in the combination of other characteristics: one had darker, longer stems and shorter racemes and another was much more upright but had shorter, more poorly branched laterals. Possibly it was a hybrid or the result of somatic segregation, so it was removed for further study (Burt and Beattie, unpublished data).

The remaining groups, 5 to 20, have been examined in similar detail but, for the sake of brevity, the results will not be presented here. Some of the groups included accessions that shared a similar geographic background, in others accessions had previously been shown to be similar (Edye *et al.*, 1974) and some groups were mono-specific. These have been left unchanged. The remaining groups were usually separated in dichotomies on the basis of characters, and combinations of characters of dubious biological significance. The resultant groups often contained mixtures of species and accessions from different geographical backgrounds. Many of these groupings were bulked and their significance examined on the Minimum Spanning Tree (MST). The exceptions were groups 8 and 9, which contained tetraploid *S. hamata* accessions from coastal South America. As noted earlier, these two groups were separated on the basis of minor characteristics and, even in these, they showed much overlap. Thus the groupings are of limited utility for purposes of description or for documenting the characteristics responsible for the patterns of variation so efficiently displayed in an MST analysis (Burt and Williams, 1979b). However they are retained as separate groups here for illustrative purposes in a later section.

Table 3.3. Geographic provenance of the groups.

Group no.		Used in minimum spanning tree analysis			Excluded from minimum spanning tree analysis		
Initial	Final	Species	No.[a]	Origin[b]	Species	No.[a]	Origin[b]
1	1 2 3	*S. seca* *S. seca* *S.* sp. aff. *hamata*.S13	1 3 2	Sao Paulo, Brazil Bahia, Pernambuco, Brazil Florida, USA	*S. seca*	5	Pernambuco, Brazil
2	4	*S.* sp. aff. *hamata*	2	Pernambuco, Brazil	*S.* sp. aff. *hamata*	2	Pernambuco, Brazil
3	5 6	*S.* sp. aff. *hamata* H *S. sympodiales*	1 2	Honduras Ecuador	*S. sympodiales*	4	Ecuador
4	7	*S. sympodiales*	1	Ecuador	*S. sympodiales*	1	Ecuador
5	8	*S. humilis*	2	Bogotá, Colombia (1) Australia (1)	*S. humilis*	5	Caruraru, Brazil (2), Crato, Brazil (2), Australia (1)
6 7	9	*S. calcicola* *S. hamata* *S. humilis*	1 2 3	Yucatan, Mexico Paraguana, Venezuela (1) Santa Marta, Colombia (1) Dominican Republic (1) Paraguana, Venezuela (2)	*S. humilis* *S. subsericea*	4 1	Costa Rica (2) Mexico (1) Calabozo, Venezuela (1) Guatemala
8	10	*S. hamata*	10	Maracaibo, Venezuela (9) Barranquilla, Colombia (1)	*S. hamata*	1	Barranquilla, Colombia
9	11	*S. calcicola* *S. hamata*	1 12	Yucatan, Mexico Maracaibo, Venezuela (4) Paraguana, Venezuela (2) Santa Marta, Colombia (6)			

10 11	12	*S. humilis* *S. hybrid*	3 1	Petrolinea, Brazil (1) Paraguana, Venezuela (1) Barinas, Venezuela (1)	*S. humilis* *S. subsericea*	2 3	Mexico (1) Sobral, Brazil (1) Mexico (1) Honduras (1) Guatemala (1)
12	13	*S. hamata*	18	Antigua (5) [13, 16][A] Barbuda (2) Curaçao (20) Guadeloupe (1) Nevis (4) [14][A] St Lucia (1) Florida, USA (2) [9][A] Paraguana, Venezuela (1)	*S. hamata*	10	Antigua (4) [14, 16, 17][A] Cuba (1) [14][A] Nevis (1) [15][A] St Lucia (1) Florida, USA (3)
13	14	*S. hamata*	30	Antigua (14) Barbuda (4) Barranquilla, Colombia (2) Curaçao (3) Nevis (4) Barinas, Venezuela (1) Maracaibo, Venezuela (1) Paraguana, Venezuela (1)	*S. hamata*	19	Antigua (14) Barranquilla, Colombia (1) Curaçao (1) Nevis (2) St Lucia (2)
14 15	15	*S. hamata*	18	Antigua (2) Crato, Brazil (1) Fortaleza, Brazil (1) Salvador, Brazil (1) Barranquilla, Colombia (5) Curaçao (1) Florida, USA (5) Barquisimeto, Venezuela (1) Margarita, Venezuela (1)	*S. hamata*	2	Antigua (1) Nevis (1)

Continued

Table 3.3. Continued.

Group no.		Used in minimum spanning tree analysis			Excluded from minimum spanning tree analysis		
16 17 18	16	*S. hamata* *S.* sp. aff. *hamata G*	12 2	Antigua (2) Barbuda (2) Barranquilla, Colombia (2) Santa Marta, Colombia (1) Guadeloupe (1) Nevis (1) Maracaibo, Venezuela (1) Paraguana, Venezuela (2) Guatemala	*S. hamata* *S. subsericea*	12 1	Antigua (7) Barranquilla, Colombia (1) Curaçao (1) Nevis (1) St Lucia (1) Chiquimula, Guatemala
19 20	17	*S. hamata*	19	Antigua (2) Bahamas (2) Cuba (2) Dominican Republic (2) Puerto Rico (1) Florida, USA (11)	*S. hamata*	4	Antigua (2) Florida, USA (2)

[a]Number of accessions in group. [b]Country and province, area or nearest town.[A]Square brackets, for new group 13 only. When an accession/s from a country were found to be mixed, the group numbers into which other members of the mixture were placed is shown. A single accession could have members placed in up to 3 different groups.

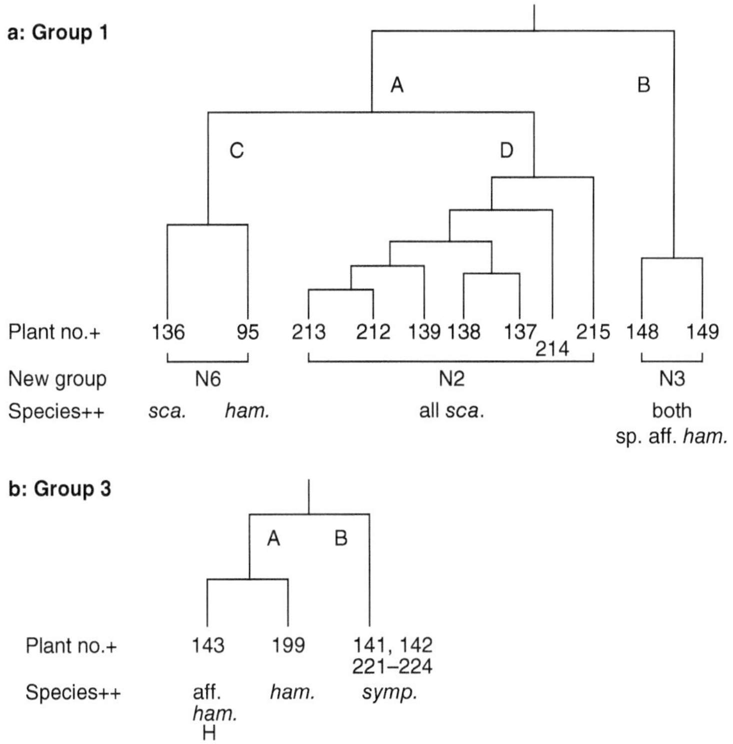

\+ as listed in tables 3.7 and 3.8

++ *ham.- S. hamata; sca.- S. scabra; symp.- S. sympodialis*

Fig. 3.2. Complete hierarchical breakdown of initial groups 1 and 3.

When the hierarchical classification produces well-defined groups, it can be used to nominate the characteristics of various zones of accessions on the MST. But with poorly defined groups this is not possible and other methodologies used previously only allow the use of numerical data (e.g. ordinations and principal components analysis). This problem is discussed in a later section when results from the MST and hierarchical classification are compared.

3.5 Preliminary discussion

3.5.1 Choice of accessions to be studied in the MST analysis

Observations taken during the course of the experiment, and confirmed by the inspection of the raw data, showed that many of the accessions of *S. hamata* contained mixed seed lots. These were renumbered, with the most commonly occurring type within an accession labelled A, and subsequently regarded as different accessions in the hierarchical classification. The *S. hamata* collection then numbered 226 entries and it needed to be reduced to 150 for the MST analysis.

The classification had revealed, however, that there were overlaps between some groupings and that some of these groups had been separated on the basis of quite minor characteristics or combinations of characteristics. This confirmed a pattern of continuous variation, nominated from earlier studies with fewer accessions (Burt and Williams, 1979b). The exclusion of some elements from the accessions known to contain mixed seed lots was not likely to affect this overall pattern of variation: the most common elements, labelled A, were therefore included in the analysis and those labelled B to D generally omitted. In some instances, however – when one of the latter elements was considered to be an interesting form – they were included.

S. subsericea accessions appeared in different, widely separated groups in the analysis, and often in association with other species, particularly *S. humilis* and *S. hamata*, which it clearly resembles. While one of the accessions is of particular interest because of the environment in which it was collected, the species is currently of little agronomic interest because of its susceptibility to anthracnose. Its relationship to other species has been described in previous publications (Burt *et al.*, 1971; Edye *et al.*, 1974; Burt and Williams, 1979b). It has, reluctantly, not been included in the data set for MST analysis; but comparisons between accessions, using the raw data, will be drawn in later sections.

S. humilis, like *S. subsericea*, has fallen from favour because of its susceptibility to anthracnose and, like *S. subsericea*, its overall relationship with other *Stylosanthes* species has been well documented. However, as it is a parent of the tetraploid *S. hamata*, interest in it continues due to its interrelationships with that species; as well it has been relatively recently discovered to have the ability to grow on clay soils (the cv. Khon Kaen). The latter accession was collected on the Paraguana Peninsula, in Venezuela, an area that also furnished an accession of *S. hamata sensu stricto* adapted to similar soils. The area is poorly collected and is generally regarded as being very dry and environmentally unsuited to *S. humilis* (Fig. 1.7 in Maass and Sawkins, 2004). Plant 129 similarly comes from a very dry region, in Brazil. A small range of carefully selected accessions, including those from the Paraguana Peninsula and plant 129, have thus been included in the reduced accession set for analysis.

S. sympodialis is another vigorous annual/biennial that is generally similar to *S. hamata*. It has shown considerable promise on clay soils, but is again very susceptible to anthracnose. Its overall similarity to other *Stylosanthes* accessions has not, however, been illustrated and a representative sample, embracing the variation depicted in the classification, has been chosen for MST analysis.

The final reductions omitted some accessions of *S. scabra*, including the variants of cv. 'Seca' – whose characteristics have already been documented – together with two accessions of *S.* sp. aff. *hamata* W. The relationships between these two species and other *Stylosanthes* species of interest here has been previously documented (Burt *et al.*, 1980b; Burt *et al.*, 1983). The final selections, and their origins, are shown in Tables 3.7 and 3.8 in Appendix 3.1.

3.5.2 Positioning of the various species, and types within species, in the classification

Any classification of a genetic resource must have a sound basis in classical taxonomy, for without it recognition and communication is impossible (Burt, 1983). But a sound taxonomist uses 'stable characters' and 'deals with features of leaves, flowers, seeds etc.' as 'the form of the whole is taxonomically irrelevant because it is so enormously variable' (Harper, 1977, quoted in Burt, 1983). Similarly, genetic or phylogenetic studies are not concerned with overall plant form; 'the evaluation of genetic diversity in a sample is a problem different from the evaluation for its agronomic merit' (Brown, 1978 as quoted in Robinson *et al.*, 1980). Many studies of this nature have recently been carried out on *Stylosanthes* and are reviewed by Maass and Sawkins (1984). In taxonomic and genetic/phylogenetic classifications, species that are very similar in agronomic and ecological features, e.g. *S. scabra* and *S. fruticosa*, may be placed in widely separate groupings and their overall similarity is difficult for users to recognize. And the reverse is true: species that are quite different agronomically and ecologically, such as *S. humilis* and the woody perennial *S. viscosa*, may be placed together (e.g. in Maass and Sawkins, 1984).

Similarly there is a major problem at the sub-specific level, i.e. at the level of greater interest to would-be users of genetic resource collections. 'Much of the published work in this area (taxonomy/phylogeny) has addressed questions relating to taxonomy rather than focusing on detailed analysis of variation within species' (Maass and Sawkins, 2004). For *S. hamata*, Cook *et al.* (2005) list herbaceous annuals/biennials, and woody perennials, but the short-beaked type from Florida (herein listed as *S.* sp. aff. *hamata* SB) has no overall description for the reader, though it has performed very well in evaluation trials (Chapter 8, this volume). Although this type has been shown to have chemotaxonomic similarities with the woody perennial *S. scabra* (Robinson *et al.*, 1980), it is much more similar overall to the herbaceous forms of *S. hamata*.

There are basically two forms of *Stylosanthes*. One form is erect, woody and perennial. These plants are usually thick-stemmed and sparsely branched, and are typified by *S. scabra*.

The other form is composed of lower growing and heavily branched herbaceous annuals/biennials. These are sometimes almost cushion-like when grown as spaced plants. They include such species as *S. humilis* and *S. hamata sensu stricto*. It is important to note that the two forms are quite different agronomically and ecologically, a feature often lost when considering and using the results of hierarchical classifications. The relevance of the two forms to plant collecting philosophies will be considered here.

Table 3.4 illustrates the behaviour of the two plant forms in relation to disturbance levels. The information for *S. capitata* and *S. scabra* was obtained from collection sites in north-east Brazil, and for *S. hamata* sites in Venezuela. The data for *S. humilis* came from sites around Townville, Australia. At the time of collection, *S. capitata* was very poorly known, *S. scabra* and the allotetraploid *S. hamata* had not been used commercially and the broad general characteristics of *S. humilis* were becoming apparent. *S. humilis*, which was first noticed around Townsville, has since become adventive in northern Australia.

The first two species, the woody perennials *S. capitata* and *S. scabra*, were found under similar climatic conditions in Brazil on sands and yellow earth soils (Burt *et al.*, 1979); however, the soils at the *S. capitata* sites were much more acid than those of *S. scabra*. Both species behaved similarly in response to disturbance and were well distributed throughout the collecting areas. They were found only as scattered plants and were never dominant.

S. humilis and the allotetraploid *S. hamata*, which contains the cultivars selected for use in Australia 'Verano' and 'Amiga', were found on similar soils and also grew on disturbed sites. However, they only became widespread in the collecting area at higher disturbance levels and then they were often dominant. These are the 'weedy' plants nominated by Baker (1974), typical of early stages of seral development and some form of disturbance may be necessary for their establishment and maintenance (as noted by Gillard and Fisher, 1978 for *S. humilis* and see Chapter 8 this volume for *S. hamata sensu stricto*). *S. scabra* is more typical of a plant from a later seral stage. It is also of interest to note, in this context, that both *S. humilis* and *S. scabra* maintained this typical form of behaviour when introduced into an area in which they are not native. Unsuitable environments or management systems can, however, cause *S. scabra* to become more dominant; in mixed pastures, if the associated grass is more palatable, it may be overgrazed and the legume is favoured. The situation can be remedied through lower grazing pressures, sowing with a more grazing-tolerant grass or spelling to allow the grasses to seed. And, unlike other leguminous genera such as *Centrosema* and *Macroptilium*, which require relatively high levels of potassium to persist in mixed pastures,

Table 3.4. Behaviour of various legume species in relation to disturbance level.[a]

	Disturbance level[b]				
Species	1	2	3	4	5
S. capitata[c]	Not found	4.0 (3.5)	3.0 (2.0)	3.5 (3.0)	3.0 (2.5)
S. scabra	4.0 (1.3)	3.0 (1.8)	3.8 (2.0)	3.3 (2.6)	3.8 (2.3)
S. hamata	Not found	2.0 (1.0)	3.0 (2.7)	4.0 (3.6)	4.0 (3.8)
S. humilis[d]	1.0 (2.0)	2.2 (4.0)	3.0 (3.0)	2.3 (4.2)	3.0 (4.0)
M. martii	Not found	2.0 (1.0)	2.8 (3.8)	2.3 (3.8)	3.0 (4.0)
M. bracteatum	4.0 (2.5)[e]	4.0 (3.0)	3.7 (3.7)	3.3 (3.0)	4.0 (3.0)

[a]Mean values of data recorded in Brazil and Venezuela. The first figure in each cell refers to plant distribution at the collecting site: 1 = barely present to 4 = well spread throughout. The figure in brackets refers the relative dominance of the species where it occurs, 1 = only one or two plants with low ground cover to 4 = several to many plants with high ground cover. [b]Increasing values indicate increasing levels of disturbance: 1, uncleared, lightly grazed; 2, uncleared, heavily grazed: 3, cleared and grazed: 4, road edges, disused gardens etc.; 5, recently graded roadsides and cuttings etc. [c]Requires a very acid soil, not commonly found in this part of Brazil. [d]Data for *S. humilis* recorded in typical collecting situations around the Townsville area, Queensland, for comparative purposes. [e]Higher dominance rating than a comparable perennial species, such as *S. scabra*, reflects a sprawling habit.

Stylosanthes does not, and to the contrary is favoured by high levels of exchangeable potassium (Burt *et al.*, 1979). This feature has been demonstrated for *S. humilis* in both pot and field experiments (R.L. Hall, 2007, personal communication). The distinctions between the woody perennial and shorter lived herbaceous forms are very important to potential users of genetic resource collections.

In the hierarchical classification these two forms have generally been well separated. For instance, the more herbaceous types of *S. hamata*, with the exception of a misclassification, have been placed with other annuals/biennials, away from the perennials, while the woodier forms, *S.* sp. aff. *hamata* from Honduras and *S.* sp. aff. *hamata* from Brazil, have been located near *S. scabra*. However, the more herbaceous *S.* sp. aff. *hamata* SB, from Florida, and the generally similar *S. sympodialis* have been placed in the same branch of the classification (B in Fig. 3.1). Similarly the perennial *S. calcicola*, admittedly less robust than *S. scabra*, has been linked to the annual/biennials in section C.

3.5.3 Treatment of the annual/biennial species in the readjusted groups (N 8 to 17) *S. subsericea*

Accessions of *S. subsericea* have been placed by the analysis in three widely separated groups (Fig. 3.1), where they are linked with the generally similar species *S. humilis*, *S. hamata sensu stricto* and the *S. hamata sensu stricto*/*S. humilis* hybrid (e.g. Edye *et al.*, 1974; Burt and Williams, 1979b). Inspection of the raw data confirmed that the accessions in group 12 (Fig. 3.1) are very similar to each other and it is known that they originated in relatively wet, high-altitude areas in Central America (Edye *et al.*, 1974). The single accession in group 16 (plant 219), however, is from a very different environment in Guatemala, a restricted area of low rainfall and on a very alkaline, sandy soil (RLB). In the classification, it is placed with accessions of *S. hamata sensu stricto* and (atypically) with *S. humilis* from alkaline soils: its nearest neighbour is an accession of *S. hamata sensu stricto* from the island of Nevis.

In Table 3.5 we compare the dryland *S. subsericea* (plant 219) with the *S. hamata* accession from Nevis (plant 196) and an accession of *S. subsericea* from a highland situation (plant 218). The dryland *S. subsericea* is quite different from that found at higher, wetter elevations. In seed characteristics, some of which are much used as 'stable characters' in classical taxonomy, it shows a shorter beak, lower beak/pod ratio and has less heavy, polymorphic seed. In this it is similar, or intermediate to, the *S. hamata* accession. This situation extends to characteristics of the leaflets; they are rounder and have more rounded leaf tips, than those found at higher altitudes.

In chemotaxonomic/genetic terms the dryland *S. subsericea* is again quite distinct from the higher altitude type (in which only one accession is not identical with others, and even then in only one band); there are several different bands and band strengths. The accession of *S. hamata* contains several bands not found in *S. subsericea*.

The taxonomic provenance of the dryland accession (plant 219) warrants investigation, as does its genetic relationships with other species (Maass and Sawkins, 2004). And, bearing in mind the variation depicted herein, and its agronomic interest (Burt *et al.*, 1970) and ability to grow on alkaline soils, a search for anthracnose resistance in this species could be profitable.

S. humilis

Group N8 accessions are of the large-leaved form (MA1), while those in N9 are the small-leaved type (MA2) (Edye *et al.*, 1974). The latter are of particular interest as they contain two accessions from the very dry, alkaline Paraguana Peninsula, one of which became the cv. 'Khon Kaen'. An accession of *S. hamata sensu stricto* from the same area is also in N9. Another, *S. humilis*, from this area is placed in group N12, where it is linked with the sole accession from semi-arid Brazil, a plant noted by one author (RLB) to be somewhat atypical. These latter groupings, with the smaller leaved types, are of considerable interest and their relationships will be explored further after the MST analysis.

3.5.4 The allotetraploid *S. hamata sensu lato*

S. hamata sensu lato is an agronomically important grouping, having produced both the ubiquitous cultivar 'Verano' (see information quoted in Chakraborty, 2004) and the cv. 'Amiga'. The

Table 3.5. Comparisons between selected accessions of *S. subsericea* and *S. hamata sensu stricto*.

Attribute[a]	*S. subsericea*[b]		*S. hamata*[c]
Plant[d]	218	219	196
Flower			
Colour (076)	Cream	White	Cream
Pod			
Length of beak (N1)	4.0	2.8	2.0
Length of pod (N3)	3.2	3.0	2.7
Beak/pod ratio (N5)	1.25	0.93	0.71
Seed			
Polymorphic (Q2)	No	Yes	Yes
Pointed embryonic end (Q4)	Yes	No	No
Weight 10 seeds (mg)	30.4	20.0	13.0
Leaf			
Length/breadth ratio of last expanded leaflet (N6)	3.69	2.83	2.86
Length stipule/length awn on stipule ratio (N22)	1.17	1.77	1.50
Leaf apex, acute (1) to round (9) (DOM7)	1	8	7
Stem			
Colour (DOM2)	All green	Lower nodes red	All red
Crown branch no. (N14)	25.8	12.3	20.5
Dry matter yield			
g/plant	9.87	2.52	8.94
Esterase bands[e]	A (M) B (M) 22 (M), 24 (M) D (M) E (M) F (M) 4 (S) 12 (S) 14 (M)	B (M) C (M) 21 (M) 24 (M) D (M) E (M) F (M) 4 (M) 12 (M) 13 (M) 14 (M)	B (M) C (M) 22 (M) 24 (M) D (M) E (M) F (M) 1 (M) 5 (S) 8 (W) 9 (S) 11 (S) 13 (M) 14 (M)

[a]See Table 3.3. [b]From Guatemala. [c]From Nevis. [d]See Tables 3.7 and 3.8 in Appendix 3.1. [e]As used by Robinson *et al.* (1980), Burt *et al.* (1983). Bands initially labelled A to C, 20 to 25, D to F, 1 to 6 and 7 to 15 in that sequence and this order preserved here. Capital letters in brackets refer to band strength W (weak), M (medium) and S (strong).

former cultivar is often taken to be representative of the species, but this is clearly not the case as the analysis shows that the two groupings concerned (N8 and N9, Fig. 3.1) are well separated from the majority of the *S. hamata* accessions.

Because *S. hamata sensu lato* is found only in adjoining coastal and sub-coastal areas in Venezuela and Colombia and these are linked by road, it is not surprising that groups N8 and N9 merge. The range of the grouping extends roughly from the base of the Paraguana Peninsula in Venezuela to just past Barranquilla in Colombia (Fig. 3.3). As far as can be ascertained, the accessions in groups 10 and 11 were collected in lowland areas with the same general climate.

These results contrast with the unpublished findings of Edye and Rebgetz (as quoted in Edye and Maass, 1997). Their analysis required six MA groups to accommodate their collection of 171 accessions. The groupings were geographically based, with three being of strictly Venezuelan origin, two Colombian and only the remaining one mixed. In the absence of published data, the reasons for this apparent difference cannot be ascertained.

S. hamata sensu stricto

With the exception of the two accessions placed in group 9, accessions of the *S. hamata sensu stricto* form of *S. hamata* have been placed in section D

Fig. 3.3. Origins of some accession groups.

of the classification and are thus well separated from those of *S. hamata sensu lato* (Fig. 3.1); and, with exception of group N19, the groups are monospecific (Fig. 3.1). All non-*S. hamata* accessions are located in group N16, itself a composite of three poorly defined, mixed groupings at the 20 group level, and the species with which they are associated are all of the herbaceous annual/biennial form. The particular accessions included are enigmatic: the sole accession of *S. humilis* is of the small-leaved type, atypical in that it is from a very alkaline region in Florida, USA (Williams *et al.*, 1980; Robinson *et al.*, 1980). This could well be the result of troop movements from the British naval base in Antigua, together with horses and hay, to Florida, when under British control, or from the establishment of a British naval base at Pensacola in 1814. This suggestion agrees with the information presented by Kretschmer and Brolmann (1984), who show that the species is confined to a narrow coastal strip of alkaline soils but is spreading into disturbed areas. This species has been shown elsewhere to be favoured by disturbance (Table 3.4 and associated text) and is found in such environments and situations in Antigua (Appendix 3.4).

Similar considerations apply to many other regions because there were strong trade routes between, for instance, cotton- and sugar-producing regions; thus a mixture of South American, Central American and Asian legumes were moved between continents and eventually many were found to be useful worldwide in plantation agriculture. Such movement of germplasm via trade could account for the *S. hamata sensu stricto* diploids in Brazil and the apparently Venezuelan-related *S. humilis* in the Philippines (as quoted in Maass and Sawkins, 2004). The Paraguana Peninsula in Venezuela, noted herein for the diversity of the few accessions that it has yielded, has strong trading links with and has similar soils and climate to such Caribbean islands as Curaçao, in which *S. hamata sensu stricto* is endemic. The Peninsula is also close to and well connected with Maracaibo, which is noted for the occurrence of the tetraploid form. The small-leaved form of *S. humilis* also occurs in the Peninsula and it is close to areas with the large-leaved acidophilic form, one of which is the parent of the allotetraploid (with *S. hamata sensu stricto*). Although the cv. 'Verano' is suited to acid soils, and has the appropriate symbiotic relationships with strains of *Bradyrhizobia*, there are some accessions of this type that, like *S. seabrana*, function most effectively with rhizobia strains associated with plants from alkaline soils (Date *et al.*, 1979). In view of the proven agronomic utility of accessions from this area, further collection would be warranted.

3.6 The minimum spanning tree: results

A conventional topological representation of the minimum spanning tree (MST) is presented in Fig. 3.4. In interpreting the MST, the first step is to incorporate the independent taxonomic identification of the accessions.

With the sole exceptions of *S. calcicola* and the Guatemalan accessions of *S.* sp. aff. *hamata* G, the various species and types within species are clearly separated, and the amount of variation within each group, which is very large in *S. hamata sensu stricto*, is very evident.

The main axis moving from left to right generally represents an increase in perenniality. It begins with the small-leaved, strictly annual *S. humilis* and this joins onto the annual/biennial allotetraploid form. This is in turn linked to the more perennial *S. hamata sensu stricto*. The *S.* sp. aff. *hamata* SB, a tetraploid off-type from Florida, comes next, placed at some distance from *S. scabra*. Finally, at some distance from *S. scabra*, the woody form of *S. hamata sensu* Cook *et al.* (2005), which is labelled *S.* sp. aff. *hamata* W.

3.6.1 The MST: preliminary discussion

Positioning of the species and species types

In the sense of overall similarities, some of the plant groupings appeared to have been misplaced in the hierarchical classification (Fig. 3.1): *S. sympodialis* and *S.* sp. aff. *hamata* F were grouped with woody perennials, whereas they are more similar overall, ecologically and agronomically, to the annuals/biennials (Chapter 7, this volume). In the MST, this situation is much improved: the woody *S.* sp. aff. *hamata* W is located

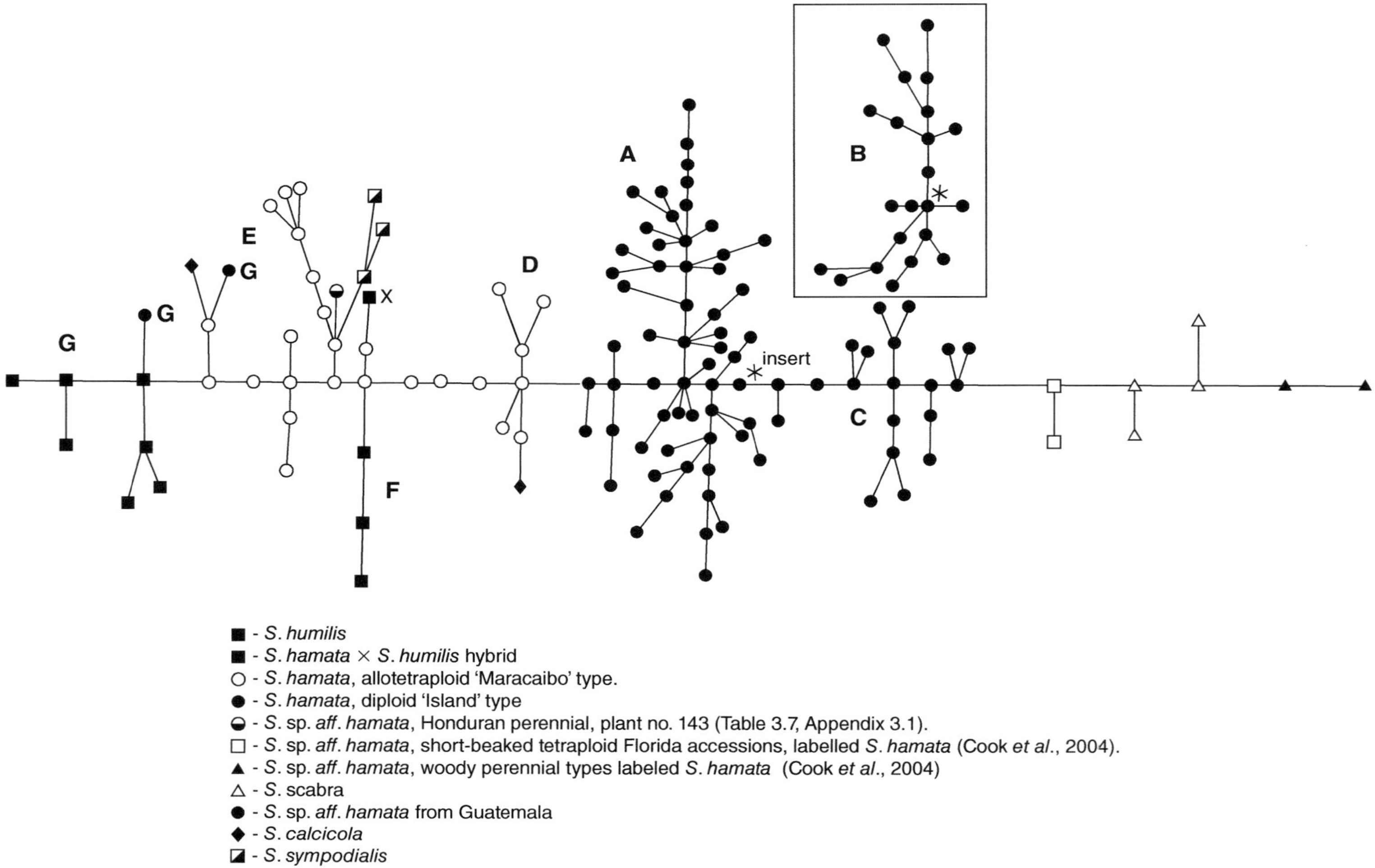

Fig. 3.4. Minimum spanning tree results: conventional topographic representation.

on one side of *S. scabra* and *S.* sp. aff. *hamata* SB is placed on the other side, next to *S. hamata sensu stricto*. This positioning, when considered with the chemotaxonomic studies of Robinson *et al.* (1980), is suggestive of hybridization. The locations of the two accessions of *S.* sp. aff. *hamata* G, often labelled simply *S. hamata*, confirms this distinction, and their separation. Their placement with *S. humilis* and the allotetraploid *S. hamata* shows their variability. Similar variability is shown by other accessions such as *S. subsericea*, from Guatemala.

The MST clearly provides a much more useful, holistic overview of the collection than does the hierarchical classification and the user is freed from difficult decision making, such as which groupings should be accepted, fused or divided and how best to position the groupings obtained relative to each other. In Fig. 3.1, for instance, the major groupings could have been shown in order C, B, D, placing the perennials between two groups of annuals/biennials. With poorly known plant material, the MST approach is important because it provides important, hypothesis-generating situations, suggesting areas where further study might best be directed and the material that might be used. For example, previous studies suggested that the allotetraploid form of *S. hamata* might be the result of hybridization between *S. humilis* and *S. hamata sensu stricto* (Burt and Williams, 1979b), and so it has proved to be (Curtis *et al.*, 1995). The MST provides a convenient framework on which to add information from other sources, to develop what Harlan (1983) refers to as a 'body of theory and general principles' to aid with the efficient utilization of genetic resource collections. Examples of this potential utility are provided subsequently.

Comparing the results of the MST and the hierarchical classification: misclassifications in the latter

[In the following sections the results of the MST are used for a variety of purposes. Should readers wish to do likewise, the positions of each accession on the MST are shown in Figs 3.11 and 3.12 in Appendix 3.2.]

With the perennial species there was good agreement between the MST and the hierarchical methods in the constitution of the groups *per se*, but as just noted, not in their positioning. Both had clearly designated and separated *S. scabra* and *S.* sp. aff. *hamata* W; both had problems with *S. calcicola*; and both had clearly defined the allotetraploid *S. hamata* and separated it from the other annuals/biennials. However, *S. hamata sensu stricto* had sometimes been badly misclassified and had often been placed with other species. This is illustrated in Fig. 3.5, where the accessions from the various hierarchical groups are nominated on the MST.

There is some general agreement between the two classifications with regard to *S. hamata sensu stricto*: in the MST (Fig. 3.5) the parts of groups A and B above the axis are dominated respectively by the hierarchical subgroups N17 and N13. They differ however in that accessions of N13 also appear in A and C, and in observant accessions of *S. humilis* and *S* sp. aff. *hamata* G from the hierarchical classification N16, having been transferred to their relevant species groupings in the MST in groups E, F and G (Fig. 3.4). In addition, two accessions of *S. hamata sensu stricto* previously placed with the hierarchical subgroups 14 and 16 have been correctly included in the section of the MST containing the allotetraploid accessions, groups A, B and C (Fig. 3.6; for detailed information see Appendix 3.3).

The same situation applied with *S. humilis*. Two groups were defined in the MST, the large-leaved types in F and the smaller leaved form in G (Fig. 3.6). They are shown as being associated with the allotetraploid *S. hamata*, but are attached at different locations. The individual accessions within groups are widely separated with, for instance, the dryland form of *S. humilis* from Brazil (plant 129) being separated from the two more typical accessions (from Colombia and Venezuela, N8) and placed in an intermediate position next to the allotetraploid *S. hamata* (see Appendix 3.2, Fig. 3.12). Three of the four *S. humilis* accessions from Paraguana, also from an environment not usually considered to be suitable for the species (see Fig. 1.7 in Maass and Sawkins, 2004), are well separated from the others and placed in a terminal position (plants 130, 131 and 132 in zone H) (see Appendix 3.2, Fig. 3.12). This is a much better arrangement than that afforded in the hierarchical classification (details will be presented subsequently) and one that can be much more readily visualized by a would-be user.

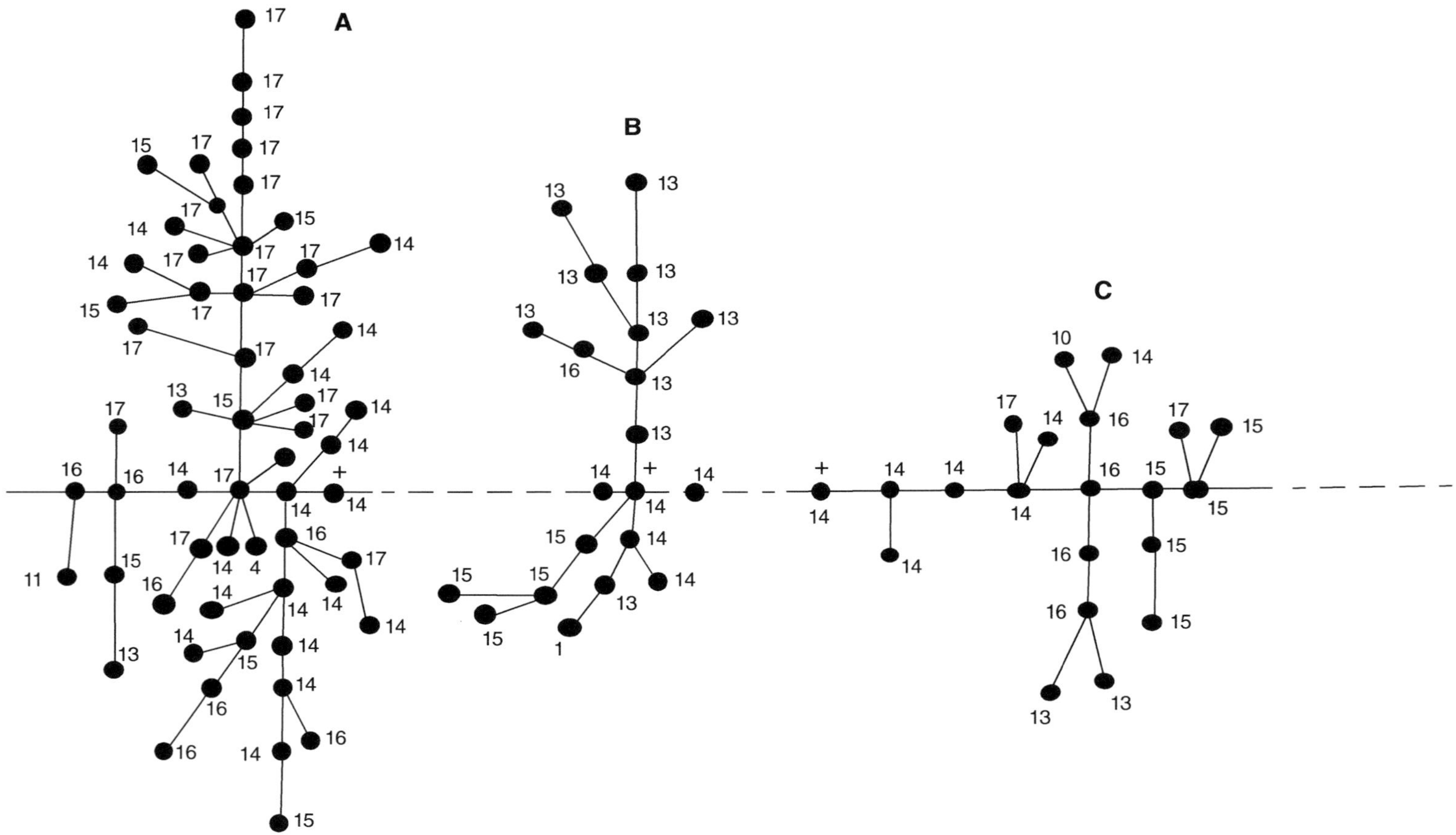

Fig. 3.5. Minimum spanning tree results for the *S. hamata sensu stricto* accessions: group numbers from the hierarchical classification (as on Fig. 3.1) superimposed. Zones A, B and C as shown on Fig. 3.4.

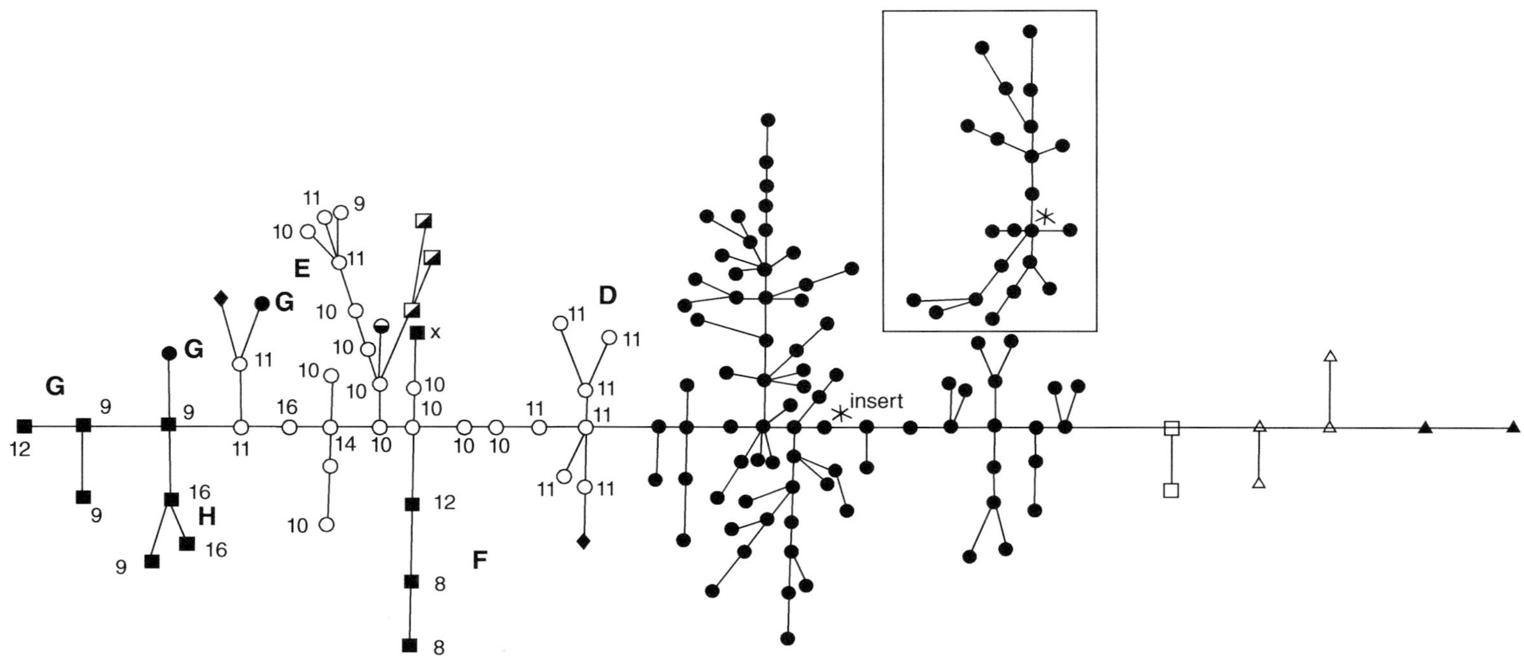

Fig. 3.6. Minimum spanning tree for *S. humilis* and the allotetraploid *S. hamata* accessions. Group numbers from the hierarchical classification (as on Fig. 3.1) superimposed. Zones D to G as in Fig. 3.4.

Using the MST: some worked examples

There are many instances when someone wishes to choose a range of material to represent a species. In agronomic evaluation, representatives of the different types of plant available can be used to avoid the initial testing of the complete collection, comprised of many virtually identical accessions: Burt *et al.* (1974) did so using MA groupings. But, with presence of larger collections and the existence of continuous variation, groupings based on hierarchical classifications, as we have seen, may no longer be suitable for this purpose. Failure to choose a truly representative range of types of a species can produce misleading results in such important studies as the determination of the potential distribution of a species (Maass and Sawkins, 2004). The type of results presented herein, especially if combined with other information, could well be used to select representative types of a species and some simple examples follow.

First, selection for agronomic potential on the basis of similarity to plants of proven value (Fig. 3.7).

The *S. humilis* cultivar 'Paterson' was selected for use on acid, sandy-textured soils and 'Khon Kaen', collected from the Paraguana Peninsula is one of the few accessions adapted to heavy clay soils. When selecting plants for testing for heavy clay soils it would be reasonable to assess other accessions within the MST group G. These include other accessions from the Paraguana Peninsula, one accession from the Dominican Republic (plant 127, labelled 8 in Fig. 3.7) and add to these the enigmatic *S.* sp. aff. *hamata* G, from Guatemala. In the tetraploid *S. hamata* the ubiquitous cv. 'Verano', often included in evaluation work as the sole representative of the species, is clearly atypical of even the allotetraploid form: it is placed in a terminal position away from the main body of accessions. But, if similar plants are required, then plants 34 and 36 (labelled as 6 and 7 in Fig. 3.7) should be selected: they are both from the Paraguana peninsula, typified by alkaline soils, and may have retained their ability to form effective symbiotic relationships with *Bradyrhizobia* from alkaline soils (Date *et al.*, 1979).

In *S. hamata sensu stricto* no cultivars have been made commercially available to date, but one, plant 4 (Fig. 3.7), has been nominated as promising on heavy clay soils in Indonesia (Cook *et al.*, 2005) (but some others will be nominated subsequently). It is atypical of the species in form and origin (the Paraguana Peninsula): the plants placed closest to it are from Antigua and Florida (as noted earlier, they may share a common origin) with those from Antigua coming from soils that are all alkaline but of contrasting textures (a coarse coral sand, a marl and a heavy clay). Their agronomic evaluation would be informative (see later examples).

Second, considerations about *S. hamata sensu stricto*.

As indicated in Edye and Maass (1997), this is a very variable type found over a relatively large geographical range, but the range is probably somewhat greater than that indicated by the authors (e.g. Stehle, 1956 for general information about the Caribbean Islands and Arnaldo, 1964 for Curaçao) and it is also probably adapted to a wider range of soil types and climates than those listed. Although there are some general relationships between the geographical origins of the plants and their type (Williams *et al.*, 1980; Robinson *et al.*, 1980; Edye and Rebgetz as quoted in Edye and Maass, 1997), this is far from complete (Fig. 3.8). Accessions from Barranquilla (in Colombia, Fig. 3.3) are found throughout. As noted in the discussion of the results of the classification, accessions from a given location can contain mixtures of widely different plants. Accessions from Antigua (A on Fig. 3.8) are like those from Barranquilla they can occur throughout the range (although many continue to be associated, in grouping A, Fig. 3.8, with those from Florida. The situation in Antigua is explored further in Fig. 3.9).

As noted in Appendix 3.3, Antigua has many different soil types, and the type of soil from which accessions were collected and the classificatory group in which they had been placed, is shown on Fig. 3.9 and listed in Table 3.6. Bendal's clay had contributed most accessions and the most common classificatory group for these was 14, but some accessions from this soil were placed in groups 13 and 15. Group 14 accessions were to be found on a range of soils from coarse coral sands to marls to heavy clays. The same is true for other soils and plant groups; examination of Fig. 3.9 shows the same dissociation between plant group and the soil from which it was collected.

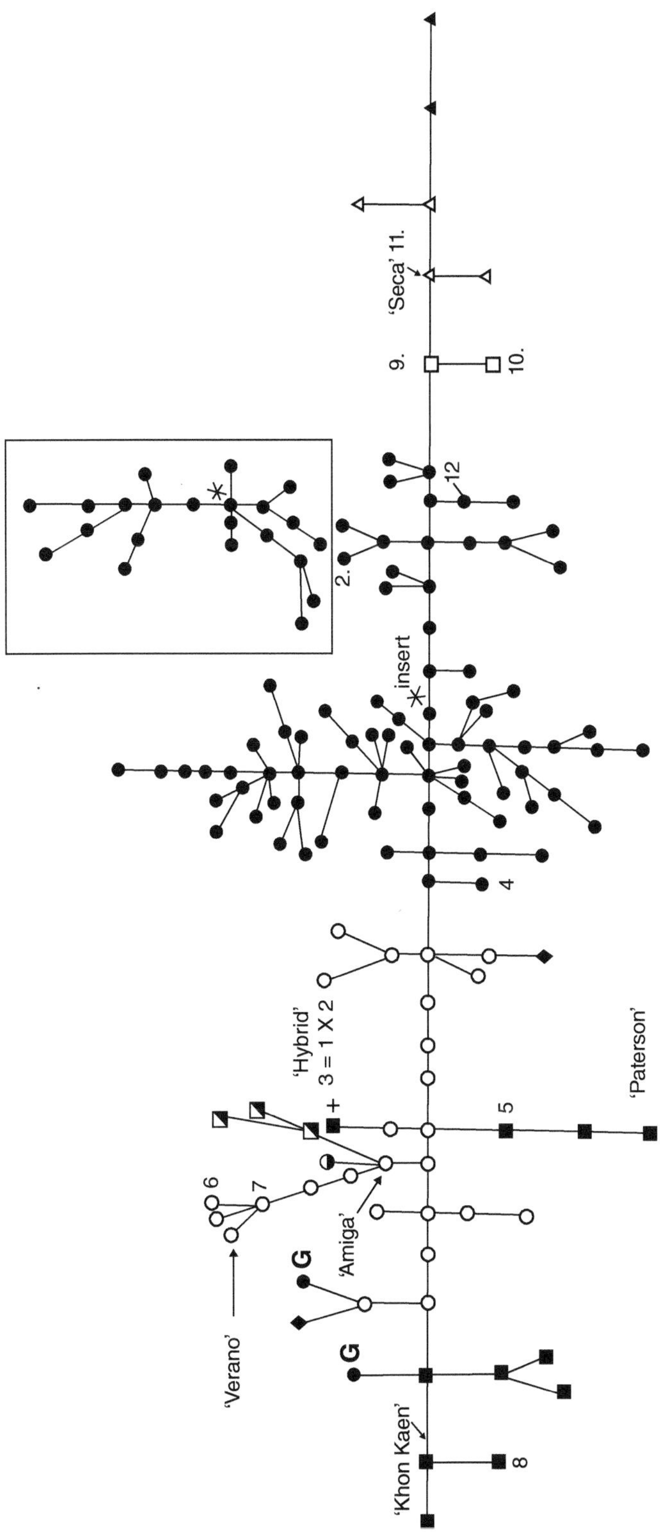

Fig. 3.7. Location of the various cultivars and other specified accessions on the MST. See section 3.6.1.

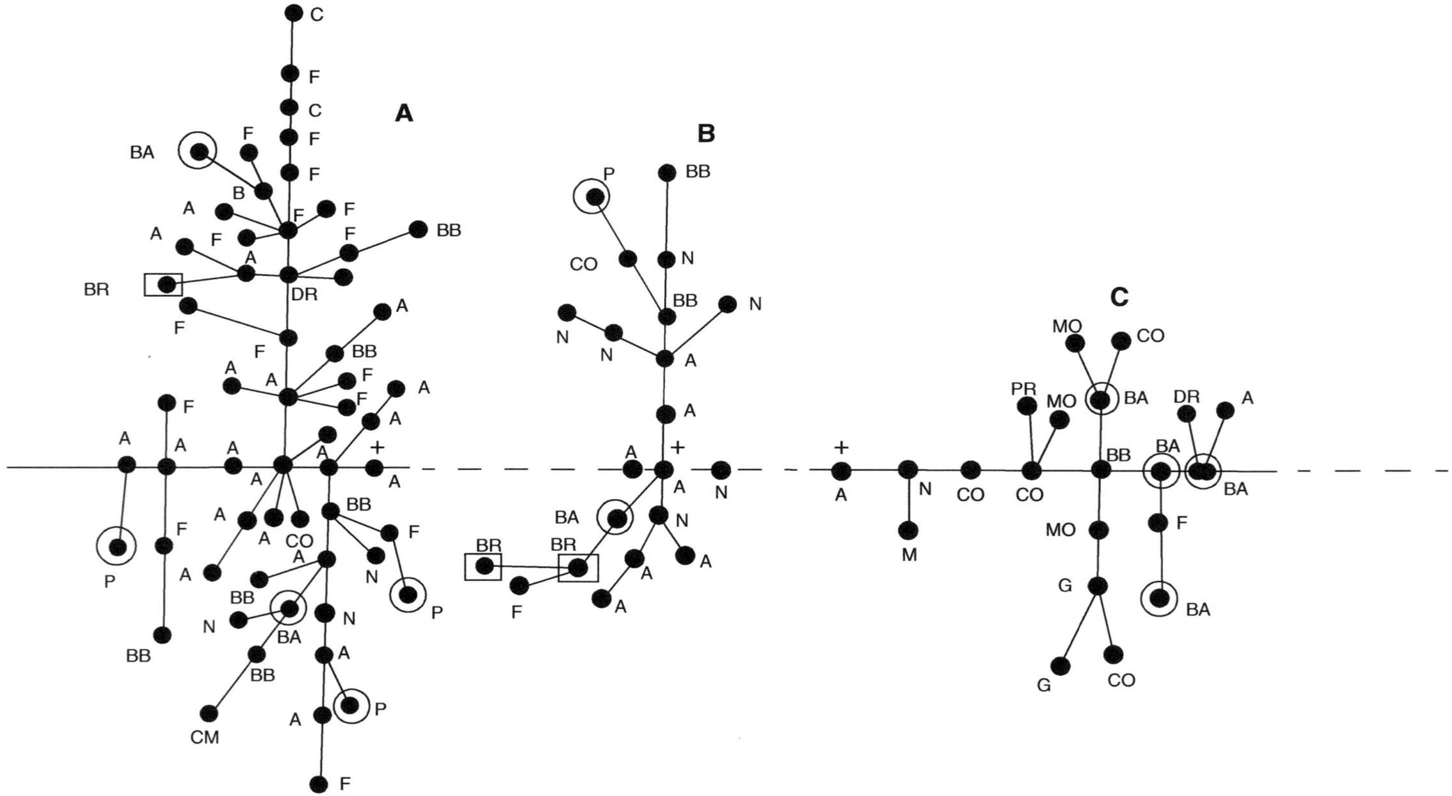

A Antigua
B Bahamas
BA Barranquilla, Colombia
BB Barbuda
BO Barquisimeto,Venezuela
PR Puerto Rico
BR Brazil
C Cuba
CM Colombia
CO Curaçao
DR Dominican Republic
F Florida, U.S.A
G Guadeloupe
M Margarita,Venezuela
N Nevis
P Paraguana Pen., Peninsula

Fig. 3.8. The geographical origins of the *S. hamata sensu stricto* accessions and their position on the MST.

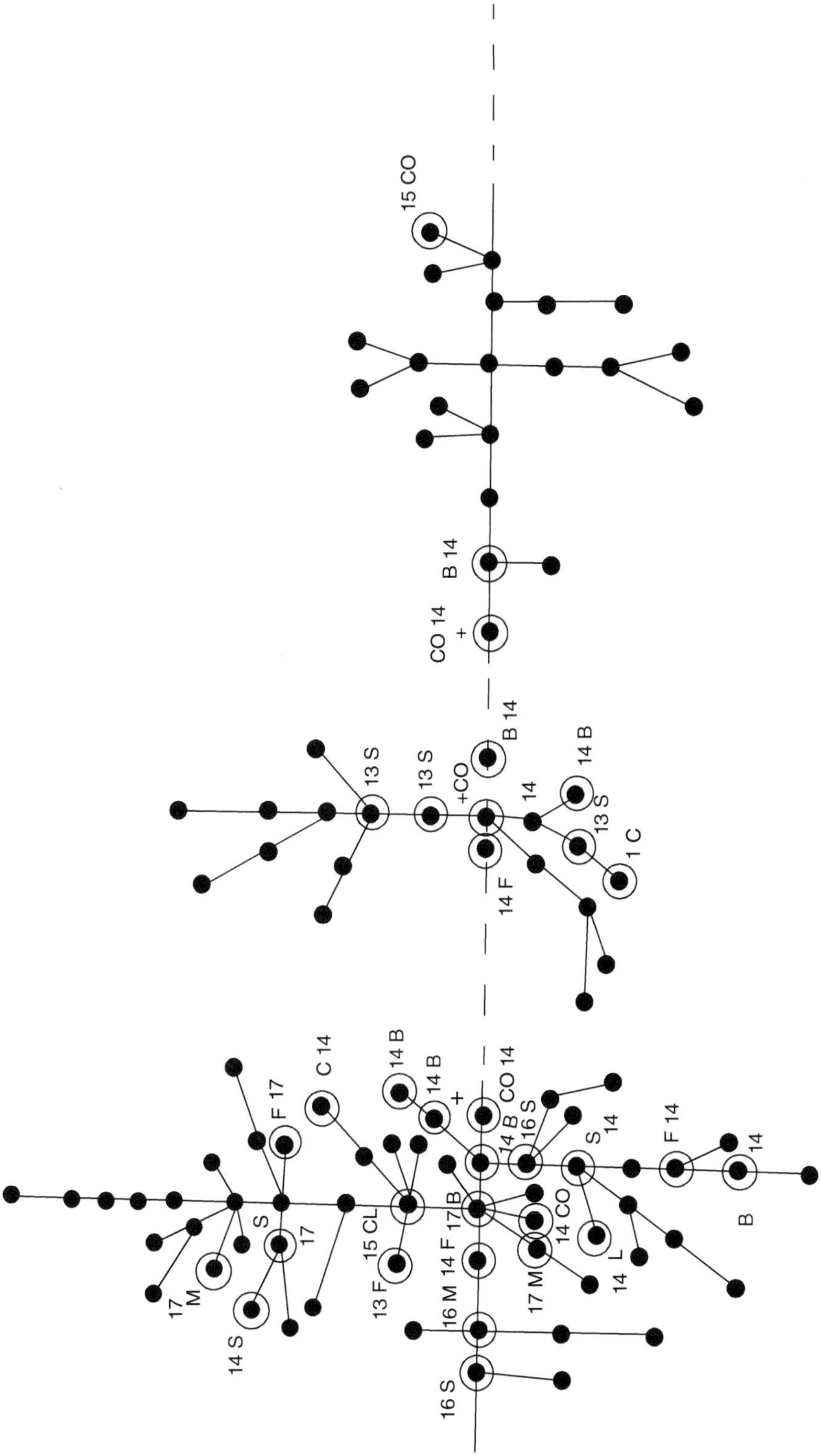

+Soil type as defined by Hill (1966): +B - Bendal's clay; +CO - Coolidge clay; +F- Fitche's clay; S - Shelly sand; C - clay; CL – clay loam; M - Marl; L - ex. Railway line mixed soil.

Fig. 3.9. Antiguan accessions of *S. hamata sensu stricto*: their classificatory group (as in Fig. 3.1) and soil type at the point of collection.

Table 3.6. Distribution of types of diploid *S. hamata* on Antiguan soils.

	Classificatory group[a]						
Soil	1[b]	13	14	15	16	17	Total
Coral sand		4	4		1		9
Clay loam				1			1
Marl			3		2	1	6
Clay	1						1
Bendals Clay[c]		1	10	1			12
Coolidge Clay[c]			2	1			3
Fitches Clay[c]		1	2	2	1		6
Total	1	6	21	5	4	1	38

[a]As defined in Table 3.3 [b]Before reallocation to group 13 or 14. [c]As defined by Hill (1966).

Plant populations are of mixed types, as has been noted for *S. humilis* (Burt *et al.*, 1973); the allotetraploid form of *S. hamata* (the cv. 'Amiga') was selected as being different from other plants from the same population (also Chapter 8, this volume). Such a mixture of plant forms would confer population plasticity and survival in an unpredictable environment (Burt *et al.*, 1973) and this approach is applied in farming practice by the sowing of mixtures of cultivars and species (e.g. *S. scabra* cv. 'Siran' is a mixture of three registered cultivars).

The third example continues this investigation. The backgrounds of various accessions that have been found to be best adapted to heavy clay soils in dry tropical environments were examined; after a comprehensive long-term evaluation study the best rated accessions produced well, survived and spread by seedlings in a disturbed situation (Chapter 8, this volume).

The best adapted accessions were plants 62, 110 and 83 (these are the *S.* sp. aff. *hamata* SB form from Florida, USA) (Fig. 3.10); they were placed in widely spaced locations on the MST, were from different classificatory groups, from a range of soil textures and from very different types of climates – Florida is subtropical, Antigua dry tropical and Curaçao semi-arid tropical. The only constant environmental feature is that they were all found growing on alkaline soils, as was the short-beaked form of *S.* sp. aff. *hamata* from Florida. Those accessions whose performance was good, but not as outstanding, were similarly diverse in their environmental background but were again from alkaline situations.

Double lines link accessions on the main axis of the MST. Single lines link accessions placed on side branches of the MST.

It would appear that pre-adaptation to alkaline soil is a prerequisite for the survival of introduced material of this form (the annual/biennial type). Almost certainly this involves the ability to form effective symbiotic relationships with the local *Bradyrhizobia*, for introduced strains do not persist for any length of time. The climatic background is of far less significance as is indicated by plant 32, which thrives in the wet tropics of Indonesia (Cook *et al.*, 2005), although it is an accession from the semi-arid Paraguana Peninsula. This is in contradiction to hypotheses used in previous work, e.g. by Burt *et al.* (1976), about the types of climate most likely to furnish useful plant material of *S. hamata* for use in the dry tropics, and it should influence the choice of material for use in future testing programmes, selection of areas or vegetation types for plant collecting missions among other plant introduction activities. However, such a revision of attitudes may be applicable only to the 'weakly' annual/biennial types (*sensu* Baker, 1974), where disturbance is a key feature in their establishment and survival. All of the evidence available with the long-lasting, woody perennials, such as *S. scabra*, continues to suggest that pre-adaptation to climate, and possibly to ecological conditions/associate species, is of importance (Burt *et al.*, 1974; Burt, 1983; Burt *et al.*, 1983; information quoted in Chakraborty, 2004; Cook *et al.*, 2005). Further review of this latter situation is unwarranted here.

The examples presented demonstrate how suitable graphic presentations, such as the MST, can be much more amendable for use than a 'standard' hierarchical classification and can be used to provide a holistic overview of genetic resource data. In validating the meaningfulness

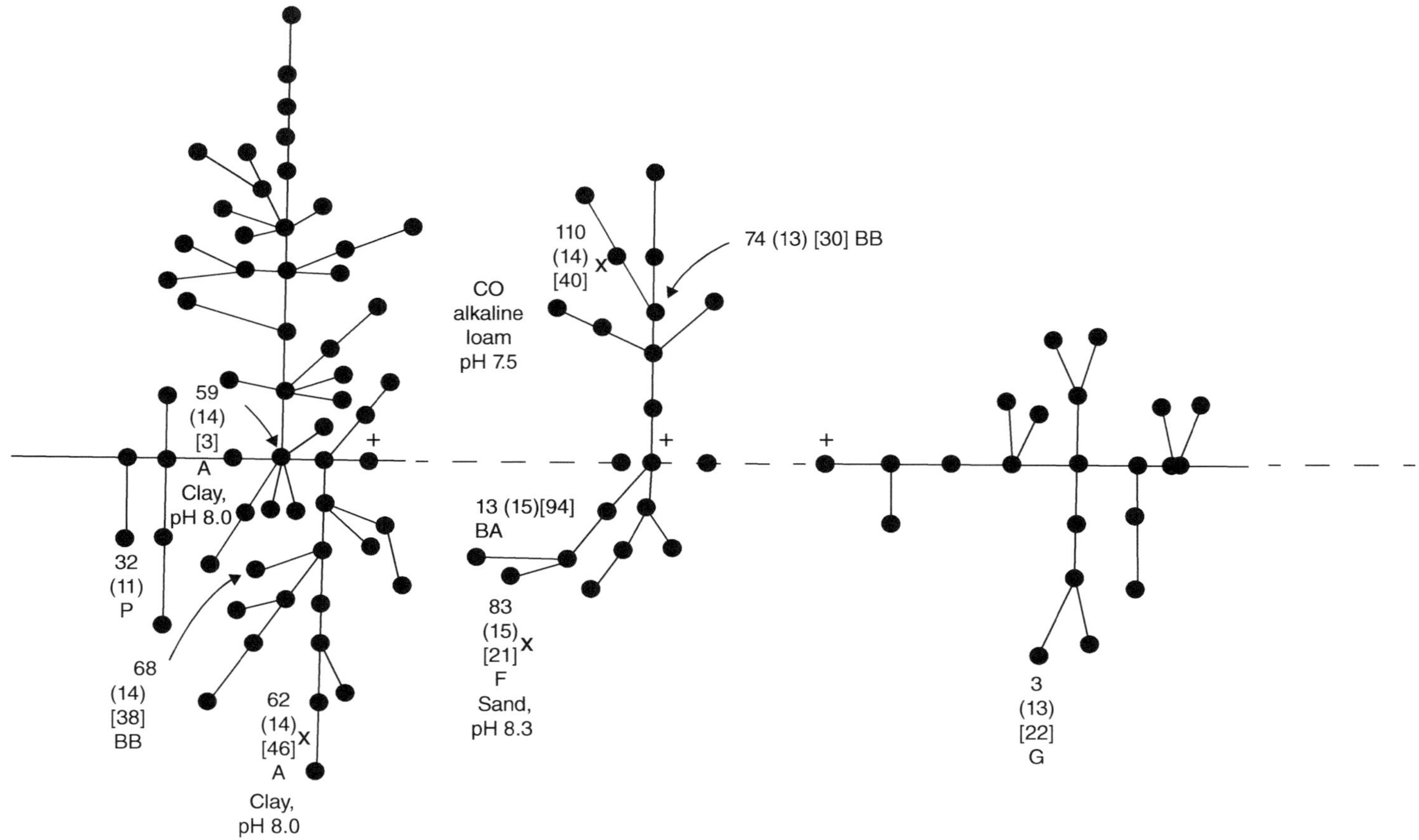

+ Accessions most outstanding generally.

Place of origin: A, Antigua; BA, Barranquilla, Colombia; BB, Barbuda; CO, Curaçao; F, Florida, U.S.A.; G, Guadeloupe; P, Paraguana Peninsula,Venezuela.

N.B.: Also performing well on the clay soil after 3 years were plant numbers 148 and 149, short beaked tetraploid *S. sp.aff. hamata* from alkaline sands in Florida, and, for a short period, plant 8 (*S. hamata* cv.'Verano'.)

Fig. 3.10. The positions on the MST and background information for promising accessions on heavy clay soils. Accessions given plant numbers (Table 3.7, Appendix 3.1) and classificatory group numbers (as in Fig. 3.1) in round brackets. Numbers in square brackets are for plant survival/m^2 after 3 years of growth on a clay soil in the dry tropics (Chapter 8, this volume).

of such a classification, the addition of information from other disciplines assists in efficiently providing the 'body of useful information and general theory' (Harlan, 1983) that aids the would-be users of genetic resource collections. The more recent graphic method, described by Burt *et al.* (1983), was demonstrated by the authors to be even more informative than the MST. The further development of this methodology, or its incorporation into a system of classification and description, would greatly aid work such as that described here.

3.7 General discussion

The need for tropical pasture and ley legumes was documented in the introduction of this chapter. This need can be readily confirmed on a worldwide basis, particularly for legumes adapted to important clay soils, by reference to relevant chapters in Chakraborty (2004). Global warming and the energy crisis make the demand for such plants, and the loss of the plant biodiversity necessary to meet this demand, a matter of some urgency.

The main provider of suitable pasture legumes for use in the dry tropics has undoubtedly been the genus *Stylosanthes* (Maass and Sawkins, 2004; Cook *et al.*, 2005). But, as has been shown herein, our knowledge is less than adequate for those species of major importance, and, as documented by Maass and Sawkins (2004), 'new' species and centres of diversity are being discovered and more esoteric approaches developed. Some of the more practical issues of significance at the present time are discussed further below, while more detailed studies are presented in the Appendices. The points raised are almost exclusively directed at the genus *Stylosanthes*, but, where directly relevant, comparisons will be drawn with other genera.

3.7.1 The adequacy of genetic resource collections with especial reference to *S. hamata* and *S. scabra*

Plants new to an area may arrive from a variety of sources and there is always the chance of serendipity, of something good 'turning up'. 'Hass', the most popular variety of avocado, was found as a seedling growing in an orchard, as was R2 E2, a very large-fruited variety of mango. In tropical forage legumes, the *S.* sp. aff. *hamata* SB is an 'off-type' detected in Florida (Brolmann and Sonoda, 1981) that has performed very well in dry tropical situations (Chapter 8, this volume). However, there is increasing dependence on genetic resource collections, and in *Stylosanthes* 'new' species are still being found (Maass and Sawkins, 2004), 'new' types of variation are being discovered (e.g. *S.* sp. aff. *hamata* G and *S. subsericea* from Guatemala) and the possibility of developing 'new' traits (such as anthracnose resistance) would be welcomed.

Some of the species and species types mentioned are very poorly represented in genetic resource collections. For example, *S. sympodialis*, the only *Stylosanthes* species to be found in a phytogeographically interesting area, geographically removed from other *Stylosanthes* species (and associated with *Desmanthus*, a genus of major importance for clay soils and that also merits collection from this region; Pengelly and Liu, 2001). But for the purpose of illustration, and brevity, remarks will be confined to the two contrasting, well-collected species, *S. hamata* and *S. scabra*.

S. hamata

As noted earlier, the genetic resource collections of this species in the mid- to late 1960s were totally inadequate and very biased towards forms most easily obtained by correspondence; the same is true of *S. scabra*. Thereafter, Australian collections were rapidly increased by visits to Venezuela (Burt *et al.*, 1979), the Paraguana Peninsula by Smith in 1972, Venezuela and Colombia (Edye, 1988) and the Caribbean Islands, by Keoghan and others involved with the IDRC project there (see Appendix 3.3 for a fuller appreciation of the significance of this project). To this collection could be added increasing amounts of material from what were then new sources in Florida (Kretschmer and Brolmann, 1984) and from South and Central America generally through CIAT, Colombia (Schultze-Kraft *et al.*, 1984). By 1997 the Australian collection contained at least, and probably more than, 139 accessions of *S. hamata sensu stricto* and 171 of the allotetraploid. Classifications and descriptions of the collections were undertaken but not published (Edye and Maass, 1997).

The allotetraploid has a very limited pattern of distribution in a coastal and sub-coastal region of Colombia and Venezuela, and dedicated collecting missions, especially that of Edye in 1986, would appear to have covered most of the area (Edye and Maass, 1997). But, as noted previously, the Paraguana Peninsula in Venezuela has largely escaped attention (apart from the efforts of Dr R. Smith; Burt *et al.*, 1979), and it differs from the rest of that region in terms of its more arid climate, alkaline soils and vegetation (the latter as attested to by Andrade-Lima, 1971, personal communication). Indeed, on climatic grounds it is sometimes considered to be an unlikely source of *S. humilis* (Maass and Sawkins, 2004). Two of the very few accessions from this area have, however, been shown to be of major significance. The first, *S. humilis* cv. 'Khon Kaen', outperformed other *Stylosanthes* species and accessions on clay soils in Thailand. The second, *S. hamata sensu stricto* (CPI 61670) showed much promise in Indonesia and was shown to be the most consistent performer, better than the best accessions of *S. seabrana* and *S. scabra*, across a range of environments and soil types in India (Ramesh *et al.*, 2004).

By contrast, *S. hamata sensu stricto* has a much wider geographical range, extending from the coastal ranges just mentioned – where it appears to be more common than the allotetraploid in more subtropical, drier climates (Edye and Maass, 1997) – through many of the Caribbean Islands (Stehle, 1956) and into the higher latitudes of the Bahamas and Florida, USA. Collections from many of these islands appear to be very limited, although on an island such as Puerto Rico, which has a very wide range of soils and climates, the species is known to occur and should be easy to access. Is further collection, apart for a search for new sources of anthracnose, warranted?

It has been shown that this is a very 'wide' species, but one in which the different forms are broadly distributed across the islands and, on a given island (Antigua, Appendix 3.4), widely distributed over the different textured (alkaline) soils. It is favoured by disturbance but, provided it can form effective symbiotic relationships with the appropriate rhizobia, is capable of growth over a range of climates. Similar species, the allotetraploid *S. hamata* and *S. humilis*, are capable of producing adaptive variants (Burt and Beattie, unpublished data; Burt *et al.*, 1973). The ability of *S. humilis* to adapt has enabled a chance introduction of a plant, whose flowering behaviour is under strict day length control, to become adventive over a wide range of latitudes and day lengths. Such adaptive changes occur frequently, especially under stressful conditions such as those imposed by growth on infertile soils (Gottlieb, 1977). They can occur by somatic segregation without even the necessity for outcrossing and often produce agronomically useful plants (Snaydon, 1978). Although anthracnose-resistant plants would be welcome, these considerations suggest that further targeted collections for the species need not be given priority.

S. scabra

S. scabra is a South American species, primarily from Brazil but occurring also in Venezuela and Colombia (Liu, 1997); those accessions attributed to Belize were almost certainly collected in Brazil and taken to an IDRC/University of West Indies project, based in Belize, by one of the authors (RLB). In 1997 the Australian collection contained some 600 accessions (Liu, 1997).

S. scabra is a very different plant from the more usually encountered forms of *S. hamata*, *S. hamata sensu stricto* and the tetraploid. These differences should affect considerations relating to its collection and evaluation. A strong, long-lasting woody perennial, it is found in the later stages of the ecological sere (see Table 3.4 and associated text); in grazed, mixed communities it has been observed as being well spread throughout, but not dominant, and plants in a given collecting area were very similar overall – this was confirmed by studies in Australia (Burt *et al.*, 1983). Plants from different climates, and different disturbance levels within them, shared many characteristics. In the more stressful, drier areas – where rainfall patterns are irregular and unpredictable – the individual plants show considerable plasticity, as previously shown for species from this ecological stage (Baker, 1974). For instance in controlled environment studies when grown in different temperature and light regimes, they can produce flowers at very different positions on the plant, and over a greater range of times, while forms from other, more predictable environments, could not (R.L. Burt and J.L. Davidson, unpublished results).

The species relies on persistency and plasticity for its survival, rather than the population heterogeneity requirement of colonizing forms such as *S. hamata sensu stricto*. With the sole exception of the cv. 'Seca', the *S. scabra* accessions studied do not produce natural variants (Liu, 1997).

S. scabra is a very diverse species and is spread over a wide range of latitudes and altitudes in South America. The occurrence and distribution of some of the MA types, those denoted earlier, is related to the climatic types found in the varying locations (Burt, 1975a) as well as to the different ecological conditions within them (Burt *et al.*, 1983). And the five different types accepted by Liu (1997), on the basis of DNA markers and a hierarchical classification, are geographically based. These groupings, as with those designated by Maass (as shown in Liu, 1997), are very wide: Liu's group 2, recognized by Maass as being the 'coastal (Brazilian) type' – as opposed to the 'continental type' – contains the MA groups (sensu Edye *et al.*, 1973) 17, 18 and 30 in addition to those defined in later studies (Burt *et al.*, 1983). And in Liu's analysis types 17 and 18, which have yielded all of the cultivars listed in Cook *et al.*, 2005), or provided the breeding material utilized for cvs. 'Jecuipe' and 'Feira', are represented by only a very small segment in his group 2.

All of his material was collected in a narrow range of latitudes, from 7°S to 13°S, in north-east coastal and sub-coastal Brazil. Liu (1997) shows few collections from this area. In contrast, there are many from more southerly latitudes and from inland, continental areas. Where comparisons can be drawn, these correspond to MA type 19 (Edye *et al.*, 1973); this form, although widely tested, showed little agronomic promise (Burt *et al.*, 1974), has yielded no cultivars and has not been useful for other purposes, for instance as a source of resistance to anthracnose. This situation is reviewed further in Appendix 3.4.

The S. scabra–S. sp. aff. hamata W–S. sp. aff. scabra–S. seabrana–S. viscosa complex

These are all woody perennials that, with the sole exception of *S. viscosa*, are of potential agronomic value. *S. scabra* and *S. seabrana* are already of importance throughout many parts of the tropical world (e.g. Chakraborty, 2004; Cook *et al.*, 2005) and future improvements and new cultivars in both should be sought. As noted above, and in Appendix 3.4, the genetic resource collections of the more valuable forms of *S. scabra* are limited and this is almost certainly true of *S. seabrana*, as will be shown shortly.

All of the material in this complex is related. It has recently been shown that the tetraploid *S. scabra* is the result of hybridization between the diploids *S. seabrana* (separated as a 'new' species in 2002; Maass and t'Mannetje, 2002) and *S. viscosa*. Here we note for future reference that while all accessions of these species were collected on sandy-textured, acid soils, *S. seabrana* has subsequently been found to be best adapted to heavy, alkaline soils in more subtropical conditions (Cook *et al.*, 2005). Prior to its delineation as *S. seabrana*, the species was labelled *S.* sp. aff. *scabra*, a species that it resembles in general form. The situation is, however, confusing to the reader because it is difficult to ascertain if all of the *S.* sp. aff. *scabra* accessions, such as those mentioned in Jansen and Edye (1996), and as nominated separately by Maass (1989) and Burt and Williams (1979b), have been elevated to species status as *S. seabrana*. Certainly, in the latter case, they have been retained as *S.* sp. aff. *hamata*, the woody form from Brazil, and these differ from *S. seabrana* in their ability to nodulate effectively with *Bradyrhizobia* from acid soils. However, the experimental work described herein shows that they are much more similar overall to *S. scabra* and, in geographical terms *S. seabrana* and *S.* sp. aff. *hamata* and *S.* sp. aff. *hamata* W were first found, and are most common, along a road transect through the Chapada Diamantina area in tropical Brazil. Here they are intermixed with *S. scabra* and *S. viscosa*.

The MST analysis provides a useful holistic overview of some of this complex, and interrelationships with *S. viscosa* have been considered previously using it (Burt *et al.*, 1983). Unfortunately accessions of the newly designated species *S. seabrana* were not available at that time. Some of the problems confronting would-be users of this part of the collection are discussed in the next section when their taxonomic identification is considered. Here we are concerned only with the development of the genetic resource collection, further details being presented in Appendix 3.4.

Much of the material contained in this complex was virtually restricted to a road transect through the Chapada Diamantina, in Brazil (Fig. 3.15). This is an anticlinal area with many types of rock formations and associated soil types, with varying altitudes and climates. It is an internationally recognized centre of biodiversity and a noted refuge area (Andrade-Lima, 1982). Several previously uncollected species were found around its margins (Burt *et al.*, 1979). Although almost certainly collected from acid soils along the road transect, where it occurred with acid-loving *S. scabra* and *S. viscosa* (Burt *et al.*, 1979), *S. seabrana* has been found to be adapted to alkaline, heavier soils and to more subtropical conditions than these species. Andrade-Lima (1971, personal communication) believed that a heavy, clay soil was present near the town of Lencois, in the Chapada Diamantina area, and it is possible that this is the source of the species *S. seabrana*, which is named after the town Seabra in that region. In this floristically rich area, and with such a diverse range of soils and climates, more detailed collecting missions could well yield 'new' variation.

3.7.2 Classification of plants in genetic resource collections

The 'major hurdle to unlocking the secrets (of our genetic resource collections) has been our inability to satisfactorily classify this variability' (Krull and Borlaug, 1970). This topic has been taken up on many occasions (as mentioned, for instance, in Burt, 1983). Here we will only emphasize that it must be a 'satisfactory' classification, one suited to the purpose in hand, and this includes plant breeders (who may be interested in only, for instance, seed colour as a 'marker character'), agronomists (interested perhaps in combinations of persistence, yield and flowering time), or those wishing to choose a range of material to represent the species. The significance of a character may not be recognized by the classifier, as was the case, for some time, with short straw in cereals, the character that finally allowed the production of the 'miracle rice' varieties.

Classical taxonomy

The basic taxonomic work for *Stylosanthes* is the revision of the genus provided by Mohlenbrock (1957). In recent years this has attracted much attention and comment, especially since 'new' species have been designated and the phylogeny of the genus considered in more detail (Maass and Sawkins, 2004). For our present, illustrative purposes, we shall confine our discussion to the classification of the more problematic accessions contained in our collection. These will be examined using the criteria set out by Jansen and Edye (1996), dealing with the delineation of *S.* sp. aff. *scabra* from *S. hamata* and *S. scabra*, and those of Maass and t'Mannetje (2002), who renamed some or all of *S.* sp. aff. *scabra* as *S. seabrana* and in doing so found it necessary to distinguish it from *S. scabra* and *S. hamata* (see Box 3.1).

The material listed in Box 3.1(vii) is part of a very important gene pool as *S. scabra*, *S. seabrana* and the tetraploid form of *S. hamata* and *S. hamata sensu stricto* are already in widespread use, and both *S.* sp. aff. *hamata*, the woody *S. hamata* (*sensu* Cook *et al.*, 2005) and the off-type short-beaked form from Florida have shown considerable promise. The enigmatic *S.* sp. aff. *hamata* H, a woody perennial from Honduras, and the herbaceous *S.* sp. aff. *hamata* G from Guatemala are part of a group of species, some of which are newly described, from Central America (see information quoted in Maass and Sawkins, 2004) and – as shown earlier – these also are of considerable interest.

The application of the keys provided have, however, failed to distinguish between the various types of *S. hamata*. And some listed as belonging to that species, for instance *S.* sp. aff. *hamata* W and *S.* sp. aff. *hamata* SB, do not appear to belong to *S. hamata*, *S. scabra* or *S. seabrana*.

The situation is particularly confusing for the accessions 146 and 147, labelled here as *S.* sp. aff. *hamata* W and as the perennial woody form of *S. hamata* by Cook *et al.* (2005). These form part of a complex, possibly a 'compilospecies', comprising *S. scabra*, *S. viscosa*, *S. seabrana* and accessions 146 and 147; they were collected from the same area in the Chapada Diamantina and these species are found growing together.

In the first instance these accessions were identified (for Burt *et al.*, 1979) as *S. hamata*, but in general form and behaviour they were much more like the sympatric *S. scabra* and to avoid confusion were re-labelled *S.* sp. aff. *hamata* and *S.* sp. aff. *scabra* and their relationships illustrated on a minimum spanning tree (Burt *et al.*, 1980b). Further detailed studies, and chemotaxonomic relationships, were provided by Burt *et al.* (1983) and Robinson *et al.* (1980). The name *S.* sp. aff. *scabra* was later applied to a group of accessions that had, in morphological-agronomic studies, been intermixed with those of *S. scabra* (Jansen

Box 3.1. Keys for the *Stylosanthes* species of concern here.

First, using Maass and t'Mannetje (2002).

i. Plant 59: *S. hamata* CPI 70360 from Antigua. Centrally placed in the A grouping on the MST (Fig. 3.2) and in the classificatory group 14 (Fig. 3.1).
   - Beak ≥ pod, leaflets without bristles – *S. hamata*.
   - A diploid, *S. hamata*, *sensu stricto*.

ii. Plant 22: *S. hamata* CPI 55825 from the Maracaibo region, Venezuela. Centrally placed on the tetraploid grouping on the MST (Fig. 3.2) and in the classificatory group 10 (Fig. 3.1).
   - Beak ≥ pod, leaflet without bristles – *S. hamata*.
   - A tetraploid *S. hamata*, *sensu lato* (or *S. hemihamata*, as in Maass and Sawkins, 2004).

iii. Plants 148 and 149: CPI 70522 and 70529. The short-beaked *S. hamata* (Cook *et al.*, 2005): listed herein as *S.* sp. aff. *hamata* SB., an off-type from Florida.
   - Beak ≤ pod, not *S. hamata*.
   - Leaflets without bristles, not *S. scabra*.

iv. Plants 146 and 147: CPI 55811A and 55871. The perennial woody *S. hamata* as listed by Cook *et al.* (2005): herein listed as *S.* sp. aff. *hamata* W. Sympatric in north-east Brazil with *S. seabrana* *S.* sp. aff. *scabra* and *S. scabra*.
   - Beak ≤ pod, not *S. hamata*.
   - Leaflets bearing bristles, possibly *S. scabra* or *S. seabrana*.
   - Leaflets narrowly elliptical, not *S. scabra* (length × breadth ratios 1.8 to 2.5 for accessions of *S. scabra* and 3.2 to 4.1 for accessions 146 and 147).
   - Bristles all over lower leaflet surfaces, not confined to margins or midribs, not *S. seabrana*.

v. Plant 143: *S.* sp. aff. *hamata* H, CPI 34148 from Honduras.
   - Beak ≥ pod, not *S. scabra* or *S. seabrana*.
   - Leaflets with bristles, not *S. hamata*.

vi. Plants 144 and 145: CPI 47587 and CPI 46588, *S.* sp. aff. *hamata* G, from Guatemala (and see Table 3.5 and associated text for Guatemalan accessions of *S. subsericea* and associated accession of *S. hamata sensu stricto*).
   - Beak ≤ pod, not *S. hamata*.
   - Leaflets without bristles, not *S. scabra* or *S. seabrana*.

Second, using the information provided by Jansen and Edye (1996) to separate *S.* sp. aff. *scabra*, from *S. hamata* and *S. scabra*, and subsequently employed by Maass and t'Mannetje (2002), to elevate some (or all) of this material to species rank, *S. seabrana*.

vii. Plants 146 and 147, shown previously (iv) not to be *S. hamata*, *S. seabrana* or *S. scabra*.
   - Leaflets more elongate than *S. scabra* not *S. scabra*.
   - Hairs on upper leaflet surface, hairs not confined to one side of the stem, not *S.* sp. aff. *scabra* *sensu* Jansen and Edye (1996).

and Edye, 1996). The authors, recognizing the heterogeneity of the groupings, chose to distinguish *S.* sp. aff. *scabra* on the basis of various morphological characteristics, many of them as used to separate *S.* sp. aff. *hamata* W herein, but did not present reasons for this choice (cf. the 'unanalysed entities' problems described by Heywood, 1973). Part, or the whole, of this collection of accessions was then elevated to the species level as *S. seabrana* (Maass and t'Mannetje, 2002). Initially deemed to be most similar genetically to *S. scabra* (Date *et al.*, 1996), at least one accession of *S. seabrana* is more recently placed closest to *S. hamata* (Liu, 1997), a species adventive to the area and one with many similar morphological characteristics and, as noted for accessions 146 and 147, listed herein as *S.* sp. aff. *hamata* W. These accessions do not belong to the species *S. hamata*, *S. scabra* or *S. seabrana*.

Further clarification, and illustration, of the relationships within this important gene pool are called for.

Numerical taxonomy

If taxonomic classifications are to be used for communication, and hopefully to also provide information of predictive value (e.g. that all members of a certain grouping are likely to be poisonous or unpalatable), then it is important that they possess integrity, that all members of the group are indeed similar. While this appears to be the case in some herbaceous tropical legume genera, it is not so with others. An example of the former is, conveniently, drawn from north-east Brazil, home to many of the groupings that have caused problems in *Stylosanthes*. *Macroptilium martii*, a colonizing annual species, found in the same regions as *S. scabra*, is similar to *S. hamata* and *S. humilis* (Table 3.4 and associated text; Burt *et al.*, 1979) but is much more restricted climatically and much less variable. Almost the only variable feature was the ability of some accessions to produce a few flowers near the crown. It is clearly separated from other species, such as the perennial *M. bracteatum*. The latter species is, like *S. scabra*, a plant favoured less by disturbance (Table 3.4) and is found over a much wider range of climates than *M. martii*. *M. bracteatum* also shows relatively little variation and those species closest to it are readily separable by the usual general-purpose type of taxonomic classification, possibly coupled with some performance characteristics such as flowering time and seed yield. This is quite different from the situation encountered in *Stylosanthes*, as we have seen, and where even species from different continents, such as *S. fruticosa* (syn. *mucronata*, syn. *bojeri*) and *S. scabra* are placed in different sub-sections of the genus, but can readily be mistaken for each other and have similar ecological and agronomic characteristics (see information contained in Cook *et al.*, 2005).

In such situations numerical taxonomy using computer-aided pattern-seeking methodologies has been found to be useful. For our present purpose with an agronomic/ecological bias, this has proved to be the case with the taxonomically difficult *Stylosanthes* genus (see information quoted in Liu (1997), where such difficulties are deemed to lead to admixture of species collections). We have earlier commented on the utility of hierarchical classificatory methodologies (such as those employed by Burt *et al.*, 1971; Jansen and Edye, 1996) and in those using genetic markers such as Pengelly and Liu (2001) with *Desmanthus* and Liu (1997) with *S. scabra*. With the chemotaxonomic studies of Robinson *et al.* (1980), network analyses proved to be more appropriate, while with morphological/agronomic data in *Stylosanthes* the MST (Burt and Williams, 1979b) and a graph theory multi-nearest neighbour technique were much more informative, especially with collections that are continuously variable. The use of such methodologies provides a clear holistic view of the patterns of variation present, indicates the characters responsible for the discontinuities that they impose between groupings and provides a framework onto which 'outside data' can readily be appended for validation. Such a process provides an integrated 'picture' of the various groupings, useful as a provider of theory and general information with which to aid the efficient utilization of genetic resources (Harlan, 1971). Further information on this topic, with examples, is provided in Appendix 3.3.

Use of other methodologies, such as chemotaxonomy

The cultivation and characterization of plants grown from seed is relatively demanding of time and resources. Chemotaxonomy, often based on laboratory analysis from a single seed, is much less so and efforts have long been made to use it to complement or replace the more traditional approaches. In recent years, much use has been made of DNA sequencing techniques (e.g. Maass and Sawkins, 2004). Here the main emphasis in the discussion will again be placed on *Stylosanthes*.

In earlier studies, at the protein level, it was possible to recover much of the classification at species level (Robinson and Megarity, 1975) and to partially predict likely responses of the various forms of *S. guianensis* (syn. *guyanensis*) to strains of *Bradyrhizobia* (Robinson *et al.*, 1976). It was also possible to link the various isozyme bands with different morphological/agronomic characteristics (Robinson *et al.*, 1980) and, because the latter were in turn often indicative of various species, or types within species, to consider their possible adaptation. It was not however possible to use the information to define MA 'Morphological/Agronomic' types directly or to indicate accession performance in field conditions.

DNA sequencing techniques have been much more definite in separating the various species of *Stylosanthes* and have proved to be very useful in phylogenetic studies (e.g. the information quoted in Maass and Sawkins, 2005). But they have been less successful with the leguminous genus *Desmanthus*. For instance, the agronomically important, individual accessions of *D. virgatus* are placed in each of the first three major groupings and some are most closely linked to accessions of *D. pernambucanus* and others to *D. bicornutus* (Pengelly and Liu, 2001).

The work of Liu (1997) is particularly important for the present purpose. Although initially designed to explore relationships within *S. scabra*, in selecting material to cover the geographical range of the species, he found that some accessions had been misidentified and, with great serendipity, these included accessions from some of our 'problem' groups, *S. seabrana* (shown by Liu as *S.* sp. aff. *scabra sensu* Jansen and Edye, 1996 and as *S.* sp. aff. *scabra* by Maass in Edye and Maass, 1997), *S. viscosa* (parent, with *S. seabrana*, of *S. scabra*); *S. hamata sensu stricto* and *S. hamata sensu lato* (or the tetraploid *S. hamata* of *S. hemihamata*, Maass and Sawkins, 2004).

In the analysis (Liu, 1997, Fig. 1) the two accessions of *S. hamata sensu stricto* are most closely related to each other and these are closely similar to a tetraploid *S. hamata*. The grouping is shown to be most similar to three accessions, one taxonomically unidentified and the others of *S.* sp. aff. *scabra* and *S. seabrana*. All are well separated from *S. scabra* and *S. viscosa*. This again raises the question about the possible significance of *S. hamata* in this complex.

In *S. scabra* Liu (1997) delimited five major groupings that are, with only slight admixtures, geographically based. These generally conform to the two types *S. scabra*, 'continental' and 'coastal' as defined by Maass on the basis of 26 agronomic and 9 morphological features (quoted in Edye and Maass, 1997). Both classifications are, however, very 'wide', for instance, a very small segment of the Liu group 2, which forms only a part of the Maass 'coastal' type, contains at least two very different morphological/agronomic types. One of these became the ubiquitous cv. '*Seca*' and the other the now rarely used cv. 'Fitzroy'. Almost certainly a third 'ecotype', as noted in Burt *et al.* (1983), was also contained therein. In *Desmanthus virgatus*, a single grouping based on DNA sequencing was noted to contain a great deal of agronomic variation (Pengelly and Liu, 2001).

In reviewing the role of isozyme studies in relation to population structure and genetic conservation, Brown (1978) concluded that 'the evaluation of genetic diversity in a sample is a problem different from the evaluation for its agronomic merit'. There is, however, in *S. scabra* (but not *S. hamata*), a strong link between geographic distribution and agronomic merit (Burt *et al.*, 1976; Burt *et al.*, 1983), and work of the type presented by Liu could be used to explore the former, and reduce the range of material for test. As such it complements the evaluation work, and, in *Stylosanthes*, guards against taxonomic misclassifications. Certainly it could be used to bias 'core collections' towards material most likely to be useful in a given region.

3.7.3 Using an ecologically based approach to the collection and evaluation of genetic resource material

In the 1960s and early 1970s, when the domestication of many of the tropical pasture legumes for dryland areas began, very little was known about the characteristics of the plants concerned. Anecdotal evidence, such as the description of *S. scabra* as the 'Alfalfa do Nordeste' in a semi-arid region of Brazil, was only just becoming available and its Bayesian significant unappreciated. Despite their now apparent differences, accessions, of, for instance, *S. hamata* and *S. scabra* were subjected to the same evaluation regimes in which yield under cutting was the main criterion (as in Burt *et al.*, 1974). The same methods appear to be in general use today and, in many cases, they may be appropriate. Yield estimates over short periods may be adequate when the purpose is to evaluate plants for use in a 'ley' or 'alley' system in cropped areas, where re-sowing, environmental modification by land preparation and fertilizing, and inoculation with *Bradyrhizobia* is feasible.

But, if the feeding of animals is involved in the system, then other information could well be of importance. In experiments with *S. hamata* there were large differences in the rate of regrowth after cutting and in the mineral content of the herbage produce (McSweeney and Burt, 1987); accessions of the tetraploid form of *S. hamata* showed two-fold variations in concentrations of P, while those from the Caribbean Islands, *S. hamata sensu stricto*, did not. In *Desmanthus*, long known as the 'frijollilo de burro' (the bean of the donkey) because of the nutrient value of the seeds, there were large differences between accessions in such factors as crude protein and sodium concentration (Schlink and Burt, 1993). All these factors are likely to influence feeding value (as demonstrated by Rangel *et al.*, 1996 for wool growth on sheep) and could be important in ley feeding and pasture systems. Strategic research and measurements before commercial release could be used to improve cultivar selection.

Introduced fodder legumes often fail, and if they are to be used for long-term pastures they must meet the requirements of being able to establish and persist, often with little or no land preparation or application of fertilizers. Competition from associated grasses and the impact of grazing animals force even the perennials to re-establish from seed for persistence. Ability to utilize local *Bradyrhizobia* efficiently is essential for their survival and the production of high-quality feed. This latter ability can be readily measured using a method devised by Dr Don Norris, founder of the famous 'CB' (Cunningham, Brisbane) collection of *Bradyrhizobia.* He found that readily available soil nitrogen in an evaluation area can be 'locked' up by the addition of a high carbon source, for instance bagasse from a sugar cane mill, and legumes that fail to nodulate effectively simply die (Norris, 1968, personal communication). This character could be readily assessed and incorporated in special-purpose classifications or onto a MST. While dry matter yield is a plant character commonly used in plant selection, other characteristics of importance in assessing ecological and agronomic potential that can be readily observed in early evaluation studies can be stressed. These include ability to produce seedlings in disturbed and non-disturbed situations, resistance to invasion by aggressive species such as Indian Couch (*Bothriochloa pertusa*), persistence of individual plants (Chapter 8, this volume) and germination response to fire and high temperatures (Rangel and Gardiner, 1996). When first evaluating the potential of 'new species' poorly represented in germplasm collections, accumulative dry matter yields were sufficient to distinguish the very large differences found between species and species forms (e.g. Burt *et al.*, 1974). But with the advent of larger collections, and the resulting very much smaller differences in yield encountered within species collections, this has become more difficult. And there is a need to become more discriminatory for both agronomic and ecological purposes, through the use of more effective distinguishing characteristics such as the feeding value (of *Desmanthus* seed) or the ability of plants to spread at various disturbance levels. In view of earlier comments about colonizing and non-colonizing species, the recognition that mixtures of different genotypes may be desirable in species such as *S. humilis* (Burt *et al.*, 1973) and *S. hamata* should be considered at all levels of evaluation, and during cultivar development.

Collecting methodologies, like evaluation strategies, can be improved with hindsight, particularly reducing the need for exploratory road transects. Road transects can be convenient, for at times 'it is virtually impossible not to step on a *Stylosanthes* plant when the car stops at a new collecting site' (Maass and Sawkins, 2004). However, the species encountered are often different from those found in more natural environments nearby. They may be of different levels of dominance and even when considering the same species may differ in form (as with *S. scabra*; Burt *et al.*, 1983), response to minerals and so on. Although the collections *per se* are of value, they do not truly represent the climate, soil or disturbance level of the area, and provenance records for them, for instance those relating to climate of nearby settlements, must be used cautiously in prediction of, for instance, zones of adaptation.

The data recording card devised by Myers (1968) has been widely adopted and its mandatory data collection requirements rigorously upheld. It has done much to ensure that geographical and provenance data were recorded on collecting missions. However, completing the forms can be problematic because some of the information required takes considerable effort to obtain, if it is available at all. It is often not possible to readily assess such information as parent rock, depth of soil to clay, or even – to the level of consistency of a trained pedologist – the texture of the soil surface. The classification of the soil can be problematic as well, even to a trained pedologist (e.g. Isbell, in Burt *et al.*, 1979). As well, even the environmental information recorded at the site of collection (sown pasture, grassland, forest, roadside etc.) is rarely utilized in analyses of collection data. The card can undoubtedly be improved in light of subsequent knowledge.

When specialists from other disciplines have been in the collecting team it has proven to be of considerable value in acquiring environmental information. During the 1971 collecting mission to Brazil and Venezuela, Mr Ray Isbell, an accomplished field pedologist, was able to characterize the soils and, later in laboratory studies, explore their chemical and physical characteristics. In combination with simple ecological observations at the collecting sites, such as dominance, distribution and associated grass species, the various collections could be categorized in terms of the soils, climates and disturbance levels in which they were found (Burt *et al.*, 1979; Burt *et al.*, 1983). From this hypothesis-generating situation it was possible to suggest, for instance, the mineral nutrient requirements that favour the various genera and species when they grow in competition with grass. Many of these suggestions have been proven to be correct in subsequent evaluation and commercial usage: for instance, *Centrosema* and *Macroptilium* require higher levels of P than *Stylosanthes*, whereas the latter when in competition with grass (but not in pure culture) is favoured by high levels of exchangeable K (Hall, 1975, personal communication). *S. capitata* thrives on soils of higher acidity than *S. scabra*, and *S. humilis* and *S. hamata* behave as colonizing species (for further information see Cook *et al.*, 2005).

In any collection in 'targeted' areas (see Appendix 3.4), 'new species' or 'new types' of existing species could be found. Ideas about the potential adaptation of such material, and its possible utility, could well be developed by using a more ecological approach to its collection and by use of anecdotal information provided by the existing users.

Appendix 3.1

Accessions used in the experiment

Table 3.7. Accessions studied in the MST analysis (see 'Analytical Methods' in text).

Species, Plant no.	CPI no.[a]	Origin and soil type[b]
S. calcicola		
1.	73524	Yucatan Pen., Mexico
2.	73525	″ ″
S. hamata		
3.	33205	Anse Bertrand, Guadeloupe
4.	33231	Isla Verde, Puerto Rico
5.	36046	Big Pine Key, Florida, USA
6.	37037	Santiago, Dominican Republic
7.	37038	″ ″
8.	38842 A	Maracaibo, Venezuela
9.	38843	Miami, Florida, USA
10.	40264 A	Salvador, Brazil
11.	40268	Fortaleza, Brazil
12.	40275	Crato, Brazil
13.	49080	Barranquilla, Colombia
14.	50997	Barquisimeto, Venezuela
15.	50998	″ ″
16.	51391	Bahamas
17.	55812	Maracaibo, Venezuela[xx]
18.	55820	[acid sands and earths]
19.	55821	″ ″
20.	55822[2]	″ ″
21.	55824	″ ″
22.	55825	″ ″
23.	55826	″ ″
24.	55827	″ ″
25.	55828	Maracaibo, Venezuela
26.	55830	″ ″
27.	55831	″ ″
28.	56211 A	Guadeloupe
29.	61623 A	Maracaibo, Venezuela
30.	61624	″ ″
31.	61669	Paraguana, Venezuela
32.	61670[6]	[alkaline]
33.	61671 A	″ ″
34.	61671 B	″ ″
35.	61672 B a	″ ″
36.	61672 B b	″ ″
37.	62160	Colombia
38.	62162	Isla Margarita, Venezuela
39.	65361	Barranquilla, Colombia
40.	65363	″ ″
41.	65364 A	″ ″
42.	65365 B	Maracaibo, Venezuela
43.	65367	Barranquilla, Colombia
44.	65368	″ ″
45.	65369	″ ″
46.	65370	″ ″
47.	65371 A	″ ″
48.	65372	Barinas, Venezuela
49.	65962	Santa Marta, Colombia.
50.	65963	″ ″
51.	65965	″ ″
52.	65966	″ ″
53.	68837	″ ″
54.	68838	″ ″
55.	68840	″ ″
56.	68841	″ ″
57.	70358 A	Antigua [Fitches Clay[xxx]]
58.	70359	Antigua [Bendals Clay[xxx]]
59.	70360	″ ″
60.	70361 A	″ ″
61.	70362	″ ″
62.	70364	″ ″
63.	70365	″ ″
64.	70366 A	″ ″
65.	70367	Antigua [marl]
66.	70368 A	″ ″
67.	70369	Cuba
68.	70370	Barbuda
69.	70371 A	″ ″
70.	70372	″ ″
71.	70373 A	″ ″
72.	70374	″ ″
73.	70375 A	″ ″
74.	70376	″ ″
75.	70520	Florida, USA
76.	70521	″ ″
77.	70524	″ ″
78.	70525	″ ″
79.	70526	″ ″
80.	70527	″ ″
81.	72850 A	Florida, USA
82.	72851	″ ″
83.	72852	″ ″
84.	72853 A	″ ″
85.	72854	″ ″
86.	72855 A	″ ″
87.	72856	″ ″
88.	72857 A	″ ″
89.	72858	″ ″
90.	72859	″ ″
91.	73484 A	Antigua [Coolidge Clay[xxx]]

Continued

Table 3.7. Continued.

Species, Plant no.	CPI no.[a]	Origin and soil type[b]
92.	73486	Antigua [marl]
93.	73487	Antigua [Coolidge Clay[xxx]]
94.	73488	" "
95.	73489 B	Antigua [clay]
96.	73490 A	Antigua [coral sand[xxx]]
97.	73491	" "
98.	73492 A	" "
99.	73494 A	" "
100.	73495	" "
101.	73497	" "
102.	73498	Antigua [coral sand[xxx]]
103.	73499	" "
104.	73500 A	" "
105.	73501	Antigua [Fitches Clay[xxx]]
106.	73502	" "
107.	73503	" "
108.	73504	Barbuda
109.	73506	Curaçao [alkaline]
110.	73507	" "
111.	73508	" "
112.	73509	" "
113.	73510	" "
114.	73511 A	Cuba
115.	73512	Nevis [alkaline]
116.	73513	" " [acid at surface only]
117.	73514 A	
118.	73515	[alkaline]
119.	73516 B	" "
120.	73517	" "
121.	73518	" "
122.	73519A	" "
123.	73520 A	" "
124.	73521	" "
125.	73522 A	St Lucia
126.	73523	Curaçao [alkaline]
S. humilis		
127.	37036	Dominican Republic
128.	38753	Bogotá, Colombia
129.	55797	Juazeiro, Brazil
130.	61667	Paraguana, Venezuela
131.	61668	[alkaline]
132.	61672 A	" "
133.	61674[3]	" "
134.	65372	Barinas, Venezuela
135.	cv. Paterson[4]	Australia [acid earth]
S. scabra		
136.	51392	Sao Paulo, Brazil
137.	55805	Itaberaba, Brazil [acid earth[xx]]
138.	55870	Anage, Brazil [acid earth[xx]]
139.	40292[5]	Gravata, Brazil
S. sympodialis		
140.	65958	Ecuador
141.	65961	" "
142.	67704 B	" "
S. sp. aff. hamata		
143.	34148	Las Mesas, Honduras
144.	46587	Guatemala
145.	46588	" "
146.	55811 A	Vitoria da Conquista, Brazil (W) [acid earth[xx]]
147.	55871	Itambe, Brazil [acid earth[++]] (W)
148.	70522	Florida, USA [alkaline sand] SB)
149.	70529	" "
S. hamata × S. humilis		
150.	plant no. 4 × plant no. 135	

[a]Plant no.: for ease of reference in the text, tables and figures. [b]Origin and soil type: origin given as the country, region (Yucatan, Paraguana) or nearest large town (Maracaibo, Santa Marta, Barinas). CPI no.: Commonwealth (of Australia) Plant Introduction number; W: woody.

[x]Accession found to contain seed mixture, this component denoted A or B, or as numeric superscript etc. 41 mixtures were separated before this experiment.

Cultivars: [1, 2]Of *S. hamata*: [1]cv. Verano; [2]cv Amiga; [3, 4]Of *S. humilis*: [3]cv. Khon Kaen; [4]cv. Paterson; [5]Of *S. scabra*: [5]cv. Seca; [6]*S. humilis*: regarded as promising on heavy clays in Indonesia.

Soil type: When known or stated by collector (e.g. 'alkaline') or as defined in: [xx]Burt *et al.* (1979); [xxx] Hill (1966).

Table 3.8. Accessions excluded from the MST analysis (see 'Analytical Methods' in text).

Species, Plant no.,	CPI no.[a]	Origin and soil type[b]
S. hamata		
151.	65364 B	Barranquilla, Colombia
152.	65365 B	Maracaibo, Venezuela
153.	65371 B	Barranquilla, Colombia
154.	70358 B	Antigua [Fitches Clay]
155.	70358 C	" "
156.	70358 D	" "
157.	70359 B	Antigua [Bendals Clay]
158.	70361 B	" "
159.	70361 C	" "
160.	70361 D	" "
161.	70366 B	" "
162.	70368 B	Antigua [marl]
163.	70368 C	" "
164.	70368 D	" "
165.	70373 B	Barbuda
166.	70371 C	" "
167.	70371 D	" "
168.	70373 B	" "
169.	70375 B	" "
170.	72850 B	Florida, USA
171.	72853 B	" "
172.	72855 B	" "
173.	72857 B	" "
174.	73484 B	Antigua [Coolidge Clay]
175.	73484 C	" "
176.	73485 A	" "
177.	73485 B	" "
178.	73489 A	Antigua [clay]
179.	73490 B	Antigua [coral sand]
180.	73492 B	" "
181.	73492 C	" "
182.	73493 B	" "
183.	73494 B	" "
184.	73494 C	" "
185.	73494 D	" "
186.	73496 A	" "
187.	73496 B	" "
188.	73500 B	" "
189.	73503 B	Antigua [Fitches Clay]
190.	73508 B	Curaçao [alkaline]
191.	73511 B	Cuba
192.	73512 B	Nevis [alkaline]
193.	73515 B	" "
194.	73516 A	" "
195.	73519 B	" "
196.	73520 B	" "
197.	73522 B	St Lucia
198.	73522 C	" "
199.	'Verano' variant	Found on heavy clay soil, near Townsville, Australia.
S. humilis		
200.	33503	Calabozo, Venezuela
201.	33829	Vera Cruz, Mexico
202.	33830	" "
203.	33979	Las Canas, Costa Rica
204.	34001	Buenos Aires, Costa Rica
205.	40265 A	Caruaru, Brazil
206.	40265 B	" "
207.	40271	Sobral, Brazil
208.	40275 A	Crato, Brazil
209.	40275 B	" "
210.	55852	Chitre, Panama
211.	cv. 'Gordon'	Queensland, Australia
S. scabra		
212.	40292 A++	Gravata, Brazil
213.	40292 B	" "
214.	40292 C	" "
215.	55814	Vitoria da Conquista, Brazil [acid earth]
S. subsericea		
216.	33940	Palmares, Brazil
217.	37274	Tegucigalpa, Honduras
218.	41217	Tapuchula, Mexico
219.	67643	Chiquimula, Guatemala
S. sympodialis		
220.	65959	Ecuador
221.	65960	" "
222.	67702	" "
223.	67703	" "
224.	67705	" "
S. sp. aff. hamata		
225.	55804	Itaberaba, Brazil [acid earth]
226.	55813	Vitoria da Conquista, Brazil [acid earth]

[a]Notations used as in Table 3.1; [b]naturally occurring variants of the cv. 'Seca'.

Appendix 3.2

Location of the various accessions on the MST

This figure is presented to provide further detail on the MST groupings, specifically the location of the individual accessions listed in Table 3.7 in Appendix 3.1. The numbers in the figure correspond to those provided for each accession in the first column in the table.

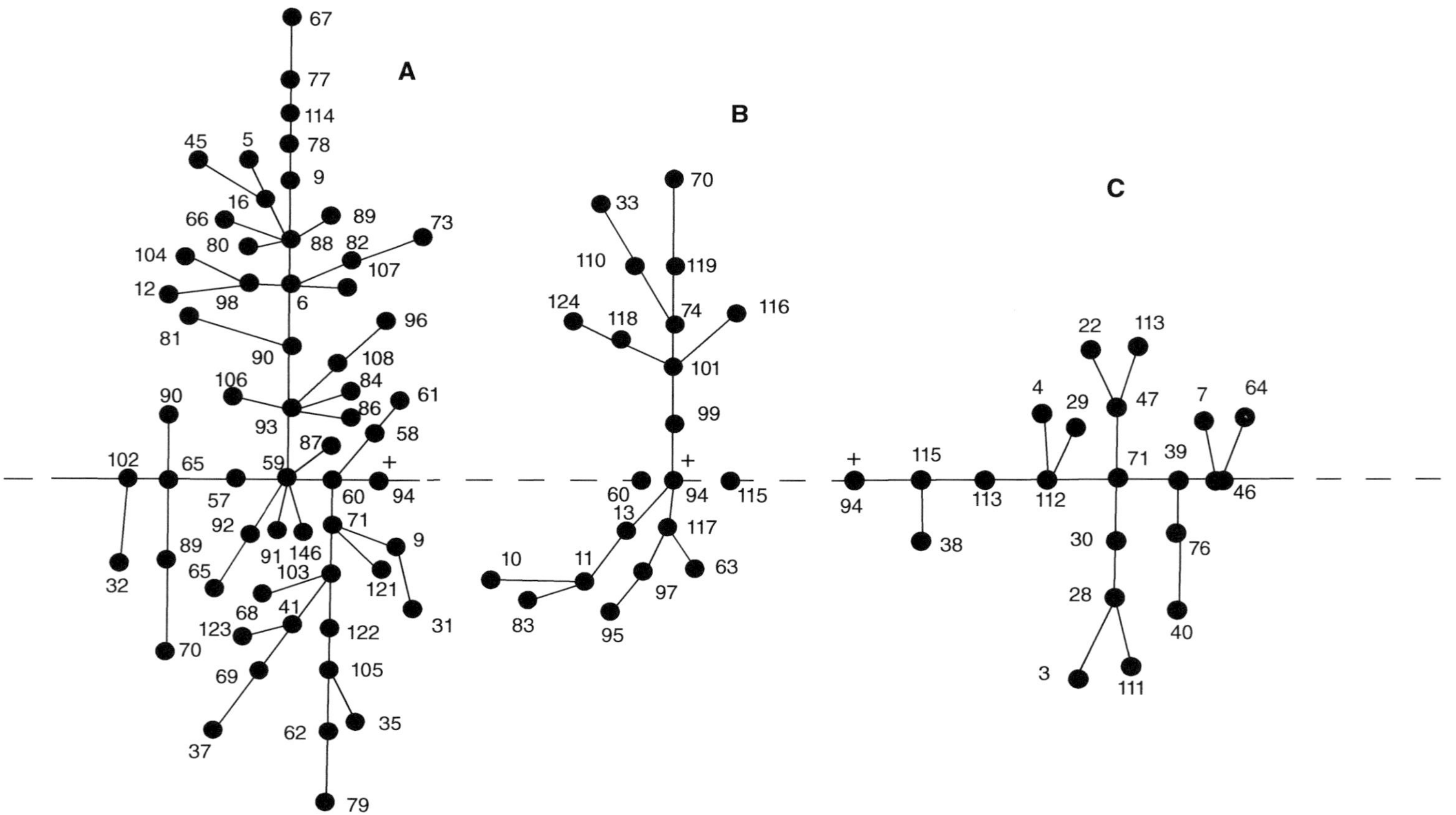

Separated into cones A, B and C for ease of reference in the text.

Fig. 3.11. The diploid accessions of *S. hamata sensu stricto*. Separated into cones A, B and C for ease of reference in the text.

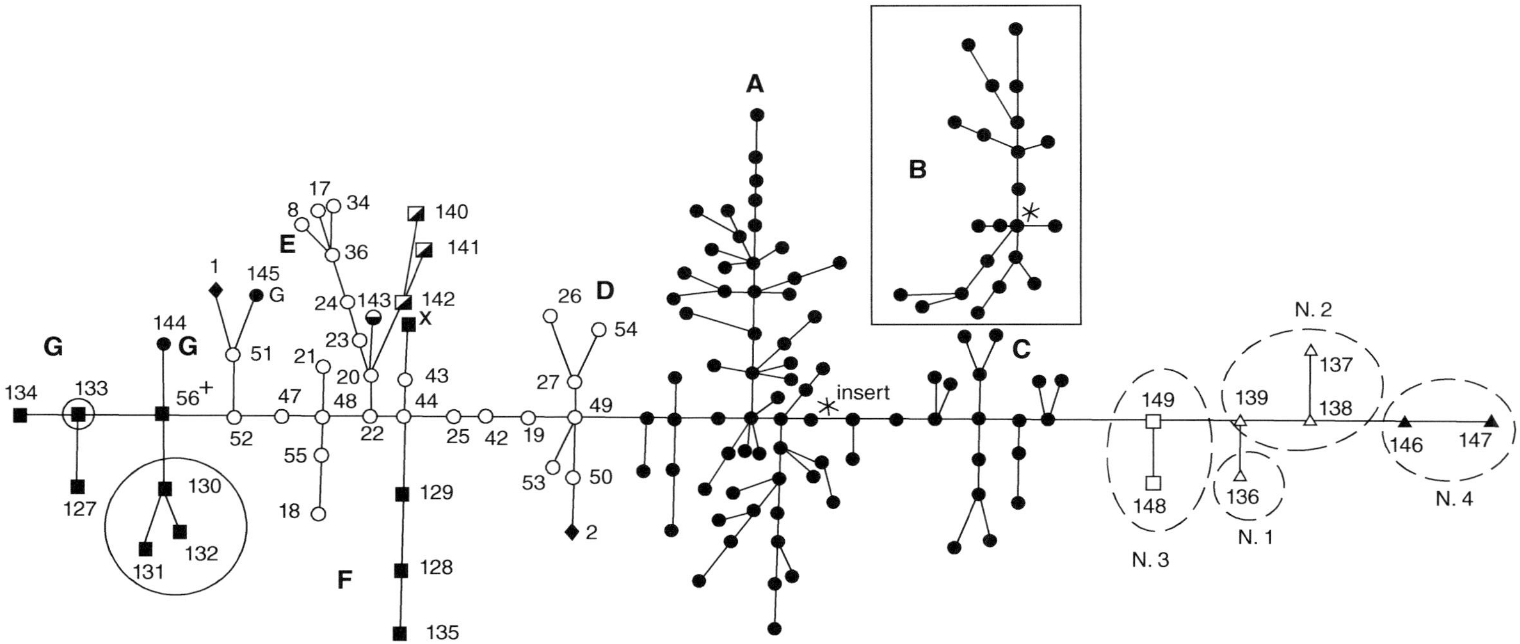

Zones D to G nominated for ease of reference in the text. Symbols as in Figure 3.4

Fig. 3.12. Accessions of *S. humilis*, *S. hamata sensu lato*, *S. sympodialis* and associated species as shown in Fig. 3.4.

Appendix 3.3

Comparing results based on the hierarchical classification and the MST

A worked example is given here that shows some of the considerations employed in the definition of the groups.

When classifying a population it is first necessary to seek discontinuities between groups of individuals. These can then be described and, if necessary, a key drawn up to aid their rapid distinction. But, in many instances, difficulties may be experienced in finding the discontinuities, and differences of opinion may occur. What one classification system/researcher readily accepts may be deemed by another to be based on characteristics too minor to warrant such distinction – and even 'stable characters' such as the presence or absence of an axis rudiment in the flower of *Stylosanthes* plants or the presence of one or two loments in the fruit may only be defined as 'usual' (Mohlenbrock, 1957). As well, separation may be made on the basis of subjective decisions, for instance, leaflets being 'more elongate' or 'narrowly elliptical'. Problems may also arise when the integrity of the groups is being examined; one classifier may accept a relatively heterogeneous group but another may not. For instance, Luckow (1993) in her review of *Desmanthus*, bulked three species previously recognized by an earlier authority (Turner as quoted in Burt, 1993a), despite each having seemingly distinct geographic and ecological characteristics. In this instance she regarded the three species as one very wide, variable species with a considerable ecological and climatic range. The situation is reviewed in Burt (1983) and topics related to species delineation have been covered in Maass and Sawkins (2004).

Pattern-seeking, computer-aided methodologies can be used in the process of classification and it is usual to employ strongly clustering methodologies to seek the discontinuities. But the problems just mentioned still prevail: misclassifications can occur and, as shown previously, the computer may provide boundaries that are based on meaningless associations of characters. In addition, apparently homogeneous groups may have occasional accessions that differ markedly in perhaps only one of the many characters studied, but this characteristic, for example the persistence of a species valued for its perenniality, may be very important. Validation of the meaningfulness and potential usefulness of the groups, and their relationships with each other, is essential.

The meaningfulness and homogeneity of the groups may be checked in various ways. It is possible, for instance, to use a less intensely clustering strategy, one that quickly separates out plants that are very different in only one or two characteristics, then by constructing a two-way table of the results from the two types of classification attention can be drawn to 'aberrant' plants. These can then be studied more intensively and possibly reallocated (Burt *et al.*, 1971). The same method can be employed using a classification based, for instance, on purely agronomic data, or on chemotaxonomic studies and so on. In the present instance, as in the case of *S. subsericea*, the species grouping did not coincide with those found in the hierarchical classification, and a manual examination of the raw data, combined with some chemotaxonomic information was used to investigate the situation. In other instances, 'aberrant' accessions have been delineated by comparing the results obtained from the hierarchical classification and the MST and, if necessary, their positions reallocated with the assistance of 'outside' information.

Our example is drawn from the allotetraploid *S. hamata* grouping (Appendix 3.2, Fig. 3.12). On the main axis, between the two groupings of *S. humilis* (E and G), plants 47 and 48 have been located in between 52 and 22. Both accessions were placed in the hierarchical classification with accessions of *S. hamata sensu stricto*: they were most closely related, in groups 14 and 16 respectively, with plants 14 and 102 respectively. The pod characteristics of these plants are shown in Table 3.9. in relation to those of plant 22, an accession unquestionably of the tetraploid form of *S. hamata*/*S. hemihamata* from Maracaibo. These pod characteristics are regarded as 'stable' characters, often used by taxonomists to separate species (e.g. Maass and t'Mannetje, 2002). From the data presented it is clear that accessions 47 and 48 are better placed with the allotetraploid *S. hamata* grouping, as shown on the MST, than with the *S. hamata sensu stricto* groups in the classification.

The geographic relationships between accessions on the relevant part of the MST are shown in Fig. 3.13. The 'aberrant' plant 48 from near Barinas, in Venezuela, is joined (to the right) with plants 21 and 22, also from Venezuela. To the left it is linked to other allotetraploid *S. hamata* accessions, 47, 52 and 51, and from these to plant 56, an accession variously attributed to *S. hamata* and *S. humilis*. From the latter, located near a coastal trading post, they are linked to accessions of *S. humilis* from another coastal trading area, the Paraguana peninsula in Venezuela. This in turn is closely linked to the island of Curaçao, which has a similar arid and alkaline environment, and on which *S. hamata sensu lato* is endemic. The 'aberrant' accessions have been well placed in the MST and the resultant groupings are meaningful and hypothesis-generating.

Table 3.9. Pod characteristics of plants considered for reallocation.

	Plant no.[b]	14	48	22	47	102
Character[a]	Group no.[c]	14	14	10	16	16
N1 Length of beak (mm)		2.1	3.8	3.3	3.5	2.7
N2 Length of pod + beak (mm)		4.7	6.6	6.6	7.1	5.7
N3 Length of pod		2.6	2.8	3.3	3.6	2.0
N5 Beak/pod ratio		0.81	1.36	1.00	0.97	0.90
Origin: Country		Venezuela			Colombia	
Origin: Area		Barquisimeto	Barinas	Maracaibo	Barranquilla	

As listed on: [a]Table 3.3; [b]Table 3.1; [c]Table 3.4.

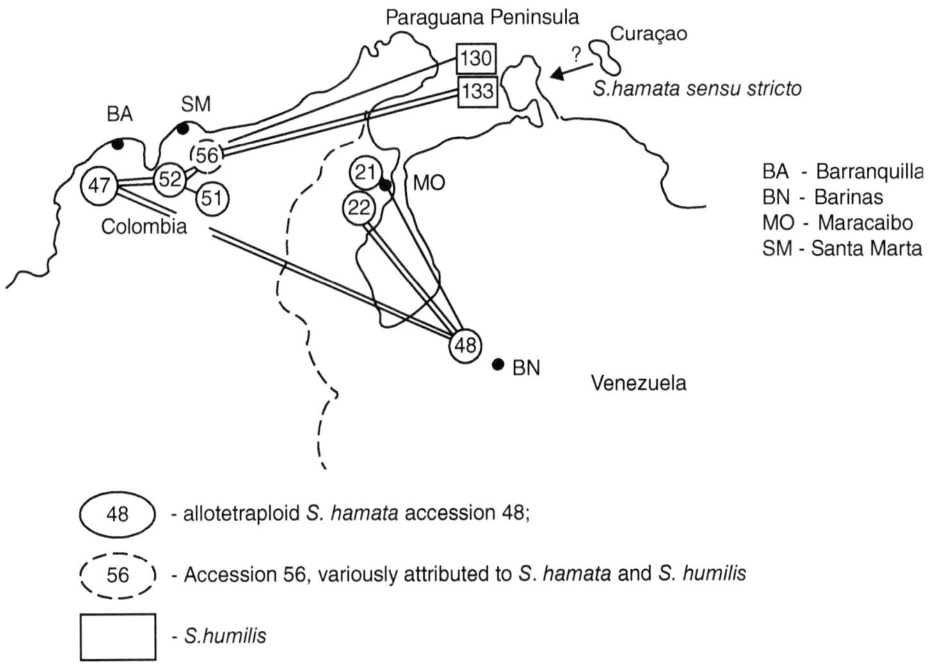

Fig. 3.13. Geographical relationships of 'aberrant' accessions with others as shown on the MST.

Appendix 3.4

Developing the *S. scabra* genepool: a brief review and suggestions for targeted collection

S. scabra, 'Shrubby Stylo', is well known to graziers in the tropical, semi-arid regions of north-east Brazil as the 'alfalfa do nordeste', the 'lucerne' of the north-east. Like lucerne (*Medicago sativa*), it is a perennial that provides valuable animal feed and has a related annual species that is quite different ecologically and agronomically. But, whereas lucerne is a well-studied temperate species from western Asia, transported to both Europe and Mexico in the 16th century, *S. scabra* is a species from South America, which has had little research attention, and the first cultivar was not made available until 1997. Their associated data bases reflect these facts both practically and theoretically. Lucerne and its relatives are so well known that they are used as examples in, for instance, the role of breeding systems in population structures and plant adaptation (Adams and Pipoly, 1980). On the other hand the origins of *S. scabra* were not delineated until 1997 (Liu and Musial, 1997), its relationship with other species is still unclear (Maass and Sawkins, 2004), and new species continue to be designated within the genus (Maass and Sawkins, 2004). *S. scabra* was probably first collected by I.R.I. in the Brazilian subtropics (Hymowitz, 1971). Genetic resources are relatively limited because worldwide accessions number only about 600 accessions, there is some duplication and further collection has ceased (Maass and Sawkins, 2004). The characteristics of the accessions held are generally poorly documented (Liu, 1997; Edye and Maass, 1997).

Most of the original accessions in the Australian collection were from subtropical environments and belonged to MA (morphological/agronomic) type 19 (Edye *et al.*, 1974), which performed poorly over a range of environments (Burt *et al.*, 1971; Burt *et al.*, 1974). A few accessions came from more northerly regions: MA18 and 30 from Minas Geraiu, MA17 from Hahia and more accessions of type 18 from the state of Pernambuco further north. Accessions of types 17 and 18, from latitudes between 8° and 12°S, furnished the cultivars 'Fitzroy' and 'Seca', the first domesticants of the species. Both came from the collections made by Mr R.J. Williams, who was based at the CSIRO Cunningham Laboratory in Brisbane, Australia. To these were added the collections made by Burt and Isbell in 1971 in north-east Brazil (Burt *et al.*, 1979). These later accessions provided yet more variation, ecotypes adapted to different climatic conditions, and within the drier climates different levels of disturbance and associated grass species (Burt *et al.*, 1983). The germplasm has provided a source of resistance to the disease anthracnose additional to those already identified from the same area of north-east Brazil. Later collecting missions by CIAT (in Colombia) and CENARGEN and EMBRAPA (in Brazil) have added more material. Many of these latter accessions appear to have been obtained at higher latitudes, longitudes, altitudes and rainfalls than those that have yielded the best-adapted genotypes for dry tropical areas. Only a small proportion of the Australian collection is from north-east Brazil (see data mapped by Liu, 1997), and most interest has understandably been placed on the more acid-tolerant species *S. guianensis* (Edye and Maass, 1997).

The early collecting missions, such as that of Hartley (in 1947–48), were largely exploratory: even the choices of target genera were uncertain and were largely biased towards the genera *Centrosema*, *Desmodium*, *Pueraria* and *Calopogonium*, which had come to be used as cover crops in plantation agriculture and whose use had occasionally spread into tropical pastures. Such knowledge as was available about plant distribution was difficult to access and often enshrined in local publications. In addition, taxonomic reviews of relevant species were lacking, that for *Stylosanthes* being first produced in 1958 (Mohlenbrock, 1957) and that for *Desmanthus* in 1993 (Luckow, 1993). Thus the same species would appear under different names in the local floras. The choice of the climates in which to collect was equally difficult. It was generally assumed, as de Candolle had in the 19th century (quoted by Darwin in 1859), that the plants most likely to be adapted to a given recipient area would be found in a similar climate elsewhere, but relevant climatic data and suitable systems of climatic classification were lacking. Hartley (1954 and 1963) did much to alleviate this problem when he was able to define biologically similar areas in terms of the composition of their grass floras. The long-term results of the introduction of exotic grasses and legumes to Australia has demonstrated that this methodology has worked extremely well (Williams and Burt, 1982) and could be developed further.

The only feasible approach for collection was an exploratory one based on road transects. From the data shown by Liu (1997) and by plotting the data presented by Maass and t'Mannetje (2002), it is clear that this is still, in some situations, an adequate strategy (see also, for the replotted information, Fig. 3.1).

The identification of successful cultivars *S. hamata* cv. 'Verano' and *S. scabra* cv. 'Seca' and cv. 'Fitzroy' reduced the difficulty of selecting collection zones because the climates that had yielded the adapted plants would be of higher potential (Burt *et al.*, 1976). Newly devised methods of climatic classification, like Hartley's, based on the climates' abilities to support different suites of plants, could then be used to nominate the regions of the world in which these climates could be found (Reid *et al.*, 1976). Newly published floras have made it possible to identify species of potential likely to be found therein (Burt and Reid, 1976). Local knowledge from the areas of origin should also be collected and utilized in the screening process. For example, in Brazil *S. scabra* was regarded as the 'alfalfa do nordeste'; *Desmanthus leptophyllous* is known locally as 'frisolillo de burro' (donkey bean), and is more valued for its nutritious seeds than its vegetative growth. Today such information is now available in computer-driven search engines, and is an aid to the identification of plants of potential.

In more recent times the development of computer-generated models and geographic information systems (GIS), and the greatly increased availability of environmental information, have allowed researchers to select homologous areas, areas similar to those known to contain species or forms of interest, from overlays of vegetation and climate. Further, detailed environmental data from collecting sites (including GPS data) can be classified to predict further areas for collection (Edye and Maass, 1997; Maass and Sawkins, 2004): it is deemed possible to map areas in terms of probabilities of finding target species based on the degree of climate similarity.

While such methodologies are of value, problems remain. In the data presented by Maass and Sawkins (2004, Figs 1.6 and 1.7), accessions are shown to be present in environmental areas that bear 'no similarity' to those nominated for the species concerned. The Paraguana Peninsula yielded the exciting new *S. humilis* cultivar 'Khon Kaen', which is able to thrive in areas where other introductions failed, but the environment is classified as showing 'no similarity'. *S. hamata* accession 61670 from the same area was noted in India (Ramesh *et al.*, 2004) to be the most consistent performer across a range of sites, including heavy clays and alkaline soils. An area inland of Recife, Brazil is also classified as 'no similarity', although the map shows the species to be present and one such accession is included in our collection herein (plant 129).

Maass and Sawkins (2004) note that these divergences occur because, among other reasons, it is not always possible to nominate a range of accessions that 'adequately represent the known distribution of the major species'. In the analysis of *S. humilis*, for example, the cv. 'Khon Kaen' was not included. In addition, 'other subspecies or varieties [may have been included] despite best efforts', as happened with *S. guianensis*, when the potential distribution of the subspecies *guianensis* was investigated. Another reason could be the absence of disturbance level as a factor in plant distribution (see Table 3.4 and associated text). Divergence also becomes apparent due to the increase in knowledge from published work relating to legume distribution in north-east Brazil.

In this region legume distribution was generally a function of soil × climatic mosaics, and disturbance levels (Burt *et al.*, 1979; Burt *et al.*, 1983). *S. capitata* was only found on soils of very high acidity, an *Arachis* sp. was only found on a heavy clay soil, *Centrosema* and *Macroptilium* were favoured by high soil P levels and *Stylosanthes* by high exchangeable K. But, with the exception of the heavy clay (as classified by Mr R.F. Isbell, a field pedologist whose expertise cannot be doubted), the distribution of the various species could not be limited to any soil grouping of the two classification systems he used. A special classification, using factors that specifically relate to plant growth, is more likely to be useful for the present purpose, such as the Fertility Capability Classification (FCC) developed by Sanchez, Couto and Buol from 1965 to 1982 (Sanchez *et al.*, 1982). In 1997 the FCC system was integrated with WISE (World Inventory of Soil Emission Potentials) by the University of Hawaii at Manao and the National Resource Conversation Service of the USDA. It is possible, with such a system, to incorporate new 'limiting edaphic factors' for specific plant types. Similar considerations apply to the use of general purpose climatic classifications (Reid *et al.*, 1976).

But, even with suitable methods of classification for climates, soils or vegetation, there remain problems associated with scale. North-east Brazil, the area that has furnished the *S. scabra* cvs 'Fitzroy' and 'Seca', and provided breeding material for the cultivars 'Fera', 'Jecuipe', 'Siran' and 'Recife' (the only cultivars listed by Cook *et al.* (2005), is shown in the map provided by Papadakis (1966) (Fig. 3.14) to have six types of climate, with the material mentioned coming from a narrower range of climates. But, within each climate there are many 'niches'. Roadsides are often in 'run-on' situations for rainfall and have different levels of disturbance and soil minerals. Each factor can help determine the species and types of species to be found. In *S. scabra*, for instance, accessions from disturbed sites are of a type different from those found in other situations (Burt *et al.*, 1983). In the semi-arid tropical region around Juazeiro (J in Fig. 3.14), *S. humilis*, *Centrosema pascuorum* (parent of the cv. 'Cavalcade') and *C. brasilianum* (to become cv. 'Oolloo', the first domesticant of that species), were found only on a run-on situation near an air strip.

In the same region a perennial *Stylosanthes* was unexpectedly found on a coarse sandy soil north of Juazeiro. Although overgrazed and affected by drought, it was almost certainly *S. scabra* or *S. viscosa*. It was noted to be in caatinga 'scrub 4' to 6' high' 'in transition to cerrado'. After discussions with Andrade-Lima (1971, personal communication), and particularly after reference to his publications (Andrade-Lima, 1982), it would seem likely that this was one of his stipulated 'brejos' (refuge areas). The flora of the area has evolved over a long period and has suffered numerous invasions and retractions. Refuge areas are relatively small and isolated sites where plants have survived when they died out in intervening areas. In many instances (see references quoted in Andrade-Lima, 1982) they have evolved very different forms. He lists three types of refuges. First, slightly elevated positions that receive slightly more rainfall than the drier climate barriers that surround them. The species therein will tolerate high temperatures. Second, refuges that because of their local topography are more humid with lower temperatures than the regions around them and the species contained thrive at lower temperatures. Third, small areas of drier climates surrounded by wetter regions containing species usually found in semi-arid climates. Many of these areas are relatively small and are often situated away from towns that provide most of the climatic data used in climatic mapping. We note that these areas are often avoided by builders of major roads and thus are not featured in road collection transects. A retrospective view of the results of collecting missions in the region is of interest.

S. humilis, an annual species favoured by disturbance, was found spreading rapidly under the influence of overgrazing by Williams in the most northerly of these (Fig. 3.15), the hot-tropical climates of Ceara (Fig. 3.14.). He also found, not unexpectedly, the perennial *S. viscosa*, but not *S. scabra*, a species often found in association with it (Burt *et al.*, 1979). Surprisingly, near Juazeiro (Fig. 3.14), *S. scabra* (together with many other genera and species) was only abundant in a 'dry' refuge area to the west of Senhor do Bofim. Two of the accessions of the 'new' species *S. seabrana* were collected to the south of this, in or near a refuge, at a somewhat higher altitude and higher than expected rainfall (data listed in Maass and t'Mannetje, 2002, plotted there on Fig. 3.15).

The same species, related to a sympatric with *S. scabra*, *S. viscosa* and *S.* sp. aff. *hamata* W and *S.* sp. aff. *scabra* (Burt *et al.*, 1979), was most commonly found around the Chapada Diamantina, part of which is listed by Andrade-Lima as a low-temperature refuge. *S. seabrana* has been found to be adapted to more humid, subtropical conditions than *S. scabra* accessions collected from the same geographical region (Cook *et al.*, 2005). The first introduction of *S. scabra*, cv. 'Fitzroy' (MA type 17), was collected in or near another designated refuge area.

Plant breeding is increasingly seen as an option for the provision of new cultivars of *S. scabra* and these are certainly needed due to the need to identify resistance to anthracnose attack, the need for slightly different agronomic forms (Jones and Clem, 1997), and the worldwide importance of the species (as documented in

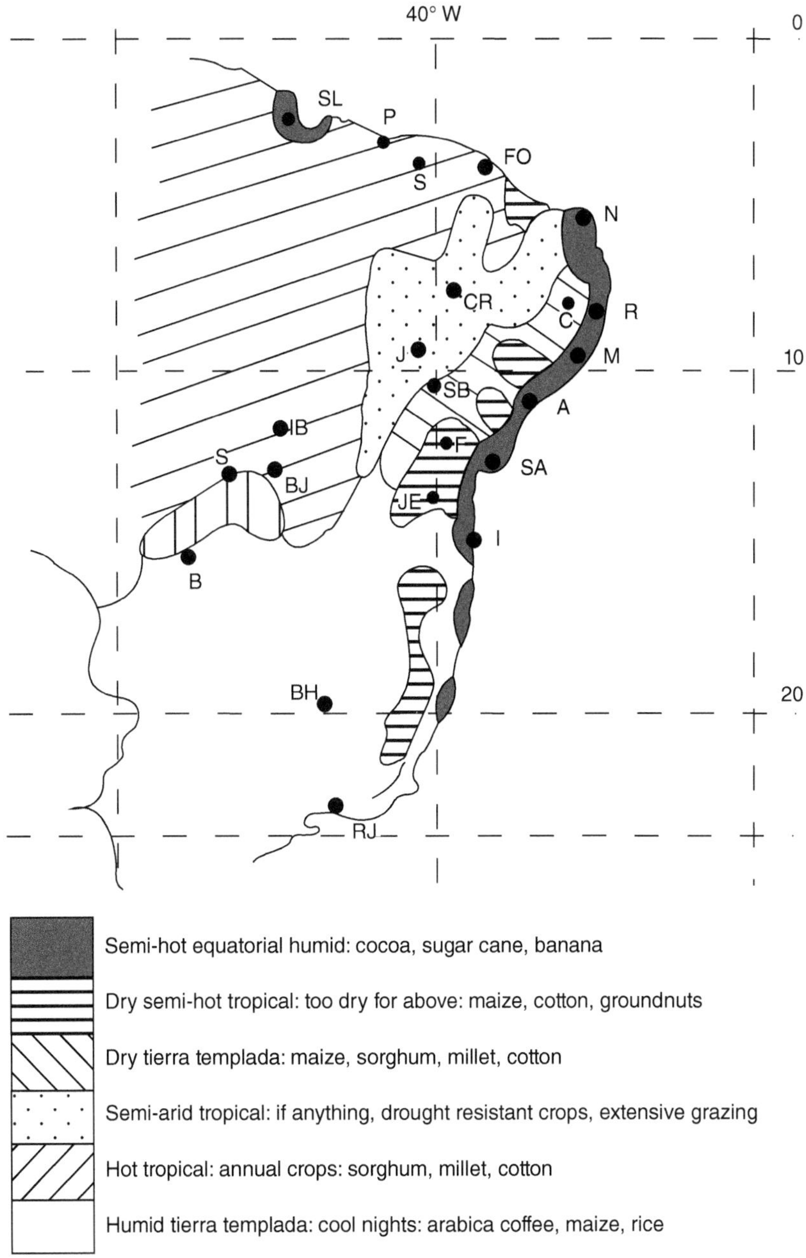

A - Aracaju
B - Brazilia
BH - Belo Horizonte
BJ - Bom Jesus de Lapa
C - Caruaru
CR - Crato
F - Feira de Santana
FO - Fortaleza
I - Ilheus
IB - Ibotirama
J - Juazeiro
JE - Jequie
M - Maceio
N - Natal
P - Parnaiba
R - Recife
RJ - Rio de Janeiro
S - Santana
SA - Salvador
SB - Senhor du Bofim
SL - Sao Luis

Fig. 3.14. Climatic types in north-east Brazil after Papadakis (1966).

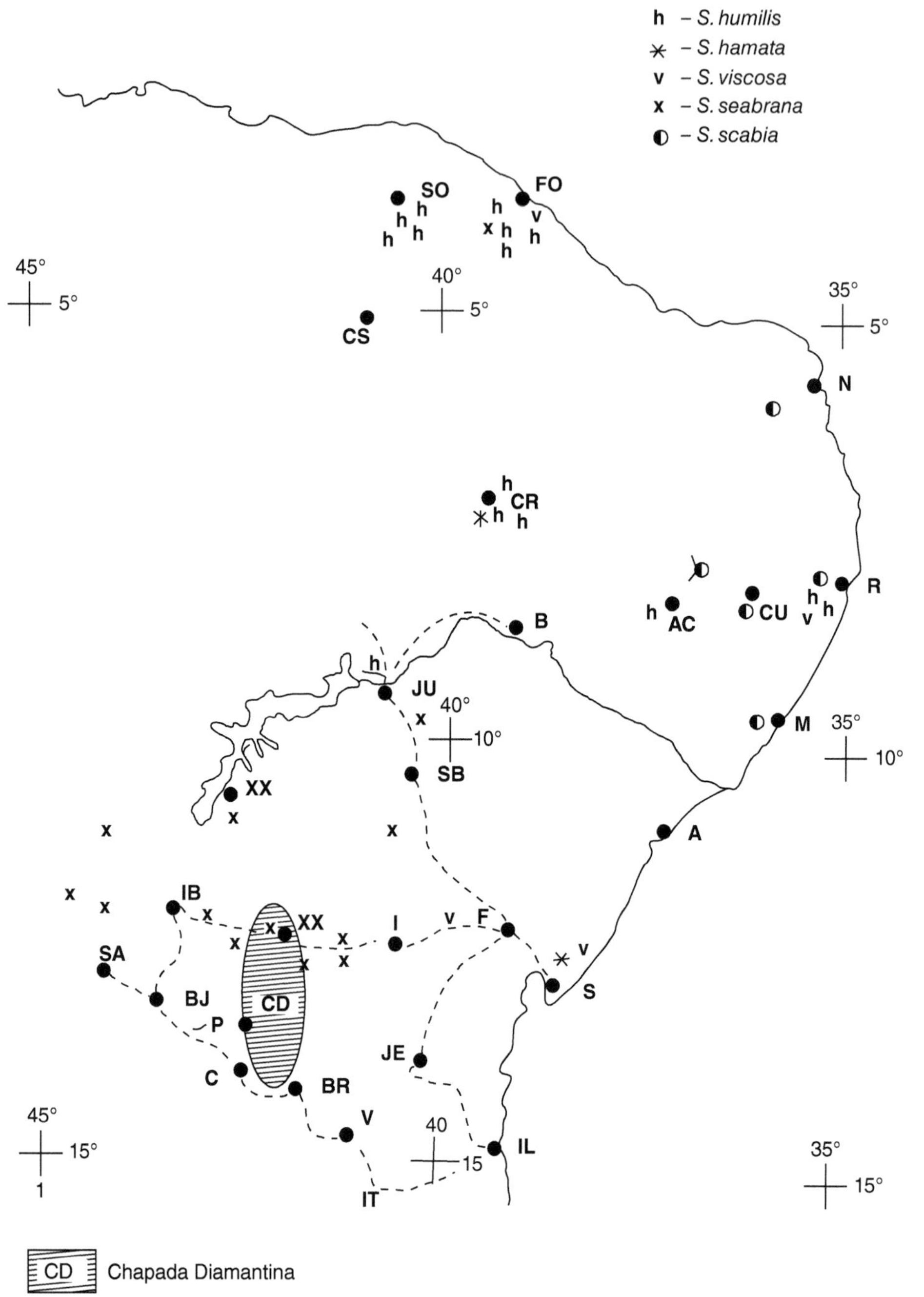

CD Chapada Diamantina

A - Aracaju; **AC** - Arcoverde; **B** - Belem de Sao Francisco; **BJ** - Bom Jesus de Lapa; **BR** - Brumado; **C** - Caetite; **CR** - Crato; **CS** - Crateus; **CU** - Cararu; **F** - Feira de Santana; **FO** - Fortaleza; **I** - Itaberaba; **IB** - Ibotirama; **IL** - Ilheus; **IT** - Itapetinga; **JE** - Jequie; **JU** - Juazeiro; **M** - Maceio; **N** - Natal; **P** - Paramarim; **R** - Recife; **S** - Salvador; **SA** - Santana; **SB** - Senhor du Bofim; **SE** - Seabra; **SO** - Sobral; **V** - Vitoria da Conquista; **XX** - Xique Xique

Fig. 3.15. Collecting zones in north-east Brazil.

Chakraborty, 2004). But data presented by Liu (1997) shows a paucity of follow-up collections from the very region that has provided our most important accessions, including cv. 'Seca'. This is in contrast to the situation for *S. hamata*, where following early successes missions were dedicated to the collection of further material (Edye and Maass, 1997). Refuge areas should clearly to be targeted as a potential source of novel variation, pre-adapted for use as cultivars. An ecological approach to collection, with relevant soil analysis, could well be used to explore the potential utility of any material collected (and see Appendix 3.5).

One area of outstanding interest is the Chapada Diamantina region, an area listed as a world centre of biodiversity. For tropical legumes of agronomic interest, this biodiversity is now well documented. Road transects in the region furnished a number of species referred to previously, including *Stylosanthes scabra*, *S.* sp. aff. *hamata* W, *S. seabrana*, *S.* sp. aff. *scabra* and *S. viscosa*, as well as the first collections of *S. capitata*, the only accession of *M. gracile* (syn. *M. longepedunculatum*) found on the 1971 collecting mission, and novel forms of *Centrosema* sp. This diversity could be a reflection of the numerous soil types encountered, for the area is anticlinal with a great mixture of rock types, climates and ecological niches, many of them inaccessible by road. One of the species located in the region, *S. seabrana*, is ecologically quite different from many other elements in the flora; it is adapted to cooler, more subtropical climates and thrives on less acid, heavy soils. It is highly valued throughout the world because of this (e.g. the information presented in Chakraborty, 2004). The species also requires inoculation with *Bradyrhizobium* strains that are typically found on alkaline soils, not the acidic sandy textured soils from which it was collected. A nearby area is reputed, however, to contain heavy clay soils (Andrade-Lima, personal communication) and plants growing on such an area might well be the source of the alkaline response pattern found in *S. seabrana*.

Recent surveys of the useful plants in north-east Brazil, including some of the likely refuge areas mentioned by Andrade-Lima (1982), have been made by the Brazilian authorities in association with the Royal Botanic Gardens, Kew. Such work is clearly of great interest and it is entirely possible that 'new species' and 'species types' may be forthcoming (as in Mexico and Central America, see information quoted in Maass and Sawkins, 2004). In view of the increasing pressures placed on biodiversity, and the rapid changes of climate involved in global warming, the survey and collection of such plants particularly in refuge areas already under considerable pressure is a matter of urgency.

3.8 Appendix 3.5

Effects of soils and soil characteristics on the distribution and growth of tropical herbaceous legumes on the island of Antigua: with special reference to *S. hamata sensu stricto*

The development of tropical pasture legumes for use in drier areas is a relatively new phenomenon. Much of the early work was primarily aimed at producing cultivars for use in Australia on soils of medium acidity and sandy textures. Work in South America followed, with much of it by or in association with CIAT. These studies concentrated on similar soils with greater acidity and as a result different species were developed. There are, however, extensive areas of clay soils in dry tropical conditions (notably the Republic of the Sudan 40 m ha, Australia 70.5 m ha and India 60 m ha, Dudal and Bramao, 1965), which compared with those with sandier textures are often more fertile, and more critically, retain their water content better into the dry season. Such soils are of huge significance for both cropping and pasture production; even small areas may be of great strategic significance both at farm and regional levels. However, there is a very real dearth of pasture legumes adapted to these situations (e.g. information quoted in Cook *et al.*, 2005) and presently only a narrow range of legume accessions are available for evaluation (see data in Chakraborty, 2004; cf. Chapter 6, this volume).

Although some evaluation of legumes in tropical clay soils in dry situations has undoubtedly taken place (A.H. Bunting, 1957, personal communication and Bunting and Lea, 1962 for results obtained in the Tozi irrigation area, Sudan), it has been largely limited to legume species native to Africa and India, and to dual purpose pulse-forage plants such as *Phaseolus trilobus* and

Clitoria ternatea. Such programmes furnished, for instance, one of the components of the *Clitoria ternatea* cultivar 'Milgarra', one of the few cultivars that is useful on heavy clay soils (as noted in Cook *et al.*, 2005). However germplasm of legume species with fodder potential from South America, the main source of tropical legume genetic variation, was not available until quite recently, and clay soils that are the most relevant source of potentially adapted legumes for such soils elsewhere are of very limited occurrence there. But they do occur commonly throughout the Caribbean islands and these have a rich, often endemic, legume flora as well as species that may have been introduced from South America, particularly those utilized in plantation agriculture (Stehle, 1956; Loveless, 1960; Howard, 1973). *S. hamata sensu stricto* is a long-valued indigenous pasture plant, common throughout the islands, known locally as 'Mother Segal' and somewhat more formally as 'Caribbean Stylo'.

The Caribbean islands comprise two phytogeographic zones (Howard, 1973). The island of Antigua is situated at the junction of these zones and has a tropical, moist monsoon climate with a well-defined, somewhat variable dry season (e.g. Ahmad, 1980). Temperatures vary little across the island and differences in legume distribution can be largely attributed to variations in soil and disturbance levels (Loveless, 1960). The island possesses soils of both volcanic and sedimentary origin, and these have been well documented (Hill, 1966). Prompted by the need to develop pasture systems for the region, the University of the West Indies in conjunction with IDRC (the International Development Research Centre of Canada) developed a sustained collection and evaluation programme and chose Antigua as one of the bases due to its unique phytogeographic situation and its representative soils (Spurgeon, 1977). An ecological approach to the collection and evaluation of legumes was developed, initially centred on Belize and Antigua, and applied to other areas in the region and, in doing so, it provided valuable germplasm, including collections of *S. hamata sensu stricto*, which were of value to research programmes in other centres such as CSIRO and CIAT.

There are five major soil groupings on Antigua and these are related to their origins, volcanic or sedimentary, and to their position on the island (Hill, 1966). An initial survey of the soil groups delimited locations of particular interest, in terms of their economic importance, floras, areas occupied by the given soil types and so forth. At each location 500 m line transects were laid out and 16, 1 m × 1 m quadrats located at equidistant intervals along them. In each quadrat the legumes present were listed and their dominance assessed (further details, if required, may be obtained from JK).

There were many interesting findings involving potentially useful legumes: for instance *Desmodium incanum* was restricted almost entirely to soils of volcanic origin and *S. hamata sensu stricto* tended to occur, like *S. humilis*, in legume-dominant patches throughout the sward. For the present purposes, however, the data presented will be for two legumes for use on clay soils: *S. hamata sensu stricto* and *Desmanthus virgatus* (for comparative purposes) (Burt, 1993a; Cook *et al.*, 2005).

In previous work of this nature, general purpose soil classifications proved to be of very limited value (Burt *et al.*, 1979). For instance, a yellow earth as defined by Stace *et al.* (1968), in a given type of climate, carried three different legume genera (of the limited range collected) and four different species. Subsequent scientific studies and commercial practice have shown them to be of different adaptation and utility (Cook *et al.*, 2005). This potential usefulness of the germplasm was much more apparent from studies in which the variation present in the different soils was displayed using a principal coordinate analysis, and the distribution of the various legume species and genera, in mixed swards, was then superimposed. This is the approach used herein, using the soil information provided by Hill (1966).

The first three vectors (using the analytical methods described previously in Burt *et al.*, 1979, remove 48%, 17% and 10% respectively of the total variation (the characteristics used are listed in Table 3.10). The third vector is concerned almost entirely with exchangeable Na, a factor that is linked to salinity, which figures in the previous vectors, and will not be considered further. The properties of the soils as displayed in Figs 3.16 and 3.17 follows a logical pedological sequence, from shallow hill soils of volcanic origin through to poorly drained, low-lying sedimentary soils. Onto this framework has been appended information showing the distribution and the dominance of the two legumes in question.

S. hamata is predominantly a plant to be found on sedimentary, more alkaline soils: in these situations it is favoured by high calcium carbonate

Table 3.10. Edaphic adaptation study of *Stylosanthes hamata sensu stricto* and *Desmanthus virgatus*. Characteristics of the surface soils at sites of collection used in the principle coordinate analysis.

1. pH	**5. Drainage (at 0.5 m)**	**9. Total N**
a) at soil surface	a) good	
b) at 0.5 m	b) medium	
	c) poor	
2. Mechanical analysis (surface soil)	**6. Surface soil structure (%)**	**10. Other analyses**
a) sand	a) stones	a) extractable P
b) sandy loam	b) sand	b) exchangeable K
c) loam	c) silt	c) exchangeable Mg
d) clay	d) clay	d) caption exchange capacity (CEC)
3. Depth	**7. Total P**	**11. Total organic carbon**
a) shallow (<0.5 m)		
b) medium (0.5 – 1 m)		
c) deep (> 1 m)		
4. Drainage (surface soil)	**8. Total K**	**12. Salinity**
d) good		a) surface soil
e) medium		b) at 0.5 m
f) poor		

concentrations and more freely draining situations and, as noted for *Stylosanthes* species previously, is favoured by high levels of exchangeable K in mixed swards.

Although the addition of superphosphate is generally advocated in low-fertility situations, and in pure culture P improves the growth of *Stylosanthes*, Hall (1979, personal communication) has convincingly demonstrated that the legume/grass balance in swards with *S. humilis* is favoured by the addition of K.

As *S. humilis* is a colonizing species favoured by disturbance and capable of growth on soils of very different textures, it can be found in most situations. However, it was never present in large quantities and its long-term persistence and utility under less disturbed conditions is open to question.

Desmanthus virgatus follows a generally similar pattern but it has been more dominant on heavier, often waterlogged soils and including those that were saline at depth.

Evaluation work was integrated with these earlier studies, and among others included accessions acquired from research bases in Belize and Antigua (Keoghan and Lazier as quoted in Ahmad, 1980), which had been tested over a wide range of climates and soil types. As noted earlier, there was a good correspondence between the environmental features at the sites of collection and those testing sites where the accessions proved to be better adapted. *S. scabra* from acid soils, for instance, failed in the more alkaline situations on which a range of *S. hamata sensu stricto* accessions thrived, and *Neonotonia wightii* (syn. *Glycine wightii*), a legume from Africa, was reported in earlier evaluation work to have thrived on the calcium-saturated soils of Barbados.

Although not large by international standards, this IDRC-UWI research programme provided a great deal of valuable germplasm for use throughout the world: one accession, of *Desmanthus pubescens*, has become one of the first cultivars of that genus, cv. Uman; others could well follow, especially for use on clay soils. The ecological approach used has provided information for what Harlan (1971) has described as 'a body of theory and general knowledge' to aid the efficient utilization of genetic resources. In doing so it has already been shown that the hypotheses developed in the course of the work have been of predictive value. For instance, the adaptation of *S. hamata sensu stricto* to heavy clay soils has indicated its ability to survive in competition with the domineering *Bothriochloa pertusa* ('Indian bluegrass in Australia, syn. *Andropogon pertusus*, 'Seymour grass' in Antigua). More detailed, integrated studies, useful for training purposes, could be equally rewarding. For undisturbed soil core studies (Keoghan, Ahmad and Burt, unpublished data) revealed that the silt:clay ratio was the major soil factor influencing legume yield and

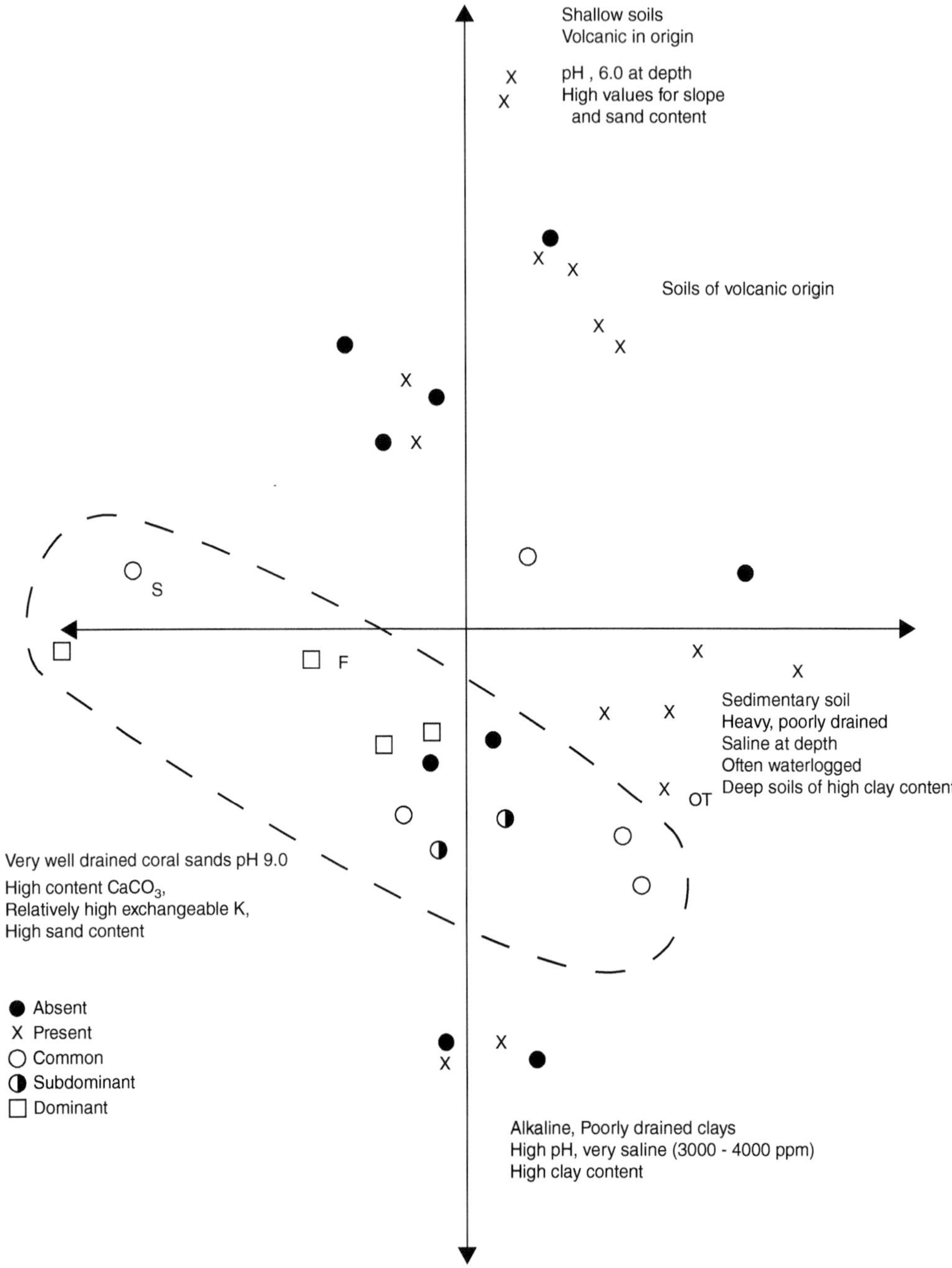

Fig. 3.16. Principal coordinate analysis of 35 soils and the distribution of *S. hamata sensu stricto* on Antigua. For detailed explanation see the text in Appendix 3.5.

was the major discriminatory factor between clay soils from northern Australia (and one likely to be of importance in 'silt fertility') (Isbell, personal communication; Burt, unpublished data). Incorporation into a special purpose classification of soils, such as the FCC system of Sanchez *et al.* (1982), might well increase the predictive value of such a system.

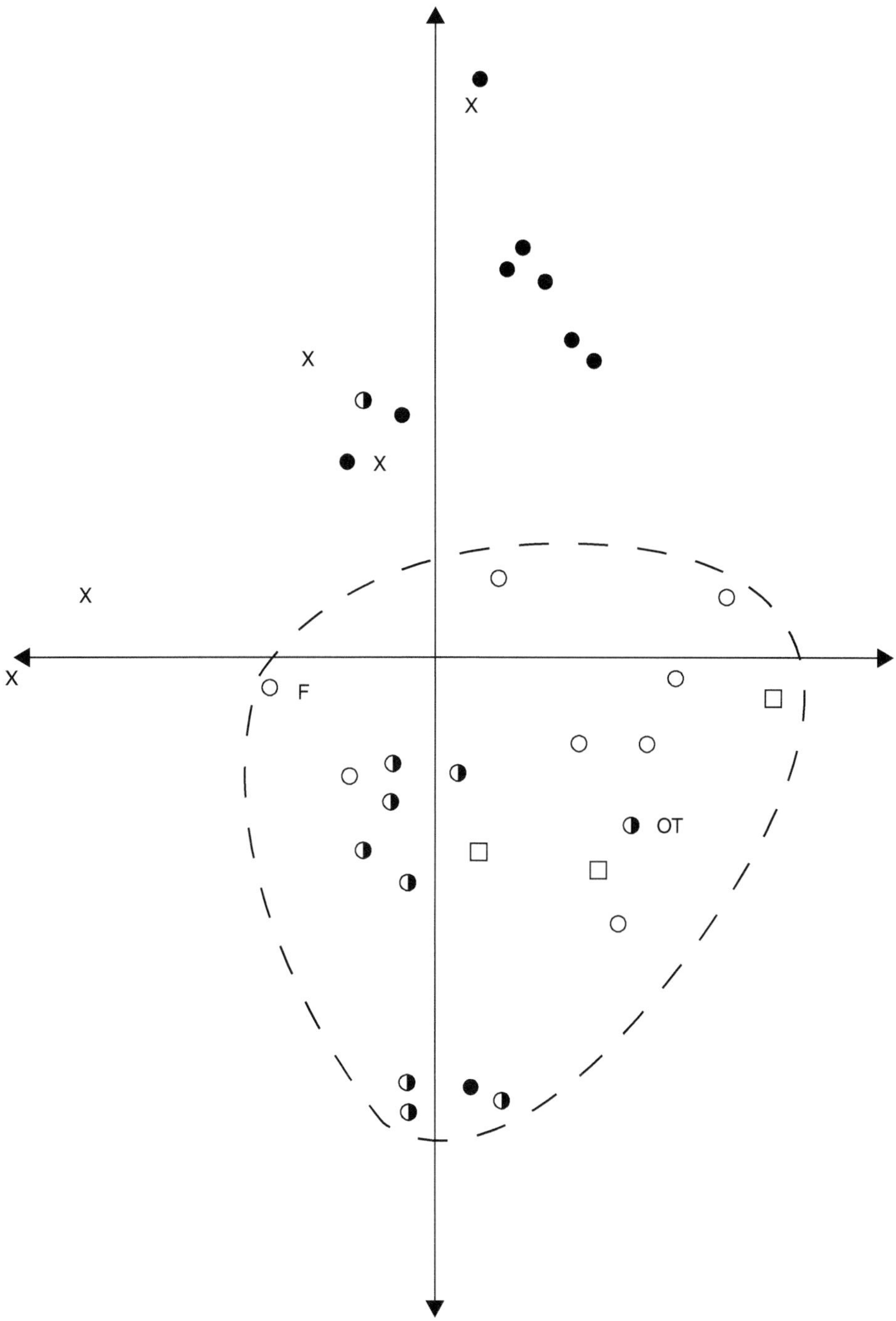

Fig. 3.17. Principal coordinate analysis of 35 soils and the distribution of *Desmanthus virgatus* on Antigua. For detailed explanation see the text in Appendix 3.5 and key in Fig. 3.16.

4 *Desmanthus*, a Tropical and Subtropical Forage Legume: Developing Germplasm Resources for More Subtropical and High Altitude Environments

R.L. Burt† and J.R. Lazier*[1]
**Formerly International Livestock Centre for Africa*

Abstract

The 24 species of the genus *Desmanthus* are widely distributed because they occur in environments ranging from tropical to temperate. Little collected or studied before the Belize-UWI research project, accessions of a number of species from the Yucatan and Belize showed considerable promise as forage plants. A review of *Desmanthus* species is presented that highlights their potential for and current success in a range of environments, particularly tropical and subtropical heavy clay soils. Further collection and research is recommended for a number of species with forage characteristics.

4.1 Introduction

Particular interest was stimulated in examining the potential of the genus *Desmanthus* because of the collections made by the IDRC/UWI forage programme in the Yucatan and Belize in the 1970s: a number of the accessions demonstrated excellent growth and productivity on heavy clay soils in introduction plots and in an initial trial (see Chapter 2, this volume). *Desmanthus* is a New World leguminous genus composed of some 24 species (Luckow, 1993), some of which have been deemed to be of agronomic value, particularly in tropical areas. Before 1993 and Luckow's taxonomic revision, there had been much confusion about nomenclature and it was common to use the specific name *D. virgatus* as a 'taxonomic hold-all'. An examination of the literature indicates that a *Desmanthus* species, now thought to be *D. pernambucanus*, was long ago transferred from the West Indies to the East Indies – possibly with other legumes to be used for plantation agriculture. It appears in 19th-century floras of India and has become naturalized there as well as in Fiji and Mauritius, among other countries. In India it was utilized as a 'hedge legume' or 'velimasal' because it was found to be well adapted to heavy clay soils where many of the more commonly used tropical pasture legumes fail to persist, and was especially useful in small farm systems as fodder for cattle. As communications improved, anecdotal evidence grew that the genus is utilized as fodder in many parts of the tropical world, as indicated by the common name 'donkey bean' ('frijillo de burro'). In 1977 Lazier showed that another species, *D. pubescens*, was also vigorous and equally well adapted to wet tropical conditions (Chapter 2, this volume).

The genus differs from the well-known genus *Stylosanthes*, source of many of the commercialized

[1] E-mail: john.lazier@sympatico.ca

tropical pasture legumes, in that *Stylosanthes* is primarily a tropical genus while many *Desmanthus* species are subtropical or even temperate. One species, *D. illinoensis*, is found at latitudes of over 40°N (Luckow, 1993). Many of its species are largely restricted to Central and North America, although it is also found in South America.

To date most of the *Desmanthus* accessions present in the major germplasm collections were collected while searching for material adapted to tropical, usually dry tropical, conditions (see information presented in Chapter 5, this volume), and in South America, an area favoured by those collecting legumes for use in the tropics. As a result, the major collections lack germplasm from cooler environments and are thus woefully inadequate. The collection of the Australian Tropical Forages Genetic Resource Centre (ATFGRC), for example, which is also housed in Colombia, contains some 442 accessions, many of which are reintroductions and lack even one example of 10 of the 24 named species (Luckow, 1993). In contrast, the US National Temperate Forage Legume Germplasm Collection accessions of *Medicago*, *Lotus* and *Trifolium* total over 13,000 and dwarf the tropical collections of *Desmanthus* and *Stylosanthes*. These facts indicate that there is much potential and valuable variation remaining to be collected in these two important tropical forage genera.

Many of the *Desmanthus* species that are poorly represented in genetic resource collections are those from more temperate climates, either from higher latitudes or from 'tierra fria', high altitude regions (non-frostless tropical highland) in the tropics. The potential value of legumes from such environments has long been recognized, especially those from the more accessible areas in Africa. *Neonotonia wightii* (syn. *Glycine wightii*), a higher altitude African species, was one of the early introductions into Australia and has been widely used in the subtropics and high altitude areas in the tropics (see information presented subsequently). There is a continuing need for similarly adapted material in Australia (Pengelly and Conway, 2000) and in such environments as 'tierra fria' in, for example, Ethiopia. The temperate New World *Desmanthus* species have the potential of being as valuable as, or more valuable than, currently utilized fodder legumes in subtropical and higher altitude environments.

Hartley (1947–48) and W.T. Atkinson (1965) were early collectors who sought new species of legumes with forage potential in South and Central America for subtropical Australia. Hartley sought to nominate potentially useful collecting areas by defining homologous collection zones, those judged to be similar to the areas for which the plants were required. Areas were considered to be homologous if they had similar grass floras. Much subsequent evaluation and plant introduction has confirmed that he was remarkably successful (Williams and Burt, 1982). However, most researchers have chosen to follow the early lead of de Candolle (as quoted by Darwin in *On the Origin of Species*, 1859) and define homologous zones in climatic terms (for further discussion on this aspect, see Chapter 3, this volume). A modification of the latter approach is to define the climates that have yielded adapted material and, using a suitable system of climatic classification, locate similar areas elsewhere. The floras of these areas can then be examined to see if they contain genera or species likely to be of value (Burt *et al.*, 1976; Burt and Reid, 1976). Geographic information systems (GIS) are now widely used for such purposes (e.g. Maas and Sawkins, 2004; Jones *et al.*, 1997). All methods currently used have their strengths and weaknesses (Chapter 3, this volume).

In view of the known global problems of loss of biodiversity, and the effects of climate change, further focused collection of useful phyto-genetic variation is a matter of urgency. Thus the present study uses a variety of methods to explore the adequacy of the ATFGRC *Desmanthus* germplasm collection. This is the collection most commonly used as a resource for tropical areas and the authors are familiar with its assembly and evaluation. On the basis of the information so gained, the study will assess the importance of the acquisition of additional genetic variation of *Desmanthus* species, and areas that warrant collection.

4.2 Environmental tolerance of individual genotypes: examples drawn from *D. virgatus*

The best-known tropical forage legume genus is *Stylosanthes* and it is tempting to apply the strictures learned from the study of this genus to

Desmanthus. However, as has been noted, the geographical distribution and overall climatic adaptation of *Desmanthus* differs markedly from that of *Stylosanthes*. *Desmanthus* is also much more likely to be adapted to heavier soils and alkaline conditions, similar in its edaphic adaptation to the temperate genus *Medicago* rather than its more acidophilic counterpart *Trifolium*. Another major difference is that all *Desmanthus* species are perennial (Luckow, 1993), while *Stylosanthes* contains annual and perennial species, both of which have yielded valuable cultivars. The annuals such as *S. humilis* and *S. hamata* rely on population heterogeneity to survive over a range of climates, or in climatic/management conditions that may vary from year to year. Perennials such as *S. scabra* have distinct 'types' for different climate conditions and, in the more severe climates, rely on 'plasticity' of the individual genotypes (see information quoted in Chapter 3, this volume).

Desmanthus species often have a strong woody taproot and some are tremendously variable 'prostrate, decumbent or erect herbaceous perennials to 1.5m tall' (Luckow, 1993). That such variation exists and is not geographically based (Burt 1993a,b; Chapter 5, this volume) are important considerations in the planning of collection missions.

Although many accessions of *D. virgatus* remain largely untested (Jones and Clem, 1997), three have been grown in a wide range of climates in Australia. Fortunately, for the present purpose, these were from two very important collection areas in Argentina and Mexico, and from very different latitudes, altitudes and rainfall regimes. Two were from Argentina, the cultivar 'Marc' and the accession CPI (Commonwealth Plant Introduction) 78382; the third, CPI 85178, is from north-east Mexico. They were collected respectively at 23°S, 24°S and 27°N, in mean annual rainfalls of 650, 700 and 400 mm and from altitudes of 300, 1000 and 220 m. All were from neutral to alkaline clays or clay loams.

The initial evaluation of the *Desmanthus* collection was carried out from the CSIRO Davies Laboratory in Townsville, Queensland, in dry tropical conditions and at latitude 19°S. Subsequently, representatives were evaluated further inland in a drier, hot semi-tropical environment and also at higher latitudes (24°S) in conditions verging on semi-desert. The study was then extended into subtropical conditions using a narrower range of plants at sites up to 28°S. More detailed information on the evaluation programmes is presented elsewhere (Chapter 8, this volume).

All three widely tested accessions proved to be remarkably resilient in tropical and subtropical conditions (as defined only on the basis of latitudes). They were very persistent and regenerated by seedlings at the Lansdowne site (near Townsville, mean annual rainfall 869 mm) and in the much more demanding climate of Isisford (latitude 24°S, mean annual rainfall 446 mm). At the latter site, over an extended evaluation period (1989 to 2003), yearly rainfalls ranged from 164 to 873 mm and the evaluation plots were subjected to flooding, drought, fire and overgrazing (see information in Chapter 8, this volume). Clem and Hall (1994) had similar results in north-eastern Queensland from three accessions of *D. virgatus*, with the Argentinean accession CPI 78373 cv. 'Marc' outperforming those from lower latitudes.

When the same three *D. virgatus* accessions were evaluated sown with a pasture grass and under various grazing regimes, on a clay soil site in sub-coastal subtropical Queensland (latitude 26°S, longitude 151°E, mean annual rainfall of 717 mm with a strong autumn/winter component and a mean minimum temperature of −4.7°C), there were large differences between the accessions (Jones and Brandon, 1998). The Mexican accession (CPI 85178) virtually died out and accession CPI 78382 produced higher plant numbers, yields and soil seed reserves than cv. 'Marc'. However, in the search for tropical legumes to complement the currently utilized *Medicago* and *Trifolium* species in even more temperate conditions, *D. virgatus* CPI 78382 outperformed a large range of other tropical material.

Accession 78382 was collected at a much higher altitude (1000 m) than the other two accessions (220 and 300 m) and from a higher latitude (24°S vs 23°S) than cv. 'Marc'. This is consistent with the belief that plants best suited for high altitude conditions in the tropics, and for highest latitudes, are most likely to be found in similar conditions elsewhere. In the case of *Desmanthus* such cooler temperature adapted germplasm is to be found in countries of the western hemisphere.

4.3 High altitude/latitude collections of *Desmanthus* in the ATFGRC germplasm bank

Desmanthus has a centre of diversity in Mexico (Luckow, 1993) and a peculiar bi-centric distribution, with high altitude/high latitude collection sites in more southerly situations in South America and in Central and North America. Relatively few species have been found in Brazil and Venezuela, the prime countries for most tropical pasture legume collection (e.g. Coradin and Schultze-Kraft, 1990).

4.3.1 South America

Major collecting zones

Although Brazil, Colombia and Venezuela have been the countries of major focus in the search for tropical pasture legumes, they have furnished only 55 of the 442 *Desmanthus* accessions currently held in the ATFGRC collection (data abstracted from the Australia Plant Genetic Resource Information Service, 2008). In contrast, Argentina has provided 79 accessions and Paraguay 65. These two countries are floristically rich in *Desmanthus*, but lack two primarily tropical species, *D. leptophyllus* and *D. pernambucanus*. They contain *D. hexapetalus* (and are the only two countries to do so), as well as *D. acuminatus* and *D. tatuhyensis* var. *brevipes*. The latter was formerly held to be a separate species and is noted for its ability to thrive on clay soils (see information presented in Burt, 1993a). Although *D. tatuhyensis* can be found in Brazil as well as in Argentina and Paraguay, it is only found in the south-eastern corner. In searching for plants adapted to subtropical regions in Australia, the early collectors Hartley and Atkinson introduced accessions from both Argentina and Paraguay. Atkinson, for instance, collected *D. acuminatus* from over 34°S. However, lacking a taxonomic review these accessions were listed as *D. virgatus*. *D. tatuhyensis* was often treated similarly and this renders a review of the literature from those times difficult. Such is the variation in *D. virgatus* that separation of these three species, especially material from more southerly latitudes, is taxonomically and agronomically difficult (e.g. Luckow, 1993; Chapter 5, this volume).

Later collecting missions were undertaken to provide plants for more tropical conditions (Burt and Reid, 1976; Reid, 1983), with emphasis increasingly placed on finding plants adapted to clay soils. The Gran Chaco, a region of deep alluvial soils bounded to the west by the Andes, was of major interest. The main collecting zones are shown in Fig. 4.1, labelled A to H. Lack of collection in the area is due to difficulties of access because it is poorly drained and contains numerous swamps and flooded areas.

Zone A is the Chaco Boreal in Paraguay, where flooding and salty groundwater are major problems. Here even slight differences in soil texture or relief have great effects on the type of vegetation found. The provenance data associated with the collections from the zone indicate that they were found at low altitudes of 150 to 300 m, with mean annual rainfalls of 800 to 1000 mm and usually from clay soils of pH 6 to 9. The provenance data are scanty however, and in view of the overriding effects of relief and soil texture such climatic data as are available cannot sensibly be used to predict areas of adaptation elsewhere. In view of the relatively low latitudes and altitudes the material from it would seem to be unlikely to be adapted to more subtropical conditions. Zone C is similar.

Zones B and D lie in the Chaco Central and the Chaco Austral respectively and, for our present purpose are again of little interest. Accessions from B were from relatively low altitudes (250 to 280 m), relatively high rainfall (1100 to 1250 mm mean annual rainfalls) and sandy loam or clay soils of pH about 7. Zone D is similar.

The more westerly zones E to H are of considerable interest. They border on the Andes, with H near the Sierras de Cordoba, and often have higher altitudes or, in the case of H, much higher latitudes. They are also climatically diverse. Zone E, which yielded the very promising *D. virgatus* accessions cv. 'Marc' and CPI 78382, contains a wide range of altitudes. Oran at the northern end has a hot, semi-tropical climate (*sensu* Papadakis, 1966), while Salta is in the low tierra fria. Zone H is separated from the more northern zones by a hot, subtropical desert (around Catamosea and La Rioja) and, because it is more temperate, is defined as a 'peri-pampeau'.

These differences in climate are reflected in the *Desmanthus* flora. *D. virgatus* is found throughout the lower latitude zones (A to F), but

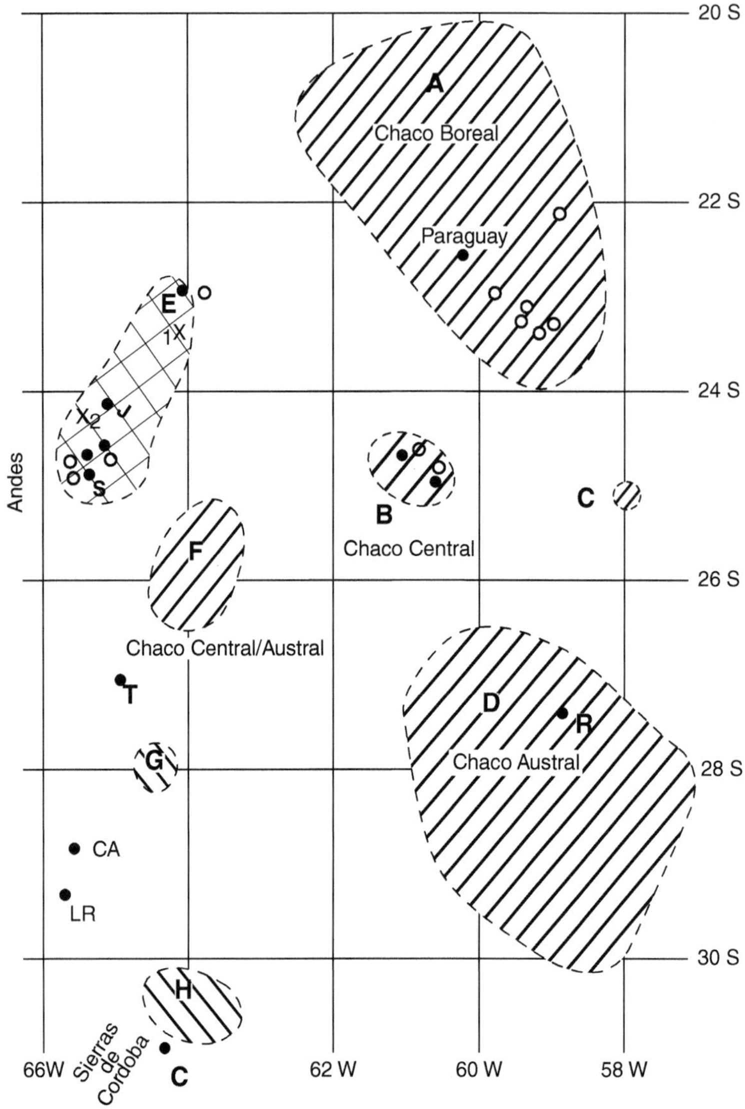

+ As listed for the Australian Plant Genetic Resource collection and abstracted on 5/2/2008

○ *Desmanthus paspalaceus* collection sites

● *Desmanthus tatuhyensis* collection sites

X_1 Collection site of CPI 78373, *D. virgatus* cv.'Marc'

X_2 Collection site CPI 78382, *D. virgatus*

C - Cordoba; CA - Catamarca; J - San Salvador de Jujuy;
LR - La Rioja; O - Oran (beside the E); R - Resistancia; S - Salta; T - Tucaman

///// *D. virgatus* \\\\\ *D. acuminatus* XXX *D. acuminatus* + *D. virgatus*

Fig. 4.1. Geographical location of collecting zones for *Desmanthus* in Argentina and Paraguay.

only *D. acuminatus* is found in the more southerly upland areas (G and H). The latter species can also be found with *D. virgatus* in the upland areas of zone E. This same zone is particularly rich because it also has *D. tatuhyensis* and *D. paspalaceus*. These species are also found in similar latitudes at more easterly locations (A and B). These distributional patterns are in accordance with those shown for the area by Luckow (1993). We note for future discussion that Luckow regards *D. tatuhyensis* as being primarily a species of south-east Brazil, an area close to zones A and B. Soils in that area are primarily acidic non-clays and this could explain the unusual (for *Desmanthus*) ability of introductions of this species to survive on such soils in Australia (Cook *et al.*, 2005). A more detailed examination of species/environment interrelationships follows.

Species/environmental relationships in higher altitude/latitude zones

The higher altitude/latitude zones are E to H (Fig. 4.1). Provenance records are most complete for E and are lacking for F. The records for zone E are presented in Table 4.1 and locational information is shown in Fig. 4.2.

D. virgatus has been collected from a wide range of latitudes and altitudes. Two of the accessions noted previously, cv. 'Marc' and CPI 78382, are from quite different environments, with CPI 78382 being much more subtropical. Should more subtropical tolerance be required, then there are accessions of *D. virgatus* available from higher latitudes and altitudes such as south of Salta (S in Fig. 4.1), where they are to be found with *D. acuminatus*, a species that occurs at yet higher latitudes where *D. virgatus* has not been

Table 4.1. Accessions of *Desmanthus virgatus* and *Desmanthus acuminatus* found in zone E+.

Species	Number[a]	Latitude (°S)	Altitude (m)	Longitude (°W)	Soil texture[b]	Soil pH[c]	Land use[d]	Mean annual rainfall (mm)
D. virgatus[e]	79373	23.25	300	64.08	clay loam	–	GNP	650
" "	319442	23.28	380	64.17	–	–	RS	–
" "	78371	23.30	400	64.17	sandy loam	8	GNP	700
" "	78372	23.33	400	64.25	sandy loam	–	GNP	680
" "	319445	23.50	330	64.12	clay loam	–	–	–
" "	78370	23.67	500	64.60	sandy loam	8	GNP	750
" "	319441	23.70	360	64.58	sandy loam	7	RS	1000
" "	319446	23.87	455	64.95	clay loam	–	RS	500
" "	319440	24.10	500	64.82	loam	–	GNP	–
D. acuminatus	78369	24.20	1000	65.67	clay loam	–	GMP	700
D. virgatus[f]	78382	24.20	1000	65.67	clay loam	–	GNP	700
D. acuminatus	78383	24.22	1660	65.82	clay loam	–	GNP	1000
D. virgatus	319439	24.27	700	64.90	loam	–	RS	–
" "	319447	24.32	426	63.98	loam	–	RS	–
" "	319436	24.38	1036	65.10	loam	7	RS	700
" "	92595	24.58	–	65.08	–	–	–	–
D. acuminatus	78381	24.58	1250	65.83	loam	–	GNP	650
D. virgatus	319435	24.67	731	65.03	loam	7	–	600
D. acuminatus	78379	24.77	1325	65.47	clay loam	–	GNP	690
" "	78380	24.77	1330	65.48	clay loam	–	GNP	690
" "	319427	24.80	1260	65.28	clay loam	–	RS	650
D. virgatus	319432	24.80	1260	65.28	clay loam	–	RS	650
" "	319437	24.83	1220	65.48	loam	7	RS	800
D. acuminatus	319428	24.83	1220	65.48	loam	7	RS	800
D. virgatus	319448	24.90	645	64.42	clay loam	7	RS	–
D. acuminatus	319426	24.95	1390	65.60	–	–	RS	–

E+ Accessions from the Australian Plant Genetic Resource Collection found in zone E, Fig. 4.1. [a]Australian TRFC number in the above collection. [b]Soil texture: Texture of surface soil as noted by the collectors. [c]Soil pH: of surface soil. [d]Land use: GNP, grazed native plants; RS, roadside collection. [e]and [f]: *D. virgatus*[e] cv. 'Marc', compared with CPI 78382[f] in text.

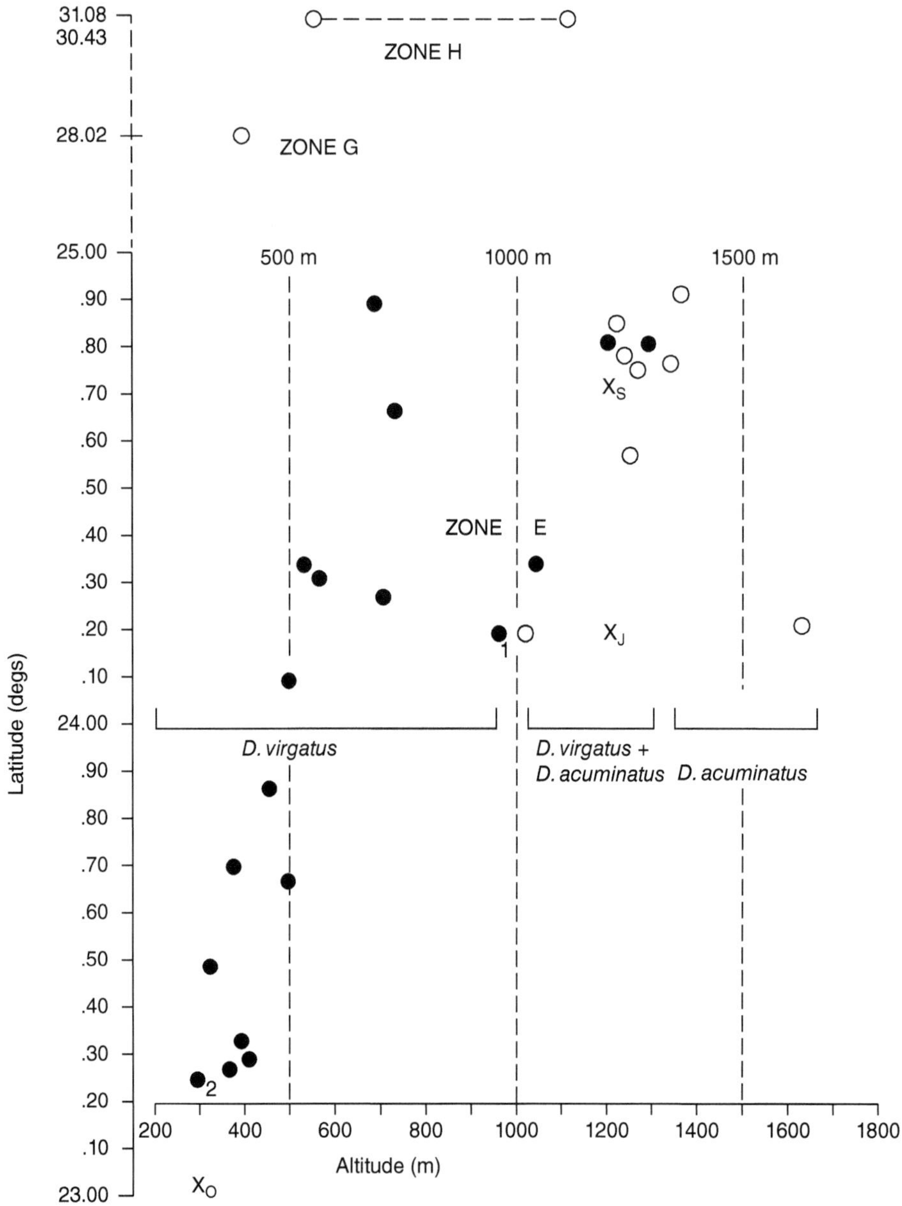

+ As listed for the Australian Plant Genetic Resource collection (2008) and found in zone E (Figure 4.1) in Argentina

○ *D. acuminatus*; ● *D. virgatus*

X_O Oran; X_S Salta; X_J San Salvador de Jujuy;

$●_1$ Collection site CPI 78382, *D. virgatus* (see text)

$●_2$ Collection site CPI 78373, *D. virgatus* cv.'Marc' (see text)

Fig. 4.2. Geographical location of collecting zones for *Desmanthus virgatus* and *D. acuminatus* in zone E: location zones G and H.

recorded by collection, or reported by Luckow (1993). *D. acuminatus* is found in grazed native pastures, usually on clay loams (although Luckow locates it primarily on sands), and it might well be useful as a pasture/ley legume in more subtropical environments. This very valuable species is described as 'prostrate to decumbent perennial herbs to 3 dm tall, much branched from the base' with a woody taproot, characteristics that could well predispose it to be a useful grazing plant. It is represented in the ATFGRC collection by only 13 accessions. No mention is made of the species in Cook *et al.* (2005) or Jones and Clem (1997), possibly because evaluation of even these accessions has been very limited.

4.3.2 Central and North America

Although Central America is floristically rich, and has yielded maize, avocado and major legume crops (common bean, lima bean, tepary bean and runner bean), it has received only scanty attention by the earlier collectors of tropical pasture plants. Later phytogeographic studies, however, drew attention to the potential importance of Mexico as a source of plants for dry tropical regions (Burt and Reid, 1976), and the same studies noted the potential importance of a small area of clay soils in an atypically dry spot in Guatemala. A visit to the latter area revealed that the only herbaceous legume species growing on these soils was *D. virgatus*. Collecting missions in Mexico (Reid, 1983) confirmed the importance of this genus on clay soils of that country. At about the same time the IDRC-UWI programme in Belize, which helped to sponsor the visit to Guatemala mentioned earlier, was making collections of native legume germplasm, mainly in Belize and the Yucatan Peninsula. The Belizean programme identified *D. pubescens* as a new shrub legume of considerable potential utility (Chapter 2, this volume). The germplasm collections were added to those of ATFGRC. One of these *D. pubescens* accessions became the cultivar 'Uman', part of the trio of the first domesticants of the genus *Desmanthus*.

These few investigative collection activities are the only ones to have taken place despite the fact that Mexico, which contains 16 species, is almost certainly the centre of diversity for the genus; the USA is also species-rich. As a result, the ATFGRC collection contains only 130 accessions from Mexico, 29 from Belize and 11 from the USA (and several of these were originally collected elsewhere). This is discussed in a later section.

4.3.3 Mexico

The sites of collection of the *Desmanthus* species found at high altitude/high latitude sites in Mexico are shown in Fig. 4.3. These are labelled A to I. Zone A lies along a transect in which altitudes vary. In this instance the locations of collecting sites at lower altitudes are also shown, including the collection site of CPI 85178 mentioned in a previous section. Provenance data are listed in Table 4.2.

As is particularly evident from the transect in the north-east of Mexico (Zone A, Fig. 4.3), *Desmanthus virgatus* can be found over a wide range of both latitudes and altitudes, but at the highest latitudes and altitudes it gives way to *D. velutinus*, a species that extends into the USA from New Mexico and Texas to a latitude of over 30°N. The same situation occurs again in Zone C, where at higher altitudes and latitudes *D. virgatus* yields to *D. glandulosus*, a species that also extends into the USA (Texas, New Mexico and Arizona). *D. leptophyllus*, a more tropical species, is found more commonly at lower latitudes from 17° to 22°N in Missouri, Kansas, Oklahoma and Texas, and its range extends into South America. *D. painteri* and *D. pumilis* were located at altitudes of 1850 m and 2150 m respectively. The occasional presence of *D. leptophyllus* and *D. covillei* in such situations is unexpected (see also information in Luckow, 1993). They may have been recent roadside introductions or may have been in niche situations. *D. painteri* is restricted to Mexico, where it is only found at high altitudes, from 1900 to 2400 m (Luckow, 1993). It is sometimes found with *D. pumilis*, a species also restricted to Mexico and to high altitudes.

The accession that was chosen to represent the range for *D. virgatus* from north-east Mexico, CPI 85178, was primarily selected for trial on clay soils in semi-arid regions in northern Australia. It had not figured in morphological/agronomic evaluation trials (*sensu* Chapter 8, this volume), but was chosen from provenance records as growing on heavy clays in dry regions. It performed well in the trials in north-east

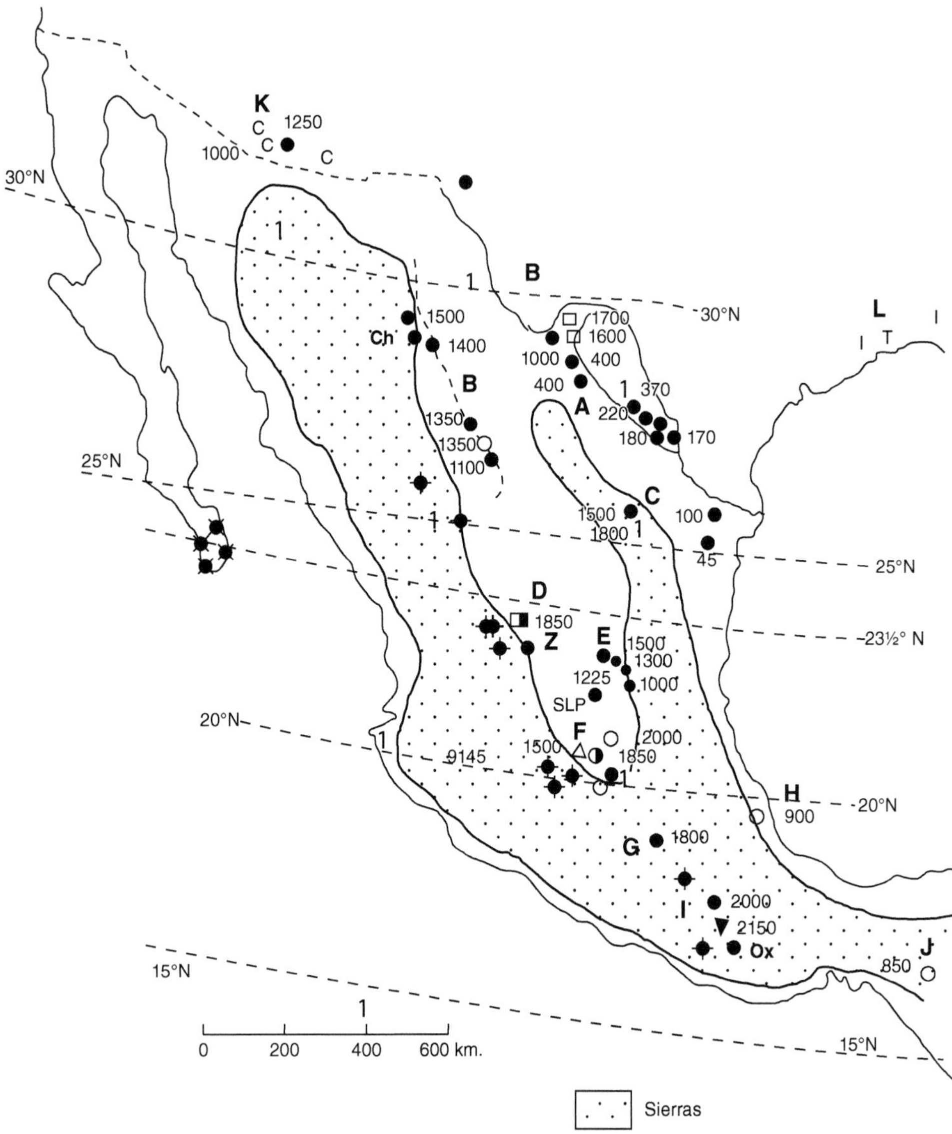

1 *D. glandulosus* | I *D. illinoensis* | O *D. leptophyllus* | **B** *D. obtusus* | ◑ : *D. painteri*
▲ *D. paspalaceus* | ▼ *D. pumilis* | △ *D. bicornutus* | **T** *D. tatuhyensis* | ◨ : *D. covillei*
□ *D. velutinus* | ■ *D. spp* | • *D. virgatus* | **X** *D. acuminatus* | c *D. cooleyi*
Not in collection: ✦ *D. interior;* ✱ *D. oligospermus*

• **Ch** Chihuahua City; • **Ox** Oaxaca City; • **Q** Queretaro; • **Z** Zacateca City

*APGR – Australian Plant Genetic Resources. Incorporates the Australian Tropical Forage Genetic Resource Collection (ATFGRC) mentioned elsewhere.

Fig. 4.3. *Desmanthus* accessions in the APGR collection from high altitudes and latitudes: Mexico.

Table 4.2. High altitude/high latitude *Desmanthus* accessions from Mexico in the APGR collection: provenance data and suitability for clay soils rating.[a]

Latitude (°N)	Altitude (m)	*Desmanthus* species	Annual rainfall (mm)	Clay soil rating	Surface soil, pH
Zone A[b]					
29	1700	*D. velutinus*	400	–	–
29	1600		400	–	–
29	1000	*D. virgatus*	400	3	–
28	400		400	2	clay 8.0
28	400		400	2	–
28	400		400	8	cracking clay
28	400		400	–	–
28	370	*D. glandulosus*	400	–	–
28	380	*D. virgatus*	250	–	–
27	220[c]		440	5	clay 8.5
27	180		420	6	clay loam 7.0
27	220		400	–	loam 8.0
27	170		440	2	clay loam 7.5
26	100		425	–	clay loam
26	280		500	5	clay loam
25	45		550	–	clay loam
Zone B					
28	1500	*D. virgatus*	300	–	loam
28	1400		300	7	loam
27	1350		320	–	loam
27	1350	*D. leptophyllus*	320	7	clay
26	1100	*D. virgatus*	230	7	loam 8.0
Zone C					
26	1500	*D. virgatus*	400	–	–
26	1800	*D. glandulosus*	400	3	–
Zone D					
23	1850	*D. covillei*	400	–	–
Zone E					
22	1500	*D. virgatus*	700	–	loam
22	1300		600	–	clay 8.0
22	1625		600	–	clay
22	1000		550	–	clay loam
Zone F					
21	2000	*D. leptophyllus*	500	–	–
21	1850	*D. painteri*	500	–	–
Zone G					
19	1500	*D. virgatus*	1200	–	clay
Zone H					
19	900	*D. leptophyllus*	1000	–	–

Continued

Table 4.2. Continued.

Latitude (°N)	Altitude (m)	*Desmanthus* species	Annual rainfall (mm)	Clay soil rating	Surface soil, pH
Zone I					
18	2000	*D. virgatus*	470	–	loam
18	2150	*D. pumilis*	470	–	–
Zone J					
17	850	*D. leptophyllus*	1000	–	–

[a]Adaptation to clay as listed by the APGR Information Service (which kindly supplied accession listings). [b]As shown in Figs 4.1 and 4.2. [c]Accession CPI 85178.

Queensland for which it was chosen (Gardiner and Burt, 1995), but poorly in cooler more subtropical zones (see data quoted earlier from Jones and Brandon, 1998; Jones and Clem, 1997). Reference to Fig. 4.3 and Table 4.2 shows that it was collected in a relatively high latitude (27°N), but at lower altitudes; several are listed in Table 4.2. Those from the highest latitudes may be better adapted to more subtropical conditions.

Of the 16 species that occur in Mexico, three, *D. balensis*, *D. covillei* and *D. fruticosus*, are essentially small, drought-tolerant deciduous trees seemingly of little direct agronomic value; *D. balensis* is rare and *D. covillei* is a common roadside weed. Another woody species is *D. oligospermus*. It has long woody underground stems. Such characteristics suggest that it might be a potentially 'invasive' plant. It possesses a feature unusual in the *Desmanthus* genus, indehiscent pods, the positive significance of which for crop plants has been stressed by Smartt (1976). *D. illinoensis* is currently being developed in the USA as a perennial pulse crop with major potential for low-input sustainable farming systems. Other species of interest and not represented in the Mexican portion of the ATFGRC collection are *D. interior* (a high altitude, intensively grazed and probably threatened species) and *D. cooleyi* (which is reported as good browse, grows at high altitudes and extends to latitudes of 37°N in the USA). Three of the species of interest in the existing collection, *D. glandulosus*, *D. painteri* and *D. pumilis*, are each represented by only one or two accessions.

4.3.4 The USA

It has long been known that states such as Arizona are floristically homogeneous with some subtropical environments (Hartley, 1963). Fifteen *Desmanthus* species are found in the USA and data provided by Luckow (1993) and Burt (1993a) show the potential value of some of them. In addition to those mentioned in the previous section, they include *D. reticulatus* (blackland prairie soils to 500 m altitude and 30°N), *D. leptolobus* (or 'prairie mimosa' on clays sands and loams to 38°N), *D. tatuhyensis* var. *brevipes* (formerly *D. brevipes*, on clay soils in the USA) and *D. acuminatus* (on a range of soil types and ecological situations in Texas and discussed above in the section on South America). *D. illinoensis* is a species of major potential importance as a perennial pulse crop (e.g. Glover and Wilhelm, 2005). The potential of some of these species, *D. illinoensis*, *D. leptolobus*, *D. velutinus* and *D. acuminatus* (together with the almost entirely Mexican *D. bicornutus*), in revegetation of blackland prairie areas in Texas is now well documented (Muir *et al.*, 2005). They could well be of use in subtropical or 'tierra fria' situations. Two of these are found at much higher latitudes: *D. illinoensis* is found from the eastern seaboard to the mid-west, and north to near the Canadian border, while *D. leptolobus* occurs in Missouri, Oklahoma, Kansas and Texas. Three species are reported to have selections commercialized for revegetation and for forage for goats, sheep and wildlife, *D. velutinus* 'Hondo', *D. virgatus* 'Rayado' and *D. illinoensis* 'Sabine' and 'Reno' (Muir and Pitman, 2004; Wynia, 2008). However, of the 15 *Desmanthus* species only four are recorded in the USDA Germplasm Resources Information Network and these are represented by very few accessions: *D illinoensis* – 61, *D. virgatus* – 47, *D. cooleyi* – 5 and *D. velutinus* – 3. It is thus important that variation in these species be collected and included in a basic germplasm bank for the genus.

4.4 Homologous high altitude/high latitude situations in Argentina, Mexico, Australia, Ethiopia and the USA

In the first parts of this volume an attempt has been made to delimit those *Desmanthus* species that could be useful as pasture or ley legumes in subtropical or 'tierra fria' situations. Some representatives of these species may already be present in existing germplasm collections. The ATFGRC collection has been widely used for tropical areas, potentially useful species in it have been noted, and it has sometimes been possible to suggest which accessions are likely to be adapted to specific climates. In seeking situations in which specific plants are most likely to be adapted, it is usual to delimit homologous collecting and receptor areas in terms of their climates (e.g. De Candolle as quoted in Darwin's *On the Origin of Species*, 1859). However, many other factors can be involved, including type of soil, grazing pressure, and over or under abundance of a single soil nutrient (see, for 'new' tropical species, Burt *et al.*, 1979). This is illustrated by the famous Aberystwyth-developed strains of pasture grasses and legumes, which were from different 'niches' within the same climate. Such complex environmental interactions are mirrored in the suites of species to be found growing in them, and both agronomists and ecologists have long used such suites of species to classify environments. Common guinea grass (*Panicum maximum*) is to be found in wet tropical areas. When in association with Centro (*Centrosema pubescens*) it is on a fertile soil, when with Stylo (*Stylosanthes guianensis*) the soil is less fertile. If homologous areas can be defined in these terms, especially in conjunction with some indication of their suitability for the different crop plants, this greatly aids the agronomist, especially when dealing with 'new species' whose environmental requirements are virtually unknown.

We begin by denoting the general types of climate for the various collecting zones mentioned (Figs 4.2 and 4.4, using the climatic classification devised by Papadakis (e.g. Papadakis, 1966)). This system, which separates zones in terms of their ability to support the growth of suites of crop plants, has been proved useful for the present purpose in earlier work (Burt and Reid, 1976), and demonstrates the wide range of environments involved. Floristic and long-term plant evaluation data (*sensu* Hartley as reanalysed in Williams and Burt, 1982) are then used to more closely define homologous regions. Examples presented here are drawn from Ethiopia, the USA and Australia.

4.4.1 Comparisons based on the use of suites of crops

The climates of the various sites of interest are listed in Table 4.3 (see Figs 4.1 and 4.4 for collecting zones, as defined by Papadakis (1966)). As far as is practicable, the order in which they are presented follows the north–south transect through the collecting zones of particular interest in Argentina (Fig. 4.2). This proved to be informative in terms of the distribution of the *Desmanthus* species. For the sake of brevity, only occasional comments are appended about the suites of crop plants suited to the environments mentioned. If further details are required, they can be obtained from Papadakis (1966).

The *Desmanthus* collections came from very different temperature regimes, ranging from subtropical and 'tierra templada' through various types of 'tierra fria' (non-frostless tropical highland) to 'Pampean'. There were no collections recorded from subtropical desert regions in Argentina. More moderate 'temperate templada' regions, also at high altitudes, are of particular interest for floristic reasons (see next section). With the sole exception of the desert regions, each of these major climatic types contains both collection and global potential receptor areas identified on the basis of climatic data alone. As some genotypes of *D. virgatus* have a proven ability to thrive over a range of environmental conditions (see section 4.3), it seems likely that at least some of the material identified will be useful in new receptor areas.

4.4.2 Floristic/agronomic comparisons

Using floristic data from highland conditions in tropical East Africa

Tropical East Africa is itself an important source of both tropical forage grasses and legumes. It is, for instance, a centre of diversity for true clovers (*Trifolium* spp., Adams and Pipoly, 1980), and also for the subtropical species *Neonotonia wightii* (syn. *Glycine wightii*, Glycine), one of the first tropical legumes to be developed for commercial use as a pasture species. It has provided several cultivars. The highest altitude zones of interest in tropical East Africa are located in the more temperate

Table 4.3. Climates of origin[+] of accessions deemed to be of interest.[++]

		Country					
		Acquisition sites		Receptor sites			Acquisition sites
Climatic type		Mexico	Argentina	Ethiopia	Other African countries[+++]	Australia[+++]	USA
1. Tropical							
1.7 Humid Tierra Templada							
	1.77			Ambo, Axum, Bako			
1.8 Dry Tierra Templada							
	1.81				Moshi, Tanzania		
	1.84			Harar			
	1.85				Dodoma, Tanzania		
2. Tierra Fria (non-frosted highlands)							
Semi-tropical	2.12				Eldoret, Kenya		
	2.14	Oaxaca City[a]					
	2.15	Queretaro[a]					
Low Tierra Fria	2.21		Salta[b]				
	2.25					Toowoomba[c]	
	2.26		San Salvador de Jujuy[b]				
	2.27		Tucuman[b]				
	2.28	Chihuahua City[a]					
	2.29	San Luis de Potosi[a]					
Medium Tierra Fria	2.32			Addis Ababa			
	2.35			Asmara			
High Tierra Fria	2.43	Zacatecas City[a]					
3. Desert							
Hot, subtropical	3.26	[+++]	Catamarca[b], La Rioja[b]			Longreach[c], Isisford[c]	Winslow[d]
4. Subtropical							
Humid	4.123						Galveston[d]
Monsoon	4.14					Lismore	
	4.22					'Havilah'[c], 'Trafalger'[c]	
Hot, semi-tropical	4.31	[+++]				Hughenden[c], Blackall[c]	

Continued

Table 4.3. Continued.

		Country					
		Acquisition sites		Receptor sites			Acquisition sites
Climatic type		Mexico	Argentina	Ethiopia	Other African countries[+++]	Australia[+++]	USA
Semi-hot	4.33		Oran[b]			Charters Towers[c]	
Semi-hot, semi-tropical	4.44					Atherton, Gympie	
Semi-Mediterranean	4.51					Brisbane	
5. South Pampean							
Subtropical	5.36						Corpus Christi[d]
	5.38						Austin, Houston, San Antonio
Monsoon peri-pampean	5.61		Cordoba[b]				
Semi-arid peri-pampean	5.76	+++					Laredo[d], San Angelo
6. Steppe							
Warm	9.12						Fort Worth[d], Wichita Falls
Semi-arid continental	9.71						Lubbock
	9.72						Amarillo

[+]As defined by Papadakis (1966). [++]As shown in Tables 4.3 and 4.4 and noted under Mexico and Argentina. Data for other countries are used in preliminary and Discussion. [+++]Climatic data for collecting zone A in Mexico (Fig. 4.3) are scanty and the transect taken passes from low to high altitudes (Table 4.2) and various rainfalls. It starts in subtropical climate 4.31, passes into a much drier, higher zone (climate 3.26) before encountering a semi-arid and peri-Pampean climate in Texas.

[a,b]For location in relation to collection zones see Figs 4.1 and 4.3. [c]Evaluation sites for *Desmanthus* as described in Chapter 7 (this volume). [d]Areas in which various *Desmanthus* species are to be found in Texas, USA (Luckow, 1993).

'tierra templada' situations, while others are in the non-frostless 'tierra fria' (see Table 4.3).

Ethiopia has large areas of tropical highlands, often with clay soils, and these are of tremendous agronomic significance. The provision of adapted pasture legumes for use in both pasture and cropping systems is seen to be of major importance. At the lowest altitudes below 1800 m, using data extracted from Mengistu (1985), both tropical and subtropical species are adapted (Table 4.4) (see next section and information presented in Cook *et al.*, 2005). *Neonotonia wightii* is commonly found in this region. The pasture grasses recommended for use (Table 4.4) are African, but are under-utilized throughout the tropical world. All of the pasture legumes recommended – excluding *Cajanus cajan*, which is primarily a pulse legume – are introduced cultivars, originally from South or Central America but developed for use commercially in Australia, with *Desmodium uncinatum* and *Macroptilium atropurpureum* being best adapted to subtropical conditions (Cook *et al.*,

Table 4.4. Highlands of Ethiopia: native and sown grasses and legumes at various altitudes.[a]

Native pastures		Recommended for sowing	
Grasses	Legumes	Grasses	Legumes
<1800 m		<2000 m	
Aristida spp. *Bothriocloa* spp. *Chloris* spp. *Cynodon* spp. *Heteropogon* spp. *Hyparrhenia* spp. *Setaria* spp.	*Indigofera spicata* *Neonotonia wightii* *Acacia nilotica*	*Cenchrus ciliaris* *Chloris gayana* *Panicum coloratum* *Panicum maximum* var. *trichoglume* *Pennisetum purpureum*	*Desmodium* *uncinatum*[b] *Leucaena* *leucocephala*[b] *Macroptilium* *atropurpureum*[b] *Stylosanthes* spp.[b] *Cajanus cajan*[c]
1800–2000 m			
Occur here and at lower altitude			
Chloris spp. *Cynodon* spp. *Hyparrhenia* spp. *Setaria* spp.	*Indigofera spicata* *Neonotonia wightii*		
Occur here but not at lower altitude			
Andropogon spp. *Cenchrus* spp. *Cymbopogon* spp. *Eleusine* spp. *Eragrostis* spp. *Paspalum* spp. *Pennisetum* spp.	*Crotalaria* spp. *Desmodium* spp. *Medicago* spp. *Rhynchosia* spp. *Sesbania* spp. *Stylosanthes fruticosa* *Trifolium rupellianum* *T. semipilosum* *T. steudneri* *Vigna* spp.		
2000–3000 m		2000–2400 m	
Occur here and at lower altitudes		Suitable for here and lower altitudes	
Andropogon spp. *Cynodon* spp. *Eragrostis* spp. *Paspalum* spp. *Pennisetum* spp. *Setaria* spp.	*Medicago* spp. *Trifolium rupellianum* *T. semipilosum* *T. steudneri*	*Chloris gayana* *Panicum maximum* *var. trichoglume*	*Cajanus cajan*[c]
Occur here but not at lower altitudes		Not suitable for lower altitudes	
Brachiaria spp. *Festuca* spp. *Sporobolus* spp.	*Lotus* spp. *Trifolium burchellianum* *T. decorum* *T. polystachyum* *T. quatianum* *T. tembense*	*Festuca arundinacea*[d] *Phalaris* spp. *Setaria anceps*	*Medicago sativa*[c] *Sesbania* spp.
		2400–3000 m	
		Suitable for here and lower altitudes	
		Festuca arundinacea[d] *Phalaris* spp.	
		Not suitable for lower altitudes	
			Medicago spp.[c] Indigenous *Trifolium* spp. *Vicia* spp.
>3000 m		>3000 m	

Continued

Table 4.4. Continued.

Native pastures		Recommended for sowing	
Grasses	Legumes	Grasses	Legumes
Occur here and at lower altitudes			
Andropogon spp. *Festuca* spp.	*Trifolium burchellianum* *T. polystachyum* *T. tembense*		
Occur here but not at lower altitudes			
Agrostis spp. *Poa* spp.	*Trifolium acaule*		

[a]Data abstracted from Mengistu (1985). Species originally from: [b]South or Central America; [c]Asia; [d]Europe.

2005). Initial screening and seed multiplication of lines of *Desmanthus* spp. in 1987–88 at 1600 m elevation, and at Soddo at 1950 m on an acid nitosol (pH 5.5 to 6) under rainfed conditions (1050 mm) identified seven *Desmanthus* accessions with particular promise and of these four had been collected in the Yucatan, Mexico and Belize (*D. leptophyllus* CF 764, and the *D.* spp. CF 448, CF 711, CF 896) and one in Curaçao.

In the next higher altitude range (1800–2000 m), *Neonotonia* continues to be found naturally but to it is added a range of cryophilous legumes, notably various species of *Trifolium* and *Medicago*. These latter species also persist at high altitude (2000–3000 m) where a wide range of other *Trifolium* species also occur. Lucerne (alfalfa, *Medicago sativa*) is a more temperate species whose environmental requirements, and those of the temperate European grass *Festuca arundinacea* (tall fescue), are well documented in the literature. The well-known temperate pasture grasses *Lolium perenne* (ryegrass) and *Dactylus glomerata* (cocksfoot), both of European stock, persist up to 3000 m. *Neonotonia wightii* did not occur naturally at these elevations and was not recommended for use in areas above 2400 m. Rhodes grass (*Chloris gayana*), a native species, was not used as a sown pasture grass above 2400 m or found occurring naturally above 3000 m. These grasses are simply not adapted to the higher elevations. The significance of these observations in relation to the potential adaptation of selected *Desmanthus* species is explored below.

Using plant introduction/evaluation data from Australia

Low fertility soils are characteristic of most Australian environments and much of the agriculture has depended on nitrogen from leguminous crops. The commercially available forage legumes are almost entirely those developed from introduced species. For tropical and subtropical regions this began before the Second World War, mainly with African grasses and South American legumes via plantation agriculture elsewhere. Among these were the subtropical legumes *Neonotonia wightii* (Glycine) from Kenya and *Desmodium uncinatum* (Silverleaf Desmodium) from Brazil. Other legumes introduced later included *Desmodium intortum* (Greenleaf Desmodium) from Central America, *Trifolium semipilosum* from Kenya, and an expanded range of *Neonotonia* germplasm from Tanzania, Malawi and South Africa. These plants have long since passed into commercial usage and it is thus possible to characterize the various environments in which they are planted in Queensland in terms of the suites of introduced grass and legume forage species that are adapted there (Williams and Burt, 1982). These can then be equated with similar species groupings in Africa and thus denote homologous areas.

Much of the work on *Desmanthus* has also been carried out in Australia (Chapter 6, this volume) and as noted earlier attempts have been made to develop material for use in more subtropical conditions (Cook *et al.*, 2005; Jones and Rees, 1997) where *Trifolium* and *Medicago* species were the main genera to be used (the latter on more alkaline, heavier soils). This again relates to the changes in legume floras, which occur with increasing altitude in Ethiopia, and homologous situations can again be sought. Relevant information is presented in Fig. 4.4.

In the 'tierra templada' situation around Atherton in Queensland (see Table 4.4, A in Fig. 4.4), the main adapted exotic legumes are *Neonotonia wightii*, *Desmodium uncinatum*, *D. tortuosum*, *Trifolium repens*

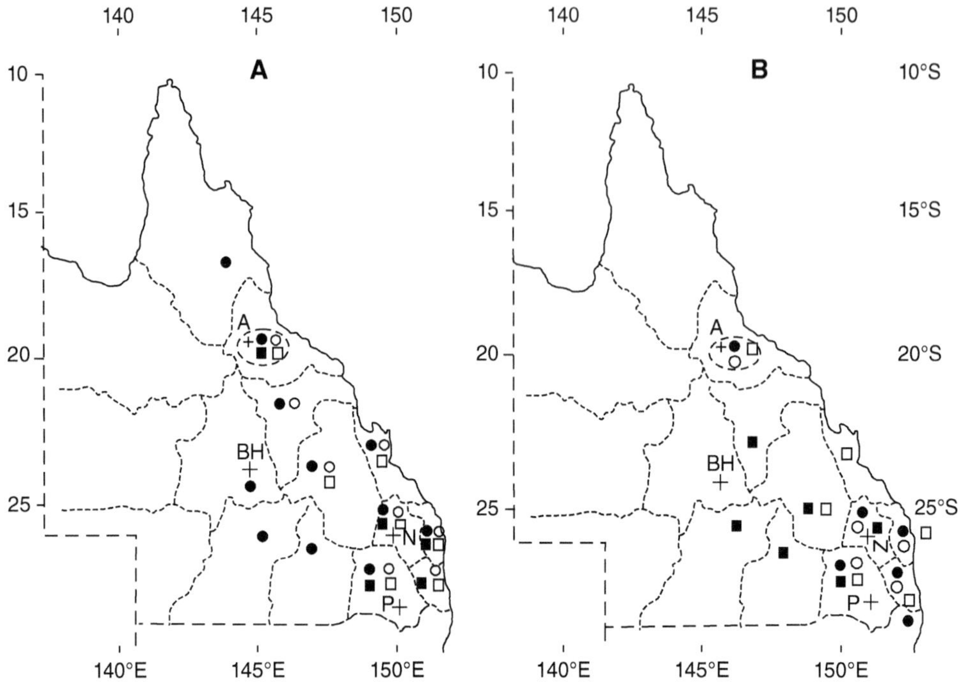

A. Distribution of sown pasture grasses in the various pastoral districts

- ● *Cenchrus ciliaris* 'Buffel Grass'
- ■ *Festuca arundinacea* 'Tall Fescue'
- ○ *Chloris gayana* 'Rhodes Grass'
- □ *Lolium perenne* 'Perennial Ryegrass'

B. Distribution of sown pasture legumes in the various pastoral districts

- ● *Neononia wightii*, 'Glycine'
- ○ *Desmodium uncinatum* and *D. intortum*, ('Silverleaf' and 'Greenleaf Desmodium')
- ■ *Medicago truncatula* 'Barrel Medic' and *M. scutellata* 'Snail Medic'
- □ *Trifolium repens* 'White Clover'
- +A Atherton – Centre of the Atherton Tablelands with a climate similar to that found at higher latitudes (see Table 4.3). The cryophilous grasses and legumes in the pastoral district in general are found on this tableland.
- +BH Blackall, in the southern portion of the hot semi-tropical region (Table 4.3)
- +P Pitsworth, near Toowoomba in a 'terra fria' climate (Table 4.3). The testing site for Jones and Rees (1977).

Fig. 4.4. The adaptation to and use of selected pasture legumes and grasses in Queensland, Australia. Data abstracted from Williams and Burt (1982) and originally taken from Anon (1979).

and occasionally *Medicago* sp. Exotic grasses include *Cenchrus ciliaris* (Buffel grass), the more subtropical *Chloris gayana* (Rhodes grass), and the temperate species *Festuca arundinacea* (Tall Fescue) and *Lolium perenne* (Ryegrass). The same suites of species are encountered at lower altitudes but higher latitudes (around 25°S) in coastal regions north of Pittsworth (P) and west of Narayen (N). These sites are very similar to the Ethiopian situation below 2000 m (Table 4.4) where *Neonotonia wightii* is native and the recommended exotic forage species include *Desmodium uncinatum* from South America combined with the grasses *Cenchrus ciliaris* and *Chloris gayana*. In these regions *Desmanthus virgatus* has shown great promise on clay soils (see section 4.3), more especially CPI 78382 from a higher altitude/latitude situation, while the accession from lowest altitudes failed.

At high elevations in Ethiopia, where the environments are more temperate, *Medicago* and

Trifolium species become more important in the native vegetation (Table 4.4) and grasses and legumes from Europe (*Festuca arundinacea, Lolium perenne* and *Medicago sativa*) are recommended for use. These species are also used in coastal and sub-coastal Queensland, where the more temperate *D. virgatus* accession CPI 78382 has proved to be well adapted. Accessions from these situations (see section 4.4) merit attention in Ethiopia. Those from highest altitudes and latitudes, which include Pampean type climates (Table 4.4), could be pre-adapted to even higher altitudes (>2400 m, Table 4.4). Some of the species concerned, *D. illinoensis, D. leptolobus, D. velutinus* and *D. acuminatus*, have been tested for use on blackland prairie soils on Pampean environments in the USA (Muir *et al.*, 2005).

4.5 Discussion

The genus *Desmanthus* contains a very wide range of plant forms and these, combined with the high nutritional value of its seed (e.g. Schlink and Burt, 1993), give it a wide range of potential uses. *D. pubescens* and *D. virgatus* have been referred to as 'tree legumes' or 'browse plants'. Other species and phenotypes are useful as 'hedge' lucernes or 'protein banks', 'ley legumes' in cropping systems and as persistent components in long-term pastures. *D. illinoensis* or 'prairie mimosa' is seen as a potential perennial pulse crop, the development of which would reduce tillage and fertilizer use (Kulakow, 1999; Kulakow *et al.*, 1990). Many members of the genus, moreover, thrive on clay soils, areas of high importance for crop production, but with few adapted tropical legume cultivars (Chapter 6, this volume, and see information presented in Cook *et al.*, 2005).

Despite its potential, *Desmanthus* has received less attention than *Stylosanthes* for example, a genus that has provided the bulk of the newly domesticated species for use in the tropics (Burt and Williams, 1975; Chakraborty, 2004). In *Advances in Legume Science* (Summerfield and Bunting, 1980), which reports on the proceedings of an international legume conference at Kew Gardens, there are only two references to *Desmanthus* and neither of these is known to be entirely correct. A comparison of Google searches for *Stylosanthes* forage and *Desmanthus* forage yielded 21,700 vs 4300 sites respectively. In addition, the lack of taxonomic review of the genus until 1993 (Luckow, 1993) greatly hindered meaningful communication and successful development of the genus. As noted earlier, germplasm collections are inadequate. Although some effort has been made to alleviate this deficiency for tropical lowland areas, it would seem that this is not the case for higher altitudes or latitudes.

The potential of the genus has been demonstrated in recent years as promising accessions and cultivars from the available germplasm have been spread internationally. Alan Robertson (New South Wales, 2009, 2014, personal communication) reported that in Ethiopia *D. leptophyllus* and *D. pernambucanus* were looking very promising in Afar on occasionally irrigated sites. It has been excellent around Alaba, on the west side of the Rift, on the way to Soddo, in West Gojam, and on the north side of the Blue Nile gorge, in Welo, where some stands have persisted over a long period. *D. leptophyllus* has great potential in Eritrea and the Sudan around Gezira, in Cambodia and Thailand where it is productive and widely naturalized on roadsides and common land. In India it is being actively promoted and is common in the south. *D. virgatus, D. leptophyllus* and *D. pernambucanus* have some promise in Nepal up to 1000 m in lower rainfall areas. It is also successful on black clays in Uganda (Emmanuel Zziwa, Uganda, 2013, personal communication).

Early indications of the productivity and nutritive value of *Desmanthus* were obtained under the IDRC/UWI forage programme in Antigua over 1974–1982 under about 1100 mm annual rainfall when 14 accessions of *Desmanthus* were evaluated. During that period a simple trial to assess the potential of promising forage species and accessions was undertaken over one year under 830 mm rainfall which involved four harvests on shallow and deep high pH calcareous clays. The yield data which was analyzed using the diagnostic programme Grouper gave 10 groups with *Leucaena leucocephapla* cv. Cunningham in the highest ranking group, and *Clitorea ternatea, D. leptophyllus* CPI 55719, *D. virgatus* CF 649 and *D. pubescens* CF 543 in the next most productive group. While *Desmanthus* was subsequently recommended for fodder production *Leucaena* remained the option utilized by farmers due to its usually higher productivity and multiple uses (John Keoghan, New Zealand, 2013, personal communication).

It is indisputable that plant biodiversity is being eroded (e.g. Maxted and Guarino, 2006) and geographic information systems are doing much to delineate areas for conservation. Luckow (1993) shows that some *Desmanthus* species of great potential value are at imminent risk. For instance, *D. oligospermus* is restricted to a very small geographic range, the tip of the Baja California peninsula, and is said to be uncommon. It was not sighted or collected by Reid (1983), despite his successful collection of other species. Unusually in the genus, *D. oligospermus* and some *D. illinoensis* phenotypes (Kulakow, 1999) possess indehiscent seeds, a feature of huge importance in the development of perennial pulse crops (Smartt, 1976). *D. interior* 'widespread but uncommon ... is intensively grazed which probably contributes to the rarity of the species' (Luckow, 1993). The loss of such variation, due especially to the overgrazing of palatable plants and the expansion of cultivated land, is now being exacerbated by climate change.

In the initial stages of development of forage germplasm collections for use in tropical areas little was known of the species or genera of potential value (e.g. Burt and Williams, 1975; Chapter 6, this volume) or the regions or climates in which to collect them. But this is no longer so, and in the case of *Desmanthus* specific collecting areas can be identified and preliminary examination of the areas concerned be conducted on computer-based imagery such as Google Earth. In addition, in the USA, plant distribution is well-documented and relevant bodies established for conservation and seed exchange. The development of more comprehensive collections of the genus and of relevant characteristics and evaluation programmes is a matter of great importance to better utilization and productivity of tropical soils. This in turn will result in better livelihoods for smallholder farmers in such areas.

5 A Numerical Analysis of Variation Patterns in the Genus *Desmanthus*: An Exploratory Study

R.L. Burt† and J.R. Lazier*[1]
**Formerly International Livestock Centre for Africa*

Abstract

Desmanthus is the genus with most potential for having productive, persistent genotypes for development as forages for clay soils. An exploratory pattern analysis of the morphological and agronomic characteristics of 35 accessions in a collection of *Desmanthus* species from a wide range of latitudes was undertaken. The resulting groupings seemed to correspond to species or groups within species. The results were discussed with relation to variation within and between species, geography and the environments in which the collections were made. Recommendations are made as to the species of potential requiring further collection and development (e.g. *D. leptophyllus*, *D. tatuhyensis*, *D. virgatus* among others), and the approach to be used.

5.1 Introduction

Desmanthus is, arguably, the most likely source of pasture legumes for use on clay soils in tropical and subtropical environments (Chapter 4, this volume; Burt, 1993a; Cook *et al.*, 2005), however it is poorly known. In 1978, when *Stylosanthes* had already yielded cultivars of the 'new' species *S. hamata* and *S. scabra*, an authoritative conference on legume science (Summerfield and Bunting, 1980) contained only two brief references to *Desmanthus*. One of these had incorrect information, stating that it had the 'ability to nodulate effectively with a wide range of [rhizobial] strains'. The other paper referred to *D. virgatus*, which the plant in question may not have been. Initial collections of the genus were undertaken in the Yucatan, where the species of recognized Australian potential were collected in a range of environments by the University of the West Indies/International Development Research Council of Canada project in the 1970s (*D. pubescens*, *D. virgatus* and *D. leptophyllus*) (Chapter 2, this volume), as well as by Reid about the same time.

Indeed a taxonomic revision of the genus was not published until 1993 (Luckow, 1993). There is still some confusion about the genus because it is viewed variously as a tree legume, a small shrub to be used in alley farming or as a grazing-resistant plant to be used in semi-arid rangeland areas. One species, *D. illinoensis*, is being considered for development as a pulse crop. Each potential use requires different evaluation strategies.

The genus lacks general-purpose descriptions on which agronomists or others interested in developing or utilizing germplasm collections of the genus can draw (e.g. Burt, 1993a; Pengelly and Liu, 2001). This is a problem commonly encountered when dealing with poorly known

[1] E-mail: john.lazier@sympatico.ca

tropical genera, such as the now familiar *Stylosanthes*. In the latter case, pattern-seeking methods, integrating both morphological and agronomic plant attributes, provided a solution (Burt, 1983). As germplasm collections improved and hierarchical classificatory systems became less appropriate, the minimum spanning tree was developed (Burt *et al.*, 1983). As the *Desmanthus* collection increased from 69 accessions in 1979 to 272 in 1984, three such analyses were carried out to aid plant evaluation. The values of some agronomic attributes were dependent on seasonal conditions, so these were not consolidated into one analysis. Nor, in the absence of a taxonomic revision, was it sensible to allocate species names (although an in-house key was compiled and later published; Burt, 1986).

Of the three analyses, the first, undertaken in 1980, is of most interest here. It contains reasonable representation of the species of major interest agronomically and illustrates the type of variation contained and the problems faced by would-be users of the germplasm collections. In validating the meaningfulness of the various accession groupings (Burt, 1983; Burt *et al.*, 1983), the relevant information can be drawn from taxonomic identification (based on Luckow, 1993) by a variety of authors such as Pengelly and Liu (2001) and Gardiner *et al.* (2004) or from the complete data sets assembled for the later classifications (Burt and Beattie, unpublished data). The utility of this particular part of the collection, and the need for further studies, is considered in section 5.4.

5.2 Materials and methods

5.2.1 Experimental methods

The accessions used in the experiment are listed in Table 5.1. At the time most of these were placed in the *D. virgatus* hold-all (e.g. Turner, 1950; Luckow, 1993), and the subject of their taxonomic identity is discussed subsequently.

Seed was sown into small peat pots in December 1980 and the seedlings were raised in a shade house of the CSIRO Davies Laboratory, Townsville, Australia. A mixture of *Bradyrhizobia* strains was watered onto the seedlings. In January 1981 they were transplanted into a spaced plant field experiment located at the Lansdown Field Station, near Woodstock in the dry tropics (latitude 19.5°S, longitude 146.8°E). The experiment was sown into a prepared seedbed into which adequate levels of super phosphate had been incorporated; spray irrigation was used to ensure good survival and growth. The accessions were replicated three times, each replicate consisting of four spaced plants, and individual replicates were randomly located within each of three blocks.

One of the major features of agronomic interest was persistence, then poorly documented for the genus as a whole. Previous experience with *Stylosanthes* had shown this to be a useful discriminatory factor both within and between species. However, unlike *Stylosanthes*, all *Desmanthus* species are perennial, with a well-developed taproot system to ensure survival. Under defoliation, different types of crown and productive structures could be developed (Burt, 1986). To promote differences in persistence and crown development, half of the plants in each replicate were twice trimmed near the end of the first growing season (170 and 190 days after sowing). On each occasion any growth more than 7.5 cm from the point of emergence from the main stem from the soil was removed, dried and weighed (N_5 and N_6 Table 5.2). Other agronomic attributes were measured on the un-defoliated plants from which pressed specimens were taken for morphological measurements (Table 5.2).

This first analysis was exploratory, seeking to find patterns of variation at a time prior to the publication of a taxonomic review. Only characteristics reflecting readily observable differences were recorded and the total number, 21 (Table 5.2), is much less than that for *Stylosanthes* (which had 40). The characteristics were of mixed type, so pattern-seeking methods rather than statistical methods were appropriate for clarifying the patterns and the extent of variation present. As is usual, mean values were used in the analysis.

5.2.2 Numerical methods

Details of the methods used and of the interpretational requirements necessary are presented elsewhere (Chapter 3, this volume). Briefly, the data were first subjected to a hierarchical 'cluster

Table 5.1. Accessions used in the experiment.

Treatment no.	CPI no.	Origin Country	Lat.	MAR	Treatment no.	CPI no.	Origin Country	Lat.	MAR
1	CQ1139	–	–	–	36	68882b	Brazil	–	–
2	16672	–	–	–	37	68883	Brazil	–	–
3	18005	–	–	–	38[a]	70338	Antigua	17 N	–
4	21825	–	–	–	39	71012	Brazil	16 S	800
5	25840	Uruguay	–	–	40	73463	Antigua	17 N	–
6	27732	Brazil	–	–	41	73464	Curaçao	12 N	–
7	28909	via Malaysia	–	–	42	73465	Curaçao	12 N	–
8	30205	via India	–	–	43	73466	Antigua	17 N	–
9[a]	33462	Guyana	07 N	–	44	73467	Antigua	–	–
10	33486	Venezuela	10 N	–	45	76052	Mexico	21 N	940
11[a]	37143	Mexico	19 N	–	46[a]	76053	Mexico	21 N	900
12	37481	Uruguay	34 S	–	47	76054	Mexico	20 N	1350
13	37483	Uruguay	34 S	–	48	76055	Mexico	20 N	1150
14	37484	Uruguay	34 S	–	49[a]	76057	Belize	18 N	1400
15	37485	Uruguay	34 S	–	50	76061	Mexico	18 N	1250
16[a]	37538	Argentina	31 S	–	51	76064	Mexico	20 N	1150
17[a]	38351	Venezuela	09 N	–	52[a]	78369	Argentina	24 S	700
18	38403	Venezuela	–	–	53	78370	Argentina	24 S	750
19	38820	Venezuela	–	–	54	78371	Argentina	23 S	700
20[a]	40071	Brazil	–	–	55[a]	78372	Argentina	23 S	680
21	40072	Brazil	–	–	56[a]	78373	Argentina	23 S	650
22	40073	Brazil	31 S	–	57	78374	Argentina	29 S	1150
23[a]	49788	Brazil	–	–	58	78376	Brazil	–	–
24	51382	Venezuela	10 N	–	59	78377	Brazil	–	–
25	55718	Brazil	11 N	–	60	78378	Brazil	–	–
26[a]	55719	Venezuela	11 N	586	61	78379	Argentina	29 S	1170
27	55975	Mexico	–	600	62[a]	78380	Argentina	25 S	690
28[a]	57960	Curaçao	12 N	750	63	78381	Argentina	24 S	650
29	60474	Brazil	–	–	64[a]	78382	Argentina	24 S	700
30	60794	via India	–	–	65	78383	Argentina	24 S	1000
31	63453	Brazil	05 S	–	66	787385	Argentina	28 S	1100
32	65947	Ecuador	02 S	500	67[a]	79653	Cuba	20 N	600
33[a]	67642	Guatemala	15 N	550	68	79654	Cuba	20 N	1200
34[a]	67643	Guatemala	15 N	500	69	79839	Colombia	–	–
35	68882a	Brazil	–	–					

CPI, Commonwealth Plant Introduction number; Lat., latitude in degrees north or south; MAR, mean annual rainfall (mm). [a]Used in later evaluation programmes (Chapter 9, this volume).

analysis' using a strategy that has been proven to be of value for this purpose and whose characteristics have been well studied (Abel and Williams, 1985). This showed rapid fragmentation of the resulting plant groups below the eight group level and it was therefore truncated at this number. The characteristics of the groups were then documented (Table 5.3) and the relative importance of the attributes that had contributed to their delineation determined (by using the programme GROUPER). In the second part of the analysis the accessions were located on a minimum spanning tree (Burt and Williams, 1979b), because the more useful programme NEBALL (Abel and Williams, 1981; Burt *et al.*, 1983) was not available then.

5.3 Results

The dendrogram of the hierarchical classification has been presented previously and does not

Table 5.2. Summary of the data recorded. Type of character: N = numerical; Q = quantitative (binary); M = multi-state.

Morphological characters				**Agronomic characters**			
1. Seed characters				Growth from first defoliation	g/plant	N_6	
Pod length	mm	N		Regrowth from second defoliation	g/plant	N_7	
Pod width	mm	N		Time to first anthesis	days	N_5	
Beak on pod		Q	0 absent; 1 present				
Length of beak	mm	N		Persistence	%	N_4	
Seeds/pod		N					
Seed colour		N	1 light; 2 dark				
Weight/10 seeds	mg	N					
2. Leaf				**Productive structures**			
No. pairs of leaflets per pinna		N	A leaf consisting of paired pinnae, each with numerous pairs of leaflets or pinnules	Habit		M	1. Erect 2. Spreading/ ascending 3. Prostrate
No. pairs pinnae		N		Position of lateral branches		M	1. From near the base of the main stem 2. Evenly spread on main stem
Stipule length	mm	N					
3. Stem							
Thickness	mm	N_1					
Length	mm	N_2					
Crown branching below 5 cm		N_3					

warrant representation (Burt, 1993a). It is sufficient to note that the seven dichotomies used to provide the eight groups required only nine attributes, five of them relating to the pod and none to crown branch number, persistence or yield. Two of the groups were separated from each other on the basis of a red mite infection, obtained in the shade house. The accessions concerned were very similar in other characteristics. These have been bulked up and the attribute means of the remaining seven groups listed in Table 5.3. The results of the minimum spanning tree (MST) analysis are shown in Fig. 5.1.

There is very good general agreement between the results obtained by the classification and those illustrated on the MST. Beginning on the right hand side of the main axis, there is a decline in plant height and change in plant habit until the prostrate plants of groups IV to VI are encountered. Group VII is a non-conformist group, whose members are different from everything else and each other. They are suitably isolated from each other and from other members of the collection. The prostrate groupings V and VI tend to overlap and there is some species admixture. *D. acuminatus*, *D. virgatus* and *D. tatuhyensis* appear to be very similar and this is in accordance with the observations of Luckow (1993). Similarly *D. paspalaceus* is noted to be similar to some Argentinean specimens of *D. virgatus* and

Table 5.3. Attribute means for the classificatory groups.

	Group						
Attribute	I	II	III	IV	V	VI	VII
Habit*	1	2	3	4	5	5	5
Height of main stem (cm)	133	75	97	65	5	5	72
Days to first anthesis	96	81	92	138	78	120	92
Pinnae (no. of pairs)	4.7	3.4	8.0	7.5	3.9	3.0	3.7
Leaflets (no. of pairs)	23	20	31	24	18	9	30

*Habit: 1 main stem erect, laterals evenly spaced on main stem; 2 ascending, laterals basal; 3 ascending, laterals evenly spaced on main stem; 4 main stem erect, laterals basal; 5 prostrate, laterals basal.

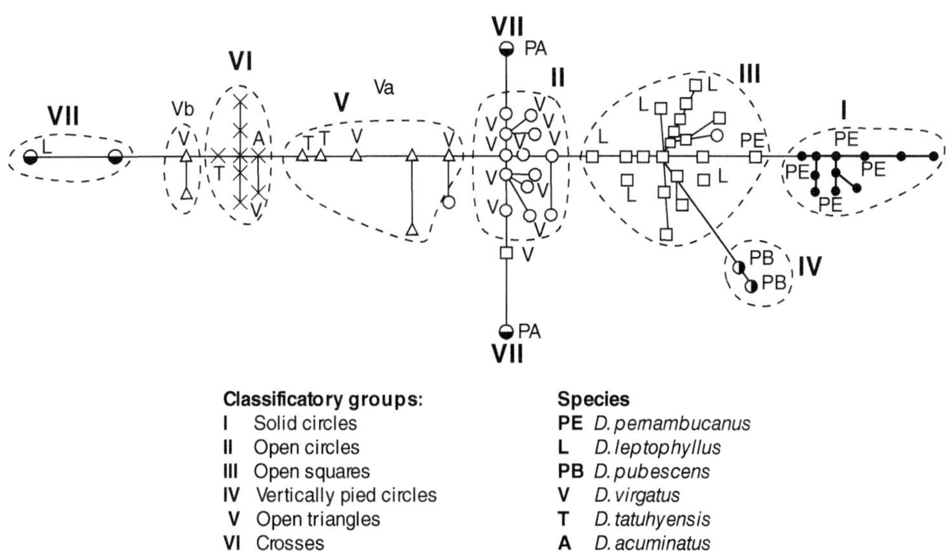

Fig. 5.1. Locations of the accessions on the minimum spanning tree, classificatory groups superimposed.

Luckow suggests that many of the specimens identified as *D. virgatus* are in fact *D. paspalaceus*. The two accessions included in this collection are most similar to, but quite different from the type II *D. virgatus*. Such differences as do occur between the results of the classification, the MST and the taxonomic identification of accessions, are themselves informative and examples of these are discussed in the following sections.

5.3.1 Validation and reinforcement of the groupings

After any objects have been classified and split into groups of like individuals, it is necessary to validate the classification to ensure that the groupings are meaningful. This is accomplished by the use of 'outside information', information that has not been used in the classification *per se*. Homogenous groups have characteristics shared by their members, for instance species of *Indigofera* tend to be poisonous and this gives them predictive value. In such a process the imposition of outside information gives a more holistic overview of the plants concerned (e.g. Burt *et al.*, 1983 for relevant genetic resource work with *Stylosanthes*).

It is impractical to document this process for all of the collection and for all of the information used, but some examples follow.

Taxonomic studies

The first exploratory study reported herein used only 20 plant attributes and was undertaken when no taxonomic review of the genus was available. Rather surprisingly, many of the accession groupings seemed to correspond to species or, in the case of *D. virgatus*, types within the species (Fig. 5.1). There were exceptions however, and each of these has been examined.

In her review, Luckow (1993) makes frequent use of the leaf characters, petiole length and number of pinnae and leaflets, especially when separating some of the taxonomically difficult species found in southern Brazil, Uruguay and northern Argentina. On the MST, *D. virgatus*, *D. acuminatus* and *D. tatuhyensis* from these regions often 'overlapped'. Similarly an accession *of D. pernambucanus* was placed in an intermediary position between groupings of *D. leptophyllus* and *D. pernambucanus* (Fig. 5.1). Using the data set collected for the latest classifications of the new parts of the *Desmanthus* collection (much improved, covering some 72 plant attributes), it was possible to explore the taxonomic status of some of the accessions in overlapping groups. It is impractical to present all of the relevant data, but for the purposes of illustration, information for seven of the accessions (the positions of which on the MST are shown on Fig. 5.2) is superimposed on a relevant set abstracted from Luckow (1993). To the leaf attributes mentioned above have been added that for staminate length, an apparently useful discriminatory character for the species in question.

First, the prostrate, low pinnae members (Table 5.3) from groups V and VI (Figs 5.1 and 5.2). These are plants numbered 1 (*D. virgatus*), 2 (*D. acuminatus*) and 3 (*D. tatuhyensis*). These accessions, which are bracketed for ease of comparison on the figure, overlapped on all of the leaf characters. Staminate length clearly separated only *D. tatuhyensis* and *D. acuminatus*. The values for individual attributes often lay at the extreme ends of the ranges, or even outside of them for the species, as observed by Luckow (1993). These species or species forms are indeed very similar and are difficult to meaningfully separate. In this the MST is in accordance with the findings of Luckow (1993) and those of Pengelly and Liu (2001), who based their findings on genetic similarities. There are, however, large differences between the plants in such agronomic features as flowering time (Table 5.3), and it is important that such differences be illustrated in classifications to be used in, for instance, selecting material for trial or commercial cultivar development (see later examples).

Finally plant 6 (Fig. 5.2), placed between groupings of *D. leptophyllus* (with plant 5) and *D. pernambucanus* (with plant 7), Luckow suggests that '*D. pernambucanus* differs from the other species of the *D. virgatus* complex (which includes *D. leptophyllus*) in having fewer pinnae and fewer, larger leaflets than either *D. leptophyllus* or *D. virgatus*.' As shown on Fig. 5.3, however, there is no definite distinction between the species in terms of the numbers of pinnae or leaflets, and this is also true when leaflet sizes are compared. When staminate length is considered (a feature that should clearly separate the species) the putative absence of staminodes would favour the

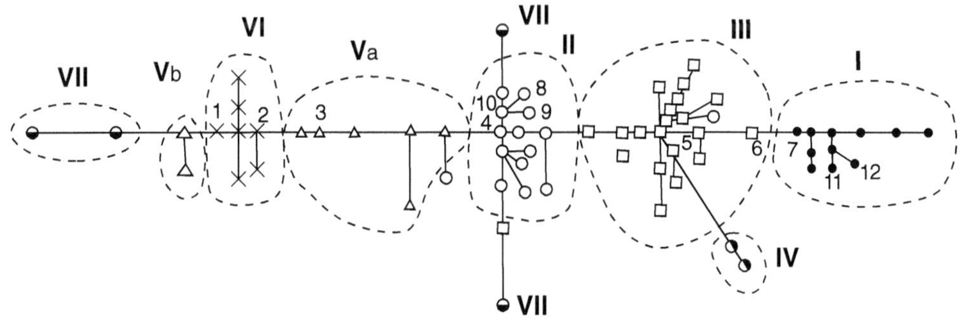

Fig. 5.2. Location of individual accessions considered in more detail in the text Symbols the same as for Fig. 5.1.

placement of plant 6 with *D. leptophyllus*. The MST analysis, which did not have access to information on staminodes, has drawn attention to an unusual form of *D. pernambucanus*.

Individual comparisons of accessions

In searching for pasture legumes for use on clay soils in dry tropical areas, particular attention was paid to three zones, all with semi-arid climates and clay soils. Three of these zones, B, C and G (Fig. 5.4), yielded accessions of *D. virgatus* and those from B and G will be considered subsequently.

Zone C, a small, atypically dry and isolated area in Guatemala, yielded only two accessions (plants 8 and 9, Fig. 5.3). It was surprising to find that they were deemed to be more similar to plant 10, collected from a very different soil in a much wetter area in the distant northern area of the Yucatan peninsula.

The three accessions have been compared using the 72 attribute data set available for the latest classifications (Burt and Beattie, unpublished data). Differences are listed in Table 5.4. Of the 72 plant attributes available for comparison, only 20 differed, many of them only slightly. The accession from the Yucatan differed primarily in pod characteristics. They had shorter peduncles, but were longer and less curved. However, all were of median length of pod and peduncle for the specific range given by Luckow (22–88 mm and 10–52 mm respectively). The accessions are indeed very similar.

The two Guatemalan accessions were, as indicated by the MST, more different from each other than they were from the Yucatan collection. Plant 8 morphologically had highest values for all of the numeric values associated with the legumes and also had differences in the shape and positioning of the nectary glands. Although Luckow records that *D. virgatus* is 'much branched from the base', it is clear that the accessions differ in this respect. Accession 8 has many lateral branches arising from the base, while accession 9 has fewer basal lateral branches, instead having them spaced more or less evenly up the main stem. Such differences can lead to different productive structures following grazing (Burt, 1986). It is noted for future reference that the basally branched plant was common and found interspersed with grass in a grazed pasture. The plant with less basal branching was from an area some 50-km away with a slightly higher rainfall, but from an ungrazed roadside.

GEOGRAPHICAL VARIATION. The previous section showed that very similar plants could be found in very different geographical and environmental situations. This is quite different for the situation reported for perennial species of *Stylosanthes* on the basis of morphological/agronomic studies (Williams *et al.*, 1980; Burt, 1983), and with chemotaxonomic methodologies (Robinson *et al.*, 1980; Liu, 1997). The geographic backgrounds of some of the plant groupings are explored further in Fig. 5.5.

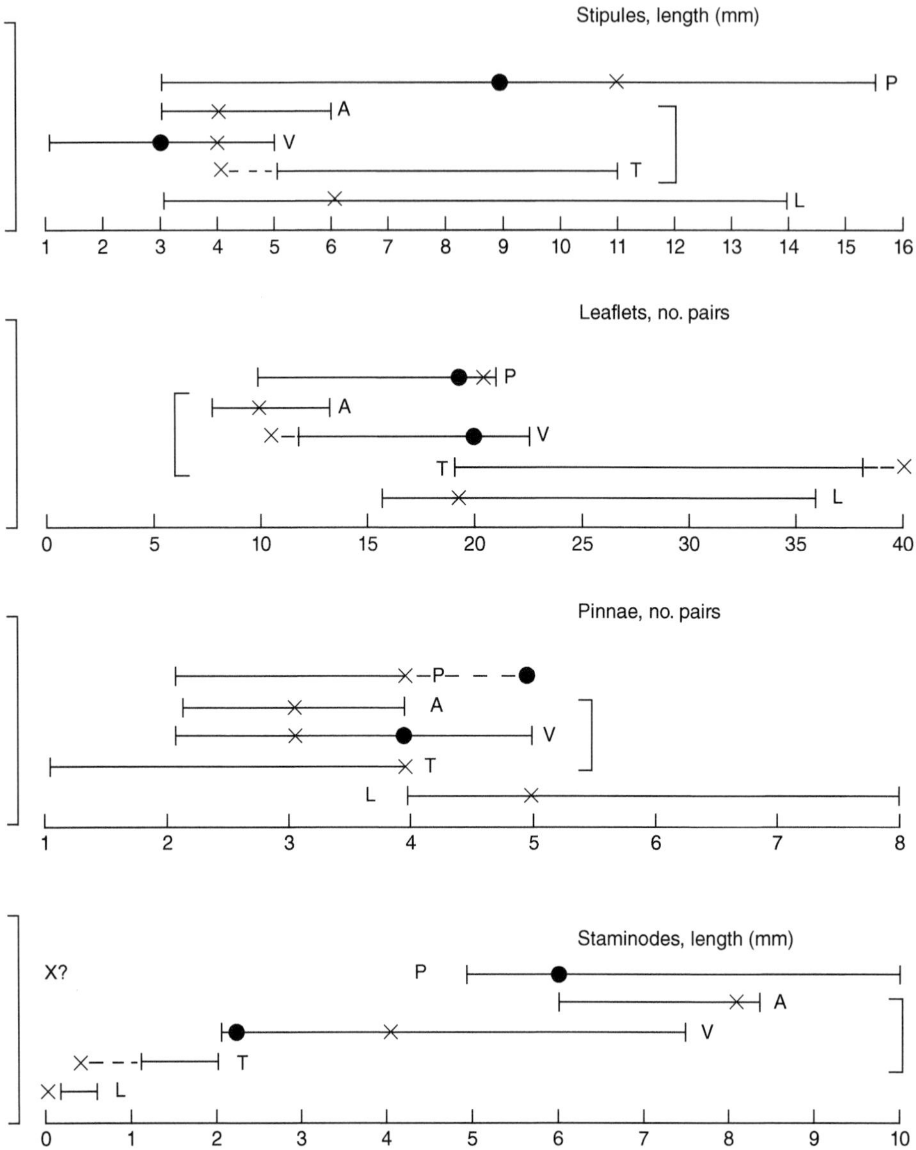

Species

PE *D. pernambucanus*
L *D. leptophyllus*
PB *D. pubescens*
V *D. virgatus*
T *D. tatuhyensis*
A *D. acuminatus*
P *D. pernambucanus*

Values for individual accessions indicated by x or ●

For plants (see Fig 5.2) *D. leptophyllus* - plant 5; *D. pernambucanus*- plant 7; *D. virgatus* - plant 1 (x), - plant 4●; *D. tatuhyensis* - plant 3; *D. acuminatus*- plant 2; *D. pernambucanus*- near *D. leptophyllus*, plant 6

Fig. 5.3. Species' limits for discriminatory attributes (quoted from Luckow, 1993) and values recorded for individual accessions

Fig. 5.4. Areas deemed most likely to yield legumes adapted to dry tropical conditions. To the original areas (A to H, Burt *et al.*, 1976; Burt and Reid, 1976) have been added the Balsas Depression, Zone Z in Mexico.

In group I *D. pernambucanus*, the bulk of the accessions shown were introduced from countries where they are not native, so they are given the suffix [1] (Fig. 5.5). The Antiguan accession was collected from an area near an airport runway and though the species is widely distributed throughout the Caribbean islands it could conceivably have been introduced into the area. All of the remaining accessions of the group were introduced from Brazil and, where relevant provenance records are available, were from coastal and sub-coastal areas in the north-east of that country. The Brazilian accession placed closest to that of the Antiguan one was found in a roadside area where it was favoured by a local grazier. It was likely to have been sown there and the species was not located again in the region. Luckow notes that the species is generally found in moist situations. The species has long been naturalized in India, South-east Asia and various islands in the Pacific and Indian Oceans, together with wetter tropical areas in Guyana, Surinam, Brazil and southern Africa. The distribution pattern is suggestive of purposeful plant transfer for use in the East and West Indies. The species contains little morphological/agronomic variation (Fig. 5.6) and is genetically less diverse than other groupings (Pengelly and Liu, 2001).

The next grouping, III, is of *D. leptophyllus* accessions. Unlike the situation with *D. pernambucanus*, these originated over a wide geographical range, from Brazil in the south to Mexico in the north. The central core is Brazilian, an area favoured for collection and seed exchange,

Table 5.4. A comparison of similar accessions of *D. virgatus*[a]

Attributes	Accession no.[b] 9	8	10	Attributes	Accession no.[b] 9	8	10
Morphological				Morphological			
Leaf				Pod			
Gland shape[c]	0	1	0	Peduncle length (mm)	34	30	16
Gland position[b]	1	3	1	Length (mm)	45	52	60
Stipule vestiture[e]	1	1	0	Curvature[g]	3	2	1
Stipule length (mm)	5	6	5	Colour[h]	1	3	1
Petiole length (mm)	2	3	3				
Pinnae (no. pairs)	3	4	3	Agronomic			
Pinnae, rachilla length (mm)	14	22	17	Habit[i]	3	2	2
Rachis, length (mm)	4	6	4	Crown branch no.[j]	12	19	12
Pinnules (no. pairs)	13	17	15	Stem thickness (mm)	6	6	5
				Time to flowering (days)[k]	79	80	82
Floral[f]				Seed weight (mg/100 seeds)	48	51	44
Florets (no./head)	11	8	4				
Stamens, length (mm)	4	5	3				

[a]For selections see text. [b]See Fig. 5.2. [c]Gland (foliar nectary): shape – 0 orbiculate; 1 elliptic. [d]Gland position: 1 between lower pinnae only; 3 between upper and lower position. [e]Stipule vestiture: 0 absent; 1 present. [f]Stomatas: recorded as being vestigial on accession 76055 only: awaits confirmation. [g]Pod curvature: 1 linear; 2 slight; 3 marked. [h]Pod colour: 1 pale; 2 medium; 3 nearly black. [i]Habit: 1 prostrate; 2 main stem erect with lateral branches from the base; 3 main stem erect with the lateral branches evenly distributed on the main stem. [j]Crown branch number: number of branches arising from the crown, below 5 cm on the main stem. [k]Time to flowering, days on the Julian calendar.

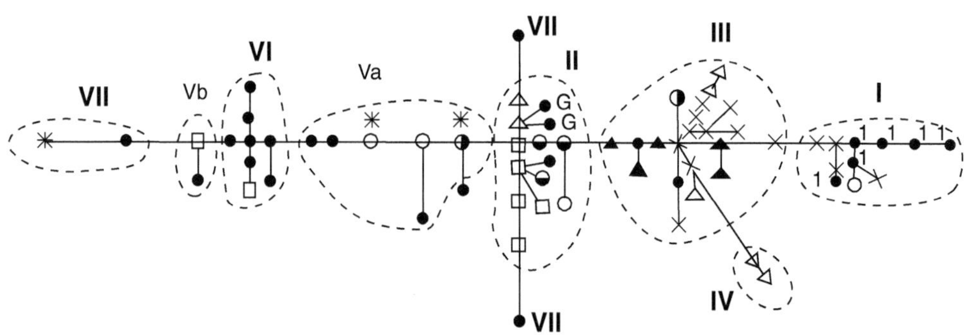

Fig. 5.5. Countries/regions of origin for some accessions. Classificatory groups I to VII as before. Caribbean Islands: Antigua, open circles; Cuba, vertically pied circles; Curaçao, horizontally pied circles. Central America: The Yucatan Peninsula, Mexico and Belize, open triangles; Guatemala, G. South America: Venezuela, closed triangles; Brazil, crosses; northwest Argentina, open squares. Countries of origin not presented: small closed circles; [1] Accessions so labelled introduced from a country to which they were not native.

but the most different accessions – those placed peripherally – are usually from the Yucatan Peninsula of Mexico, Venezuela, a Caribbean island (Cuba) and Ecuador (in group VII). The overall similarity of the Brazilian accessions, which is echoed in their genetic similarity (Pengelly and Liu, 2001), may be at least partially attributed to their source. Many were obtained by correspondence from research centres that well may have exchanged seed stocks. Without better information about the distribution of the species in Brazil, this country cannot be ruled out as a potential source of useful variation. In the existing collection (as shown by Pengelly and Liu, 2001), the Venezuelan material

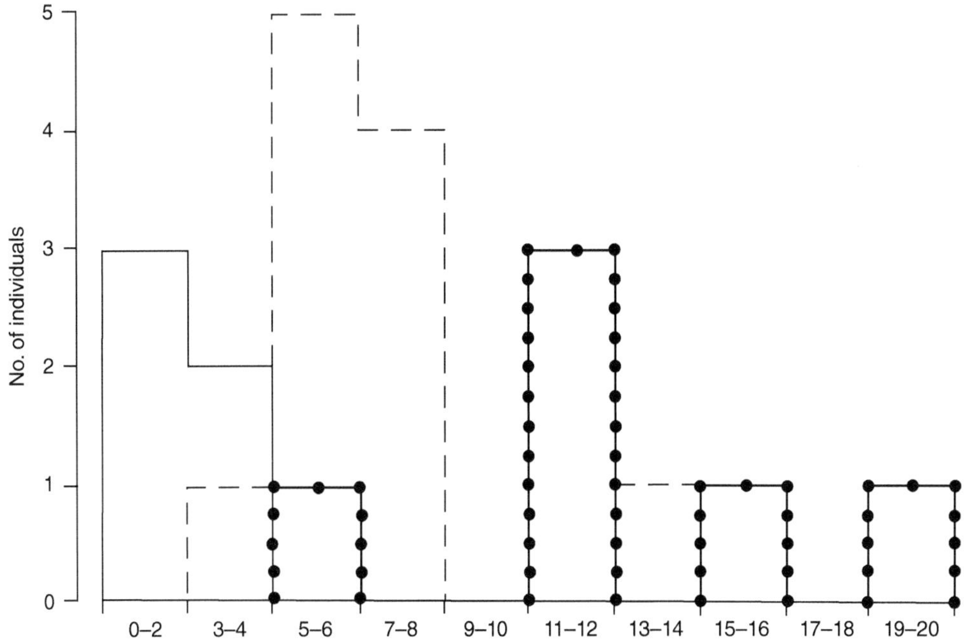

Fig. 5.6. Crown branch number in selected plant groups of *D. virgatus* and in *D. leptophyllus* and *D. pernambucanus.*

is quite variable, while the single accessions from the Caribbean islands and Ecuador are highly so.

In regard to *D. virgatus*, the most diverse species, there are three zones to consider: groups II, V and VI. In the core grouping II, which has produced the most promising material in field studies (Chapter 8, this volume), there are accessions from the Caribbean, Argentina, the Yucatan and Central America. Very similar accessions as shown in the previous section can be found in widely separated locations: Curaçao and Argentina, Guatemala and the Yucatan. Group V members link the core type II and the prostrate type VI – possibly what was called *D. depressus*. In group V, the *D. virgatus* accessions from Antigua and Cuba (asterisks in Fig. 5.5) are quite different from each other and the core group II. This reflects the genetic differences noted by Pengelly and Liu (2001).

Finally there are the groupings VI and Vb, both of which contain Argentinean accessions.

Unlike the situation in *Stylosanthes*, there is little geographical pattern between or within some of the more important groupings in this collection (compare *D. virgatus* group II with the narrow geographic range for the Brazilian coastal type of *Stylosanthes scabra* as shown in Liu (1997) and Edye and Maass (1997)). A relatively narrow geographical range in north-west Argentina has yielded four species of *Desmanthus* and very different forms of *D. virgatus*. A great range of variation was also evident in the Caribbean material. It follows that if genetic material is to be selected for trials, for use as 'core collection', or 'representative range', it cannot be done solely on the basis of geographic or environmental provenance data. A meaningful classification is required.

Variation within selected groups of D. virgatus

In seeking to find legumes adapted to dry tropical conditions, it was necessary to select areas most likely to yield such plants. The areas delineated were areas with climates similar to those that had yielded *S. hamata* cv. Verano and

S. scabra cv. Fitzroy (Burt *et al.*, 1976; Burt and Reid, 1976). The location of these zones is shown on Fig. 5.4. Collecting missions to B, C, H, F and G all yielded *D. virgatus* accessions (zone E in Ecuador awaits collection). Information on those from C (Guatemala), H (Curaçao) and G (Antigua) has already been presented. Accessions from B, northwest Mexico, have not. In later classifications, and in final field evaluations, these have been shown to be somewhat different morphologically from the Argentinean material and to be well adapted to clay soils in semi-arid conditions in Australia (Chapter 9, this volume).

The characteristics of plants from these two geographical areas are shown in Table 5.5. Data are again drawn from the full data set. The Mexican material used data only from those accessions tested in field studies (Chapter 9, this volume). It was collected along a transect stretching northwest from Nuevo Laredo in Coahuila. The Argentinean material is the upright form of the plant and is located in zone II on the MST. An accession from a more southerly, wetter climate in Mexico (labelled C in Table 5.5) is also included for comparison with the Mexican material from the semi-arid climate (B in Table 5.5). It has shown promise in evaluation tests in central and southern Queensland (Cook *et al.*, 2005), in wetter areas than those in which the suite B has been so promising. An accession from Costa Rica (D), which has also shown promise in the dry tropics (Chapter 9, this volume), is the final accession inclusion. To make these comparisons, the attribute listings for each of the accessions involved was scanned manually and where differences occurred the characters were listed, and ranges and mean values calculated. Only 15 of the 72 plant attributes were implicated.

The accessions from north-east Mexico are from drier environments than those from Argentina. They have shorter, narrower pinnae, fewer, shorter pinnules and shorter pods and stamens. The accession from the wetter more southerly zone C is similar but it has longer pods and leaf characteristics, near the upper range of those from zone B. The Costa Rica accession has longer, darker coloured pods and fewer pinnules.

A feature of major importance however is in the differences to be found in agronomic characteristics between the various accessions. The group II accessions from Argentina show much variation in crown branch number, stem diameter and flowering time. The plants may or may not show much basal branching. The Mexican accessions from the semi-arid zone are thinner stemmed, very uniform and early in their flowering time but have wide variation in their crown branch numbers. The accession from the wetter zone in Mexico and that from Costa Rica have the same habit and flowering time as the dryland Mexican type but have very high crown branch numbers. The variation in this attribute within plant groups (I, II and III, Fig. 5.1) is shown in Fig. 5.6. *D. virgatus* is the most basally branched and *D. pernambucanus* least so. Within *D. virgatus* and *D. leptophyllus* there are individuals with atypically high values.

In *D. virgatus* there are various 'types' of plant and the use of the complete list of attributes to validate and complement the original MST has confirmed that they are homogeneous for many morphological attributes. They contain, however, much agronomic variation that is not geographically or climatically based. A study of the geographical locations from which the plant material was collected cannot alone be used to select 'representative ranges', 'core collections' or individual accessions for test. Although DNA-based molecular tests seem to offer more possibilities than older chemotaxonomic methodologies, for instance seed protein studies, they are still not entirely suited to this purpose. In *Desmanthus* for instance, most accessions of *D. virgatus* are simply placed in one group and large agronomic and morphological differences are undetected (Pengelly and Liu, 2001). At the present time 'the evaluation of genetic diversity in a sample is a problem different from its agronomic merit' (Brown, 1978).

Variation within D. pernambucanus

In their work with *Desmanthus*, Pengelly and Liu (2001) use DNA-based techniques to suggest that there is much variation in *D. virgatus* but little in *D. pernambucanus*. To illustrate this further, two accessions were chosen for comparison, one from Antigua and one from Brazil (numbers 37 and 23 respectively), abstracting information from the full data set as before. To maximize the chances of finding differences (and the aberrant position of plant 6 (Fig. 5.3) has already been discussed), these accessions are from very different geographical locations, and are field collections rather than institutional accessions to guarantee this difference. The accessions were different in only 11 of the 72 characteristics measured and many of these

Table 5.5. Examples of geographical variation in collections of *D. virgatus*.

	Region[a]			
Attributes	A	B	C	D
Provenance				
Country	NW Argentina	NE Mexico	SW Mexico	Costa Rica
Latitude (°)	23°–24°	26°–28°	16°	
Altitude (m)	400–1000	400–1350	20	
Mean annual rainfall (mm)	650–750	250–500	1600	
Morphological				
Leaf				
Petiole, length (mm)	1.2–5 (3.2)	1–3 (2.0)	2	4
Stipule, length (mm)	4–7 (6.2)	2–6 (4.5)	3	3
Rachis, length (mm)	14–23 (17.4)	11–19 (16.5)	19	21
Pinnae, no. pairs	3–4 (3.4)	2–5 (4.0)	5	3
Pinnae, length (mm)	24–38 (30.6)	14–22 (18.0)	22	18
Pinnae, breadth (mm)	9–11 (9.4)	7–8 (7.5)	7	7
Pinnules, no. pairs	17–20 (18.4)	11–16 (13.3)	16	12
Pinnules, length (mm)	5–7 (5.8)	4–5 (4.3)	5	5
Floral				
Stamen, length (mm)	4–5 (5.3)	2–3 (2.8)	3	4
Pod				
Colour[b]	1, 2	2, 3	2	3
Length (mm)	48–70 (57.6)	40–52 (46.8)	60	65
Agronomic				
Crown branch no.[c]	5–12 (7.6)	8–16 (11.0)	16	26
Stem diameter (mm)	5–8 (6.4)	3–5 (3.8)	7	6
Time to first anthesis (days)	66–86 (78.2)	70	70	70
Seed weight (mg/100)	34–49 (42.4)	26–46 (37.5)	46	38
Growth habit[d]	2, 3	2	2	2

[a]Mean and range for selected accessions: A, north-west Argentina, CPI accessions 85173, 85178, 90751 and 91181. B, north-east Mexico, CPI accessions 78370, 78371, 78372, 78373 and 78382. C, south-west Mexico, CPI 91326. D, Costa Rica, CPI 83563. [b]1 pale, 2 medium, 3 nearly black. [c]Number of branches arising from the crown, below 5 cm on the main stem. [d]1 prostrate, 2 main stem erect with lateral branches from the base, 3 main stem erect with lateral branches evenly distributed along the main stem. Note: As the lead author had no footnotes in the manuscript those for b–d have been assumed from other tables in the chapter.

differences were slight (Table 5.6) and of little taxonomic importance. The results as shown on the MST and as verified here support the findings of Pengelly and Liu (2001).

5.4 Discussion

There is continuing need for the development of new pasture legumes (e.g. Jones and Clem, 1997; Pengelly and Conway, 2000). This need is likely to increase as energy costs increase, and fertilizer prices escalate and their polluting effect on the environment gains more attention. Global warming and its deleterious effects on rangelands and potential collecting areas alike can only exacerbate this need.

This situation is most marked in tropical areas and it is unfortunately true that the development and utilization of legume genetic resources collections in these zones lags well behind that in more temperate areas. The best developed

Table 5.6. Differences between contrasting accessions of *D. pernambucanus*.

Attribute	Accession and source	
	Antigua CPI 73463	Brazil CPI 55718
Floral		
Petiole width (mm)	<1	>1
Staminode length (mm)	6	8
Peduncle length ((mm)	32	40
Pod		
Colour[a]	2	1
Seed		
Colour[b]	2	3
Leaf		
Nectary gland shape[c]	1	0
Petiole length (mm)	7	4
Rachis length (mm)	45	34
Stipule vestiture type[d]	3	4
Pinnae length (mm)	35	31
Pinnae width (mm)	10	15
Leaflet breadth (mm)	1.0	1.5
Agronomic		
Crown branch number[e]	3	2
Time to first anthesis, days	93	99

[a]Pod colour: 1 pale; 2 dark; [b]seed colour: 2 dark; 3 black; [c]nectary gland shape: 0 orbiculate; 1 elliptic; [d]stipule vestiture type: 3 velutinous; 4 pilose; [e]crown branch number, number of branches on the crown to 5 cm up the main stem.

tropical leguminous pasture genus is arguably *Stylosanthes*, but it was not systematically utilized until the mid-1970s (Burt and Williams, 1975) and new species were still being made available for commercial use in 1996 (see information quoted in Cook *et al.*, 2005). Indeed, new potentially useful species are still being found in several tropical areas (Maass and Sawkins, 2004). The development of *Desmanthus*, probably the most important source of pasture/ley legumes for use on clay soils in the tropics (Chapter 3, this volume; Jones and Clem, 1997; Cook *et al.*, 2005), lags far behind even *Stylosanthes*. The Australian Tropical Forages Genetic Resources Centre (ATFGRC) collection, the best available, contains representative accessions from altitudes ranging from 34°S to 32°N, but has only about 300 *Desmanthus* accessions and many of these are duplicates. The equivalent collection for the tetraploid form of *Stylosanthes hamata*, found only in a small coastal area of northern Venezuela and Colombia from 9°N to 15°N, numbers about 200 (Edye and Maass, 1997), twice the number of the very valuable widely distributed *D. virgatus* (Pengelly and Liu, 2001). In contrast, the USDA *Medicago* collection has over 7500 accessions.

Most tropical cultivars are simply selections of introduced material (see information contained in Cook *et al.*, 2005). In *Desmanthus* promising new types are already on hand but have not yet been made commercially available, for instance the dryland Mexican type for drier tropical areas (Chapter 4, this volume; and as described herein). Until these various types have been delineated and evaluated and the most useful types specified, it is difficult to suggest plant breeding objectives. In *Stylosanthes* improved resistance to the fungal disease anthracnose is sought (Chakraborty, 2004), in *Macroptilium atropurpureum* it is rust resistance. In the more poorly known *Desmanthus*, a very different plant form from most other pasture legume genera, selection may be made for crown branch density, the nutritive value of the seed or, as is mentioned later, the non-dehiscence of pods. To take advantage of the variation naturally available and to avoid a costly piecemeal approach to its development, it is now widely accepted that an integrated approach to the collection, description and evaluation of the germplasm collection should be developed (e.g. Burt, 1983; Guarino *et al.*, 1999; Pengelly and Liu, 2001).

The approach adopted herein used methodologies developed for and successful with *Stylosanthes*. It was applied to only a small part of the *Desmanthus* collection and was exploratory, with no taxonomic review of the genus available, and used a limited, trial set of plant attributes. Despite this, after the process of validation it proved to provide the basis for a very useful classification and holistic overview. Where apparent incongruities occurred, as with the overlap with *D. acuminatus*, *D. tatuhyensis* and *D. virgatus*, these were informative and in the example quoted confirmed the taxonomic and genetic similarities noted by Luckow (1993) and Pengelly and Liu (2001) respectively.

In *Stylosanthes* at the same stage of development, it was found that morphological and agronomic plant attributes reinforced one another in the original morphological/agronomic classification. The various morphological/agronomic (MA) types of *S. fruticosa* (syn. *S. mucronata*) differed in terms of stem thickness, hairiness, pod colour, persistence and flowering time (Burt *et al.*, 1971). The individual MA groups were quite homogeneous and were often geographically based. In *S. fruticosa* they were from different parts of Africa. In *Desmanthus* the only agronomic feature to figure in the initial classification was habit, and individual groupings could be quite valuable for some characteristics, such as crown branch numbers. Very similar accessions could be forthcoming from different geographical regions and climates. Unlike the situation in *Stylosanthes* where prior knowledge of the soils, climates, altitude and so on has been useful in selecting collecting areas or in predicting plant adaptation, other factors may warrant more consideration in *Desmanthus*.

One such factor could conceivably be the microenvironment from which the individual accessions were collected. In *Stylosanthes scabra* accessions from the same geographical region and from the same climates and soils but from different disturbance levels differed in various morphological features (Burt *et al.*, 1983). In two morphologically very similar Guatemalan accessions of *D. virgatus* from the same region there was a difference in the pattern of crown development and lateral branching. The accession from a grazed pasture showed much more basal branching than that collected on a roadside. Several of the more grazing-resistant accessions of *Desmanthus* exhibit this trait (Chapter 4, this volume; Jones and Brandon, 1998) and its significance has been discussed elsewhere (Takahashi and Ripperton, 1949; Burt, 1986). In the United Kingdom, many of the famous pasture strains identified at the Welsh Plant Breeding Station, Aberystwyth were simply selections of plants that had been produced in such niche situations. Work at this institute showed that the imposition of different patterns of grazing, coupled with chromosome breakages and somatic segregation could rapidly change the plant habit, flowering time, and so on without even the necessity for flowering and seed production.

A search through the provenance records of the *Desmanthus* germplasm collection, possibly in conjunction with the ecogeographic survey techniques described by Maxted *et al.* (2000), might well provide useful insights into the occurrence of such traits. As had been found with *Stylosanthes* (e.g. suggestions for the edaphic requirements as given in Burt *et al.* (1979) with those reported in later studies and summarized in Cook *et al.* (2005)), such hypothesis generation in poorly known germplasm collections can be of importance in selecting relevant strategic research areas.

In assessing the adequacy of the ATFGRC collection of *Desmanthus*, a question posed by Pengelly and Liu (2001), it is important to realize that it really has three components. The first of these is concerned with the provision of legumes for use in long-term pastures on clay soils in dry tropical and semi-arid conditions in Australia (Chapter 8, this volume). The main emphasis in collection was on plants from similar conditions in South and Central America and the Caribbean. To this was added material from a University of the West Indies/International Development Research Council of Canada project concerned with wetter areas in the Caribbean and Belize, both major centres of diversity for *Desmanthus* (Spurgeon, 1977; Ahmad, 1980; Keoghan, 1980; Lazier, 1981a). From wet tropical areas in Belize and neighbouring regions, particularly in the Yucatan peninsula, came accessions of *D. pubescens*, *D. leptophyllus* and

D. virgatus. From the islands of Antigua and Curaçao, less wet monsoonal areas, came collections of *D. virgatus* and *D. pernambucanus*. Linked introduction/evaluation programmes showed that such material could be very useful for long-term pastures and as 'protein banks'.

This same collection has, however, been pressed into service for possible use as ley legumes, often in more subtropical areas (Jones and Rees, 1997; Pengelly and Conway, 2000). Although a historic introduction into India, *D. pernambucanus*, has found a use as 'hedge lucerne', it has many shortcomings (Cook *et al.*, 2005) and productive pasture/ley legumes are still required for India's large, economically very important areas of clay soils (Ramesh *et al.*, 2004). Indeed many such areas exist elsewhere in the tropics (Dudal and Bramao, 1965). Global warming and habitat degeneration, for instance in the *Desmanthus*-rich areas of Texas, USA, have prompted a conservation/evaluation programme in that area (Walker *et al.*, 2005; Muir *et al.*, 2005). There is indeed a need to review the adequacy of the collections of *Desmanthus*, particularly as plant-collecting efforts in South America have virtually ceased (Maas and Sawkins, 2004).

The MST has provided a useful, holistic overview of only a small part of the existing ATFGRC germplasm collection and a similar classification of the entire collection is clearly warranted. It does however provide a basis for discussion on this subject. *D. pernambucanus* germplasm is readily accessible and has long been exploited for use as a 'protein bank'. It has been shown to contain little morphological/agronomic variation and Pengelly and Liu (2001) show that it is genetically limited. After an exhaustive survey of the world literature, Cook *et al.* (2005) point out that it has numerous limitations that are inherent in the species. For instance, it is only well adapted to wet areas, has pithy stems, which break under grazing, and it is not compatible with grasses in pastures. Although, as in other species, there is always the chance of serendipity, of a useful plant being identified from an unexpected source, this species does not seem to warrant any emphasis in genetic resource programmes.

D. pubescens is also from wet, tropical situations (J.R. Lazier, Belize, 1976, personal communication), and Luckow (1993) notes that it is found in marshy situations within such climates. In nature it is almost confined to the Yucatan Peninsula, an area collected by the UWI/IDRC-sponsored project in Belize (Chapter 2, this volume). It has shown little morphological/agronomic or genetic variation (Pengelly and Liu, 2001) and the performance of its cultivar Uman has been disappointing in pasture situations (Cook *et al.*, 2005). It is not well adapted to drier tropical conditions (Chapter 9, this volume). It does however thrive in wet tropical conditions (J.R. Lazier, Belize, 1976, personal communication) and produces high-quality feed that greatly stimulates wool growth. It could well be useful as a 'protein bank' in such situations and the existing collections are adequate for testing such a use.

This is not the situation in *D. leptophyllus* and *D. virgatus*, both variable species distributed over a wide geographic range. *D. virgatus* in particular is found over a wide range of climates. Although both species have proven to be of value (e.g. Cook *et al.*, 2005; Chapter 9, this volume) and *D. virgatus* has been found to be of high value in pastures (Stehle, 1956) and in protein banks (Keoghan, 1980), collection is wanting. There are, for instance, virtually no collections from Ecuador, yet this region contains both *D. virgatus* and *D. leptophyllus* (Luckow, 1993) and is a well-separated distinctive phytogeographic region (see information quoted in Burt and Reid, 1976). The region is also the sole source of *Stylosanthes sympodialis* (which has the ability to grow on clay soils in dry tropical regions) and provided a very distinctive form of *D. leptophyllus*. Similarly, in the Caribbean islands, an area stretching from 60°W to 84°W, where *Desmanthus* is common throughout and it is known that the accessions of *D. virgatus* from there are morphologically and genetically diverse (see herein, and Pengelly and Liu, 2001), the ATFGRC collection contains accessions from only four of the islands. Evaluation with some of the other species in the collection, such as *D. tatuhyensis* (of particular interest because of its ability to grow on acid soils; Cook *et al.*, 2005), has been so limited as to render discussion implausible.

Another feature of some concern is the nature of the collections. First the geographic location of the major collecting zones. Although the main interest in *Desmanthus* has come from tropical areas, it is primarily a genus of the subtropics (Burt, 1993a; Luckow, 1993). Brazil, the main source of pasture legumes, contains only five species of *Desmanthus* mostly in the non-tropical areas, while the southern USA lists 12 species. Consequently the majority of accessions from Brazil are of such genera as *Stylosanthes*, *Centrosema* and *Desmodium* and the number of *Desmanthus* accessions (in Coradin and Schultze-Kraft, 1990) is too few for mention. Yet Hartley (1963), whose forecasts have proven accurate (Williams and Burt, 1983), suggests that the southern states of the USA are a likely source of adapted material for subtropical and more temperate zones in Australia. Pengelly and Liu (2001) have shown that an accession of *D. virgatus* from this region is genetically distinct from the remainder of the germplasm collection. Unlike the situation often encountered elsewhere, details of plant distribution and collections of native species are well documented in the USA (Luckow, 1993). An ecogeographical survey of the type described by Maxted *et al.* (2000) would be of considerable value in considering the future direction for plant germplasm collections. Certainly the data provided in Luckow (1993), for instance, that *D. cooleyi* is a 'common component of short grass prairies', 'occurs on sandy ... sandy-loam-basalts ... limestones and rocky substrates' and is reportedly good browse for antelope, suggests that this species, present in the AFTGRC collection, is worthy of study.

In many other areas however, plant distribution is more poorly documented, and access is difficult. For *Desmanthus*, the situation was confounded further by the lack of a taxonomic revision (Luckow believes that many of the plants identified in northern Argentina as *D. virgatus* were in fact *D. paspalaceus*). Indeed it has been deemed necessary to use more advanced techniques in Geographic Information Systems to suggest where some species of interest may be found (Maas and Sawkins, 2004; Sawkins *et al.*, 1999). Germplasm collection missions have, of necessity, often been exploratory, making use of 'main roads with a collection site selected every 30 to 50 km' (Coradin and Schultze-Kraft, 1990). However roads are often routed away from geographical features that may be favoured as centres of biodiversity a situation noted in *Stylosanthes* (Chapters 8 and 9, this volume). *Stylosanthes* and probably *Desmanthus* plants growing near the roadside may be quite different from those in other ecological situations. It may well be necessary, as has been done with *Stylosanthes hamata* and *S. seabrana*, to revisit areas for future, more detailed collections once a successful 'type' has been delineated (Chapters 8 and 9, this volume; Edye and Maass, 1997). A case in point is provided by a form of *D. bicornutus* that has proved to be adapted to heavy clay soils in semi-arid, western Queensland (Chapter 9, this volume), and had found use in grassland conservation schemes in Texas, USA, an area to which it was introduced (Ocumpaugh *et al.*, 2004; Muir *et al.*, 2005). The accession in question was one of only two collected from the Rio Balsas depression in Mexico, an area noted to contain a type of *D. bicornutus* different from those found elsewhere (Luckow, 1993). In that region it is found in a range of climates, soil types and ecological situations and is likely to show significant variation.

If the long-term utility of the collection is to be considered, then the species *D. oligospermus* warrants early inclusion. Found only in a small, semi-desert region on the southern tip of the Baja California peninsula, this species may not be suitable for agronomic trials because it has underground stems that enable it to spread, which could possibly render it invasive. However, it is almost unique in the genus in that it possesses indehiscent pods and the seeds of *Desmanthus* are of high nutritional value (Schlink and Burt, 1993). Indeed *Desmanthus* is viewed as a potential pulse crop with seeds more nutritious than soya bean. The incorporation of non-dehiscence into other, more desirable plant types could further improve the known efficiency of the genus to increase wool production/unit of forage consumed (Rangel *et al.*, 1996), a characteristic of value to those planning sensible land utilization (e.g. McIvor, 2002). The overall significance of non-dehiscence in the evolution of the pulses such as peas and beans needs no further emphasis (Smartt, 1990). The restricted geographical range of *D. oligospermus* and its rarity within that range is a matter of considerable concern (Luckow, 1993).

6 Searching for Pasture Legumes for Heavy Clay Soils in the Australian Dry Tropics and Subtropics: I. Initial Literature Reviews, Data Analysis and Choice of Material for Test

R.L. Burt† and J.R. Lazier*[1]
**Formerly International Livestock Centre for Africa*

Abstract
As a preliminary step in the selection of germplasm for heavy clay soils in the Australian tropics and subtropics a review was undertaken of known genera and species of leguminous plants with known and suspected potential. Groupings were made of the genera based on the percentage of species occurring on clay soils. Assessments were then undertaken of their interest based on the environments in which they occur and their general forage characteristics. The report concludes with brief comments about the adequacy of genetic resource collections of the genera and species that have proven to be of value.

6.1 Introduction

Almost all Australian pasture legume cultivars are plants that have been introduced from elsewhere, and all crop varieties are 'aliens', with the sole exception of the Macadamia nut, which is native to Queensland but was developed for commercial use in the USA. This is not altogether surprising because Australian flora is unique since it has been long isolated from those regions in which the seed-bearing plants developed, and thus has relied on the evolution of endemic plants to cope with the ever-changing climate and decreasing levels of soil fertility. The dry areas of Australia illustrate the adaptations that have been necessary (White, 1994). There the soils are poor and sclerophyll–xerophyte grasslands occur; termed 'spiniflex' plains, they are dominated by tussock (*Astrebla*) and hummock (*Triodia* and *Plectrachne*) grasses. Such change has been slow and adaptation not always complete. Although the climate of the southerly portion of Western Australia changed from summer to winter rainfall, a 'Mediterranean' type climate, some plants retained their summer growing characteristics (White, 1994). The arrival of man some 50,000 years ago, in small numbers and reliant on hunting and gathering, was sufficient to change the composition of the flora. This disadvantaged browsers and grazers, many of which became extinct (Prideaux and Roberts, 2007). The advent of Europeans, and their hard-footed animals, produced similar dramatic effects. In the 'Mediterranean' regions of Western

[1] E-mail: john.lazier@sympatico.ca

Australia, for instance, tall growing grasses were quickly replaced by undesirable native plants or by those accidentally introduced from Mediterranean countries. Today there is much cause for concern about the degradation of such areas as the *Astrebla* (Mitchell grass) Plains. This situation is likely to worsen with global warming and the associated droughts and changes to rainfall patterns. Australian vegetation, which barely coped with climatic change in times past, could be faced with much more rapid changes in the near future.

The sowing of native species is one way in which areas may be rehabilitated, but, even without climatic change, this can be difficult. There have been many attempts to utilize native Australian legumes and these have not been successful (e.g. by Kerridge and Skerman, 1968; Clem and Hall, 1994). For example, the native subtropical *Glycine latifolia* cv. 'Capella', despite its pre-release testing and subsequent trials, generally performed much more poorly than introduced material when tested on clay soils in regions similar to those in which it was collected. In Texas, 'varieties considered even remotely native to Texas' such as *Desmanthus bicornutus* are being considered for introduction to rangeland with heavy clay soils and a native flora that already contains several native species of *Desmanthus* (Muir *et al.*, 2005). The genus *Desmanthus* is very closely allied to *Neptunia*, which is native on the clay soil plains of western Queensland, and its potential to Australia will be discussed herein.

There are very large areas of clay soils in Australia (70.5 m ha), and in countries such as India (60 m ha) and Sudan (40 m ha) (Dudal and Bramao, 1965). These soils are especially valued in tropical and subtropical situations for their ability to 'hold water' and their generally higher fertility levels compared with the more free-draining, sandier-textured soils. Consequently clay soils are often used for cropping and ley farming, and even a small area can provide 'green pick' for grazing animals in dry periods. Whereas there are now a wide range of pasture/ley legumes available for use on tropical and subtropical sandier textured soils (e.g. the information available in Cook *et al.*, 2005), there is a very real dearth of material adapted to clays.

Attempts to meet this demand have, in fact, been made for some time, with the due regard that must be paid to the potential of any new species to be invasive, both within and outside of the various regions in which they may be seen to have agronomic and economic merit (Williams and Burt, 1982). The work reported here, for instance, commenced at the same time as the '*Stylosanthes* programme' (Burt and Williams, 1979a), which ultimately led to the domestication of 'new species' such as *S. hamata*, *S. scabra* and *S. seabrana*. But the development of plants for clay soils was hampered by logistical problems and the lack of intuitively valuable clay soil collecting sites in South America, the home of most of the 'new species' for sandier-textured soils.

The results obtained from the 'clay soil programme' are presented herein and the final components of the study, concerned with evaluation on the clay soils of western Queensland, are published subsequently (Chapter 9, this volume). We begin with describing the review that was used to select the genera and species of interest, and conclude with brief comments about the adequacy of our genetic resource collections of the genera and species that have proven to be of value.

6.2 The review

In any plant development programme there is the problem of selection, of knowing what material to use and, in many instances, of where to obtain it. This problem was particularly difficult in developing plants for use as tropical pasture legumes for at the time there was little *a priori* information on which to base such decisions, and there were a great number of species and genera from which to choose; Williams *et al.* (1976) estimated about 50 leguminous genera and some 2000 species had potential.

As such numbers are too large for detailed consideration, they were reduced to a list of 'best bets' through long-term consultation with colleagues (including those encountered on collecting missions), reference to the world's literature, visits to herbaria and various climatic studies (e.g. Burt and Reid, 1976). A re-analysis of the biological classification of Hartley (1963), in which he selected homologous areas on the basis of their similarities in grass floras, showed its relevance for this purpose and was of use later. In total, some 29 genera were identified as having greatest potential (Table 6.1).

Table 6.1. Numbers of sites at which each species of the 29 legume best-bet genera identified in the review occurred on clay and non-clay soils throughout the tropics.

Genus/species	Soil type: Clay	Soil type: Non-clay	Total sites
1. *Aeschynomene*			
americana	2	1	3
amorphoides		1	1
cristata	1		1
fluitans	1		1
indica[a]	14	1	15
sensitiva	1	1	2
Totals	19	4	23
2. *Alysicarpus*			
longifolius[a]	3		3
monilifer		1	1
ovalifolius[a]	3	2	5
rugosis	18	2	20
vaginalis[a]	6	4	10
zeyheri	1		1
Totals	31	9	40
3. *Arachis*			
diogi		1	1
glabrata	1	1	2
helodes		1	1
marginata	1	1	2
prostrata	1		1
pusilla	1		1
villosa		1	1
Totals	4	5	9
4. *Calopogonium*			
galactoides		1	1
mucunoides	2	3	5
Totals	2	4	6
5. *Centrosema*			
brazilianum		3	3
pascuorum	1	3	4
plumieri		1	1
pubescens	2	4	6
schottii	1	1	2
virginianum		3	3
Totals	4	15	19
6. *Clitoria*			
ternatea[a]	7	2	9
Totals	7	2	9
7. *Crotolaria*			
burkeana		1	1
cephalotes		1	1
crassipes[a]	1		1
dissitiflora[a]	7		7
glauca	1		1
incana	1		1
laburnifolia		1	1
linifolia[a]	2		2
medicaginea[a]	1		1
mitchellii		1	1
pallida syn. *mucronata*		3	3
pycnostachya	1		1
retusa		2	2
sagittalis		1	1
spinosa	1		1
thebacea	2	1	3
trifoliastrum[a]	3	1	4
Totals	20	12	32
8. *Desmanthus*			
illinoensis	1		1
leptolobus	1		1
virgatus	9	4	13
subulatus	3		3
Totals	14	4	18
9. *Desmodium*			
adscendens	1	1	2
axillare		1	1
barbatum	1	2	3
brachypodium[a]	1		1
campylocaulon[a]	3		3
encanum syn. *racemosa* syn. *frutescens*	5	3	8
dichotomum	3		3
muelleri[a]	2		2
polyneurum[a]	1		1
scorpiurus[a]		1	1
tortuosum	1	2	3
triflorum	3	3	6
varians[a]	1		1
velotinum	1		1
Totals	23	13	36
10. *Dolichos*			
amablis		1	1
oliveri		1	1
sericeus syn *formosus*	2	1	3
Totals	2	3	5
11. *Galactia*			
canescens	1	1	2
grayii		1	1
jussiaena		1	1
marginalis		1	1
muelleri[a]	1		1
tenuifolia	1	1	2
Totals	3	5	8
12. *Glycine*			
clandestinum[a]	1		1
falcata[a]	4		4

Continued

Table 6.1. Continued.

Genus/species	Soil type: Clay	Soil type: Non-clay	Total sites
sericea[a]		1	1
tabacina[a]	3	2	5
tomentella[a]	2		2
tormentosa[a] syn. *tomentella*	2		2
Totals	12	3	15
13. *Heylandia*			
laterbrosa	4		4
Totals	4		4
14. *Indigofera*			
arenaria		1	1
australis[a]		1	1
brachynema syn. *semitryuga*		1	1
coerulea syn. *articulata*	2	2	4
colutea syn. *viscosa*	2	1	3
comosa		1	1
cordifolia	2	2	4
costata		1	1
dendroides	1		1
doleoides	1	2	3
dominii		1	1
enneaphylla[a]	1	3	4
glandulosa	2		2
hirsuta[a]		3	3
hochstetteri	6	4	10
lespedozoides		1	1
linifolia[a]	4	4	8
mucronata	1	1	2
nigricans		1	1
paniculata	1		1
parviflora[a]	3	1	4
ruspotii		1	1
schimperi	3		3
simplicifolia syn. *tetrasperma*	1		1
spicata	2	1	3
subargentea	1		1
suffruticosa	3	1	4
tinctoria	1	1	2
trita[a] syn. *subulata*	4	1	5
vohemarensis syn. *suaveolens*		1	1
Totals	41	37	78
15. *Lablab*			
purpureus		1	1
Totals		1	1
16. *Leucaena*			
leucocephala	4	4	8
Totals	4	4	8
17. *Macroptilium*			
atropurpureum	2	4	6
geophilum	1		1
lathyroides	3	3	6
martii		4	4
Totals	6	11	17
18. *Macrotyloma*			
africanus		1	1
axillare		1	1
daltonii	1		1
Totals	1	2	3
19. *Neonotonia*			
wightii syn. *Glycine wightii*	2		2
Totals	2		2
20. *Neptunia*			
dimorphantha[a]	1		1
gracilis[a]	6	1	7
lutea	2		2
monosperma[a]	3		3
oleraceae	2		2
plena		1	1
Totals	14	2	16
21. *Phaseolus*			
adenanthus	2	1	3
elegans	1		1
linearis	1	1	2
pilosus	1		1
Totals	5	2	7
22. *Psoralea*			
cinerea[a]	7		7
cuspidata	1		1
holubis	1		1
hypogea	1		1
linearifolia	1		1
patens[a]	2	1	3
rhombifolia	1	1	2
tentax[a]	3		3
Totals	17	2	19
23. *Rhynchosia*			
albissima	1		1
americana		2	2
capitata	1		1
holosericea		1	1
minima[a]	29	12	41
sublobata		3	3
totta		1	1
badocana[a]	1		1
leucantha[a]	1		1
acutifolia		1	1
Totals	33	20	53

Continued

Table 6.1. Continued.

Genus/species	Soil type: Clay	Soil type: Non-clay	Total sites
24. *Sesbania*			
arabica	1		1
benthamiana[a]	7		7
bispinosa	1		1
brachycarpa[a]	1		1
microphylla	2		2
mosambicensis	1		1
pachycarpa[a] syn. *aculeata*	2		2
transvaalensis	1		1
Totals	16		16
25. *Stylosanthes*			
calcicola		1	1
fruticosa	3	4	7
guianensis	1	3	4
hamata	4	2	6
humilis	1	1	2
leiocarpa		1	1
mexicana		1	1
montevidensis	1	1	2
scabra		1	1
subsericea		1	1
sympodialis	2		2
viscosa		2	2
sundaica	1		1
Totals	13	18	31
26. *Teramnus*			
uncinatus	1	2	3
labialis	3		3
Totals	4	2	6
27. *Tephrosia*			
bracteolate	3	1	4
burchellii		1	1
elata		1	1
elegans		1	1
filipes[a]	1	1	2
leptoclada		1	1
linearis	2	2	4
lupinifolia	2	2	4
potovina		1	1
pumila[a]	1	1	2
purpurea[a]	5	3	8
rosea		1	1
spollinea	1		1
supina[a]		1	1
tenuis	1	1	2
uniflora	2		2
Totals	18	18	36
28. *Vigna*			
ambacensis syn. *pubigera*	1		1
filicaulis	1		1
lanceolata[a]		2	2
lasiocarpa syn. *pilosus*	3		3
luteola[a]	2		2
oblongifolia	3		3
radiata[a]	1	1	2
trilobata	4		4
unguiculata		1	1
umbellata	1	1	2
vexillata[a] syn. *angustifolia*	9	1	10
canescens	1		1
Totals	26	6	32
29. *Zornia*			
diphylla[a]	1	1	2
glochidiata		3	3
milneata		1	1
uriculata[a]	1		1
reticulata		2	2
Totals	2	7	9

[a]Species found in Australia.

Once these genera had been delimited, the information available on them was re-examined for use in the studies that followed. Emphasis was placed on data from clay soil situations, but this was relatively limited. Although edaphic and floristic information on species is commonly reported, the two aspects are not usually reported in the same publication. As well, for our present purpose, the definition of soil types needs to be reliably based. Climatic information, usually based on long-term records (as used subsequently in this study), can be obtained independently and is much more reliable. The information obtained was heuristically divided into two sets, that in which the data obtained were deemed to be relatively reliable, and that in which they were less so. The results from the two were subsequently found to be similar, and, for the sake of brevity, have been combined in this presentation.

6.2.1 Distribution of the various genera and species in relation to soil type

The basic data on the relationship between soil type and species are presented in Table 6.1, and indicate that many of the 205 species of the 29 genera can be found on clay soils (140 species). Non-clay soils were associated with 127 species. Among those on the clay soils are some now well known for their association, e.g. *Clitoria ternatea* which has yielded the cultivar 'Milgarra' and *Desmanthus bicornutus*. Australian legumes figure prominently in this listing, especially those in the genus *Glycine*, whose members are valued in natural grazing lands. Some idea of the extent to which the various genera are adapted to clay soil can be seen in Fig. 6.1. There are clear adaptation differences between the genera and these have been sorted using pattern-seeking methodologies. The six groups formed are labelled A to F in Fig. 6.1. Some genera, for instance those in group E, have a much higher proportion of their species growing on clay soil than others.

The genera in each group will now be briefly reviewed in order to create a prioritized listing of those most likely to provide pasture plants for use in dry tropical conditions. For the sake of brevity, reference will usually be restricted to material for which information is not generally available in such works as Allen and Allen (1981), and, more recently, Cook *et al.* (2005), and for native Australian legumes in particular, Hacker (1990).

6.3 General characteristics of the genera of interest

6.3.1 Group A *Indigofera* and *Crotalaria*

These two very large genera have only a small portion of their constituent species found on clay soils.

Many *Indigofera* species are palatable to some degree, but they are often toxic to grazing animals.

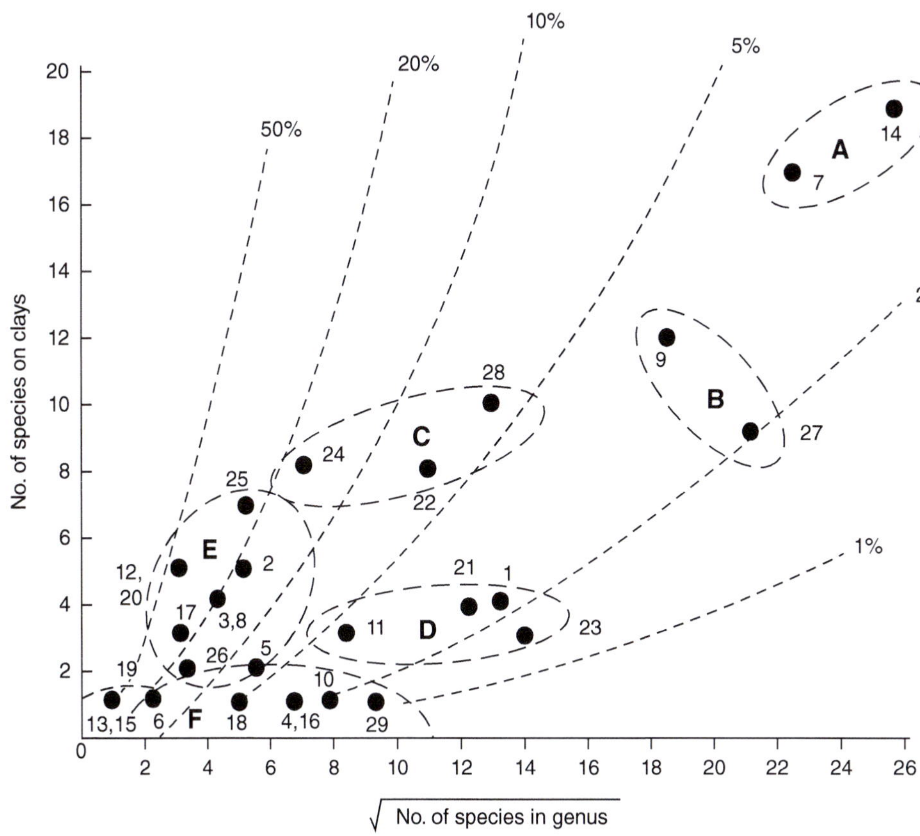

Fig. 6.1. Number of species/best-bet genus found on clay soils and total number of species/genus. Dotted lines indicate the percentages of species from each best-bet genus found in clay soils. Capital letters refer to the species groups defined in the text.

In the occasional references in the literature to their utility in West Africa, some of these toxins are said to disappear when plants 'hay off' in the dry season. In Australia *I. trifoliate* and *I. linifolia* have sometimes been considered to be useful forage plants because of their persistence, and thus much early research was carried out on the genus. Species such as *I. schimperi*, which is sufficiently variable as to warrant the taxonomic definition of three varieties, has been noted to thrive on black cotton soils (Polhill and Raven, 1971), but it is reported to be unpalatable to cattle (confirmed by Clem and Hall, 1994) and this, coupled with the possible toxicity of members of the genus, has caused interest in the species in Australia to lapse in favour of less potentially problematic genera. Cook *et al.* (2005) nominate two species in their list of forages, *I. schimperi* and *I. spicata*. However, *I. schimperi* is now considered an invasive species in Australia and is being removed from trial sites in which it has been planted (C. Gardiner, Australia, 2009, personal communication).

Crotalaria is also common throughout tropical Australia, but the various species are often unpalatable and, when not so, are toxic to grazing animals. Some species, such as *C. juncea* or 'Sun Hemp', a fibre crop, have also been widely used as green manure crops and where their lack of palatability may be an advantage. *C. pallida* and *C. lanceolata* have occasionally been mentioned as native pasture species, but not in Australia. *C. ochroleuca* has been used as a fodder in Tanzania, but its toxicity has yet to be determined (Sarwatt *et al.*, n.d.). The genus is deemed to be of little interest for Australia.

6.3.2 Group B

This group also contains two large genera, *Desmodium* and *Tephrosia*, with only a small proportion of species of each found on clays.

Tephrosia has been noted to often grow in swampy areas in East Africa (Polhill and Raven, 1971). It is common in Australia, where it is usually regarded as having little agricultural significance, with one species reported as 'sometimes grazed'. Leaves of several species have been shown to be toxic to rats in laboratory studies. *T. purpurea* has occasionally been used as a green manure crop in Australia, and members of the genus are currently being promoted for that use in East Africa, but the genus is of little interest here and Cook *et al.* (2005) list none as tropical forages.

Desmodium is well researched as a source of pasture legumes and has already provided five species that are used for fodder: *D. intortum*, *D. uncinatum* and *D. heterocarpon* (for higher altitude tropical and subtropical zones), *D. heterophyllum* (as a green manure or pasture legume for wet tropical areas) and *D. canum* – or 'Kiami clover' – a species that occurs widely on clay soils in wetter tropical areas, is well eaten by cattle and is often touted for commercial development.

Most of the species native to Australia that are palatable are generally regarded as being unproductive. *D. scorpiurus* occurs widely but is generally regarded as being of little value, even when tested in animal feeding experiments. *D. campylocaulon* or 'creeping tick trefoil' occurs on clay soils in Mitchell grass pastures in western Queensland, where it is highly palatable and probably an important component of native pastures. *D. dichotomum* is found on semi-arid clay soils in the Sudan and is considered to be a valuable element of native pastures and is of particular interest here.

6.3.3 Group C

Group C consists of three smaller genera, *Psoralea*, *Vigna* and *Sesbania*, the last of which has nearly 20% of its species found on clay soils.

Sesbania has several species in Australia, where in addition to those listed in Table 6.1, the species *S. campylocarpa*, *S. chippendale* and *S. erubescens* (Hacker, 1990) are to be found on Mitchell grass downs and floodplains on clay soils; in East Africa *Sesbania* species are also found in swampy conditions. Although Hacker (1990) notes that several of the Australian species are probably grazed when young, only *S. eruberscens*, found in standing water, is regarded as being palatable and well grazed. Members of the genera often occur as weeds in cultivated regions, especially wetter sites and in association with rice fields in Africa, and while useful sources for cut and carry fodder, they are not usually recognized as pasture plants. The shrubs *S. sesban* and *S. rostrata* are currently being promoted for short-term fallows on clay soils and have potential as fodder crops as well (e.g. Onim and Dzowela, n.d.). One Asian species is – atypically – a tree (*S. grandiflora* or 'white Dragon Tree') and has also been used for fodder.

Vigna is a well-known source of pulse crops, such as *V. unguiculata* or 'Cowpea'. Many of the *Vigna* species such as *V. radiata* (Mung Bean) and *V. umbellata* (Rice Bean) were at one time included in the genus *Phaseolus*. It is primarily a genus of climbing or twining annuals, native to Africa and Asia, where it has been domesticated for thousands of years. *V. hosei* and some forms of the very variable *V. vexillata* have been used as green manure crops and others have proven to be highly palatable and have been considered for use in wet tropical situations: *V. parkeri*, from Africa, has been made commercially available for use in subcoastal, subtropical Queensland. Another species, *V. luteola*, has produced the cultivar 'Dalrymple'.

Some *Vigna* species can be found in dry tropical regions of Australia. *V. lanceolata* ('Malaga Bean') occurs on the Darling Downs in southern Queensland and in moist areas such as creek banks in more northerly regions and *Vigna trilobata* (syn. *Phaseolus trilobatus*), a well-regarded pulse and forage plant in India, Pakistan and Sudan is found on heavy clay soils of India. *V. ambacensis*, syn. *Pubigera*, which is found in similar situations in Sudan, is also of interest.

6.3.4 Group D

This group consists of the genera *Rhynchosia*, *Galactia*, *Aeschynomene* and *Phaseolus* and each contains three or four species that commonly occur on clay soils.

Rhynchosia has one species, *R. albissima*, which is commonly found on black cotton soils in Africa and, because of this, there have been numerous *Rhynchosia* introductions and inclusions in evaluation studies, for instance of *R. minima* and *R. verdcourtii*. No accessions of potential for pastures have yet been identified.

Galactia has been of interest for some time and *G. muelleri*, or 'Muellers Pea', while more common on sandier soils in Australia, can also be found on clay soils (Hacker, 1990). In 'nursery' evaluation in the tropics, various species introduced mainly from Brazil have been found to not be productive, a situation also noted in their native environment.

Aeschynomene has five species found on clay soils and seven species are regarded as being of value as forage plants (Cook *et al.*, 2005). Most references to these species place them in marshy or waterlogged conditions and some species, such as *A. elaphroxylon* and *A. fluitans*, occur in standing water. *A. americana*, 'American joint vetch', is a well-researched, very variable species that has yielded two pasture cultivars in Australia. *A. falcata*, 'Joint vetch', is used in pastures in subtropical, coastal areas of south-east Queensland and northern New South Wales. *A. indica*, 'Budda pea', can be found on clay and black cotton soils in Africa, and has been used as an annual green manure plant. Most species are said to be moderately palatable but to drop their leaves with the onset of the dry season. They are herein deemed to be of little further interest for semi-arid conditions.

Phaseolus is a genus that has long been taxonomically intertwined with *Macroptilium* and *Vigna* and this poses problems in literature reviews. *Macroptilium* and *Vigna* are discussed here under separate headings. In Africa, *Phaseolus adenanthus* has been noted to occur on clay soils in swampy conditions as has *P. pilosus* (probably syn. *Vigna pilose*). Neither is regarded in their areas of origin as being of value as pasture plants.

6.3.5 Group E

This group consists of genera with relatively few species, many of which have pastoral value and a high proportion can be found on clay soils. The group contains eight genera, *Stylosanthes*, *Alysicarpus*, *Glycine*, *Neptunia*, *Arachis*, *Desmanthus*, *Macroptilium* and *Teramnus*. Although not native to Australia, two very important New World genera, *Stylosanthes* and *Desmanthus*, are closely related, and very similar to the widespread Australian native genera, *Zornia* and *Neptunia* respectively. Both *Stylosanthes* and *Desmanthus* species can often utilize native *Bradyrhizobia* effectively, which could greatly increase their potential utility.

Stylosanthes is primarily a genus of South America, but some species also occur in Central and North America and the Caribbean, and there are two species native to Africa. Sown pasture development in the Australian dry tropics and elsewhere is dominated by three species, *S. humilis* (Townsville Stylo), *S. hamata* (Caribbean Stylo) and *S. scabra* (Shrubby Stylo). *S. capitata* and *S. macrocephala* are of importance on more acid soils, *S. guianensis* is widely used in higher rainfall situations, *S. seabrana* (Caatinga Stylo) is becoming of wider use as a ley legume on clay soils in more subtropical conditions and *S. fruticosa* (syn. *S. bojeri*, *S. mucronata*) is a valued forage plant in the African dry tropics (e.g. Chakraborty, 2004; Cook *et al.*, 2005).

With the exception of the work described by Ahmad (1980) and Keoghan (1980), little emphasis has been placed on the diploid form of *S. hamata*. This is a small-leaved, more perennial form than the better known tetraploid type, and is found throughout the Caribbean islands and parts of coastal Venezuela and Colombia (Stehle, 1956; Burt and Miller, 1975). Unlike the tetraploid type, which yielded the cultivars 'Verano' and 'Amiga', it is adapted to clay soils. *S. sympodiales*, which is found in South America, is similarly adapted to clay soils and is poorly represented in genetic resource collections. *S. fruticosa*, which is very similar to the very successful *S. scabra*, has been collected from clay soils in Africa. These little-studied species and genotypes warrant further research attention.

Alysicarpus, 'Alsyce' or 'Buffalo clover', is highly regarded as a pasture plant in areas in which it is native. The Australian species listed in Table 6.1 are all palatable and *A. rugosus*, 'Rough chain pea', has both annual and perennial forms; the latter is very common on seasonally wet, black cotton soils in Africa. *A. monilifer* occurs in similar situations in Africa.

Neptunia, and in particular the species *N. gracilis*, is almost always found on clay soils in Australia. The plant is very palatable, has not been found to be toxic to animals and is valued as a pasture plant, especially in drought years. *N. amplexicaulis* is also valued as a drought feed in Mitchell grass areas in western Queensland, but is known to cause selenium poisoning in grazing animals. *N. monosperma* is also commonly found on the clay soils plains of tropical Australia. The native species, *N. gracilis*, has received some research attention in the past, especially in nursery studies, and was included in tests by Clem and Hall (1994). Relatively little is known about the genus, however, and fears of toxicity problems persist.

Desmanthus is a small genus native to South and Central America and the Caribbean and to southern areas in the USA that is taxonomically very closely related to the genus *Neptunia* and has often been confused with it. It is highly regarded as a pasture plant in areas to which it is native (Burt, 1993a). Its seed is particularly nutritious, and the plant is very effective in increasing wool production in sheep. *D. illinoensis* has been considered to be a potential pulse crop for semi-arid conditions.

There are many records of the occurrence of *D. virgatus* and *D. subulatus* in the tropics on clay soils, often in semi-arid areas. The very frequent occurrence of *D. virgatus* in such reports could, however, stem from the lack of a taxonomic review, for this was not available until 1993 (Luckow, 1993). The name *D. virgatus* was almost certainly used as a 'taxonomic dustbin' by agronomists and botanists alike. *D. subulatus* has now been included in the species *D. bicornutus*. The taxonomic identity of the *Desmanthus* genotypes used on the black cotton soils of India is uncertain: its value as an animal feed is, however, widely appreciated there. To avoid confusion in this review the formal specific names have been used. Since the potential agronomic importance of this genus was recognized, attention has been paid to building up genetic resource collections in Australia.

Macroptilium. This genus is native to South and Central America and the southern USA. The early research and commercial successes of the pasture species *M. atropurpureum* and the green manure plant *M. lathyroides* have prompted much interest in this genus and a large introduction programme followed. In more recent times, *M. bracteatum* has also provided cultivars for commercial use as a green manure in Australia and in Brazil where it is native; it can provide well-balanced pasture mixtures with guinea grass. *M. atropurpureum* and *M. lathyroides* have both been successfully used in pastures on clay soils in Australia.

Other very interesting species to emerge from Brazil are *M. martii*, a colonizing annual, and *M. longepedunculatum* (syns. *Phaseolus longepedunculatum*, *M. gracile*). While these are both found on sandy textured soils, *M. longepedunculatum* has been found on a nearby clay soil site. There is a good genetic resource collection of *M. longepedunculatum*, but both species failed in 'nursery' tests and although the *M. longepedunculatum* cultivar 'Maldonado' was released for commercial use, it has been little utilized.

Arachis. This South American genus is highly regarded as a source of forage plants on that continent. *A. pintoii* is a very useful pasture ley legume or cover crop in wet tropical areas, and other species have persisted well in experiments in subtropical Queensland. Another species, the annual *A. pusilla*, was the only herbaceous legume to be found in a heavily grazed pasture in semi-arid Brazil on a grey clay, very similar conditions to those found in western Queensland. However, the genus produces subterranean seed and this poses agronomic problems for seed production and plant establishment.

Teramnus. This small genus has been well tested in nursery programmes. Member species have occasionally been found to be toxic to rabbits. *T. labialis* has performed well on clay soils in the Caribbean and is regarded as being suitable for use in long-term pastures.

Glycine. The Australian members of this genus are found on clay soils with *G. tomentella* (Woolly Glycine), *G. tabacani* (Glycine Pea) and *G. clandestina* (Twining Glycine) being important constituents of Mitchell grass pastures. While *G. tomentella* is especially valuable, other members are generally poorly productive. *G. latifolia*, a perennial, native to the heavier soils of the subtropics, has shown some promise in these situations and has been made commercially available as the cv. 'Capella'.

6.3.6 Group F

This consists of small- to medium-sized genera, some species of which are adapted to clay soils. The genera are *Neonotonia*, *Heylandia*, *Macrotyloma*, *Clitoria*, *Centrosema*, *Calopogonium*, *Leucaena*, *Dolichos* and *Zornia*.

Neonotonia contains one species of interest, *Neonotonia wightii*, which was formerly known as *Glycine wightii*. It is a well-researched species that has yielded several cultivars of value in high-altitude tropical areas, and in moister areas in the subtropics. In the Caribbean islands it has proved to be well adapted to clay soils and is recommended for use in long-term pastures.

Heylandia. *H. laterbrosa* is the single species belonging to this Indian genus. It is an annual that grows on black cotton soils in semi-arid tropical conditions where it is well grazed and persistent. It does not appear to have been widely included in plant evaluation exercises but is of interest here.

Clitoria contains one species, *C. ternatea* (Butterfly pea, Kordofan pea), which is pan-tropical and has often been found on clay soils. A good collection of the species has been accumulated in Australia. *C. ternatea* is quite atypical of the genus in that it is the only species to be pan-tropical and many of the others, perennial herbs and shrubs, are poisonous. It is highly palatable to cattle, but not to horses, and has often been selectively grazed. It has been widely used for fodder and green manure for many years. A mixture of accessions known as cv. Milgarra has been made commercially available in Australia. As the standard legume for clay soils, it is often the legume against which new forages are tested. In Africa and Central America, while it has been found to be easily established and somewhat persistent, its productivity is low compared with that of many alternative species (e.g. Chapter 2, Table 2.13, this volume) and has thus been of little interest there.

Macrotyloma contains two species that have yielded cultivars. *M. axillare* cv. 'Archer' is an African species used in subtropical semi-humid pastures. *M. uniflorum*, native to India, Africa and Australia, has produced the cultivar 'Leichardt'. *M. daltonii* is found on clay soils in Africa and has been included, with little success, in many plant evaluation programmes.

Centrosema has several species that are palatable and highly valued as quality forages. The species listed below have been found on, or have been shown to be, adapted to clay soils.

C. pubescens has long been used as a cover crop and pasture legume in wet tropical areas. In at least one instance it was found growing on a clay soil, in a recently flooded area, but it failed when tested on a clay soil in Australia.

C. pascuorum is an annual found, in one instance, on a clay soil in a semi-arid tropical environment. It has provided two cultivars for use in ley farming in dry tropical conditions in Australia and for short-term pastures on clay soils in the Caribbean.

C. schottii, from wet tropical areas in Belize, was found to be adapted to clay soils in both wet and drier regions in the Caribbean and has proved useful as a short-term pasture species on clay soils.

Calopogonium has produced one perennial species, *C. mucunoides* or 'Calopo', which has been widely used as a cover crop in wet tropical areas. In Panama it is widespread and provides valuable feed in the dry season, possibly because of its relative unpalatability. *C. caeruleum* has occasionally been mentioned as being of value in similar situations.

Zornia is a genus that is closely related to *Stylosanthes*, and is very similar to it in general appearance. There are several species native to Australia, and although not occurring on clay soils, they are common in open eucalypt woodland where they are occasionally thought to be of some pastoral value. *Z. glochidiata* is an annual that is useful as a stock feed in Senegal. *Z. meridiena*, a perennial from Brazil, appeared promising in 'nursery' studies but failed in swards (Burt, 1984b).

Zornia species have long been included in evaluation studies, but they have been unproductive and poorly persistent, and pose problems in such production. Although of no direct interest, their widespread occurrence in the Australian dry tropics and *Zornia*'s associated *Bradyrhizobia* may have contributed to the success of introduced *Stylosanthes* material in forming effective symbiotic relationships.

Dolichos at one time contained the species herein noted under *Macrotyloma*. This genus has also attracted much attention, and research, because it contains *D. lablab* (now renamed *Lablab purpureus*), a forage and pulse crop domesticated since antiquity and widely used throughout the tropical world. Other *Dolichos* species are used as pulses and forage in India. *D. sericeus* (syn. *D. formosus*) is found in Africa, often on clay soils. It has been widely tested as a potential pasture species, but without success.

Leucaena is a woody genus, a number of members of which, particularly *L. leucocephala*, have long been used as 'tree legumes' throughout the tropics. *L. leucocephala* thrives on clay soils and although it is very palatable to cattle it contains a poisonous chemical, mimosine, which can reduce fertility rates in cattle and lead to hair loss in horses. It can be difficult to establish and it requires irrigation in drier areas. The genus has been well researched and it is of no further interest here.

6.4 Discussion

The tropics contain a very rich leguminous flora and Williams *et al.* (1976) have estimated that some 50 genera and about 2000 species might furnish suitable pasture legumes. But this review has shown that the number to be found on clay soils is considerably less, and other considerations, such as the palatability and toxicity of the genera and species involved, can be used to reduce the numbers further to a list of 'best bets' for future investigation. In doing so, however, we must be open to the chance of 'serendipity', of a useful plant emerging from an unlikely source. Examples of this are *Clitoria ternatea*, one of our valuable pasture legumes, which is quite atypical of the genus, the widely used 'Pangola grass' (*Digitaria eriantha*) which stems from one plant and has not been found subsequently, and the cultivar 'Recife', of *Stylosanthes scabra*, a selection of naturally occurring variants of the cultivar 'Seca'.

Other bio-geographical and environmental considerations may provide useful information to further aid the choice of material for development. This is considered in Chapter 7 (this volume).

7 Searching for Pasture Legumes for Heavy Clay Soils in the Australian Dry Tropics and Subtropics: II. Ancillary Floristic, Climatic and Edaphic Studies

R.L. Burt† and N. Ahmad†

Abstract

In order to link likely sources of potentially valuable forage germplasm and sites of utilization, 76 locations with clay soils in the dry tropics were subjected to a grouping programme based on their climates, altitudes and latitudes and inferences were made as to the value of this approach to selection of germplasm of potential value. In addition, an analysis was made of characteristics of heavy clay soils of relevant locations internationally, supported by analysis of the role of Antiguan soil characteristics on the presence of three native fodder species. A further analysis was undertaken on the distribution of relevant tropical and subtropical flora globally. The relative importance of areas identified as sources of useful germplasm for Australian target environments is discussed.

7.1 Introduction

In an earlier report (Chapter 6, this volume) it was suggested that some genera and species of tropical legumes, deemed to be of potential utility as pasture plants, are more pre-disposed to grow on clay soils than others, and that there is a very real need to identify leguminous forage plants for these soils. As introduction and evaluation is an expensive and long-term process, some prioritization in choice of material to be studied is necessary ('best bet' species and genotypes) in order to reduce the effort required. Clearly adaptation to the target soils is a primary consideration, along with such important agronomic characteristics as palatability and the presence of anti-nutritional factors and toxins.

One approach that has been found to have great predictive value in indicating the likely utility of introductions in Queensland is the 'Agrostological Index' (Williams and Burt, 1982). The Agrostological Index relates plant distribution and plant 'type' to climatic, edaphic and ecological conditions (e.g. Burt, 1975b; Burt *et al.*, 1979, 1980b, 1983). The efficacy of the use of this method in earlier studies has been confirmed using information published in Chakraborty (2004) and Cook *et al.* (2005).

In the present work, similar approaches are used to explore the distribution of plants identified as having potential on clay soils in relation to climatic, edaphic and phytogeographic factors. Particular emphasis has been placed on such studies carried out on the island of Antigua, a major source of the diploid, more perennial form of *S. hamata* or 'Caribbean Stylo', which, unlike the *S. hamata* cultivars 'Verano' and 'Amiga', is well adapted to heavy clay soils.

7.2 Climatic studies

7.2.1 Methodology

The climate in the area a plant has been collected has long been used to suggest where it is most likely to be adapted elsewhere (e.g. De Candotte as quoted in Darwin, 1859; Papadakis, 1966; Burt *et al.*, 1976; Burt and Reid, 1976). A method that has proved to be very useful in the selection of plants of potential for given environments from a collection of accessions involves exploring the plant variation present in the climates that have yielded those plants by means of a principal coordinate analysis. Such an analysis provides a pictorial display, avoiding the need to define boundaries within which there are usually sets of continuously varying data. Information about a species, or 'types' within a species, in a specific collection of plants can be superimposed onto such a display. The climatic zones depicted in the display can then be equated with those of various 'target' areas. Indeed climatic data for the target areas can be included in the principal coordinate analysis. This methodology is particularly suited to the purpose in hand, where little or nothing is known about the climatic requirements or environmental limitations of the species under study and where it is therefore not possible to use some of the geographic information systems (for further comments on the latter, see Maass and Sawkins, 2004).

The details of the computing methodologies have been amply provided elsewhere (e.g. Abel and Williams, 1981). The characteristics used in the analysis are listed in Table 7.1. They are a compromise between what was generally available for the areas in question, the amount of the data that could sensibly be included and the larger amount needed for more detailed studies. Such analyses have been useful in the past and, to be relevant for our present purpose, more detailed studies need to incorporate other characters such as whether the plant colonizes or not, water runs on or runs off of the original collection site, and disturbance levels at the site of collection (Chapter 9, this volume). The sites studied are listed in Table 7.2.

Table 7.1. Climatic data used in the analysis.

1. Latitude (degrees)[a]
2. Altitude (metres)
3. Annual rainfall: Number of months giving 85–92% of the total
4. Temperature: Highest monthly maximum during growing season (°C)[b]
5. Temperature: Lowest monthly minimum during growing season (°C)[b]
6. Temperature: Lowest monthly maximum in dry season (°C)[c]
7. Temperature: Lowest monthly minimum in dry season (°C)[c]
8. Average annual rainfall (mm)

[a]Expressed as degrees or fractions of a degree not distinguishing between N and S. [b]Growing season is defined as months during which 85–92% of the total annual rainfall is received. [c]Dry season is defined as the months in which the remainder of the rain is received.

7.2.2 Results

The first two components, axes A and B (Fig. 7.1), account for 52% and 25% of the total variation respectively. This is, heuristically, more than adequate for the purpose in hand. The figure shows that axis A, the first component, is positively associated with hot climates that have high minimum temperatures, and negatively associated with high latitudes and a rainfall that is distributed over long periods. The second component, axis B, relates positively with high rainfalls and long growing seasons, and is negatively associated with high maximum temperatures. The figure also shows the site numbers and the geographical regions in which they occur; the names of the sites are presented in Table 7.2.

Five environmental zones, labelled I to V, have been designated in the figure. Zone I represents dry/semi-arid tropical locations with high maximum temperatures, high minimum temperatures, low rainfall and very short growing seasons. Zone II has wet and humid tropical locations that have higher rainfalls distributed over longer growing seasons. Zone III sites are also dry/semi-arid tropical but are less hot than zones I and II and have relatively short growing seasons. Zone IV has subtropical sites with a limited rainfall spread over a relatively long growing season, some of which occurs in winter. Zone V has only a single entry, Fort Worth in Texas, USA. This site is at higher latitude, is non- or subtropical, and has been included in this study because of its soil types, dry conditions and presence of several native *Desmanthus* species.

Australian sites are present in all zones except the last. In zone I these range from Toorak, in western Queensland, to Victoria River Downs in the Northern Territory and to Kununurra in Western Australia. In zone II the only Australian entry is Darwin. Zone III, designated 'dry tropical', contains Lansdowne, the CSIRO research station at which some of the evaluation work to

Table 7.2. Site locations for climates associated with clay soils. Most of these sites had clay soils in the vicinity but some non-clay sites were included for comparative purposes. In the literature 'clays' may also be referred to as 'black cotton soils', 'vertisols', 'grumasols' etc. In the African sites they were often found in 'mbugas'.

USA
1. Corpus Christi – clay and caliche
2. Fort Worth – clay
3. Pahala, Hawaii

Central America
4. Cayo, Belize
5. Colima, Mexico
6. El Progreso, Guatemala
7. Merida, Yucatan, Mexico
8. Ocu, Panama
9. Orange Walk, Belize
10. Tuxpan, Mexico
11. Villahermosa, Mexico
12. Zacapa, Guatemala

South America
13. Calabozo, Venezuela
14. Concepcion, Argentina
15. Georgetown, Guyana
16. Guayaquil, Ecuador
17. Juazeiro, Brazil – clay collecting site
18. Maracaibo, Venezuela
19. Salta, Argentina

The Caribbean
20. St John's, Antigua – clay soils and testing sites
21. Willemstadt, Curaçao taken as 18, Maracaibo, Venezuela

Africa
22. Abu Naama, The Sudan
23. Accra, Ghana
24. Akjoujt, Mauritania
25. Athi Plains, Kenya
26. Bouake, Ivory Coast
27. Dagana, Senegal
28. Edwelani, Zimbabwe*
29. Gallayel, Benin*
30. Gondo, Mali
31. Kanem, Nigeria*
32. Karamoja, Uganda – clay plain
33. Khartoum, The Sudan
34. Luangwa, Zambia
35. Mogadishu, Somalia
36. Mongu, Zambia
37. Nrame, Niger*
38. Nylsvley Nature Reserve, South Africa
39. Rustenburg, South Africa – clays
40. Ukiriguru, Tanzania – clay and non-clay
41. Zambezi Valley, Mozambique
42. Zinder, Niger

India
43. Bellary – black cotton soil and loam
44. Manmad
45. Morvi
46. Pandharpur
47. Yeotmal

Australia
48. Avon Downs
49. Balfe's Creek
50. Biloela
51. Blackall
52. Charleville
53. Clermont
54. Collinsville
55. Creswell
56. Dalby – clay
57. Darwin
58. Emerald
59. Glenore
60. Hughenden
61. Kununurra – clay and non-clay
62. Lansdown Research Station, near Townsville – clay and earth soils
63. Longreach
64. Marlborough
65. Narayen
66. Roma
67. Springsure
68. Theodore
69. Toorak
70. Townsville
71. Victoria River Downs
72. Walget
73. West Kimberleys – clay

In Fig, 7.1 data only for clay soils at:

74. Darling Downs
75. Chinchilla
76. Chudleigh Park
77. Meteor Downs
78. Orion Downs
79. Rio Grande do Sul

*These sites could not be identified (Eds.)

be described later (Chapter 8, this volume) was carried out, Balfe's Creek (near Charters Towers), and Hughenden, inland from these sites. Zone IV extends from Central to Southern Queensland, contains many inland areas of Mitchell grass and Brigalow and extends to the cropping zones in Central Queensland and the Darling Downs. Zones I, III and IV are of prime concern here.

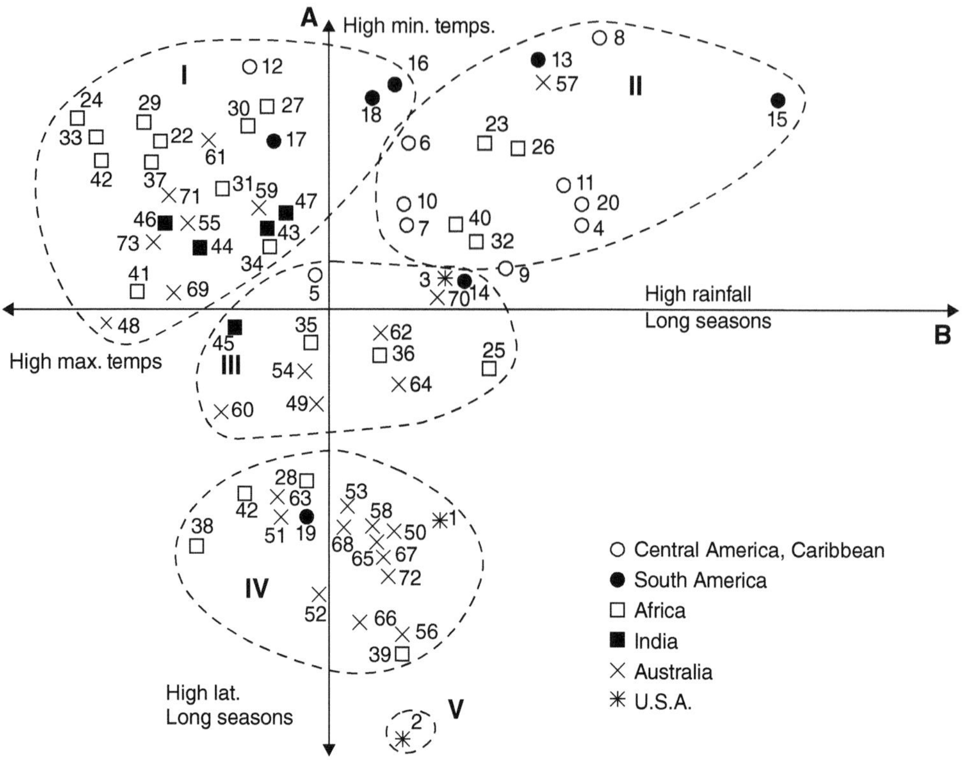

Fig. 7.1. Ordination of various sites with respect to climatic attributes. Site numbers as shown in Table 7.2.

Climatic homologues for Australian zones

The analysis presented in Fig. 7.1 may be used as a convenient framework on which to 'hang' various types of information for inspection. In zone I for instance, sites 22 (in Sudan) and 18 (Maracaibo in Venezuela) have provided, respectively, *Clitoria ternatea* cv. 'Milgarra' and *S. hamata* cv. 'Verano', arguably the two best-adapted plants for use on clays and non-clays respectively. The analysis places them near Kununurra (61) and Victoria River Downs (71), two areas in Australia where these plants have been found to be well adapted after introduction. Introductions such as *C. pascuorum* and *Arachis pusilla* from Juazeiro in Brazil (17) and *Desmanthus* from Zacapa (12) in Guatemala might also be suited. The Indian sites that are also found within the zone I group (sites 43 to 47) can provide *Vigna trilobata* and *Heylandia laterbrosa*, both of which flourish on heavy 'black cotton soils' and provide valuable grazing. Such hypotheses can be tested in the process of selecting plants for evaluation.

The Americas are a centre of diversity of the *Leguminosae*, however they lack significant areas of clay soils in climates homologous to the Australian target areas of this study (Dudal and Bramao, 1965). There have been few forage legume collections made on these South American soils. In Central America on the other hand, though the clay soils have had some collection attention they are mostly located in wetter environments (zone II). The significance of this is explored in the next section.

The possible significance of sites in the New World tropics and subtropics

The occurrence of clay soil areas in the New World tropics is very limited compared with that in similar climatic zones in Australia, India and Africa; Fig. 7.2 shows their occurrence in the data

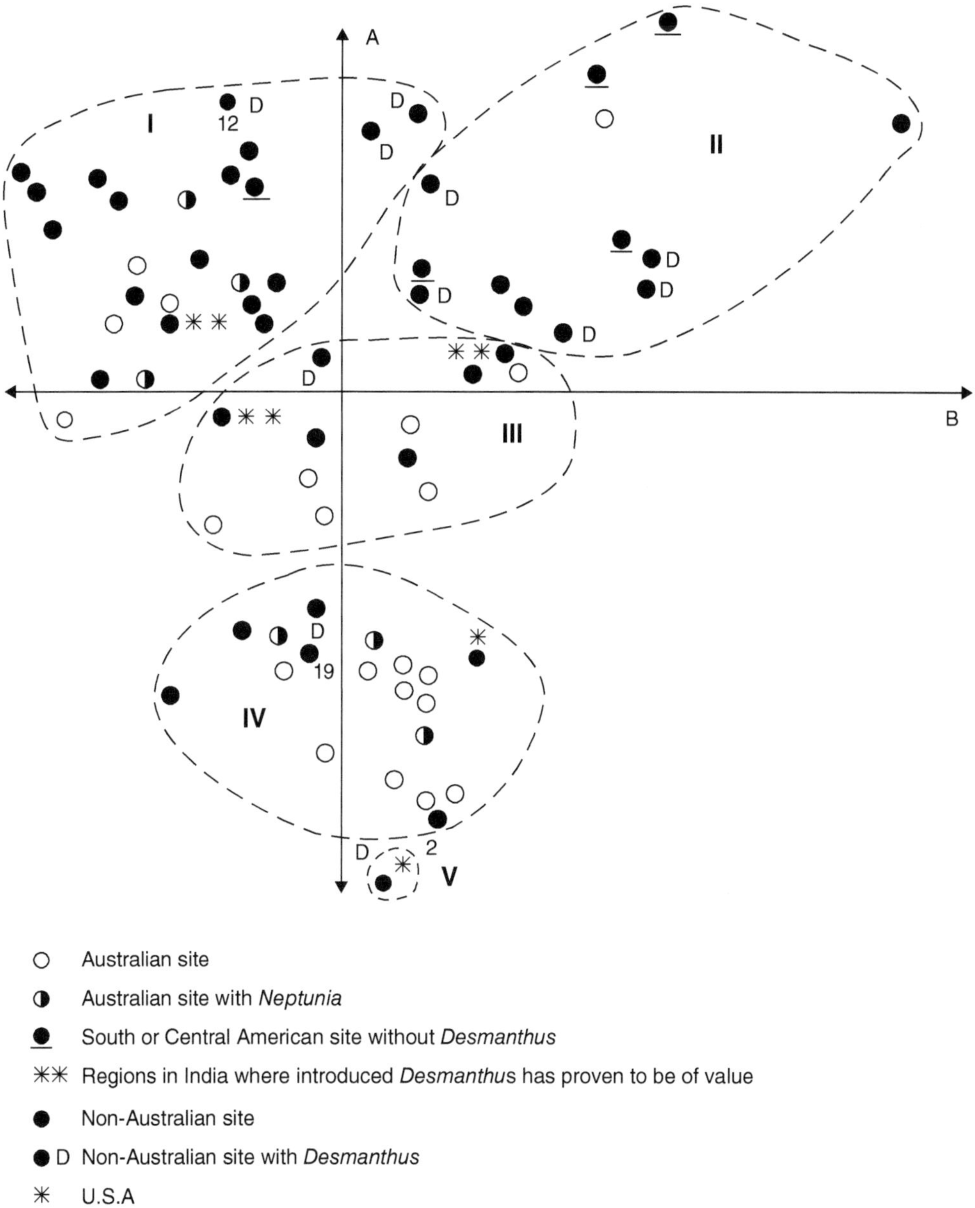

Fig. 7.2. Distribution of *Desmanthus* in relation to climates.

set used here. The genus *Desmanthus* can be found on many of these sites and these are indicated by the letter D in the figure. It occurs over a very wide range of climates, from the zone I semi-arid, clay soil situation in Guatemala (12), through to the zone III subtropical areas near Salta in Argentina (19), and even into zone V high latitudes in the USA (2). It has long been introduced to, and found to be a valuable plant in, India and Hawaii (zone I and III sites with double stars).

In all of these climates *Desmanthus* has been found to be adapted to clay soils, but different species are involved, as discussed later. Sites in Brazil without *Desmanthus* collections such as site 17 are of particular potential as sources of useful *Desmanthus* germplasm because the country has provided the bulk of tropical pasture legumes adapted to Australian conditions.

7.3 Edaphic studies

7.3.1 General edaphic study

Methodologies

Most soil classifications are 'general purpose', designed to be employed for a wide variety of uses and, as such, they may not be suited for special purposes such as explaining plant distribution (e.g. Burt *et al.*, 1979, where species adapted to widely different soil acidities were found to be growing on the same type of soil). This is not surprising because plant distribution may depend on what is, in general terms, relatively minor differences in, for instance, the level of P, K or Ca. Such differences can be important for plant growth as is shown by the long development of the soil Fertility Capability or Functional Classification (FCC) (Sanchez *et al.*, 1982, 2003). This system delineates general soil types by soil texture, pH and acidity status and so on, and is supported by numerous 'modifiers' such as nutritional elements, the levels of which can indicate adequate or restricted growth of specific plant species. This is essentially the approach developed by Papadakis (1966) for climatic classification, but in his case he was able to employ well-documented information about the climate requirements of individual crops to define boundaries between climate types.

With such poorly known species as are involved in this study, detailed information on their responses to soil parameters is not available, so the approach adopted has been to use soil data in the same manner as climatic data was used in Papadakis' study. However, detailed soil information is much less readily available than that of climates, especially in many of the climatic zones of interest here. In addition, different analytical methods are employed in determining soil characteristics, for instance for determining soil P. This has greatly reduced the number of data sets useable for analyses and compromises have been necessary between that which was available and that which would be preferred. An initial study was undertaken using clustering methods on soil analytic data from 23 locations mainly with clay soils in Australia, India and Africa. Comparison was then undertaken with more detailed results (on species and their distribution) from the island of Antigua, and other sites, where it was necessary to use only what soil characteristic information was essentially the same. A list of the characteristics included in the analysis, and the areas for which they were obtained, are provided in Tables 7.3 and 7.4.

Results

The results of the general edaphic study showed that the first two components, vectors I and II in Fig. 7.3, accounted for only 55% of the total variation, subsequent component values drifting down slowly and usually being only associated with one or two soil characteristics. The first vector was, surprisingly, concerned with silt/clay ratios which were antagonistically related to soil acidity. In the second, high cation exchange capacities were, understandably, negatively associated with high sand contents.

The data set used is primarily from sites in Australia and Africa and it shows that they are substantially different. However this difference may be of little significance in terms of plant growth and adaptation as illustrated by an accession of *Clitoria ternatea* from a site on the clay soils of Sudan (16) that has been a very successful component

Table 7.3. Characteristics of the surface soil used in the analyses.

1. pH: at surface and 0.5 m
2. Mechanical analysis: sand, sandy loam, loam, clay
3. Depth: shallow (<0.5 m), medium (0.5–1 m), deep (>1 m)
4. Drainage of surface soil: good, medium, poor
5. Drainage at 0.5 m: good, medium, poor
6. Structure of surface soil (%) - stones, sand, silt, clay
7. Soil mineral levels: total P, K, N; extractable P, K; exchangeable K, Mg
8. Cation exchange capacity (CEC)
9. Salinity: surface soil and at 0.5 m
10. Organic C: total

Table 7.4. Locations of the various soils studied.

Locations and map symbols	Latitude (approximate °)
Australia ×	
1. Darling Downs, Queensland	27 S
2. Dalby, Queensland	27 S
3. Chinchilla, Queensland	27 S
4. Chudleigh Park, Queensland	19 S
5. Lansdown Research Station, near Townsville, Queensland	19 S
6. Kununurra, Western Australia	6 S
7. Meteor Downs, Queensland	24 S
8. Orion Downs, Queensland	24 S
India ■	
9. Bellary, black cotton soil	15 N
10. Morvi	23 N
11. Manmad	20 N
Africa □	
12. Ukiriguru, Tanzania	3 S
13. Rustenburg. South Africa	25 S
14. Kanem, Nigeria	10 N
15. Mongu, Zambia	15 S
16. Abu Naama, Republic of The Sudan	15 N
17. Athi Plains, Kenya	1 S
18. Dagana, Senegal	17 N
19. Karamoja, Uganda	2 N
Antigua ○	
20. Ottos Clay	17 N
21. Bendals Clay	17 N
Brazil △	
22. Juazeiro	9 S
USA ★	
23. Corpus Christi, Texas	28 N

of the cultivar 'Milgarra' and has grown well in nursery studies at location 5 in Australia. On the basis of the main discriminatory characteristics shown (Fig. 7.3), the most similar soils are those to be found in India (sites 9 to 11) and the Caribbean (sites 20 and 21). The site in Brazil that was of interest because of its climatic similarities to others in the semi-arid tropics and the success of *S. humilis* (Townsville stylo) in the general region had a clay quite dissimilar to those found in Australia. A site in the USA (23) has an apparently similar clay soil and contains several species of *Desmanthus*. However, as noted earlier, even slight differences in soil properties can affect their suitability for various plant species, and many of these characteristics do not figure in the results.

A very interesting feature to emerge is the appearance of the silt/clay ratio as a major discriminatory factor. This factor does not appear in soils classifications but could be important to plant growth in terms of both texture and mineral availability (R. Isbell, personal communication) (and see next section for further mention). Its incorporation in special purpose soil classifications could be useful.

7.3.2 Edaphic conditions and legume distribution in Antigua

Many of the Caribbean islands are relatively small and, although many of their areas may be dominated by clay soils, they do not figure in world soil surveys, such as that used herein (Fig. 1.1, Chapter 1, from Dudal and Bramao, 1965). They possess climates and soils of interest (see Antigua, in Figs 7.1 and 7.3) and a flora that is rich in herbaceous legumes (Stehle, 1956; Loveless, 1960; Howard, 1973). Antigua, situated at the junction of two floristic zones (Howard, 1973) and with a wide variety of soils (Hill, 1966) is ideally suited as a major site for collection and evaluation. The University of West Indies, with the long-term sponsorship of IDRC (the International Development Research Centre of Canada), used Antigua as a base for such studies (Spurgeon, 1977).

Antigua is a small island of low relief and it has little climate variation. Plant distribution can be primarily attributed to differences in soil type, forms of disturbance and land use. The island had, until recently, been used primarily for sugar cane and thus the vegetation has been much altered from its original natural state.

Methodologies

An ecological approach was used in both plant collection and initial evaluation activities in Antigua. The collected germplasm was then exchanged with material originating mainly from wetter environments and different soil types in Belize, where another part of the project was situated. The information gained from studies on the exchanged material in both countries produced information of great comparative value. Principal component analysis was used on data on soil parameters (Table 7.3) abstracted from Hill (1966).

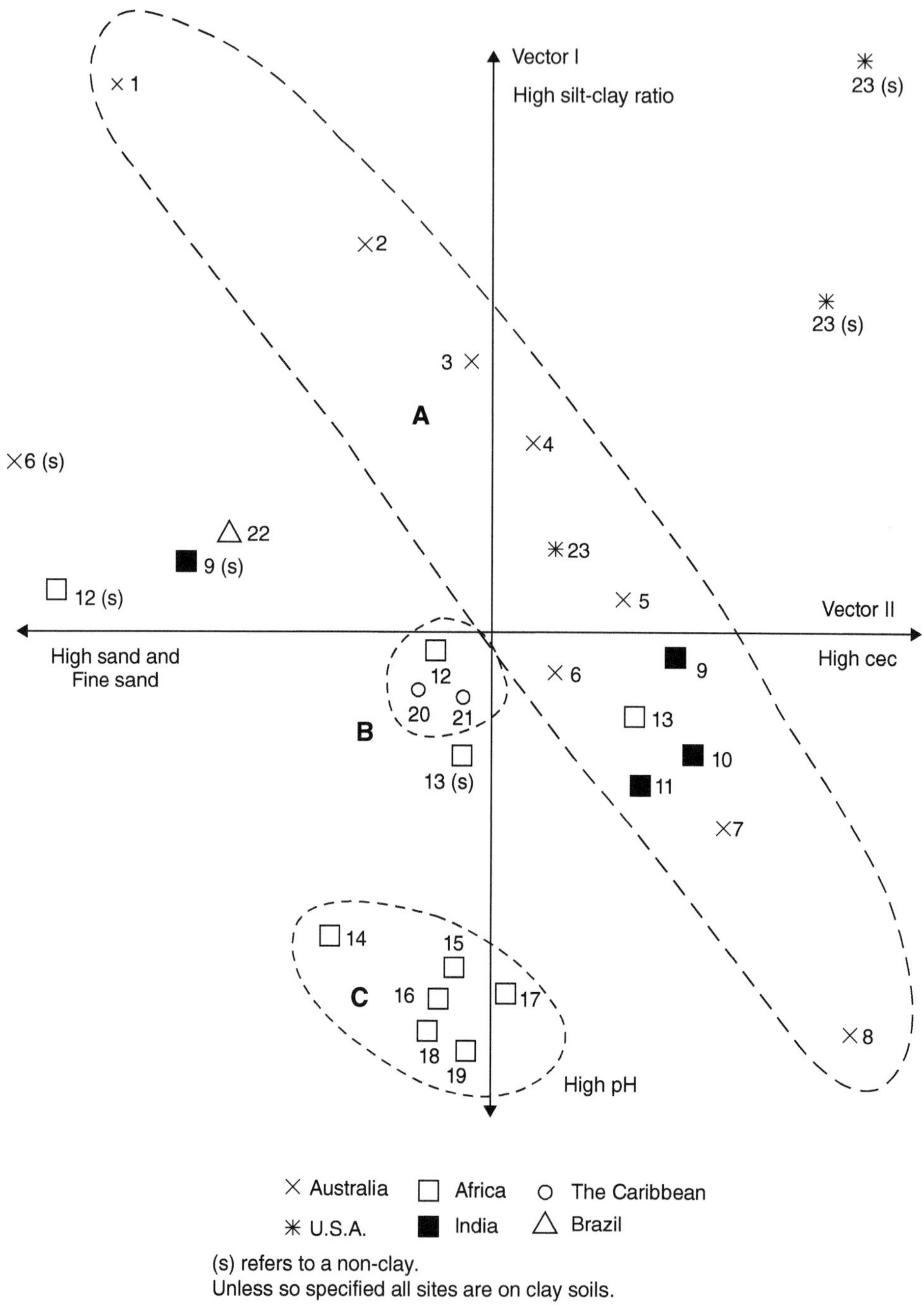

Fig. 7.3. Soil variation across a range of tropical locations.

The results are displayed in Fig. 7.3. The first two components or vectors accounted for 48% and 17% of the total variation respectively. The first vector at its positive end was concerned with high values for drainage and sand content and acidity (pH < 6.0) at depth, while the negative end was correlated with high clay content, poor drainage, salinity and high soil pH.

In the second vector high values are associated with high clay contents, waterlogging and poor drainage, and high levels of salinity, while soils located near the negative end are well drained, high in sand and high in $CaCO_3$ content, have high pH and relatively high levels of exchangeable K. The third vector accounts for only 10% of the total variation and is primarily concerned with only one factor, exchangeable Na. This is related to salinity, covered in the second vector, and has been disregarded. Results of the analysis are presented in Fig. 7.4. Decreasing values of the first vector indicate a move from shallow, volcanic, freely draining hill soils to saline, poorly drained sedimentary soils. The second vector is concerned with distinctions between more open-textured coralline sand, of high alkalinity (with relatively high levels of exchangeable K) and deep, poorly drained saline soils. The figure thus graphically portrays meaningful and important differences between a wide range of soil types.

Ecological surveys were carried out at locations representing native grasslands on soil types of major agronomic and economic significance. Data were acquired using 500 m long line transects with sixteen 1 m × 1 m quadrats located at equidistant intervals along each. The species content and dominance of each legume was noted and the names of the associated grass species recorded (for further details contact J.M. Keoghan). Mean dominance ratings for each legume were then calculated.

Legume growth in relation to soil type was investigated in a controlled experiment using undisturbed soil cores. Results of this are briefly mentioned and experimental details

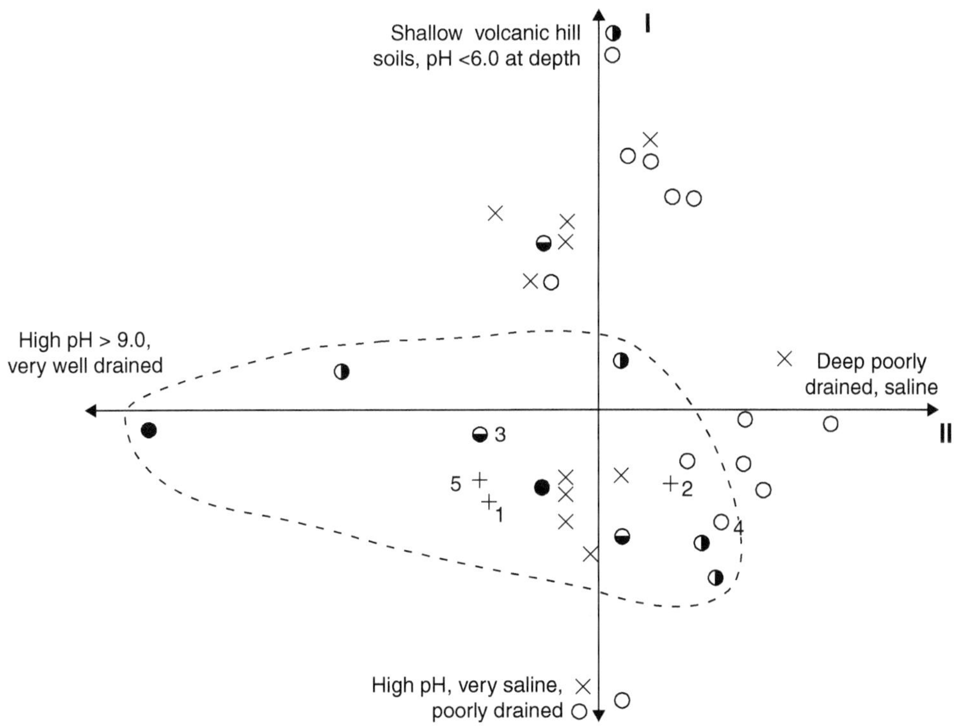

Fig. 7.4. Distribution and dominance of *Stylosanthes hamata* in relation to soil variations. For detailed explanation see text.
Legume is: × absent; ○ present; ◑ common; ◒ frequent; ● dominant; 1 and 2, Australian clays (at site 6, Fig. 7.3); 3 and 4, Evaluation sites at Antigua (sites 20 and 21, Fig. 7.3); 5, Brazilian clay (at site 22, Fig. 7.3).

may again be obtained from one of the authors (J.M. Keoghan).

Results

Early inspection of the data revealed much of interest: *Desmodium incanum*, for instance – widely regarded as a valuable pasture legume – was only ever found to be of significance on the more acid, volcanic soils. Caribbean stylo (*Stylosanthes hamata sensu stricto*) was often well distributed throughout pastures in 'clumps' in which it was dominant. Information for three species of considerable agronomic interest is shown in the following figures for comparative purposes; soil information for two Australian sites is also presented.

Stylosanthes hamata sensu stricto (Fig. 7.4) is found throughout Antigua, which is not surprising because it is an annual/biennial which pre-adapts it as a 'colonizing species' (*sensu* Baker, 1974). It is therefore more prominent in disturbed environments (Chapter 3, this volume). It is common under alkaline conditions and only dominant or subdominant in better drained, less saline conditions. As in previous studies with *Stylosanthes* species (Burt *et al.*, 1979), its presence and growth was favoured in mixed sward situations by relatively high levels of exchangeable K. Hall (1975–2007, personal communication) has explored this situation in competition studies with *S. humilis* in controlled environments, and in the field, and has achieved the same result.

Desmanthus virgatus (Fig. 7.5) was also found to be primarily restricted to more alkaline soils and is only infrequently found elsewhere. It is better suited to deeper, heavier soils and is more frequent in saline, waterlogged conditions. *Teramnus* (Fig. 7.6) has a distributional pattern generally similar to that of *D. virgatus*, but it is probably somewhat more restricted.

In the controlled soil environment study using soil cores and a wide range of soil types without grass competition, the major soil factor affecting yields of a wide range of legume species was the silt/clay ratio (for further details contact J.M. Keoghan). This ratio is a major discriminatory characteristic in a wide range of predominantly clay soils (see previous section). This contrasts with the results obtained from the field studies.

7.4 Floristic studies

In the previous sections it has been possible to nominate genera, and sometimes species, which seem to be adapted to clay soils in climates for which there have been few cultivars identified, but these studies have not delineated any geographical areas of particular interest. Because of the success of *Stylosanthes* and *Centrosema* as pasture plants on tropical sandy-textured soils, it is well known that these genera are native to South and Central America and the Caribbean. However, it is less commonly known that other important forage genera, such as *Vigna*, also occur in these areas. Given the wealth of native legumes on Australian target soils, these areas that are so similar floristically, and are in the centre of legume diversity, are likely to contain considerable material of fodder potential. This study seeks to identify geographical areas of potential as sources of legumes for clay soils in the Australian dry tropics and subtropics.

7.4.1 Methods

This study compares the legume floras of a range of tropical regions of interest (see Table 6.1, Chapter 6, this volume). The selected regions are listed in Table 7.5 and their geographic locations shown in Fig. 7.7.

For each of the regions listed, the genera present, and the number of species within each, were determined and the similarities between them illustrated by the results of a 'network analysis'. Details of the methodology are presented elsewhere (Williams *et al.*, 1980). A hierarchical classification has been used to document the reasons behind these relationships and the methods used are those which have been found to be most useful for this purpose (Burt and Williams, 1979b).

7.4.2 Results

The floras fall into zones, denoted I to V in Fig. 7.7, and are shown diagrammatically in Fig. 7.8 in order to illustrate the geographical relationships between these regions and floristic zones. Zone I contains mainly South American floras. The Brazilian

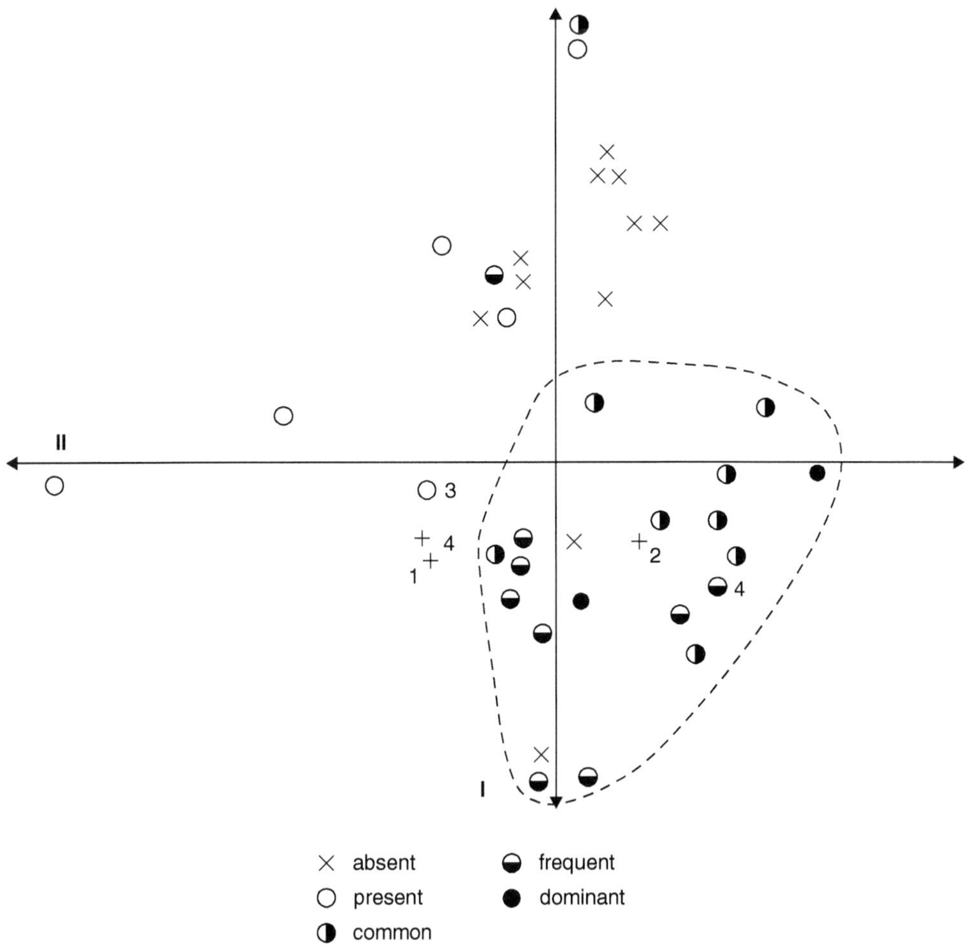

Fig. 7.5. Distribution of *Desmanthus virgatus* in relation to soil variants. Symbols as in Fig. 7.4.

flora, which is well known for its richness in legume species and is the source of many Australian cultivars, is linked through Venezuela to Peru and Guatemala; the latter three are strongly linked, and also through Venezuela to Panama in zone II and through it to other zone II floras that are mainly of islands, many of them associated through trade and many of them heavily disturbed by cultivation of sugar cane and other crops. The native and introduced legume floras of Manila and Fiji are quite similar, and the introduced floras are placed terminally here. Zone III covers northern Mexico and the southern USA, and is linked to zone II through the Bahamas. These first three zones are located distantly from the legume floras of northern tropical Australia (Northern Territory and Queensland), both of which are weakly linked to Sudan. Finally, again distantly placed, are the closely similar floras of West Africa, tropical East Africa and India.

The Sudan is a very large country with many desert and semi-arid areas, and it stretches from the tropics almost into the subtropics. Its flora is quite different from those in tropical East and West Africa. With the exception of zone II floras, which are probably the result of anthropomorphic influences and the invasion of disturbed areas, the observed differences seem to reflect occurrences

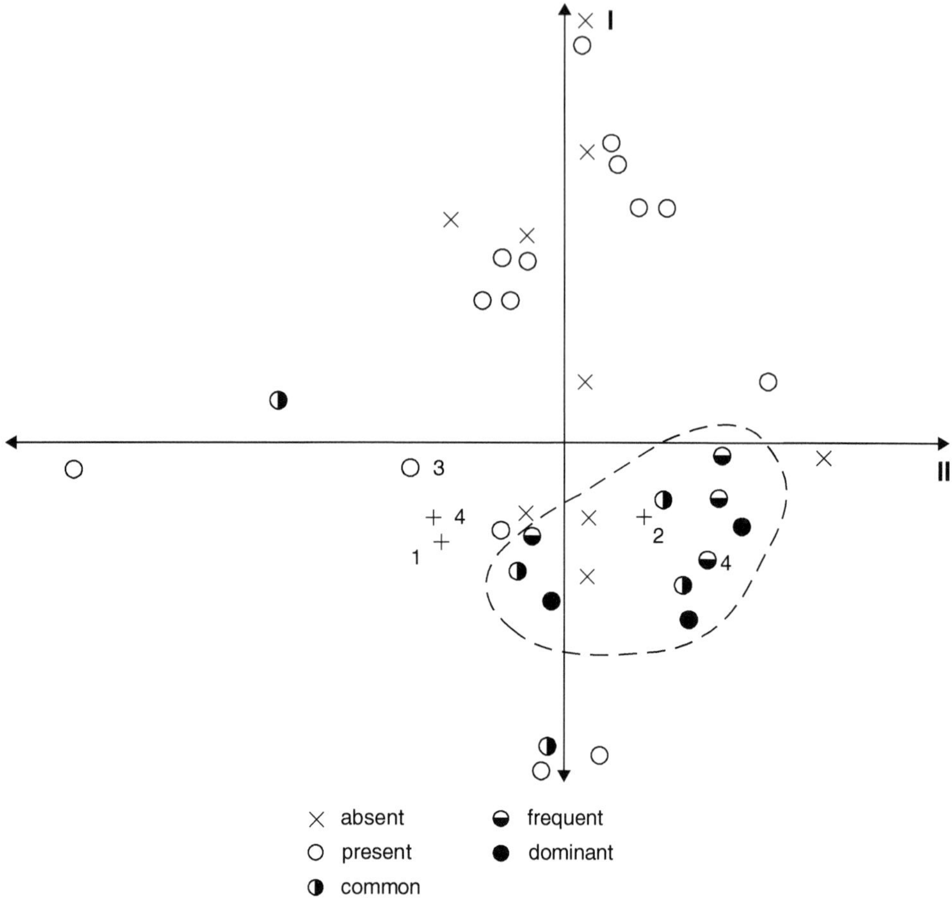

Fig. 7.6. Distribution of *Teramnus labialis* in relation to soil variations. Symbols as in Fig. 7.4.

during Gondwana (e.g. White, 1986). Australia clearly has a very distinctive legume flora that appears to be closest to the well-known Afro/Indian association. The other most distinctive legume flora, well removed from that of the other continents, is in South and Southern Central America; of these American flora, that of Brazil is the most different.

The distinguishing features of the various groupings are illustrated in Fig. 7.9, which shows the results of the hierarchical classification. (There is complete agreement between the groups designated between the two forms of analysis, and the order in which the zones are shown in Fig. 7.9 has been presented in the same order as that shown for the network analysis in Fig. 7.8).

First, members of zone V (tropical East and West Africa and India) are separated because of their high species content of *Crotalaria*, *Rhynchosia*, *Tephrosia* and *Aeschynomene* (only the latter has yielded pasture cultivars). Then the Indian flora is removed: it contains the monospecific genus *Heylandia* together with *Atylosia* and *Pueraria* – it is poorer in *Zornia*, *Indigofera* and *Sesbania*. Indian species and genotypes of *Atylosia* and *Pueraria* have provided plants of economic value, while those from Africa (Regions 2, 3) have not, and two, *Indigofera* and *Sesbania* often figure as weeds though the latter has been developed as a fertility tree for use in crop rotations, and to a lesser extent as fodder, and is being successfully promoted as such in Africa.

The South and Southern Central American zone (I) then separates. It is rich in *Canavalia*, *Phaseolus*, *Stylosanthes*, *Clitoria* and *Calliandra*, all of which have yielded very useful plants. North-east

Brazil (flora region 11) is terminally placed in this zone and has a unique flora: Brazil is the main source, for instance, of the genus *Arachis* and one section of this genus, the *Triseminalae*, is native to the north-east of Brazil (Kerridge and Hardy, 1994).

The tropical Australian/Sudanese grouping, zone IV, is then separated. It is distinguished by the presence of *Glycine*, and the relatively large number of species of the genus *Alysicarpus* (see Table 7.6) that has yielded pasture cultivars. *Glycine* is of significance as a native pasture legume and due to its relationship with *G. max*, the soybean. The zone is rich in *Indigofera*, *Tephrosia* and *Sesbania*, members of which can often be toxic and weedy.

The more northern subtropical floras of zone III, Texas and the Sonoran Desert, are separated from the largely island legume floras of zone II. The northern sites are rich in *Desmanthus*, *Desmodium* and *Leucaena* and – like Australia – in *Neptunia* and *Psoralea*. The zone II island sites (just referred to in Fig. 7.9) possess *Desmanthus*, *Desmodium* and *Leucaena* but are depauperate. *Desmanthus* is especially important because of its proven ability to grow on clay soils.

The nature of these floras is explored further in Table 7.6. There, for the sake of brevity and ease of presentation, the number of genera listed and flora covered have been reduced and, as the size of the regions covered varies greatly, the data are primarily given as percentages of the total number of species per genus covered by this study.

In the sense given above, the island floras (zone II) contain very few species and *Desmodium* is the most common genus (with *Vigna* in Manila and *Crotalaria* in Guadeloupe). In some of the islands, for instance Fiji and Manila, much of the original vegetation was removed for agricultural purposes and invasion by non-natives occurred; the numbers of invasive species belonging to those genera selected for study outnumber the

Table 7.5. Countries and areas covered by the legume floras chosen.

Country and area
1. India
2. Tropical West Africa
3. Tropical East Africa
4. Sudan
5. Australia – Queensland
6. Australia – Northern Territory
7. Fiji – native
8. Fiji – introduced
9. Philippines – native
10. Philippines – introduced
11. Brazil – the north-east
12. Venezuela
13. Peru
14. The Galapagos Islands
15. Curaçao
16. Guadeloupe
17. The Bahamas
18. Panama
19. Guatemala
20. USA – Texas
21. USA – The Sonoran Desert

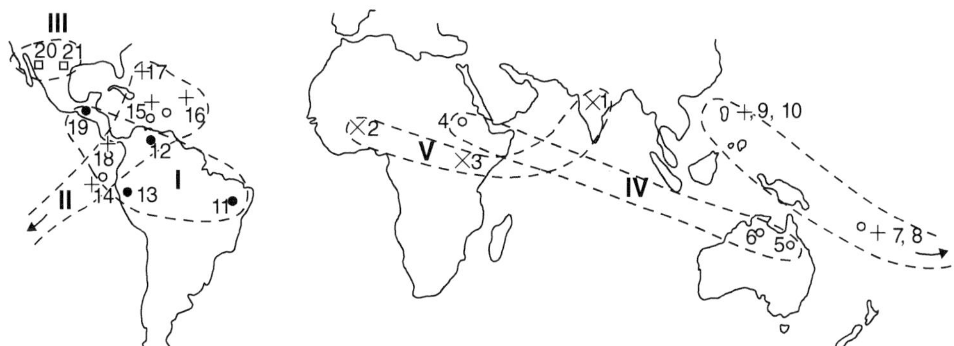

Fig. 7.7. Network analysis of the floristic data.
Numbers represent geographical areas labelled as such in Table 7.5. Roman numerals indicate the regions nominated and correspond with those in Fig. 7.9.

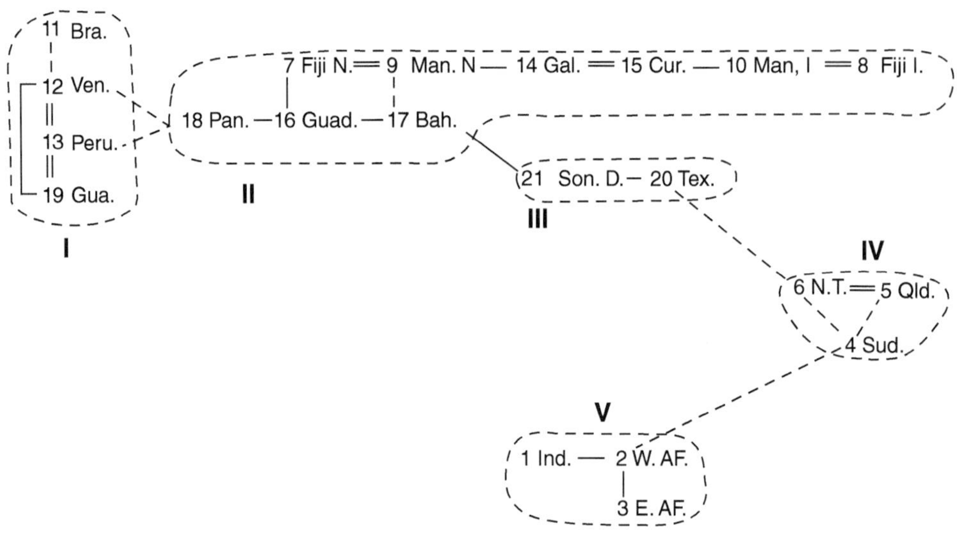

Fig. 7.8. Relationships of the various zones as shown by their floras.
Dashed lines, similar; single lines, closely similar; double lines, most similar. + as listed on Table 7.5. I – introduced species; N – native species. Zone I: 11 Brazil; 12 Venezuela; 13 Peru; 19 Guatemala. Zone II: 18 Panama; 16 Guadeloupe; 17 Bahamas; 7 Fiji native; 9 Manila native; 14 Galapagos; 15 Curaçao; 10 Manila introduced; 8 Fiji introduced. Zone III: 21 Sonoran Desert; 20 Texas USA. Zone IV: 6 Northern Territory, Australia; 5 Queensland, Australia; 4 Sudan. Zone V: 1 India; 2 West Africa; 3 East Africa.

number of native species in Fiji by 37 to 12, and in Manila by 36 to 18. However, the invasive flora is generally similar to the native, with *Desmodium* and *Crotalaria* being most prominent: undesirable additions include *Indigofera* and *Sesbania.* Purposeful introductions, *Centrosema, Stylosanthes* and *Desmanthus,* were from South and Central America (usually for use as cover crops in plantation agriculture) and the *Vigna* species were almost certainly early introductions by the Spanish galleons from India or Africa. The same pattern emerges in tropical Australia.

Finally, for future reference, attention is drawn to the rich legume floras of Central America and the southern USA (as represented by Guatemala and Texas).

7.5 Preliminary discussion

7.5.1 Climatic analyses

Most of the sites included in these analyses were chosen because they possessed clay soils. The majority of these were African and Australian and they figured prominently in all of the main climatic types designated (Fig. 7.1, I to V). Zone I represents semi-arid tropical climates and contains Australian sites ranging from Toorak in the south to Kununurra in the north. The only South or Central American sites with similar climates and clay soils and thus likely to yield 'new' pasture species are 12 (Zacapa in Guatemala), 16 (Guayaquil, Ecuador), 17 (Juazeiro, Brazil) and 18 (Maracaibo, Venezuela). The latter has yielded *S. hamata* cv. 'Verano' and cv. 'Amiga' – and these thrive in both zone I and zone II climates (Cook *et al.*, 2005). From the other sites, respectively, there are collections of *Desmanthus virgatus, Stylosanthes sympodialis* and (from 17) *Centrosema pascuorum, C. brasilianum* and *Arachis pusilla*: both *Centrosema* accessions have been used to provide cultivars and the other species are clearly of major interest. And to these can be added *Vigna trilobata* (from near sites 43 and 44 in India, and 32 in Sudan), *Heylandia laterbrosa* (from near 43) and *Desmodium dichotomum, Vigna putigera* and *Clitoria ternatea* (from 32 in The Sudan); the latter accession is a component of the cv. 'Milgara' and this has proved to be adapted to climates I, III and, marginally, IV in Australia.

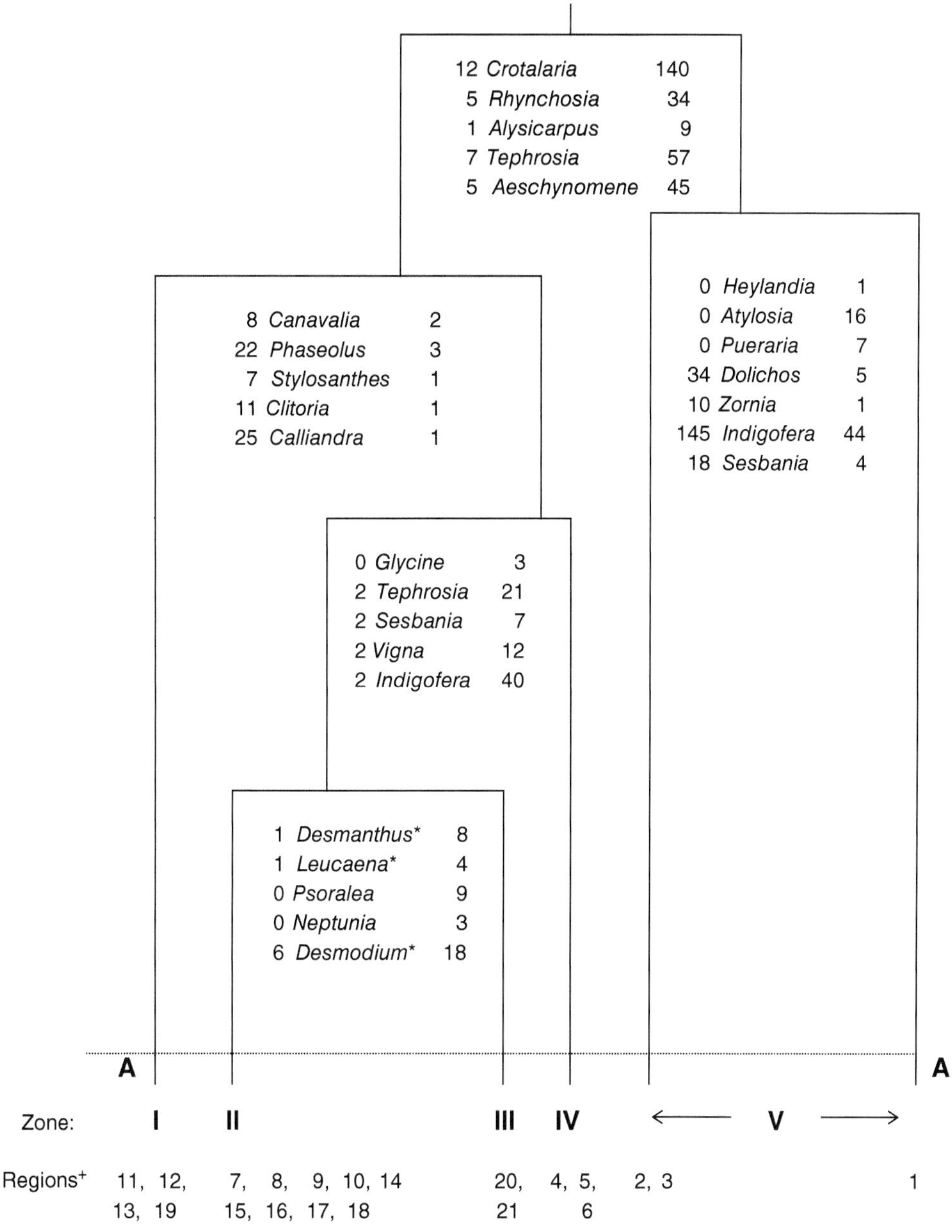

Fig. 7.9. Hierarchical classification of the floras to the 6 group level.
See text for interpretation. The attributes (genera and number of species involved) responsible for the divisions are superimposed. The figure preceding each genus name is the number of species of that genus in the segregated grouping represented by the left arm of the diagram at that level, and those following the name are the numbers of species in the grouping represented by the right arm. The geographical zones (Fig. 7.8) are shown by Roman numerals and the regions are as listed in Table 7.5. A – 6 group classificatory level; *Also appear in zone II – the islands. +Regions within each zone.

Table 7.6. Analysis of occurrence of species within selected genera in the five environments defined in this report. Given as the per cent present of the total number of species in each genus and listed in descending order of importance

	Classificatory group (zone) and region (Species present/total species in genus × 100)																			
	I		II							III		IV			V					
	Guatemala		Guadeloupe		Curaçao		Manila native			Texas		Queensland			East Africa			India		
Genus	%	A	%	A	%	A	%	A	N	%	A	%	A	N	%	A	N	%	A	Mean*
Genera of importance in Queensland																				
Crotalaria	8		19		4		11		4	5		18			31			30		13.6
Desmodium	21		14		17		16		8	21		15		1	3			16		18.3
Tephrosia	6		5		4		6		0	4		13			9			3		6.3
Indigofera	4		6		9		0		3	3		9		4	23			16		9.9
Psoralea	0		0		0		0		1	14		10			1			0		1.9
Glycine[a]	0		0		0		0		1	0		4			1[d]			1		0.8
Genera of special interest																				
Phaseolus[b]	15	27	0	0	13	3	1	6	1	6	7	1	1		1	1	3	3	9	5.0
Vigna	2	3	5	3	0	0	1	3	2	1	1	3	3	1	5	35		5	14	4.6
Alysicarpus	0	0	2	1	4	1	1	2	1	0	0	2	2		1	6		4	12	3.0
Sesbania	1	2	3	5	4	1	0	0	3	3	4	4	4		3	18		1	4	2.4
Atylosia	0	0	0	0	6	0	6	1	1	0	0	4	4		0	0		6	16	2.0
Centrosema	3	5	5	3	4	1	0	0	1	2	2	0	0		0	0		0	0	1.8
Stylosanthes[c]	2	4[c]	2	1	4	1	0	0	2	2	2	0	0	4	1	2	4[e]	1	1	1.5
Desmanthus	1	1	2	1	1	4	0	0	0	8	4	0	0	1	0	0	1	0	0	1.5
Total number of legume species**																				
	180		64		23		18			117		113			633			278		

A, actual number of species; N, number of species naturalized; *mean for native species only; **tropical species; [a]*Glycine* includes *Neonotonia* in Africa; [b]*Phaseolus* includes *Macroptilium;* [c]*Stylosanthes* does not include 'new' species found in Central America [d]see 'a'; [e]includes 'new' species *S. scabra* and *S. hamata*.

Included, marginally, in zone II – the wetter tropical type of climate – is Antigua, source of many accessions of *S. hamata sensu stricto*, *Desmanthus virgatus* and *Teramnus labialis*. It is not very different from Townsville, Queensland (70) and the testing site at Alligator Creek and this type of climate extends into Curaçao and the Paraguana Peninsula (near site 12, Maracaibo) in Venezuela. *S. hamata sensu stricto* and *D. virgatus* are found in all of these environments.

The dry tropical zone (III) contains the major Australian testing site at Lansdown (62) and extends into Hughenden (60), on the Mitchell grass plains. The only South or Central American site is Colima (5), an area in which *Desmanthus bicornutus* is common.

The most subtropical zone (IV) contains site 51 (Blackall), which figures, together with site 69 (in zone I), in further testing work in western Queensland. Most of the non-Australian sites in this zone are African, with legume floras not unlike those found in Australia. The region near Salta (in Argentina, site 19) has provided accessions of *D. virgatus*, one of which has been made commercially available as the cv. 'Marc'. The American sites 1 and 2 (the latter in zone V) possess many native species of *Desmanthus* (Luckow, 1993) and these have potential as sources of forage legumes adapted to clay soils for the subtropics (Jones and Rees, 1997).

Desmanthus can be found in all of these major climatic types (Fig. 7.2) but is largely absent from geographical areas that have been of importance for providing legumes for sandy textured soils, such as north-east Brazil (e.g. site 17, Juazeiro). It is found in climates (and soils) similar to those in Australia containing *Neptunia*, a genus it resembles closely. Although *Desmanthus* is often deemed to have 'specific' rhizobial requirements in general, its ability to form effective symbiotic relationships with *Bradyrhizobia* found on *Neptunia* should aid its adaptation to clay soils in Australia.

7.5.2 Edaphic analyses

The initial analysis using information from a worldwide selection of sites was of little value for our present purpose. The first two vectors took out only 55% of the total variation (Fig. 7.3). Though the study does suggest that the Indian clays (9, 10 and 11) are similar to those found at sites 5, 6 and 7 in Australia (5 being the Lansdown clay soil site) (Fig. 7.3), it is possible that these might be well separated on subsequent vectors. Site 16 (in Sudan), which yielded the accession of *Clitoria ternatea* adapted at site 5, is well separated from it. The analysis does, however, illustrate the wide range of variation present in clay soils in Australia. The significance of the silt/clay ratio, a major determinant here and the most significant soil factor relating to legume yield in the soil case studies in Antigua, warrants further study.

The second analysis, relating to edaphic/plant distribution studies in Antigua (Figs 7.4–7.6), proved to be edaphically meaningful and thereby useful to explore relationships with plant distribution and dominance. There are different legume floras for the different types of soil with *S. hamata sensu stricto*, *D. virgatus* and *Teramnus labialis* being well adapted to alkaline conditions and to clay soils. *S. hamata*, as with *Stylosanthes* species in Brazil (Burt *et al.*, 1979), was favoured by relatively high levels of exchangeable K in sward conditions and thrived over a range of soil textures from coral sands to clays; *D. virgatus* preferred heavier soils and could tolerate salinity and waterlogging; *Teramnus labialis* was more restricted in its distribution. All three species have been found to be suitable for long-term pastures on heavy clay soils (Keoghan, 1980).

These analyses confirm the potential importance of *Desmanthus* for use on clay soils and demonstrate that the diploid *S. hamata sensu stricto*, unlike its tetraploid derivative (*S. hamata sensu lato* with its cultivars 'Verano' and 'Amiga'), is adapted to alkaline soils where it thrives on both sands and clays.

7.5.3 Floristic studies

The similarities between the floras of India, Africa and Australia have long been known and the data presented demonstrate that this also applies to the legume genera and species that are of interest to us here. The most dissimilar legume floras are to be found in South and Central America (Fig. 7.8, zone I), with Brazil being the most different of these. The Brazilian flora is very rich in *Stylosanthes*, *Centrosema* and *Macroptilium* (Burt and Reid, 1976), but poor in *Desmanthus* (Luckow, 1993). It is this region which has furnished the most pasture legumes for acid soils and dry tropical conditions. Central America, represented here by Guatemala and Panama, belongs to a different floristic zone and like Mexico they are poor in terms of the number of species present of *Stylosanthes* and *Centrosema*, but rich in those of *Macroptilium* and *Desmodium* (Burt and Reid, 1976). They are a major centre of distribution for *Desmanthus* (Luckow, 1993).

The best known tropical pasture legume from Central America is *Macroptilium atropurpureum* and this is now generally regarded as being more suited to subtropical climates (Cook *et al.*, 2005). Species that were of interest, for instance *Phaseolus longepedunculatum* (syn. *Macroptilium gracile*) (Burt and Reid, 1976), have found little use. But *Stylosanthes*, although poorly represented in terms of species numbers, has continued to furnish useful material. The Central American type of *S. guianensis* (Edye *et al.*, 1974) has yielded pasture cultivars for the wet tropics and *S. subsericea*, prior to its attack by anthracnose, was deemed to be very promising for dry tropical regions (Burt *et al.*, 1971). Guatemala has yielded alkaline-adapted material of *S. hamata* (Chapter 3, this volume) of considerable promise and a 'new' species found in Mexico, *S. calcicola*, is probably also from alkaline situations (Lazier, 1977, personal communication). There are reports of the occurrence of the tetraploid form of *S. hamata* from Mexico as well (see information quoted in Maass and Sawkins, 2004). *Stylosanthes* material from alkaline situations, for instance *S. hamata* and *S. humilis* from the Paraguana Peninsula in Venezuela, has proved to be more useful than the same species collected from acid soils in some situations (Cook *et al.*, 2005). This Central American material warrants investigation.

7.6 General discussion

7.6.1 Introduction

The planned development of tropical pasture legumes for use in drier areas is a relatively recent and brief phenomenon. It is usually considered to have begun in the late 1960s (Burt and Williams, 1975) and by the early 1990s most plant collection and germplasm bank development had been curtailed (Maass and Sawkins, 2004). In this short period most of the research effort was placed on identifying plants adapted to sandier textured soils, for these dominate the grazing lands in Australia and South America where most of the interest was centred. Research was also restricted in several other ways. First, in the choice of 'target' genera, *Stylosanthes* figured prominently, primarily because of the earlier chance introduction of *S. humilis* to Australia and its subsequent success as a pasture legume in the dry tropics. *Macroptilium lathyroides*, introduced to Australia before 1900, had become an important fodder legume and prompted interest in *M. atropurpureum* and the development of *M. atropurpureum* cv. 'Siratro' (Cameron, 1985). From the suite of legumes used as cover crops in plantation agriculture in the wet tropics, genera such as *Centrosema* were subsequently developed as pasture legumes. The later success of 'new' species of *Stylosanthes*, first *S. hamata* and *S. scabra* and then *S. capitata* and *S. macrocephala*, intensified this interest and, as these species are primarily of South American origin, focused attention on that continent. The interest was further biased by the need to select areas deemed to be most likely to furnish adapted material and, due to the paucity of any other useful information, climatic data were used to define such areas (Burt and Reid, 1976; Burt *et al.*, 1976; Burt and Williams, 1975). This trend continues at the present time using geographic information systems to designate areas most likely to contain 'target species' (Maass and Sawkins, 2004).

But 'a plant responds to more than climate, its ability to invade disturbed ground, its response to grazing and trampling, and similar factors' may all be of importance (Williams and Burt, 1982) and affect its distribution, agronomic utility and proclivity as an invasive plant. And, if as is generally assumed, plants adapted to clay soils in Australia are most likely to be found on clay soils elsewhere, then a wider range of genera may need to be considered and this could well include species new to domestication (see Chapter 6, this volume). Many of these plants, unlike *Stylosanthes*, *Macroptilium* and *Centrosema*, are native to Africa and India rather than South America. The search needs to be widespread.

7.6.2 Choosing material for import

In India and Africa the genera of interest here for clay soils had been 'subjected to intense human selection for seed, forage and vegetable production in a variety of environments and farming systems for some time after 3500 BC' (Steele and Mehra, 1980; Jain and Mehra, 1980). *Dolichos lablab* syn. *Lablab purpureus*, one of the legumes currently of interest in Queensland and northern New South Wales, had already been developed for use and transported to Egyptian Nubia by AD 350. In *Vigna* it has long been known that *V. aconitifolia* (or 'Moth Bean') was the species best suited to the driest regions, and the utility of *V. trilobata* ('Pillepesora'), *Canavalia ensiformis* ('Jack Bean'), *Cyamopsis psoralioides* ('Cluster Bean'), *Cajanus cajan* ('Pigeon Pea', nearest relative *Atylosia*) and *Clitoria ternatea* ('Butterfly Bean' or 'Kordofan Pea') had all been well documented, even in standard text books such as Cobley (1956). To this could be added results such as those obtained from the Gezira scheme in Sudan, carried out on heavy clays in climates similar to those in the Australian dry tropics (see Figs. 7.1 and 7.3). In these studies large collections of these pulses/forage legumes were evaluated and studies of the local floras on clay soils carried out (Bunting, 1957 onwards, personal communication; Bunting and Lea, 1962). From this source, floristically different from East Africa (Fig. 7.8), came *V. trilobata*, *V. putigera* (syn. *V. ambacensis*), *Alysicarpus monilifer* and *Clitoria ternatea* 47187, an important component of cv. 'Milgara'. From his extensive collecting missions in Africa, Mr Ray Strickland provided much useful information and plant material from grazed areas, for instance the only accessions of *S. fruticosa* (syn. *bojeri*, *mucronata*) collected from clay soils. The legumes *Heylandia laterbrosa* and *V. trilobata* came from pastures on clay soils in dry tropical regions of India.

Much of this material has subsequently become readily available and has been examined in Queensland, Australia in 'nursery' studies in Brisbane (subtropics) and Townsville (dry tropics). Collections of plants such as *Clitoria ternatea*, *Dolichos* and many of the pulse crops have been examined by various authorities such as the Queensland Department of Primary Industries, CSIRO (Reid and Sinclair, 1980) and the Northern Territory Authority.

The situation was very different in South and Central America, and the Caribbean. In these areas only two leguminous genera, *Arachis* and *Phaseolus*, have been domesticated for use as major pulse crops. *Arachis hypogea* ('peanut') is South American with only one species, *A. pusilla*, in the drier tropical zone of Brazil. The *Phaseolus* species, *P. vulgaris* ('Common Bean'), *P. lunatus* ('Lima' or 'Butter Bean'), *P. coccineus* ('Runner Bean') and *P. acutifolius* ('Tepary Bean') were often from higher altitude tropical areas and it is uncertain whether the first two were from Central or South America. *P. acutifolius* is from Mexico or the southern USA. There was not a wide range of genera or information on which to draw for trials, nor are there large areas of clay soils from which to obtain pre-adapted material. Genera of particular interest, such as *Stylosanthes*, *Macroptilium*, *Centrosema* and *Desmanthus*, had not been domesticated for use at that point as sown plants. It was necessary to organize collecting missions and these proved to be very informative.

The first of these missions was to the dry tropical zones of north-east Brazil and Venezuela (Burt *et al.*, 1979) and it identified small areas of clay soils in both countries. In Brazil, in an area rich in such genera as *Stylosanthes*, *Macroptilium* and *Centrosema*, the only herbaceous legume to be found was *Arachis pusilla*. It was sub-dominant and well grazed. In Venezuela, in a zone in which the tetraploid form of *S. hamata* (which provided the cultivars 'Verano' and 'Amiga') is very common, no herbaceous legumes could be found on the clay soil in the dry tropical region. In a similar situation in Guatemala only *Desmanthus virgatus* could be found on clay soils, although other legumes, such as *Macroptilium atropurpureum*, were common throughout the area. In the drier Caribbean islands *S. hamata sensu stricto* (the diploid), *Desmanthus virgatus* and *Teramnus labialis* were very common on clay soils. Various *Desmanthus* species were subsequently collected on clay soils in Mexico and Argentina (Reid, 1983) and are seen as important species on clay soils in the southern USA (Muir *et al.*, 2005). *D. virgatus* occurs with *Stylosanthes sympodialis* on clay soils in semi-arid regions in Ecuador and with *S. humilis* and *S. hamata* on atypically alkaline soils on the arid Paraguana Peninsula in Venezuela. But, unlike the situation with many of the plants found on clay soils in India and Africa, very little is known about the characteristics and agronomic potential of these New World species. Indeed, 'new species' of *Stylosanthes* are still being designated (Maass and Sawkins, 2004) and new types of *S. hamata* found (Chapter 3, this volume). One of these species, *S. seabrana*, has been found to be adapted to clay soils in more subtropical climates (see information presented in Cook *et al.*,

2005). With little scientific evaluation of this material, it is important that it be well represented in studies on clay soils in Australia.

7.6.3 Choosing material for test: other considerations

Legumes, probably because of their ability to fix nitrogen, are often 'colonizing species' (*sensu* Baker, 1974) and this is particularly so for annuals/biennials such as *S. humilis*, an accidentally introduced plant that became adventive throughout the Australian dry tropics. On various collecting missions it was shown that *Stylosanthes hamata*, *Centrosema pascuorum*, *Macroptilium martii* and *M. lathyroides* are ecologically similar to *S. humilis* (see Appendix, Chapter 8, this volume). Early 'nursery studies' showed that these plants were well grazed by cattle, as were the annual species of *Vigna*, *Heylandia*, *Strophostyles* and *Alysicarpus* chosen for use in the experiment. Less palatable or potentially toxic genera, such as *Crotalaria* and *Indigofera*, were excluded.

In the perennials, for reasons presented previously, it was necessary to include considerable numbers of accessions from both *Stylosanthes* and *Desmanthus*. Further comment on these follows.

Alysicarpus, with a large number of species to consider and a propensity for clay soils, also warranted a range of accessions. In contrast, as *Centrosema* had shown much less promise in such situations, it could be represented by few accessions. The only species of *Macroptilium* and *Clitoria* deemed to be suitable (*M. atropurpureum* and *C. ternatea*) had previously been intensively studied (see data presented by Cook *et al.*, 2005) and could be represented by 'standards'.

The final caveats for selection were based on the dictum of Harlan (1983) that it is necessary to develop 'a body of theory and general information' to aid with the efficient development and utilization of plant genetic resources. Although it is assumed that plants for use on clay soils are most likely to be found on clay soils elsewhere, this is a hypothesis that awaits evaluation. Although earlier work found that plants adapted to a given climate in Australia had been found in a similar, but slightly more stringent, climate elsewhere (Burt *et al.*, 1976), this finding was, of necessity, based on relatively little information and further confirmation is required. It is clear that factors such as 'edaphic responses ... ability to invade disturbed ground ... response to grazing ... and similar factors' need to be considered (e.g. Burt *et al.*, 1979). In selecting the material for inclusion in adaptation trials on clay soils in the dry Australian tropics and subtropics, plants with different abilities to invade disturbed ground (see Appendix in Chapter 8, this volume) had been included, for instance *S. scabra* and *S. hamata*. Wherever possible, plants from a given species, geographic location and climatic type (as delimited on Fig. 7.1) were 'paired', with one of the pair being collected from a clay and the other from a non-clay. Similarly, although most of the plants included were from dry tropical conditions, they were from regions with different lengths of growing season, and, with the better represented species such as *Alysicarpus rugosus*, *Desmanthus virgatus*, *D. leptophyllus* and *Stylosanthes hamata*, plants were chosen from different floristic zones (Fig. 7.8) and climates (e.g. subtropical and tropical in *S. hamata*, Figs 7.2 and 7.3). The final suite of accessions is presented in Table 7.7.

Stylosanthes and Desmanthus

Both of these genera figured prominently in the review phase of this work where their ability to grow on clay soils became very evident. Little, however, was known about the various species involved. A taxonomic review of the genus *Desmanthus* was not available until 1993 (Luckow, 1993) and this greatly impeded sensible communication (Burt, 1993a,b). The variation present within *S. hamata* still awaits documentation (Chapter 3, this volume). Most of the research effort on the latter species had been placed on the tetraploid form that had yielded the cultivars 'Verano' and 'Amiga'. The majority of the accessions of this species are adapted to sandy-textured acid soils and those able to utilize *Bradyrhizobia* from alkaline conditions had escaped notice (Date *et al.*, 1979). New introductions continued to arrive during the course of this evaluation programme. To aid the selection process several *ad hoc* classifications of the seed stocks were carried out (e.g. that shown in Burt 1993a for *Desmanthus* and that for *Stylosanthes* in Chapter 3, this volume). Relevant morphological agronomic classifications for the entire CSIRO collections do not seem to be available and

Table 7.7. Places of origin and climates of plant material chosen for initial evaluation on clays in the north Australian tropics.

Species	CPI[a] no./ cultivar	Country of origin	Soil type	Latitude (°)	Altitude (m)	Mean annual rainfall	Habit
Alysicarpus							
A. longifolius	30034			18.3 N			erect annual herb
	79603	India		17.2 N			
A. monilifer	40612	Sudan	deep sand	14.1 S			erect annual herb
A. ovalifolius	50968	Senegal					
	79067	India	red non-clay	17.2 S			annual herb
A rugosus	52351	Malawi	clay loam	15.5 S	1100	850	erect, prostrate or ascending, vigorous annual or perennial, often somewhat woody
	52352	Kenya	volcanic clay loam	1.0 N	1850	1600	
	52354	Zambia	cracking clay	13.5 S	600	750	
	52355	Zambia	clay loam	13.4 S	1000	1000	
	52356	Malawi	silty clay loam	15.5 S	750	800	
	52357	Malawi	clay	16.0 S	750	800	
	56234	Chad		12.1 N			
	60167	Uganda	acid sandy loam	1.1 S	1819	1000	
	60168	Kenya	acid sandy loam	3.6 S	15	1250	
	76978	Zambia	sandy	~16 S			
	76979	Zambia	loam	~16 S			
	76980	Zambia	clay	~16 S			
	79604	India		17.2 N			
	79605	India		17.2 N			
Centrosema							
C. pubescens	common						vigorous rambling perennial
	55703	Venezuela	flooded clay	11.1 N	50	900	
C. pascuorum[b]	55697	Brazil	acid sandy soil	9.2 S	380	400	vigorous rambling annual
Clitoria							
C. ternatea	47187	Sudan	clay				vigorous rambling perennial
	49963	Brazil	sandy soil				

Continued

Table 7.7. Continued.

Species	CPI[a] no./ cultivar	Country of origin	Soil type	Latitude (°)	Altitude (m)	Mean annual rainfall	Habit
Desmanthus							
D. spp.	33426	Guyana	clay	6.5 N		~2400	perennial ranging in habit from a prostrate fine-stemmed herb to an erect ascending sometimes woody shrub
D. leptophyllus	38351	Venezuela		9.3 N	250		
	55719	Venezuela	acid sandy loam	10.3 N	80	600	
	76053	Mexico	skeletal brown	21.1 N	5	900	
	76057	Mexico	alkaline clay	18.8 N	5	1400	
D. pernambucanus	49728	Brazil					
D. virgatus	67643	Guatemala	acid clay	15.0 N	185	500	
	65947	Ecuador	alkaline clay	2.2 S	100	500	
	78369	Argentina	sandy	23.2 S	300	650	
	78373	Argentina	slightly alkaline clay loam	23.2 S	300	650	
	78380	Argentina	clay loam	24.5 S	1250	650	
	78382	Argentina	neutral clay loam	24.1 S	1000	700	
	79653	Cuba	slightly alkaline clay loam	21.6 N	5	600	
D. dichotomum	47186	Sudan	heavy alkaline cracking clay	12.3 N	450	660	herbaceous to suffrutescent
Dolichos							
D. sericeus	60218	Kenya	acid sandy loam	0.0 S	1970	1000	perennial climbing or prostrate herb
	60219	Kenya	acid sandy loam	0.0 S	1879	750	
	60220	Kenya	acid sandy loam	0.0 S	1879	750	
	60221	Kenya	acid sandy loam	0.0 S	1879	750	
Haydonia (syn. *Vigna*) (included after review undertaken)							
H. monophylla	76993	Zambia		15.4 S			erect climbing or trailing herb
Heylandia							
H. laterbrosa	31951	India					prostrate herbaceous annual

Macroptilium							
M. atropurpureum	16875	Mexico, Baja Calif.				arid	climbing or rambling perennial
	33826	Mexico, Veracruz	sand	18.0 N			
	61232	Mexico	limestone ridge	19.0 N	16	1000	
	67648	Guatemala	grey clay	15.0 N	185	500	
	73477	Mexico, Baja Calif.				arid	
	76151	Mexico	deep sandy soil	19.2 N	10	1100	
	76159	Mexico, Yucatan	sandy, roadside				
	76160	Mexico, Yucatan	shallow red soil	20.4 N	10	1250	
	76164	Mexico, Yucatan	sand on edge of sea	19.3 N	0	1100	
	76165	Mexico, Yucatan	sand over limestone	19.5 N	10	1050	
	76166	Mexico, Yucatan	shallow red clay	19.5 N	50	1000	
	cv. Siratro						bred variety
Strophostyles (included after the review was carried out)							
S. helvola	26272	USA, Texas	clay	30.0 N			twining annual
Stylosanthes							
S. calcicola	73525	Mexico	sand	20.1 N	10	1500	
S. fruticosa	41219	Sudan	sandy soil	13.0 N		609	
	41219a	segregant of the above					
	48386	Tanzania	sand				
	56211	Guadeloupe		16.5 N			
	61670	Venezuela	grey, alkaline sand	11.2 N	80	950	
S. hamata	38842	Venezuela	sandy loam	10.7 N	40	420	annual or perennial, prostrate to bushy herb
	cv. Verano						
	33205	Guadeloupe	stony calcareous soil	16.5 N	20	1000	
	49080	Colombia		11.0 N	15	847	
	50997	Venezuela	probably acid loam	10.0 N	590	509	
	70360	Antigua	heavy grey-brown clay	17.0 N	60	1346	

Continued

Table 7.7. Continued.

Species	CPI[a] no./ cultivar	Country of origin	Soil type	Latitude (°)	Altitude (m)	Mean annual rainfall	Habit
	70364	Antigua	alkaline clay pH 8.0	17.0 N	20	940	
	70366	Antigua	alkaline clay loam pH 8.7	17.0 N	20	940	
	70370	Antigua	alkaline clay loam	17.0 N			
	70376	Antigua	neutral heavy grey cracking clay, pH 7.5	17.0 N	10	1155	
	70522	USA, Florida	alkaline sand pH 8.7	26.7 N	10	1473	
	70529	USA, Florida	sand	26.5 N	10	1473	
	72852	USA, Florida	sand pH 8.3	26.5 N	10	1473	
	73507	Curaçao	loam, pH 7.5	12.2 N	10	532	
	73511	Cuba					
	73515	Nevis	sandy loam, pH 6.5; alkaline below 8 cm	17.0 N	20	940	
	73519	Nevis	heavy clay	17.0 N	100	1397	
S. scabra	40292	Brazil		8.0 S	350		
	cv. Seca						
	40205	Brazil		12.0 S			
	cv. Fitzroy						
	55870	Brazil	sandy loam pH 6.5	14.0 S	600		
S. sympodiales	67702	Ecuador	clay loam pH 7.1	2.1 S	100	500	
	67705	Ecuador		1.3 S	250		
Vigna							
V. trilobata	13671						twining annual
	47510	Sudan	clay	8.0 N			
	86179	India					
	86180	India					
	86181	India					
	86182	India					

[a]CPI: Commonwealth Plant Introduction number. [b]A collection of *C. pascuorum* from a clay soil in Ecuador was unavailable. Number of: genera 12; species 20; accessions 88.

selections of *Desmanthus* for the last part of this programme based in western Queensland (Chapter 8, this volume) were based solely on literature review and provenance data. In view of the interest in *Desmanthus* in more subtropical climates (Pengelly and Conway, 2000), and the weaknesses in existing cultivars of the genus, data from literature reviews show that other *Desmanthus* species should be of interest. *Desmanthus* is not primarily a tropical genus and Hartley (1963) and Williams and Burt (1982) have argued that some southern states of the USA, which are major sources of diversity, are likely to be a source of adapted material for northern New South Wales and southern Queensland. Several authors (e.g. Muir *et al.*, 2005; Luckow, 1993) have commented about the forage value of some of the species found there. Other sources of readily available literature on the topic are given in Maxted and Guarino (1997).

8 Searching for Pasture Legumes for Heavy Clay Soils in the Australian Dry Tropics and Subtropics: III. The Initial Evaluation of Introduced Material

R.L. Burt†

Abstract

A simple initial evaluation trial of germplasm potentially adapted to heavy clay soils was undertaken by the establishment of spaced plants of 90 accessions (12 genera involving 20 species) at three locations. The sites formed a rainfall gradient and the environments ranged from tropical moist monsoon with rainfall exceeding evapotranspiration to hot semi-tropical where the rainfall never exceeds evapotranspiration. The survival and response to environmental stresses of the plants were recorded. The best results were achieved by accessions of *Stylosanthes*, especially *S. hamata*, and *Desmanthus*. Discussions follow on the general topic of pasture legumes for clay soils in the dry tropics and on the value of provenance records in suggesting plant utility.

8.1 Introduction

In evaluating the material selected earlier in this series (see Appendix 8.1), there are two requirements: first, to examine the plants in such a way that their potential utility can be assessed and second, to add to the 'body of theory and general principles so as to aid in the efficient development and utilization of plant genetic resources' (Harlan, 1983), particularly in relation to tropical and subtropical legumes.

Methods of plant evaluation are often largely dependent upon yield estimates, usually obtained under regular cutting regimes. Such methods may be entirely relevant, especially for the more temperate, better-known species, where a relatively small difference in yield may be all that is sought. But the characteristics of tropical species, even those that are better-known, are less well documented and 'new' features may emerge. In the early 1970s, for instance, when the pastoral value of *Stylosanthes scabra* was only just being accepted, it was found to be a variable species containing several morphological/agronomic types (Burt *et al.*, 1971; Edye *et al.*, 1974). The agronomically important types were all of the 'Brazilian coastal' form and were genetically similar (see information presented in Liu, 1997). But, even within one type, there are different ecotypes adapted to different climates and ecological situations and such differences are reflected in their potential agronomic utility (Burt *et al.*, 1983). In such plants, long-term persistence, rather than yield, is of prime importance (Burt and Ebersohn, 1969), and long-term persistence is related to the ability of individual plants to survive and to re-establish from seed, often in unfertilized conditions and in competition with grass in grazed situations. The ability to utilize native *Bradyrhizobia* is essential. Moreover total yield may be of less significance than the ability to produce a smaller amount of feed at some crucial time, after frost, fire or at the onset of the dry season (Burt and Ebersohn, 1969), and this may be in the form of seed, rather than leaf or stem. Information of this type is often noted but is rarely presented for more general use. It is this same type of observation that can be useful in determining the potential agronomic utility of the plant and its possible role on an agro-ecological landscape (*sensu* McIvor, 2002). Some aspects of this are covered further in Appendix 8.2.

The second requirement, to add to a 'body of theory and general principles', involves the further testing of two hypotheses: that plants most likely to be adapted to clay soils in Australia are most likely to come from clay soils elsewhere, and from areas that are climatically homologous (or slightly more severe; Burt *et al.*, 1976). It was for these reasons that the plants chosen for this study came from a range of generally similar tropical climates and, wherever possible, with 'paired' accessions from clay and non-clay soils.

8.2 Experimental methods

This experiment sought to study the response of introduced plants to a range of environments and to determine their characteristics and likely adaptation. Such an objective is very similar to those of the ecologists who developed the concepts of 'ecotypes' within species (e.g. as described by Baker, 1974) and the system of evaluation that is used here is very similar. Basically the introduced plants were first established under favourable environmental conditions and are then transplanted into a range of environments, and nurtured initially to ensure their establishment. This range of environments is similar to the range in which the plants were initially collected in order to determine whether the plant responses fit the hypothesis that they are best adapted to homologous climates elsewhere.

The selected material (12 genera involving 20 species and 90 accessions, see Table 8.1) were grown in rows of spaced plants enabling the behaviour of individual plants to be monitored over time. The response to important environmental 'pulses' – frost, fire, grazing, invasion by other species – were observed and persistence and spread assessed. Plant features of interest were recorded as they became apparent, rather than at pre-ordained times. The characteristics measured are listed in Table 8.2 and the times at which they were taken, relative to rainfall and potential evapotranspiration at one of the sites, are shown in Fig. 8.1.

The sites at which the experiment was sown were located along a climatic gradient inland from Townsville, Queensland. The first, near Townsville, was at Alligator Creek, the second the experimental areas at the CSIRO Lansdown field station (43 km inland) and the third at Hillgrove, near Charters Towers (some 90 km further inland). The environments thus range from a tropical moist monsoon climate with 4 months in which the rainfall exceeds the potential evapotranspiration, to a hot, semi-tropical situation in which monthly rainfall never exceeds potential evapotranspiration (Papadakis, 1966). Beyond this trial the next step in broader testing of the material would be to test in a location with a more stressful environment; in Australia this would be the clay soil plains of western Queensland, which Papadakis (1966) characterized as hot subtropical desert. Such future studies would use the best adapted material obtained from this experiment augmented with more recent introductions. The Alligator Creek region is the only site of those chosen to have a climate suitable for crop production and therefore one in which the suitability of plants for use in leys or as cover crops could be considered. It is too dry for rain-fed sugar cane but sufficiently dry for cotton (Papadakis, 1966), and well-suited to mango production.

For each introduction, 100 seeds were sown into small pots, one per pot, in early December 1981. They were raised in the shade house at the CSIRO Davies Laboratory and the 60 most uniform seedlings selected for use in late February 1982. These were placed outside the shade house and the water supply reduced to 'harden them off'. They were transplanted in early March, a small amount of water being applied to facilitate survival.

At each site plants were sown onto both sand and clay soils. Prior to transplanting superphosphate was applied at the rate of 251 kg/ha and the native vegetation cut and raked off the plots. Strips 50 cm wide and 1 m apart were cultivated by rotary-hoe. Each accession was represented by two rows, each of five plants, randomly located in each of two blocks. Plants were spaced at 50 cm intervals within the strips.

Measurements taken at the end of the 1982 growing season showed a high survival rate at all sites except the Hillgrove clay. It was a drought year and pastures sown to the cultivars favoured in the area had failed. High survival, coupled with flowering and seed set, had been deemed to be necessary to ensure a fair test of the

Table 8.1. Survival of legume accessions as transplants or seedlings on 17 May 1984 and December 1985.

Species	CPI[a]	Soil at site of accession	Alligator Creek 1984		Lansdown 1984		Total surviving number/ genus 1984	No. plants/quadrat[b] Dec. 1985 Lansdown Clay
			Clay	Non-clay	Clay	Non-clay		
Alysicarpus								
A. longifolius	30034	–			x			0
	79603	–						0
A. monilifer	40612	deep sand						0
A .ovalifolius	50968	–						0
	79067	red non-clay	x					0
A. rugous	52351	clay loam	x		x[d]			0
	52352	clay loam	x		x[d]			0
	52354	cracking clay						0
	52355	clay loam						0
	52356	clay loam	x	x	x[d]	x		0
	52357	clay	x	x	x[d]	x		0
	52364	–	x					0
	60167	acid sandy loam	x		x			0
	60168	acid sandy loam						0
	76978	sandy	x		x			0
	76979	–	x	x	x			0
	76980	clay			x			0
	79604	–						0
	79605	–						0
Number surviving[c]			9	3	9	2	23	0
Centrosema								
C. pubescens	Centro		x	x	x	x		0
	55703	flooded clay	x	x	x	x		0
C. pascuorum	55697	acid sandy		x		x		0
Number surviving[c]			2	3	2	3	10	

Clitoria								
C. ternatea	47187	clay		x	x	x		1
	49963	acid sandy	x	x	x			6
Number surviving[c]			1	2	2	1	6	2
Desmanthus								
D. spp.	33426	clay	x		x	x		34
D. leptophyllus	38351	non-clay	x	x	x[d]	x		52
	55719	acid sandy loam	x	x	x[d]	x		22
	76053	skeletal brown	x	x	x	x		32
	76057	alkaline clay	x	x	x[d]	x		4
D. pernambucanus	49728	–	x	x	x	x		43
D. virgatus	67643	acid clay	x	x	x	x		27
	65947	alkaline clay	x		x			0
	78369	sandy	x		x	x		0
	78373	slightly alkaline clay loam	x	x	x	x		0
	78380	clay loam			x			4
	78382	neutral clay loam	x		x	x		54
	79653	slightly alkaline clay loam			x			25
Number surviving[c]			11	7	13	10	41	9
Desmodium								
D. dichotomum	71864	heavy alkaline cracking						0
Number surviving[c]			0	0	0	0	0	0
Dolichos								
D. sericeus	60218	acid sandy loam						0
	60219	acid sandy loam						0
	60220	acid sandy loam						0
	60221	acid sandy loam						0
Number surviving[c]			0	0	0	0	0	0

Continued

Table 8.1. Continued.

Species	CPI[a]	Soil at site of accession	Alligator Creek 1984		Lansdown 1984		Total surviving number/ genus 1984	No. plants/quadrat[b] Dec. 1985 Lansdown Clay
			Clay	Non-clay	Clay	Non-clay		
Haydonia								
H. monophylla	76993	–						0
Number surviving[c]			0	0	0	0	0	
Heylandia								
H. laterbrosa	31951	–						
Number surviving[c]			0	0	0	0	0	0
Macroptilium								
M. atropurpureum	16875	–	x	x	x			0
	33826	sand	x	x	x	x		0
	61232	limestone ridge	x	x	x	x		2
	67649	grey clay	x	x	x	x		2
	73477	–	x	x	x	x		0
	76151	deep sandy	x	x	x	x		?
	76159	sandy	x	x		x		0
	76160	shallow red	x	x	x	x		0
	76164	sea edge sand	x	x	x	x		2
	76165	sand over limestone	x	x	x	x		1
	76166	shallow red clay	x	x	x	x		2
	cv. Siratro	–	x	x	x	x		1
Number surviving[c]			12	12	11	11	46	6
Strophostyles								
S. helvola	26272	clay						0
Number surviving[c]			0	0	0	0	0	0

Stylosanthes								
S. calcicola	73525	sand			x			6
Number surviving[c]			0	0	1	0	1	1
S. fruticosa	41219	sandy				x		4
	41219a	segregate of above	x	x	x	x		3
	86	sand	x					0
Number surviving[c]			2	1	1	1	5	2
S. hamata	cv. Verano	sandy loam	x	x	x	x		90
	33205	stony, calcareous	x		x			22
	49080	–			x			94
	50997	acid loam (?)	x	x				0
	56211	–	x					0
	61670	grey alkaline sand			x	x		0
	70360	heavy grey/brown clay			x			3
	70364	alkaline clay pH 8	x	x	x	x		46
	70366	alkaline clay loam pH 8.7	x	x	x	x		0
	70370	alkaline clay loam			x			38
	70376	heavy cracking clay pH 7.5			x			30
	70522	alkaline sand pH 8.7	x	x	x[d]	x[d]		10
	70529	sand	x	x[d]	x[d]	x		64
	72852	sand pH 8.3			x			21
	73507	loam pH 7.5			x			40
	73511	–	x		x			0
	73515	sandy loam, pH 6.5 at surface						0
	73519	heavy clay			x			0

Continued

Table 8.1. Continued.

Species	CPI[a]	Soil at site of accession	Alligator Creek 1984		Lansdown 1984		Total surviving number/ genus 1984	No. plants/quadrat[b] Dec. 1985 Lansdown Clay
			Clay	Non-clay	Clay	Non-clay		
Number surviving[c]			9	6	15	6	36	11
S. scabra	cv. Seca	–	x	x	x	x		24
	cv. Fitzroy	–	x[d]	x	x[d]	x		28
	55870	sandy loam pH 6.5	x	x	x	x		83
Number surviving[c]			3	3	3	3	12	3
S. sympodialis	67702	clay loam pH 7.1	x	x[d]	x	x		0
	67705	–		x[d]	x	x	0	0
Number surviving[c]			1	2	2	2	7	0
Vigna								
V. trilobata	13671	–						0
	47510	clay			x			0
	86179	–						0
	86180	–						0
	86181	–						0
	86186	–						0
Number surviving[c]			0	0	1	0	1	0
Total surviving			50	39	60	39	188	

Totals: Genera 12; Species 20; Accessions 90. [a]CPI: Commonwealth Plant Introduction number; [b]original transplanted plants plus seedlings; [c]number of surviving accessions per taxon; [d]particularly vigorous, healthy and well-adapted plants.

Table 8.2. Observations[a] made at the trial sites.

Observation number[b]	Observation date	Season	Activity/observation	Detail
1	10/12/81		Seed sown in shade house	
1982				
2	4/3/82		Seedlings transplanted	
3	11/5/82	End of growing season	Original transplants surviving	number
			Plant height	cm
			Plant vigour	rated 1 to 5
			Maturity	1 no flowers, 2 flower buds present, 3 flowering, 4–3 or more buds, 5 ripe seed
4	28/8/82	Dry season	Green leaf	absent 0 or present 1
			Flowers	absent 0 or present 1
1983				
5	24/1/83	Growing season begins	Original transplants surviving	number
			Plant vigour	rated 1 to 5
			Flowers	absent 0 or present 1
6	20/4/83	Mid- to end season	Original transplants surviving	number
			Plant vigour	rated 1 to 5
			Ripe seed	absent 0 or present 1
7	20/6/83	End growing season	Seedling regeneration	absent 0, few 1, many 2
1984				
8	16/2/84	Growing season begins	Original transplants surviving	number
			Bothriochloa pertusa plants	absent 0 or present 1
9	17/5/84	Dry season begins	Original plants showing green leaf	number
			Plant vigour	rated 1 to 5
			Seedling regeneration	absent 0, few 1, many 2
1985				
10	27/11/84	Growing season begins	Original transplants surviving	number
			Maturity of original transplants	1 no flowers, 2 flowering, 3 green pods present, 4 ripe seed
			Bothriochloa pertusa plants	absent 0 or present 1
11	8/3/85	Mid- to end season	Original transplants surviving	number
			Plant vigour	rated 1 to 4
1986				
12	8/11/85	Growing season begins	Original transplants surviving	number
			Seedling regeneration	absent 0, few 1, many 2
13	4/12/85	Mid-growing season	Original transplants surviving	number
			Perenniating plants other than above	number
			Seedlings	number
			Maturity of perennial plants	1 no flowers, 2 flowering, 3 ripe seed
			Spread into non-cultivated native vegetation	0 no spread, 1 moderate, 3 good
14	17/12/85	Mid-growing season	Invasion of transplant plots: ground cover	0 no invasion, 2 moderate invasion, 3 severe invasion

[a]Lansdown clay soil dates are shown; all other sites observed within a few days (Lansdown non-clay and Alligator Creek until 17/5/84, Hillgrove until 24/1/83). [b]The observation numbers correspond to the numbered arrows in Fig. 8.1.

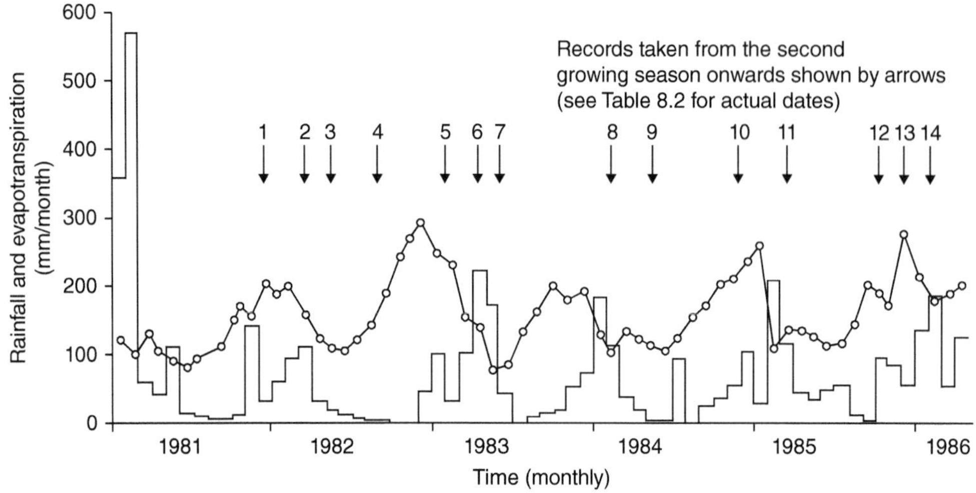

Fig. 8.1. Rainfall, evapotranspiration and recording time at Lansdown.

accessions, and this did not take place on the Hillgrove clay. This part of the experiment was declared to be null and void and further measurements were not taken.

At the remaining sites light intermittent grazing was imposed from the end of the 1983 season. With the sole exception of the Lansdown clay site, for reasons presented subsequently, the experiment was concluded in May 1984, the end of that growing season.

8.3 Results and preliminary discussion

8.3.1 From Alligator Creek, Lansdown and Hillgrove

The failure of the clay soil component of the experiment at Hillgrove rendered the result from that site useless for the present purpose. Sufficient to note that, on the sandier textured soils, various accessions of *Stylosanthes* thrived with *S. scabra* and *S. hamata* being selected for use in later experiments (Jansen and Edye, 1996).

Preliminary investigation of the data from the remaining sites revealed a very similar pattern of survival of the various species and genera. And, at the only site whose climate is deemed to be suitable for cropping (Alligator Creek), and in which legumes might be used as cover crops or in a ley system, it was species of *Centrosema*, *Clitoria*, *Macroptilium*, *Vigna* and *Stylosanthes*, which produced the high early yields needed from such plants. Such performances have been amply documented elsewhere (e.g. the information quoted in Cook *et al.*, 2005). For the sake of brevity, the presentation of results hereafter has largely been confined to those relating to persistence at the end of the third growing season (Fig. 8.1). But the performance of plants on the clay soil at Lansdown, of key concern for our present purpose, was documented for a longer period and in more detail and will be presented separately.

The number of accessions that persisted was greatest on the clay soils and highest at the drier site (Table 8.1). Some genera, represented generally by annual species, died out completely. These were from the genera *Desmodium*, *Dolichos*, *Heylandia*, *Haydonia* and *Strophostyles*. *Vigna trilobata*, which grew vigorously in the early stages of evaluation, was subjected to continual attack by insects, failed to set seed and only one accession survived (and then only at one site). Some species, such as *Macroptilium atropurpureum*, *Centrosema pubescens* and *Clitoria ternatea*, survived on both clay and non-clay sites; this is in accordance with published literature (Cook *et al.*, 2005). In other genera, such as *Alysicarpus*, *Desmanthus* and *Stylosanthes*, survival on both soil types was observed in some species and in some accessions within species. In *Stylosanthes* the woody perennial *S. scabra* or 'Shrubby Stylo' persisted on both soil types, while the closely similar *S. fruticosa* generally did not. Several of the annual/biennial accessions of *S. hamata* or 'Caribbean Stylo' and *S. sympodialis* likewise persisted well on both soils and both sites. This was particularly true for the tetraploid forms *S. hamata* cv. 'Verano' and the 'short-beaked' form from Florida (*sensu* Cook *et al.*, 2005) (whose exact taxonomic status is uncertain). Some of the smaller-leaved, more persistent diploid-type *S. hamata* were also able to do so. In the genus *Desmanthus*, *D. leptophyllus* rather than *D. virgatus* displayed this trait.

It is clear that some genera in this experiment were generally more suited to clay soils than others. This adaptation has long been suspected for a genus such as *Alysicarpus*, which is native to Australia, but is here evident in non-native genera such as *Desmanthus* and *Stylosanthes*. *Stylosanthes*, which has yielded cultivars suitable for use on sandy textured soils from the species *S. scabra*, *S. macrocephala*, *S. capitata* and one tetraploid form of *S. hamata*, has other species and species forms which are adapted to heavier soils. These data also indicate that the accessions which perform best on clay soils do not necessarily need to come from clay soils elsewhere: *D. leptophyllus* was well adapted but has never been collected from clays, nor were the very successful 'short-beaked' accessions of the tetraploid *S. hamata* (see Appendix 8.1). Accessions of particular interest failed completely, such as *Desmodium dichotomum* and *Heylandia laterbrosa*, which are both from clay soils in grazed pastures in dry tropical conditions. Further discussion on this aspect of the experiment is provided in part IV of this series of reports (Chapter 9, this volume), where the results of more detailed, extended studies are available for comment.

8.3.2 From the Lansdown clay soil site

Data are available for five growing seasons from this site because of the early onset of rains in 1985, which provided an opportunity to obtain a final set of information. In the early years survival was uniformly high and non-informative. The information available on the last four seasons, comprising data from 16/2/84, 27/11/84 and 4/12/85, have therefore been studied in more detail. The results complement those provided in Table 8.1

Survival of the original transplants

Each accession can potentially accumulate a total of 30 points, 5 (plants/row) × 2 (blocks) × 3 (years). When summarized in this way, three main groups of accessions could be discerned. First, a group with excellent survival, with counts between 25 and 30; second, a group with good survival, with counts between 20 and 24; and a group with minimal survival usually with zero points but never exceeding 2. The accession groups are shown in Table 8.3.

In terms of perenniality, four accessions were outstanding, one accession of *S. scabra*, one of the diploid forms of *S. hamata sensu stricto* and two of the short-beaked tetraploid *S.* sp. aff. *hamata* from Florida. In the 'Good' group (Table 8.3), *S. scabra* and *S. hamata sensu stricto* were again evident and were joined here with one accession of *Clitoria ternatea* and several accessions of *Desmanthus leptophyllus* and *D. virgatus*. The survival rates of all other species and genera were low. And, as noted previously, survival was not dependent upon the type of soil from which the accessions were collected.

Invasion of plot rows

Invasion of plot rows became very noticeable in 1984, the third growing season. The most invasive species for the plots, and possibly for the area in general, was the grass *Bothriochloa pertusa* (Indian Couch) and in this instance the invasion justified analysis. With other invasive species, Natal grass (*Rhyncelytrum repens*) and Rubber vine (*Cryptostegia grandiflora*), the invasion was very patchy and was virtually confined to guard rows that had remained uncultivated. The invasion by *B. pertusa* is, moreover, of especial interest because the species is widespread in the Australian dry tropics and in other tropical zones, such as Antigua (where it is known as Antigua hay grass). *A priori* information from Antigua, source of many of the accessions of *S. hamata sensu stricto* and *D. virgatus*, suggests that these forms may be able to compete with *B. pertusa*, a hypothesis worthy of test.

The effect of the *Bothriochloa* was explored over two time intervals. In the early stages these were the counts for the two survival periods 16/2/84 and 27/11/84, at which time the maximum possible survival for the transplanted plants is 10. At this stage the *Bothriochloa* was

Table 8.3. The percentage survival rates for the various accessions.

Species	CPI number	Soil texture[d]	Species	CPI number	Soil texture
Excellent: 25 to 30					
S. scabra	55870	Sandy			
S. hamata[a]	70364	Clay			
S. spp. aff. *hamata*[b]	70529 70522	Sandy Sandy			
Good: 20 to 24					
S. scabra	cv. Fitzroy	Sandy	*D. virgatus*	79653 78382 67643	Clay loam Clay loam Clay
S. hamata[a] *S. hamata*[a]	72852 73507	Sandy Sandy			
S. hamata[c]	cv. Verano	Sandy	*D. leptophyllus*	38351 76053	Sandy Sandy
C. ternatea	49963	Sandy			
Bad: usually 0 not exceeding 2					
S. hamata[a]	70360 50997 56211	Clay Sandy Unknown	*D. virgatus*	65947	Clay
Desmodium *Haydonia* *Heylandia* *Strophostyles* *Vigna trilobata*	Single accessions showed no survival after year 1				
Alysicarpus *Centrosema* *Dolichos*	All 18 accessions All 3 accessions All 3 accessions				
Others					
When the extreme cases are excluded the mean per cent survival rates for the remainder of the species are:					
S. hamata[a] *Clitoria* *Desmanthus* *Stylosanthes* spp. *Macroptilium*	8 accessions 1 accession 8 accessions 6 accessions 9 accessions	11–9 13–0 11–3 12–0 8–6			

[a]*S. hamata senso stricto*, small-leafed diploid; [b]*S.* sp. aff. *hamata*, short-beaked tetraploid 'off-type'; [c]*S. hamata sensu lato*, large-leafed tetraploid type derived from *S. hamata sensu stricto* × *S. humilis*; [d]soil texture at site of original collection.

recorded only as 'present' or 'absent'. Using only the data for *Stylosanthes* and *Desmanthus*, the information for other genera at this stage being minimal, it is possible to produce a 2 × 2 set of data: *Stylosanthes* vs *Desmanthus* and invaded vs non-invaded (Table 8.4). The number of observations in each cell of the matrix is shown in Table 8.4 column a. The raw means of the data are presented in column b and the counts after they have been subjected to a square root transformation column c. Invasion by *Bothriochloa* at this stage of the experiment was not associated with any great loss in persistence (cf. invaded vs non-invaded means).

The effect of *Bothriochloa* invasion later in the experiment can be explored by comparing the behaviour of *Stylosanthes* and *Desmanthus* on 17/12/85. As use of information from accessions that were already performing poorly at this time would be meaningless, data were used only from those accessions that had been rated as 'good' or 'excellent'. This gives 13 accessions for each of which the total number of plants originally transplanted, the number seedlings that

Table 8.4. Effects of *Bothriochloa* invasion on plant survival in *Stylosanthes* and *Desmanthus* plots.

	a: Number of observations in each cell		b: Raw means of data		c: Counts, square root transformed	
Observation	*Stylosanthes*	*Desmodium*	*Stylosanthes*	*Desmodium*	*Stylosanthes*	*Desmodium*
Invaded	17	13	5.176	2.923	2.200	1.531
Not invaded	22	12	4.909	4.083	1.857	1.774

have developed into perennial plants and the number of new seedlings can be calculated. Invasion, where it had occurred, was rated as mild or severe, and note was taken of a termite mound and of plots invaded by Rubber vine. The resulting data are presented in Table 8.5. Invasion consistently reduced plant numbers in *S. scabra* and *S.* sp. aff. *hamata*, but was less severe in the diploid *S. hamata sensu stricto* form (as found in Antigua). *Desmanthus* exhibited little or no effect. The difference between genera was almost statistically significant at the P = 0.05 level.

Seed, seedling production, ability to hold leaf into the dry season and ability to spread outside the area into non-disturbed, 'natural grassland'

Spread into non-disturbed areas was difficult to define because of the differences in plant form between the genera. Three accessions of *Macroptilium atropurpureum* cv. 'Siratro', 76165 and 76166 had spread weakly into the guard rows but had not produced seedlings therein (on 4/12/85). The only *Stylosanthes* accessions to spread into the guard rows, and then only weakly, were *S. calcicola* 73525 and *S. fruticosa* 41219. The following *Desmanthus* showed medium (2) to excellent (3) spread: *D. leptophyllus* 38351* (2), 76053 (2, sandy soil), *D. virgatus* 79653* (2, clay loam), 78382* (2, clay loam), 67643* (3, clay), 78373 (2, clay loam) and *D.* spp. 33426 (2). Those asterisked also appear in the top two listings for perenniality and must be deemed to be adapted to the clay soil site. Of these, three were collected on clays or clay loams (as opposed to sandy) and no edaphic information are available for the fourth (38351). But some collected on clays, but possibly from wetter sites, failed.

Presence of ripe seed (on 20/4/83, scored 0 = absent and 1 = present) are presented in Table 8.6 and reduced to percentages of possible maximum scores to facilitate comparison across the table (data for *Clitoria ternatea* are omitted since it only had two accessions). *Stylosanthes* and *Desmanthus* both tended to produce seed by the mid-point of the growing season: *Macroptilium atropurpureum* did not.

Seedling regeneration within the cultivated rows, as opposed to spread into the less disturbed 'natural grassland', was particularly marked in *S. hamata*, less so in perennial *S.* spp. and *Desmanthus* spp. and of very low occurrence in *Macroptilium atropurpureum*.

Green leaf retention into the dry season (17/5/84) was shown by over 40% of the accessions in *Stylosanthes* and *Desmanthus* but was considerably less in *Macroptilium atropurpureum*.

Summary

The most promising material in this study at the Lansdown clay site is almost exclusively from the New World genera of *Stylosanthes* and *Desmanthus*. This confirms observations made elsewhere, for instance in Antigua where *Bothriochloa pertusa* is a common invasive grass and *Stylosanthes hamata sensu stricto* and *Desmanthus virgatus* are endemic (Chapter 3, this volume). The difference between the genera is that the perennial *Desmanthus* showed greater ability to spread into less-disturbed grassland situations than *Stylosanthes*. However in more disturbed situations, where the grassland had been rotary-hoed, the perennial *Stylosanthes* species were able to re-establish by seedlings at about the same rate as the *Desmanthus*. The annual/biennial, 'ruderal' species, *S. hamata*, was much more adept at this strategy than the perennials. *Macroptilium atropurpureum*, in these environments, was not particularly good at either site. Although once favoured for use in dry tropical conditions, the most recent surveys suggest that it is best suited to wetter, more subtropical conditions (Cook *et al.*, 2005).

Table 8.5. Effect of invasion on plant numbers of the best adapted accessions.

Accession	CPI accession number	Block 1	Block 2	I[a]	Trans[b]
Stylosanthes					
S. scabra	55870	19[c]	64	+45	+3.641
	cv. Fitzroy	17	11[c]	+6	+0.806
S. spp. aff. *hamata*	70529	32	1[d]	+31	+4.568
	70522	8	2(T)	–	–
S. hamata	72852	17[c]	4[d]	+13	+2.123
	73507	22[c]	18	–4	–0.448
	70364	20	26	–	–
					M_s = 2.138
Desmanthus					
D. leptophyllus	38351	24	28[d]	–4	–0.393
D. virgatus	67643	14[d]	3(R)	–1	–0.136
	78382	30	24[d]	+6	+0.578
	79563	6	19[d]	–13	–1.909
					FLD = –0.465

[a]The difference of non-invaded (n) minus invaded (i). [b]Trans.: square root transformation of I. With 8 degrees of freedom, t = 2.649 at 0.05. [c]Moderate and [d]severe invasion by *Bothriochloa pertusa.* (R) Invasion by rubber vine. (T) Invasion by termite mound.

Table 8.6. Seed and seedling production and green leaf retention into the early dry season. Presented as percentages of maximum scores to facilitate comparison. Data for *Clitoria* are omitted because it had only two accessions.

Species	Ripe seed present[a]	Seedling regeneration[b] 10/6/83	Green leaf retention[c] 17/5/84
Stylosanthes spp.[d]	75.0	31.2	40.0
Stylosanthes hamata	73.5	63.2	46.5
Macroptilium atropurpureum	13.6	4.5	28.2
Desmanthus spp.	88.0	36.0	43.2
M	62.5	33.7	39.5

[a]Scored as 0 or 1 on 20/4/83; [b]scored as 0, 1 or 2 on 10/6/83; [c]number of plants showing green leaf on 17/5/84; [d]other than *S. hamata*.

8.4 General discussion: pasture legumes for clay soils in dry tropical conditions

The work that has been described in this volume was part of a long-term programme designed to provide pasture legumes for use in the dry tropical and subtropical areas of Australia and other parts of the world. For the sandier textured Australian soils, this is virtually complete largely through the development of 'new' species of *Stylosanthes*. However, as in all agricultural systems there will be a continuing need to provide relatively slight plant modifications to meet specific challenges to productivity, such as resistance to disease (Chakraborty, 2004).

In the case of clay soils, however, there is a very real dearth of leguminous plants that are persistent and productive (Clem and Hall, 1994; Jones and Rees, 1997). There are very large areas of these soils in such countries as Australia, the Republic of the Sudan and India. Even small areas, which are common in many countries, are of importance, due to their generally higher fertility levels and ability to hold water into the dry season, characteristics of great importance for the productivity of both crops and pastures. For these soils it may be necessary to develop 'new species' as was done for sandy soils, but the dual threats posed by global warming to the areas where such plants may be needed and to the potential source locations of relevant biodiversity, impose time limits to the collection and identification of such germplasm.

In developing, testing and utilizing such resources consideration must be given to how the material might be fitted most sensibly into the intended target environment. There could be need for dialogue between those concerned, particularly as there may be conflicting views. This situation is well illustrated by events in South Australia where the arrival of European settlers caused marked changes in plant cover and botanical composition (Moore, 1967). The accidental introduction of Mediterranean legumes helped to preclude the invasion of woody native species into farm land and provided the basis of a productive ley system of agriculture. However the long-term use of these annual legumes has led to soil salinity and the need for restoration. This could be achieved – in the absence of any suitable native plant – by the use of an introduced deep-rooted perennial legume (Bennett and Virtue, 2004). Similarly, to prevent woody weed encroachment and to restore native black soil grassland in Texas, there is a need to increase the competitiveness of the grass by including legumes to provide nitrogen. Although native legumes are sought for use, an exotic, *Desmanthus bicornutus* from Mexico (but accessed via Australia), is also being used (Muir *et al.*, 2005). These same accessions have persisted in areas of western Queensland where landscape restoration, as well as pasture production, is a major concern (Tothill and Gillies, 1992).

Worldwide attempts are now being made to reconcile the views of 'conservationists' and 'agriculturalists' (Maxted *et al.*, 1997; Bennett and Virtue, 2004; Reichard and White, 2007). Both should rely upon the same sorts of (ecological) information, and the discussion which follows is intended to facilitate such an exchange and to provide relevant information for the future development and utilization of genetic resources of tropical pasture species.

8.5 The initial survey

There is always the chance of serendipity, of finding a useful plant from some unexpected source. In crop plants chance seedlings, probably the result of cross pollination, provided the ever-popular 'Haas' variety of avocado and the large-fruited mango, 'R2E2'. In the experiments described earlier a plant labelled as the 'short-beaked' or 'off-type' of *Stylosanthes hamata* (Brolmann and Sonoda, 1981; Cook *et al.*, 2005) has performed well on tropical clay soils despite its subtropical origin on sands. But such cases are rare and much reliance in the development of useful genotypes is being placed on existing genetic resource collections along with plant breeding (Maass and Sawkins, 2004) rather than the collection and introduction of new pre-adapted genetic variation. As the content of most current genetic resource collections is dependent upon surveys such as that described herein, they thus warrant review.

The genera and species shown to be of interest in this report are similar to those presented recently based on an exhaustive survey (Cook *et al.*, 2005). It is clear that the long history of legume domestication in Africa and India has provided an excellent source of information on, and genetic resource collections of, relevant legumes. Many studies such as those carried out under the auspices of IBPGR (now Bioversity International), national programmes (such as that detailed by A.H. Bunting, Republic of the Sudan, personal communication), those related to the various floras (such as the Flora of Tropical East Africa) and the remarkably informative collecting missions of Mr Ray Strickland (CSIRO Cunningham Laboratory, Brisbane, Queensland) have confirmed such a belief and provided much useful information.

In the New World more recent work has provided many species new to agriculture and greatly increased genetic resource collections. Organizations such as CSIRO (Australia), IDRC (Canada), CIAT (CGIAR based in Colombia) and the Royal Botanic Gardens, Kew have worked with organizations such as ENBRAPA and CENARGEN in Brazil and the University of West Indies in Trinidad. But, unlike the situation in the dry tropical regions of India and Africa, the study of indigenous legume species is a relatively new phenomenon: of the 21 principal economic legumes listed by Adams and Pipoly (1980) only one, *Arachis hypogea* or 'Peanut', is from South America and of the 17 pulse crops listed for use in the dry tropics (Cobley, 1956), the majority (13) are from Africa and India.

In South America much of the plant collection for tropical pastures has been of an exploratory nature, road transects have, of necessity, been the norm and many areas of potential importance have not yet been visited (Maass and Sawkins, 2004). For instance, areas of major significance that yielded the all-important *S. scabra* cv. 'Seca' have not been revisited to capitalize on this important source of diversity. While clay soil areas may be relatively small and do not necessarily appear on broad scale soil survey maps (e.g. those that have yielded important cultivars such as *S. hamata* cv. 'Verano' and cv. 'Amiga'), they do exist and warrant examination. One such area, in Guatemala, yielded a distinctive form of *D. virgatus* (67643, a plant well adapted to the Lansdown clay). Another example of an area with great potential for yielding useful germplasm is in Ecuador, where *S. sympodiales* and *D. virgatus* occur, and where there is a unique semi-arid tropical phytogeographic region with clay soils.

Mention must also be made of Central America and the southern states of the USA (for the latter see Appendix 8.2). The highland areas of Central America have provided *Phaseolus vulgaris* (the 'Common Bean'), *P. coccineus* (the 'Runner Bean') and accessions of *Stylosanthes subsericea*, the latter of agronomic interest and generally similar to *S. hamata*. In lowland areas 'new' species of *Stylosanthes* are being found (Vander-Stappen *et al.*, 2006, unpublished results) in addition to variants in, for instance, *S. hamata* (Chapter 3, this volume).

8.6 Choice of material for study

When this programme was initiated little was known about many of the legumes that could be considered for use in tropical pastures because many New World species had not been introduced to Australia. The initial evaluation was essentially a 'look-see', carried out in weed-free, spaced plant 'nursery' conditions (Williams *et al.*, 1976). Such 'nurseries' contained a wide variety of species and genera and anything deemed to be 'promising' could be passed to agronomists interested in particular areas (Burt and Ebersohn, 1969) or to plant breeders interested in particular taxa. On occasions 'nursery data' were discussed, but the data were deemed to be 'descriptive' and, as such, were virtually unpublishable. Notable exceptions, such as that of Hafenrichter *et al.* (1949) – who produced very useful classifications or 'use groups' – were not favoured.

Over time it was possible to accumulate collections for some of the more obviously important taxa or plant groupings, such as those from different climates. For the tropics, that of *Stylosanthes*, assembled by Mr Ron Williams of the CSIRO Cunningham Laboratory, was to be of prime importance (e.g. Burt and Williams, 1975). Other collections were developed, of *Digitaria* (Strickland), *Macrotyloma* (Staples), *Desmanthus* (Burt) and *Rhynchosia* among others. The results from initial plantings and nursery assessments including many of the genera of interest for clay soils were occasionally discussed at meetings (such as the Queensland Department of Primary Industries Conference and Workshop on Plant Introduction and Evaluation in 1984), but the information was rarely published (Burt and Ebersohn, 1969).

The development of pattern-seeking methodologies of analysis in the 1970s and their acceptance by reviewed journals encouraged the rapid development of screening for useful germplasm through the use of descriptive plant characteristics, and allowed dissemination of the results (e.g. with *Clitoria ternatea*, Reid and Sinclair, 1980). Using these techniques, it was then possible to exclude many species of genera such as *Vigna* and *Macrotyloma* from consideration as long-term pasture plants (Chapter 6, this volume). Evaluation programmes with more recent introductions, including those conducted on clay soils in Antigua (Spurgeon, 1977; Keoghan, 1980), showed that plants such as *Macroptilium martii* were unlikely to be useful.

Other genera, such as *Neptunia* and *Indigofera*, which are native to Australia, were excluded because of possible problems with toxicity and weediness. *Leucaena leucocephala*, a small tree, which has been well studied elsewhere and was deemed to represent a special case, was excluded too. By reducing the range of genera and species of potential value, it was possible to focus research attention on genera and species of key interest, for instance *Desmanthus* and *Stylosanthes hamata*.

At the time of the establishment of this study some potentially valuable germplasm was excluded. Some material from clay soils in dry tropical climates that warranted inclusion was simply unavailable for testing (specific accessions of *D. bicornutus*, *D. virgatus* and *S. fruticosa* and the species *Arachis pusilla*). (However all were subsequently included in studies in western Queensland.) Information was lacking for others; *S. seabrana*, for example, which was subsequently shown to be adapted to clay soils in more subtropical climates, but had not then been designated as a separate species. The significance of information about other germplasm had not yet been recognized in this case for *S. humilis* and *S. hamata*, accessions from alkaline soils on the Paraguana Peninsula of Venezuela, and neighbouring areas. The latter two species are of particular interest because two accessions, one each of *S. humilis* and *S. hamata*, have subsequently been found to out-perform other *Stylosanthes* cultivars on clay soils. Also they have interbred (Maass and Sawkins, 2004), and some accessions have the ability to form effective symbiotic relationships with *Bradyrhizobia* collected from plants, such as *Neptunia*, which thrive on clays. Although not collected from clay soils, these plants should have been included in the evaluation studies.

8.7 Methods of evaluation

8.7.1 Classification

Classification is essential for communication and it must be suited to the purpose on hand (e.g. Burt, 1983). In many instances, especially after preliminary 'nursery' evaluation, it may be apparent that some taxa are relatively uniform and that 'special purpose classifications' are not necessary: such is the case, for example, with *Centrosema pascuorum*, *Macroptilium martii* and the various species of *Alysicarpus* or *Clitoria ternatea*. But other species are enormously variable and, in parts of their range, may 'overlap' with other species in certain agronomic characteristics. This is the case for *Desmanthus virgatus* (Burt, unpublished data), *S. scabra* (Burt *et al.*, 1971; Edye *et al.*, 1974) and *S. hamata* (Cook *et al.*, 2005; Chapter 3, this volume).

In the case of *S. hamata* a morphological-agronomic classification had been compiled from the results obtained in a shade house study, and together with provenance information was used to select the material for use in this study. In *Desmanthus* the problem was greatly exacerbated by the lack of a taxonomic revision of the genus, and much of the germplasm of interest was only slowly becoming available for use. A taxonomic revision using such information as was then available was therefore produced, initially for local use, but was later published (Burt, 1993b). As new material became available it was included in field studies and the resultant data classified. The first such classification is shown in Burt (1993a). The final phase of this work was undertaken in western Queensland when selection from newly arrived material was based totally on provenance records.

8.7.2 Evaluation

Evaluation at this stage of development was limited by seed supply and the requirements for seed increase, testing across a range of environments, numbers of replicates required, proposed methods of data analysis and so on. The simplest case is for ley legumes where short-term yield and associated nitrogen fixation, in an environment that can be modified to suit (with land preparation, fertilizer, *Bradyrhizobia* inoculation and weed control), is of overriding importance. But such methodologies, and the statistical methods used in their analysis (Burt and Haydock, 1968), are of limited meaningfulness when persistence is the long-term requirement (Burt and Ebersohn, 1969) and a more ecological approach is desirable (Rejmanek *et al.*, 2005; Daehler and Denslow, 2007).

The methods employed herein have proved to be appropriate for these purposes. The methods utilized should, for instance, be useful in assessing the utility of the germplasm being tested in the different parts of an agro-ecological landscape (McIvor, 2002) or in placing them in potential 'use groups' (Hafenrichter *et al.*, 1949). The methodologies depend on the presence of suitably trained observers to note and record plant responses to 'environmental pulses', rather than simply following a pre-determined protocol. There is also a need for the application of *ad hoc* statistical analyses of the resultant data.

It is now apparent that *a priori* information could well be used to aid relevant information. Indeed such an approach using Bayesian probability theory is proving to be useful in computer search engines and in scientific experimentation. For example in *Desmanthus*, the nutritive value of the seed has long been appreciated by local graziers in its native environments. *D. illinoensis*, because of the nutritive value of its seed, has been tested as a potential pulse crop for dry areas (see information quoted in Burt, 1993a). It has been shown (Schlink and Burt, 1993) that there are large differences in chemical composition, and almost certainly animal feeding value, both within and between *Desmanthus* species. At least one species, *D. bicornutus*, tends to retain its seed in the pod after dehiscence (Luckow, 1993), a feature that could preserve the seed from mould and frost and make it more accessible to animals. Another species has indehiscent pods, a feature that favoured domestication of legumes as pulse crops. Evaluation for such characteristics, prior to cultivar release, would be worthwhile.

Another prerequisite is that plants must be able to form effective symbiotic relationships with the local strains of *Bradyrhizobia* under field conditions. Dr Don Norris, founder of the CB (Cunningham Brisbane) collection of *Bradyrhizobia*, accepted this belief (1968, personal communication) and had earlier – in Brazil – developed a simple field test to measure such ability. By incorporating a high carbon material into his testing site, in his case bagasse from the local sugar mill, available soil N was 'locked up' by microbial activity. Legumes that failed to nodule effectively with the local strains of *Bradyrhizobia* simply died. Strategic research of this type would greatly add to our knowledge of the material on test.

8.7.3 Data analysis and interpretation

The sensible evaluation of plants, for whichever purpose, requires a great deal of information, much of which relates to important natural stresses that occur at unpredictable times and may be difficult to measure, such as responses to drought, fire and flood (Burt and Ebersohn, 1969). Frequently such information, if measured – usually as a rating rather than a numerical value – has not been included in plant descriptions. Additionally 'standard' methods of evaluation were heavily biased towards dry matter production of spaced plants or in small swards (Williams *et al.*, 1976). It has been argued (Burt and Williams, 1988) that this is at least partially explained by 'an over-academic attitude based on the early work of R.A. Fisher'. S.C. Pearce (1979), whose statistical qualifications and experience are not open to question, states: 'Supervisors of research and editors insist on randomized blocks and it is nearly impossible to publish a non-significant difference ... land research institutes, which once protested that the Fisherian experiment was impractical, are now so geared to it that they can conduct nothing but these grand localized "set pieces"'.

While this 'English disease', as it was often unkindly known, was very much a feature of the disciplines involved, it rarely permeated into the field of plant taxonomy (Heywood, 1973); paradoxically, agronomists have always been prepared to base their evaluations on the taxa so delineated. But 'good' taxonomic groupings are based on a multitude of factors, often ecological, although these may never be presented for examination (Heywood, 1973). It is this type of grouping that is required for the present purpose: but it is imperative that the 'unanalysed identities approach' (*sensu* Heywood, 1973) be avoided and that the meaningfulness of the groups be documented. This can be achieved by displaying the variation within plant collections using numerical methodologies and superimposing other relevant information. For instance, using a collection of perennial *Stylosanthes* accessions, it was found that there were three main types of *S. scabra*, and the addition of 'outside information' showed that they were from different climatic backgrounds, different disturbance levels and were associated with different grass species (Burt *et al.*, 1983). They were, in essence, different ecotypes and this was reflected in their agronomic performance. Such classifications were used to select material for inclusion in the present study and they could well be of value for other purposes (e.g. those of Daehler and Denslow, 2007).

The experiments described herein provide additional information to that mentioned above. Classical statistical methods were used to demonstrate the expression of individual accessions, species and genera to their environment: for instance, the ability to spread under different levels of disturbance and to survive invasion by *Bothriochloa pertusa*. There were differences in behavioural patterns between genera and between and within species. In *Stylosanthes* some accessions of *S. hamata* were able to spread into disturbed situations more readily than the perennial species, *S. scabra*: *Desmanthus* species could move into less disturbed situations and some accessions were less susceptible to invasion by *Bothriochloa pertusa*. *S. hamata* behaves like the other annual/biennial *S. humilis* (cf. information presented by Gillard and Fisher, 1978), but, in this case, it is the diploid *S. hamata sensu stricto* and it is on alkaline clay rather than an acidic sandier textured soil. Its demonstrated behavioural pattern is similar to that shown in similar situations in

Antigua (Chapter 3, this volume) and as suggested earlier in the work of Stehle (1956) and in Appendix 8.2.

The ability to predict behaviour from *a priori* information was developed long ago (1702–1761) and was popular for a while 'and has never been fundamentally challenged ... [but was] eventually overwhelmed by those of the "frequentist" school' (Anon, 2007). Its acknowledged utility in computer search engines suggests that it would be of considerable value in plant evaluation studies, for whatever purpose.

8.7.4 Use of provenance records

'A large proportion of the population enjoys eating, growing and viewing plants and prefers a diversity of species to sample. If food choices were restricted to native species, only staples such as wheat for bread and pasta or rice would not be available' (Reichard and White, 2007). There has always been a need to move plants away from the areas that have produced them and the ability to predict the areas in which they are most likely to be useful has long been accepted: Darwin, in his *On the Origin of Species by Means of Natural Selection* (1859), refers to De Candolle's views on the subject. But 'exotic' plants, like their native counterparts, can become undesirable and this is recognized in the field of 'invasive biology' and studies such as those of Hartley (1963) can be used for both plant introduction and invasive biology studies (Williams and Burt, 1982).

Results from recent plant introduction programmes have provided some equivocal answers. In terms of the suite of legume genera under study, the invasive flora of Manila and the Philippines came from the same genera as those that were native, but the introduced species outnumbered the natives and introductions of exotic germplasm of the native species have gone unnoticed. In Australia, using the data provided by Kleinschmidt and Johnson (1977) and Simon (1980) and comparing it to that provided in the floras of Australia, it would seem that many of the plants listed as weeds belong to non-native taxa. But such comparisons contain a subjective element for *Lolium perenne*, a plant widely used in pastures and amenity grasslands (e.g. in more temperate areas of Australia; Williams and Burt, 1982), is not there regarded as a 'weed' but is so labelled in Hawaii (Daehler and Denslow, 2007). The Manila/Philippines and Australia comparison are two quite different situations. In the former areas, the initial vegetation was thick forest, and the undergrowth in these situations is often quite impoverished. This could account for the paucity of native legume species and the ready invasion of exotic species to the environment once it was disturbed. But much of Australia had a relatively rich leguminous flora and invasive plants may need to have been 'different'.

From the programme described herein, and from later agronomic studies (conveniently reviewed in Cook *et al.*, 2005), it would appear that *Stylosanthes* will remain the dominant legume for use on the sandier textured, acidic soils and that *Desmanthus* will be the best-adapted legume for heavier, more alkaline soils (as with *Trifolium* and *Medicago* in more temperate climates). These are both 'exotics' in the sense of being New World genera: but both are taxonomically very close to, and often confused with, the native *Zornia* and *Neptunia* – and they can effectively use the local *Bradyrhizobia* strains associated with them.

Provenance records, made mandatory for Australian plant collections (Myers, 1968), are becoming increasingly used to predict where plants might be collected, adapted and invasive, and this subject is discussed elsewhere with relevance to *Stylosanthes* (Chapter 3, this volume). However, factors other than climatic or edaphic must be taken into account. For instance, was the plant found in a position where it can benefit from 'run-on' water or in an area where it relied on direct rainfall? In the present experiments survival was uniformly high at the least stressful site, Alligator Creek, and detailed provenance records were of no value in predicting success or failure. In contrast in the stringent environment posed by the Lansdown clay, success or failure in the 'ruderal' *S. hamata* accessions tested could only be related to edaphic factors, for all successful accessions came from alkaline soils. Accessions were from a range of environments and soil textures, and some from environments very similar to the successful accessions failed. In the perennial *Desmanthus* species, however, the best-adapted accessions (where provenance data were available) were from clay soils and dry tropical climates. Testing in more stringent environments, in western Queensland, and with a much larger range of *Desmanthus* accessions, provides more relevant information on this topic (Chapter 9, this volume).

Appendix 8.1

Table 8.7. Choice of plant material for initial evaluation on clays and non-clays in the Australian tropics: places of origin, soils and climates.

Species	CPI[a]	Soil at site of original collection	Country of origin	Latitude (degrees)	Altitude (m asl)	Mean annual rainfall (mm)	Habit
Alysicarpus							
A. longifolius	30034	–		18.3 N			Erect annual herb
	79603	–	India	17.2 N			
A. monilifer	40612	deep sand	Sudan	14.1 S			
A. ovalifolius	50968	–	Senegal				Annual herb
	79067	red non–clay	India	17.2 S			
A. rugosus	52351	clay loam	Malawi	15.5 S	1100	850	Erect, prostrate or ascending. Vigorous annual or perennial often somewhat woody
	52352	volcanic clay loam	Kenya	1.0 N	1850	1600	
	52354	cracking clay	Zambia	13.5 S	600	750	
	52355	clay loam	Zambia	13.4 S	1000	1000	
	52356	silty clay loam	Malawi	15.5 S	750	800	
	52357	clay	Malawi	16.0 S	750	800	
	52364	–	Chad	12.1 N			
	60167	acid sandy loam	Uganda	1.1 S	1819	1000	
	60168	acid sandy loam	Kenya	3.6 S	15	1250	
	76978	sandy	Zambia				
	76979	–	Zambia				
	76980	clay	Zambia				
	79604	–	India	17.2 N			
	79605	–	India	17.2 N			
Centrosema							
C. pubescens	Centro						Vigorous rambling perennial
	55703	flooded clay	Venezuela	11.1 N	50	900	
C. pascuorum[b]	55697	acid sandy	Brazil	9.2 S	380	400	Vigorous rambling annual

Clitoria							
C. ternatea	47187	clay	Sudan				Vigorous rambling perennial
	49963	acid sandy	Brazil				
Desmanthus							
D. spp.	33426	clay	Guyana	6.5 N		~2400	Perennial ranging in habit from a prostrate fine-stemmed herb to an erect sometimes woody shrub
D. leptophyllus	38351	non-clay	Venezuela	9.3 N	250		
	55719	acid sandy loam	Venezuela	10.3 N	80	600	
	76053	skeletal brown					
	76057	alkaline clay					
D. pernambucanus	49728	–	Brazil				
D. virgatus	67643	acid clay	Guatemala	15.0 N	185	500	
	65947	alkaline clay	Ecuador	2.2 S	100	500	
	78369	sandy	Argentina	23.2 S	300	650	
	78373	slightly alkaline clay loam	Argentina	23.2 S	300	650	
	78380	clay loam	Argentina	24.5 S	1250	650	
	78382	neutral clay loam	Argentina	24.1 S	1000	700	
	79653	slightly alkaline clay loam	Cuba	21.6 N	5	600	
Desmodium							
D. dichotomum	47186	heavy alkaline cracking	Sudan	12.3 N	450	660	Herbaceous to suffrutescent
Dolichos							
D. sericeus	60218	acid sandy loam	Kenya	0.0 S	1970	1000	Perennial climbing or prostrate herb
	60219	acid sandy loam	Kenya	0.0 S	1879	750	
	60220	acid sandy loam	Kenya	0.0 S	1879	750	
	60221	acid sandy loam	Kenya	0.0 S	1879	750	
Haydonia[c] (syn. Vigna)							
H. monophylla	76993	–	Zambia	15.4 S			Erect climbing or trailing herb

Continued

Table 8.7. Continued.

Species	CPI[a]	Soil at site of original collection	Country of origin	Latitude (degrees)	Altitude (m asl)	Mean annual rainfall (mm)	Habit
Heylandia							
H. laterbrosa	31951	–	India				Prostrate herbaceous annual
Macroptilium							
M. atropurpureum	16875	–	Mexico, Baja California			arid	Climbing or rambling perennial
	33826	sand	Mexico, Veracruz	18.0 N			
	61232	limestone ridge	Mexico	19.0 N	16	1000	
	67649	grey clay	Guatemala	15.0 N	185	500	
	73477	–	Mexico, Baja California			arid	
	76151	deep sandy	Mexico	19.2 N	10	1100	
	76159	sandy	Mexico, Yucatan	20.6 N	30	1150	
	76160	shallow red	Mexico, Yucatan	20.4 N	10	1250	
	76164	sea edge sand	Mexico, Yucatan	19.3	0	1100	
	76165	sand over limestone	Mexico, Yucatan	19.3	10	1050	
	76166	shallow red clay	Mexico, Yucatan	19.5	100	1000	
	cv. Siratro	bred variety					
Strophostyles[c]							
S. helvola	26272	clay		30 N			Twining annual
Stylosanthes							
S. calcicola	73525	sand	Mexico	20.1 N	10	1500	
S. fruticosa	41219	sandy	Sudan	13.0 N		609	
	41219a	segregate of above					
	48386	sand	Tanzania				

S. hamata	cv. Verano 38842	sandy loam	Venezuela	10.7 N	40	420	Annual or perennial prostrate or bushy herb
	33205	stony, calcareous	Guadeloupe	16.5 N	20	1000	
	49080	–	Colombia	11.0 N	15	847	
	50997	probably acid loam	Venezuela	10.0 N	590	509	
	56211	–	Guadeloupe	16.5 N			
	61670	grey alkaline sand	Venezuela	11.2 N	80	950	
	70360	heavy grey/ brown clay	Antigua	17.0 N	60	1346	
	70364	alkaline clay pH 8.0	Antigua	17.0 N	20	940	
	70366	alkaline clay loam pH 8.7	Antigua	17.0 N	20	940	
	70370	alkaline clay loam	Antigua	17.0 N			
S. hamata cont'd							
	70376	heavy cracking clay pH 7.5	Antigua	17.0 N	10	1155	
	70522	alkaline sand pH 8.7	USA, Florida	26.7 N	10	1473	
	70529	Sand	USA, Florida	26.5 N	10	1473	
	72852	sand pH 8.3	USA, Florida	26.5 N	10	1473	
	73507	Loam pH 7.5	Curaçao	12.2 N	10	532	
	73511	–	Cuba				
	73515	sandy loam, pH 6.6 at surface	Nevis	17.0 N	20	940	
	73519	heavy clay	Nevis	17.0 N	100	1397	
S. scabra	cv. Seca 40292	–	Brazil	8.0 S	350		
	cv. Fitzroy 40205	–	Brazil	12.0 S			
	55870	sandy loam pH 6.5[d]	Brazil	14.0 S	600		

Continued

Table 8.7. Continued.

Species	CPI[a]	Soil at site of original collection	Country of origin	Latitude (degrees)	Altitude (m asl)	Mean annual rainfall (mm)	Habit
S. sympodialis	67702	clay loam pH 7.1	Ecuador	2.1 S	100	500	
	67705		Ecuador	1.3 S	250		
Vigna							
V. trilobata	13671	–					Twining annual
	47510	clay	Sudan	8.0 N			
	86179	–	India				
	86180	–	India				
	86181	–	India				
	86186	–	India				

[a]CPI, Commonwealth Plant Introduction number. [b]A collection of *C. pascuorum* from a clay soil in Ecuador was unavailable. [c]*Haydonia* and *Strophostyles* spp. included after the review was carried out. [d]Alkaline below 8 cm.

Appendix 8.2

Provenance records: their possible use in suggesting plant utility

It has long been held that plant evaluation, for whatever purpose, begins at the time of plant collection (Burt and Williams, 1975). It was for this reason that an 'ecological evaluation at the point of collection' was proposed and with it the taking of relevant soil and distributional data (in addition to the mandatory provenance records of Myers (1968)) (Burt *et al.*, 1979). This same information may be used to aid the prediction of invisibility in pre-entry risk assessment and in resolving pertinent issues between the need for introduction and the risks involved (e.g. Bennett and Virtue, 2004; Reichard, 2004).

With the benefit of hindsight, and the widespread utilization of many of the species involved (as summarized by Cook *et al.* 2005), it can be seen that many of the hypotheses developed as the result of the study mentioned previously (Burt *et al.*, 1979) and in associated analysis (e.g. Burt *et al.*, 1980a) have been shown to be correct. *Stylosanthes capitata* has been much more useful on very acid soils than *S. scabra*, *Centrosema* and *Macroptilium* require higher fertility soils than *Stylosanthes* and, in mixed swards as opposed to mono-cultural situations, *Stylosanthes* is favoured by relatively high levels of exchangeable K in the soil (rather than the expected P). Such retrospective considerations, together with floristic studies, can be of value in predicting plant invisibility (Williams and Burt, 1983). The data presented herein may similarly be of value.

The methodology used on some plant collecting missions was developed using information from a survey of the locally adventive species, *S. humilis*. After accidental introduction into the Townsville region from South America, this annual spread throughout the dry tropics of Australia. It was widely used as a pasture species and its ecological and agronomic characteristics were well documented. It is dependent upon heavy grazing and reduced grass competition for its establishment and continued re-establishment from seed (Gillard and Fisher, 1978). If managed in this manner it is present throughout the sward in clumps, within which it is dominant, and these clumps are interspersed with grass. Over time the clumps degenerate and are re-established elsewhere (Burt *et al.*, 1973). Although *S. hamata* is capable of producing high animal live-weight gains, the management practices required to do so are not always ecologically acceptable.

Reference to Table 8.8 shows that *S. humilis*, on soils of sandy texture and medium acidity,

Table 8.8. Behaviour of various species in relation to disturbance levels.

	Disturbance level[a]				
Species	1	2	3	4	5
Centrosema pascuorum	Not found	Not found	2.0 (4.0)[b]	1.0 (4.0)	Not found
Centrosema pubescens	3.7 (2.7)	2.5 (1.5)	3.5 (2.5)	2.3 (3.0)	Not found
Centrosema brasilianum	1.0 (3.0)	2.0 (3.3	2.7 (3.3)	2.5 (2.5)	Not found
Macroptilium martii	Not found	2.0 (1.0)	2.8 (3.8)	2.3 (3.8)	3.0 (4.0)
Macroptilium lathyroides	Not found	Not found	1.5 (2.5)	2.0 (3.0)	3.0 (2.0)
Macroptilium bracteatum	4.0 (2.5)	4.0 (3.0)	3.7 (3.7)	3.3 (3.0)	4.0 (3.0)
Stylosanthes humilis	1.0 (2.0)	2.2 (4.0)	3.0 (3.0)	2.3 (4.2)	3.0 (4.0)
Stylosanthes hamata	Not found	2.0 (1.0)	3.0 (2.7)	4.0 (3.6)	4.0 (3.5)
Stylosanthes scabra	4.0 (1.3)	3.0 (1.8)	3.8 (2.0)	3.3 (2.6)	3.8 (2.3)
Stylosanthes capitata	Not found	4.0 (3.5)	3.0 (2.0)	3.5 (3.0)	3.0 (2.5)

[a]Higher numbers indicate increasing levels of disturbance: 1, lightly grazed, uncleared; 2, as for 1 but heavily grazed; 3, cleared and grazed; 4, road edges, disused gardens etc.; 5, recently graded roadsides and cuttings etc. The number in brackets refers to dominance level at the point of collection. [b]The first number of the paired data in the table refers to distribution in the area around the collecting site. It is the mean value from collecting notes in Brazil and Venezuela (Burt *et al.*, 1979) except for *S. humilis*, where the data come from preliminary studies around Townsville, Queensland, Australia. 1 = barely present, only 1 or 2 plants with low ground cover to 4 = well distributed throughout, several plants, high ground cover.

behaves as a 'ruderal; it is favoured by increasing levels of disturbance, becoming more widely distributed and dominant throughout the collecting areas. The other annuals/biennials behave similarly but require, in the case of *C. pascuorum*, *M. martii* and *M. lathyroides*, increased levels of soil fertility' (Burt *et al.*, 1979).

Ground cover of the *Centrosema* and *Macroptilium* species is high because of the sprawling habit of the plants. In contrast, plants of the perennial *S. scabra* are always well distributed through the collecting areas interspersed with grass and never dominant. They are typically plants from a later stage of the ecological sere (*sensu* Baker, 1974). Whereas the annuals rely on mixed populations for their continued survival in marginal environments (Burt *et al.*, 1973), *S. scabra* has evolved different 'types' to meet such strictures (Burt *et al.*, 1983). The types from the driest and the most disturbed situations show individual plasticity in the face of environmental variations, for instance, in the time and position of first flowering (Burt and Davidson, unpublished data). *M. bracteatum* has similar distributional and dominance patterns to *S. scabra*, although it only appears to be more dominant because of its sprawling habit, and it requires more fertile soils. Information for *Centrosema* is suggestive of a similar pattern but is based on too few occurrences to be reliable.

Since this collecting work was undertaken, and the experimental work described herein was carried out, many of the plant species concerned have been the subject of agronomic evaluation, and assessed for weed risk (e.g. by Daehler and Denslow, 2007 and Batianoff and Butler, 2002, conveniently reviewed by Cook *et al.*, 2005). It is of interest to compare the results obtained from the three approaches.

The information presented herein on the annuals/biennials suggests that these are the 'weedy' ruderals described by Baker (1974). *S. hamata*, like the adventive *S. humilis* in Australia, relies on heavy grazing early in the growing season to control competition from associated grasses, but if overgrazed can become dominant (Cook *et al.*, 2005) as the competition from associated grass is removed. *Centrosema pascuorum*, often used as a ley crop, requires heavy seeding rates and 'some soil disturbance is crucial to ensure good establishment'. *Macroptilium lathyroides*, also used as a short-term ley crop, requires both soil disturbance and fertilizer for its establishment and further disturbance to aid re-establishment. Both plants thrive in disturbed situations. Although all of these species have been used in crop and pasture systems in Queensland – *M. lathyroides* entered Australia before 1900 and has been widely used as a ley legume – none are listed as weeds in south-east Queensland (Batianoff and Butler, 2002) and none are considered to have potential as weeds in Hawaii (Daehler and Denslow, 2007).

In grazed pastures in its native environments, the perennial *S. scabra* generally occurs as individual plants interspersed with grass and rarely dominates even in disturbed conditions such as roadsides. When surface sown in Australia its seedlings are 'particularly hardy and able to establish unless grass competition is strong' and 'once established *S. scabra* can persist with many of the grasses that grow in the same environment' (Cook *et al.*, 2005). However 'under heavy grazing many valuable native grasses have not been able to stand the pressure and have been eaten out', but the grass can be favoured by ecological management strategies such as 'stocking at lighter levels', 'burning ... in early spring to check the established legume plants' or 'sowing more grazing-tolerant grasses' (Partridge, 1997). Hall (1975 and 2007, personal communication) found that the application of phosphorus favoured grasses at the expense of *Stylosanthes* (as predicted from data obtained in Brazil; Burt *et al.*, 1979). This information and the data presented herein suggest that *S. scabra* is a plant from a later stage of the ecological sere than *S. hamata* (*sensu* Baker, 1974).

In Hawaii, *S. scabra* is deemed to be very unlikely to become invasive or to have any major ecological effect (Daehler and Denslow, 2007). But in south-east Queensland it has, unusually for a non-climbing/non-weedy legume, been tentatively listed as being capable of escaping from cultivation and spreading to natural areas. With *Clitoria ternatea*, a climbing legume that survived on the Lansdown clay and is used as both a pasture and ley legume, the situation is reversed. *M. bracteatum* and *C. brasilianum* are perennial species from which cultivars have been selected and been made commercially available. For these we have reasonable amounts of collection data, and small amounts of agronomic data from pasture situations in Australia.

Neither has been regarded as invasive in Queensland or Hawaii.

Where relevant collection data are available on legume distribution and dominance (Table 8.8), and where this material has been tested herein and in agronomic usage, it is clear that the various species are behaving in Australian dryland tropical pastures in much the same way as in their centres of origin. They can be established by creating a 'gap' in the vegetation – by fire, overgrazing and so on – and maintained by other ecological means such as fire, grazing pressure and strategic application of low rates of fertilizer. The Lansdown clay soil site data show that under these conditions they are sufficiently invasive as to persist. However these same conditions do not apply away from the rangeland conditions, for the 'ruderal' *S. hamata* has not become a 'common invasive' of suburban and agricultural disturbance and *S. scabra* is rarely 'generally invasive' *sensu* Batianoff and Butler (2002). On the other hand, in more subtropical, less stressful Australian conditions, where agronomic practices such as ploughing, weed control and more intensive fertilization are carried out to suit the plant, then a plant such as *Macroptilium atropurpureum* (which failed on the Lansdown clay even when introduced from semi-arid climates elsewhere) may be widely used. Such areas are, however, subject to widespread weed invasion (Williams and Burt, 1983) and *M. atropurpureum* together with other climbing legumes utilized in pastures become major weeds and spread into natural areas (Batianoff and Butler, 2002). *M. atropurpureum* behaves similarly in Hawaii (Daehler and Denslow, 2007).

From the studies described herein, and in work published elsewhere (e.g. Keoghan, 1980; Clem and Hall, 1994; Jones and Clem, 1997; Jones and Rees, 1997), it is clear that the genus *Desmanthus* is of major importance as a provider of legumes for use on clay soils in the tropics and subtropics. Of the other possible species, *Leucaena leucocephala* is a woody plant, difficult agronomically and with weed propensities, *Glycine latifolia* is erratic in its performance, and *Indigofera schimperi* may have toxicity problems as well as weed potential. *S. hamata sensu stricto* is discussed in detail elsewhere and is not deemed to have weed potential. *Desmanthus virgatus* has been listed as being likely to be invasive in Hawaii (Daehler and Denslow, 2007), but although it is a long-time resident in Queensland it is not so listed there. *D. bicornutus* on the other hand has been noted to be 'an abundant roadside weed across its native range' (Cook *et al.*, 2005). Both species are likely to be potentially very useful pasture legumes, and are considered to have potential in regenerating degraded landscapes in Texas (Muir and Pitman, 2004) and western Queensland (Chapter 9, this volume). Under such circumstances 'Blanket condemnation of alien species in restoration efforts is counterproductive' (Ewel and Putz, 2004).

A brief review, in relation to the experiments described earlier, is relevant here. First, *D. virgatus*. For some considerable time *D. virgatus* has been used as a 'taxonomic hold all' to accommodate a wide variety of species (Burt, 1993a; Luckow, 1993) and, despite taxonomic revision by Luckow (1993) until the late 1990s many authors continued to refer to all species [of a complex] as *D. virgatus* (Cook *et al.*, 2005). In reviewing the literature, for instance for weed risk assessment, it is quite likely that the information obtained for '*D. virgatus*' will relate to other species. The *D. virgatus* complex contains, for instance, *D. pernambucanus*, a species known to be an invasive weed – and on Hawaii it is even given the same common name as *D. virgatus*, 'dwarf koa' (Cook *et al.*, 2005).

Even within *D. virgatus sensu stricto* (Luckow, 1993) there is a huge amount of variation (Burt, 1993a; Luckow, 1993; Pengelly and Liu, 2001). It is a 'Prostrate, decumbent, or erect herbaceous perennial to 1.5m tall', 'it can be a very small, prostrate herb with highly reduced inflorescences and short fruit ... it may decumbent or erect, and the inflorescences and fruits are much larger' or 'unusually robust ... and have a larger than average number of showy sterile flowers' and so on (Luckow, 1993). It seems unlikely that all of these various 'types' would have equal agronomic utility or potential for invisibility.

From such relevant collecting information as is available from Brazil, Venezuela and Guatemala, plant distribution conforms to that presented for *S. scabra*. In grazed pastures *D. virgatus* was present 'usually more-or-less evenly dispersed in stands of grass and herbs, often at 1 or 2 plants per m^2' and was not common on nearby roadsides (Daehler and Denslow, 2007). This pattern is the same as that found on the clay soil testing site at Lansdown, in the Australian dry tropics.

D. bicornutus behaves in a similar manner (Fig. 8.2). It is a highly variable species which, before the taxonomic revision of the genus (Luckow, 1993) was regarded as three separate species, each listed as coming from different geographic regions and different types of environments. The original *D. bicornutus* was largely subtropical and collected from northern Mexico, and Arizona and New Mexico in the USA on margins of streams and ponds. *D. subulatus*, bulked with *D. bicornutus*, was from more southerly latitudes and was found on hillsides and shaded

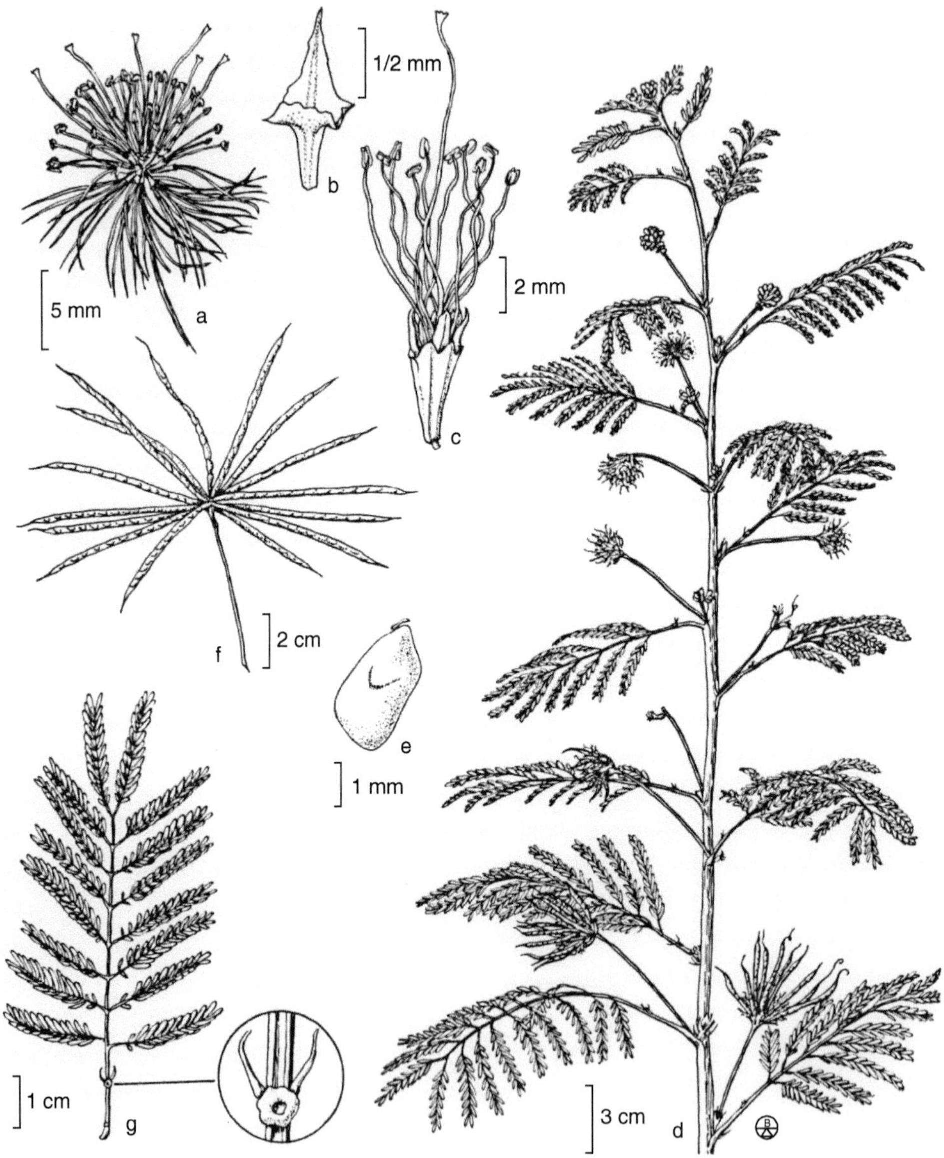

Fig. 8.2. Morphological features of *Desmanthus bicornutus*. a. Flowering head. b. Floral bract. c. Individual flower. d. Habit. e. Seed. f. Fruiting head. g. Leaf with petiolar nectary. Source: Luckow (1993). Reproduced with permission.

arroyos (wash or draw) as a scrub forest underlay, and often occurred on cracking clay soils. Luckow (1993) notes that, in this region of western Mexico, the species 'occurs as an abundant roadside weed or under story plant in deciduous thorn scrub': but at the southern end of the range, the plants are found 'in grassland or cut over thorn scrub and agricultural areas', are deemed to be 'robust' and to differ morphologically, and no mention is made of weediness. These various 'types' of *D. bicornutus* could, as with *D. virgatus*, have different agronomic utilities and potential invasiveness. In western Queensland, some 20 years after sowing, this species has not spread from trial plots but has persisted (Chapter 9, this volume).

These observations illustrate the problems faced by those attempting to sensibly evaluate the utility of this important genus. In a comparable situation, with the then poorly known *Stylosanthes* genus, it was necessary to produce a 'morphological-agronomic classification' to illustrate the types of variation occurring both between and within species (Burt *et al.*, 1971). This classification proved to be of great utility and avoided a costly piecemeal approach to the development of the genus. For instance, it delineated the numerous duplicate accessions, common in any collection and evaluation programme, and reduced evaluation costs. From the homogeneous 'groups' or 'morphological/agronomic types', it was possible to select one or two accessions as representatives. This procedure created an objectively selected 'core collection' (e.g. Brown and Spillane, 1999) that could be evaluated over a wide geographic and climatic range and the best adapted 'types' selected for study (e.g. Burt *et al.*, 1974). The domestication of the 'new' species of *S. hamata* and *S. scabra* and 'new' types of *S. guianensis* rapidly followed. The various morphological groups also proved to be ecogeographically based (Burt, 1975b; Burt *et al.*, 1983) and this greatly facilitated the choice of areas for future collecting missions.

In attempting to characterize accessions more easily, and thus expedite the evaluation process, biochemical techniques have long been employed. An attempt was made to do so in *Stylosanthes* when esterase isozymes were measured in a collection of material that had previously been used for both morphological/agronomic and ecogeographical studies (Robinson *et al.*, 1980). Although it was possible to link the isozyme banding patterns to the geographical background of the accessions, and through this to the general characteristics of the plant performance, this correlation was far from precise. This is not surprising for, as shown herein for *S. hamata*, the accessions that performed best at one site came from a wide variety of climates, soil types and geographical backgrounds. With perennial plants, such as *S. scabra*, and using molecular evidence, it was possible to more fully appreciate genetic and phylogenetic relationships (Liu, 1997), but did not distinguish between very different morphological/agronomic types that were adapted to very different climatic and ecological conditions (for a fuller discussion see Chapter 3, this volume). In *Desmanthus* this approach grouped accessions of *D. virgatus* with very different characteristics and from very widely separated geographical regions and placed other accessions closest to, for instance, *D. bicornutus* (Pengelly and Liu, 2001). As noted by Brown (1978), 'the evaluation of genetic diversity in a sample is a problem different from the evaluation for its agronomic merit'. For many tropical legumes, where most cultivars are simply selections of introduced material, sometimes slightly modified to improve disease resistance such as has been done in *Stylosanthes*, the prime concern is still with evaluation for agronomic merit.

In *Desmanthus*, which unlike *Stylosanthes* is not primarily a tropical genus (Burt, 1993a; Luckow, 1993), the search for material for more temperate regions (Jones and Rees, 1997) should consider the wealth of useful species available in the southern states of the USA (e.g. Luckow, 1993; Muir *et al.*, 2005), areas deemed by Hartley (1963) to be likely to yield adapted material (for the predictive ability of this methodology, see Williams and Burt, 1982). Consideration of the utility of these 'new species' is again hampered by the lack of a suitable classification.

9 Searching for Pasture Legumes for Heavy Clay Soils in the Australian Dry Tropics and Subtropics: IV. Evaluation in Western Queensland

R.L. Burt†

Abstract

Sixty-one accessions of 19 species belonging to nine genera potentially adapted to heavy clay soils in drier conditions were tested at six western Queensland sites in a range of climatic conditions in 13 experiments involving three trial designs: single strips (2), small swards (8) and transplants with initial irrigation (3). Growth, response to stresses and persistence of the plantings were recorded. Only *Desmanthus* species persisted at the sites, some surviving for many years. The characteristics of the *Desmanthus* species of interest and the importance of further collection of the variation in these species in order to enable their potential to be assessed and utilized are discussed, and suggestions of how this might be efficiently achieved are presented.

9.1 Introduction

There are very large areas of clay soils throughout the tropics and subtropics of the world, much of them being located in dry tropical to semi-arid and arid regions (Dudal and Bramao, 1965). These areas are of major significance for cropping, especially ley farming, and for permanent grazing. Tropical legumes are much sought after for both systems (see Chapters 6–8, this volume).

In Australia these regions with clay soils are usually referred to by the names of the native species that originally dominated, and in many cases continue to dominate each area: Brigalow (*Acacia harpophylla*), Gidgee (*Acacia cambagei*) and Mitchell Grass Plains (*Astrebla* spp.). These soils were relatively fertile and it was long held that the application of fertilizer was not necessary and, in most situations, was simply uneconomical. However with the introduction of grazing animals many of the native legumes began to disappear and this contributed to a downward spiral in soil fertility and pasture quality (Leslie *et al.*, 1987). There is currently a real danger of landscape degradation (Tothill and Gilles, 1992) and degeneration into almost unusable 'thorn scrub', as has occurred in many parts of the world (e.g. Muir *et al.*, 2005). One means of alleviating this situation is to establish suitable legumes which, because of the nitrogen that they contribute to the soil, stimulate the growth of associated grasses, which in turn compete with the woody weeds.

It has not always been possible however to re-establish native species such as *Alysicarpus*, *Rhynchosia* or *Psoralea*, despite long-term efforts to develop such species for use (e.g. Skerman, 1957; Kerridge and Skerman, 1968; Gramshaw *et al.*, 1987; Clem and Hall, 1994). The sole cultivar to come from this work, *Glycine tomentella* cv. Capella, showed little promise in the subtropical regions from which it had been collected (Jones and Clem, 1997), perhaps due among

  Tropical Forage Legumes (ed J.R. Lazier)

other factors to soil compaction, or selective grazing. However, in the face of global warming and the already evident signs of the deterioration of rangeland in many parts of the world, it would seem that reductions in productivity will be exacerbated. There is a very real need to seek possible alternatives to these native legumes.

Chapters 6–8 (this volume) were concerned with the provision and evaluation of a wide range of legume genera and species for use on Australian clay soils. The collection has been forced through an environmental sieve for increasing aridity. These final reports are concerned with their evaluation in the most stringent clay environments in western Queensland.

9.2 Methods

9.2.1 Sites and long-term climatic conditions

The earlier stages of this work were carried out at sites 1, 2 and 3 (Fig. 9.1), but a drought in the establishment year curtailed the work at the driest site (site 3), and the results from a more favourable dry tropical clay soil site (site 2) showed that few accessions were suitable for use in permanent pastures (as confirmed by Clem and Hall, 1994 at sites 12, 13 and 14). Emphasis was therefore placed on drier situations for which evaluation data were lacking but where needs (see Introduction) are likely to be very urgent (sites 6 to 11, Fig. 9.1). These areas are remote and present many logistical problems for research. It is for this reason that the Queensland Department of Primary Industries has strategically located field stations on the Mitchell Grass Downs at different latitudes at Toorak (6) and Rosebank (7) (Fig. 9.1, Table 9.1). These correspond climatically to sites such as Jodhpur and New Delhi in India (Papadakis, 1966), but rainfall varies greatly between years (Fig. 9.1). Thanks to grazier cooperation it was possible to replicate the experiment over a range of rainfalls and soil types near the more southerly site, Rosebank (7) (sites 8 to 11, Fig. 9.1, Table 9.1). Further information on some of the soils is presented in Gardiner *et al.* (2004).

9.2.2 Plant material

In the previous experiments in this series (Chapter 8, this volume), accessions of *Stylosanthes*, especially *S. hamata*, and *Desmanthus* were clearly the most promising for long-term pastures on clay soils. In *Stylosanthes*, preference was given to those accessions that had performed best in the previous studies carried out on Lansdown clay (Fig. 9.1). *S. sympodiales* (20) and *S. fruticosa* (11) (see Table 9.2) were included, although the former succumbed to anthracnose in the earlier tests. Both were from clay soils in semi-arid areas and anthracnose may not pose such a threat in the dry environments of these trials. Accessions of *S. scabra*, adapted to clay soils in moister, more subtropical conditions, would have been included as short-term ley plants, but they were not then available. However, the species failed when tested at Hughenden (Fig. 9.1, 15; Gardiner, 2003). The final additions were two accessions of *S. hamata*, which were very similar to *S. hamata* cv. Verano and derived from it, but found growing on a clay soil on which Verano would not grow. They appear to be part of a segregating population (Chapter 3, this volume).

Various species of *Desmanthus* had proved to be adapted to the Lansdown clay soil situation and it was noted that these had been collected initially from a range of climatic conditions (Burt, 1984c). As the aims of these experiments were to find suitably adapted plants and to ascertain if they were to be found in similar environments elsewhere, the range of species, types within species and environmental types from which the plants were collected was increased. Use was made again of the *ad hoc* classifications made previously (one of which was presented in Burt, 1993b). From those accessions, which had just become available for examination, a range of material was selected to cover the environments from which the species had been collected. *D. bicornutus*, for instance, is a very valuable species (Luckow, 1993) that had previously been regarded as three separate species with somewhat different ecological requirements (see data quoted in Burt, 1993a). All relevant authors had noted that it is adapted to clay soils in dry tropical conditions. More detailed information is presented in Appendix 9.1. In total the collection contained 61 accessions and represented 8 genera and 19 named species.

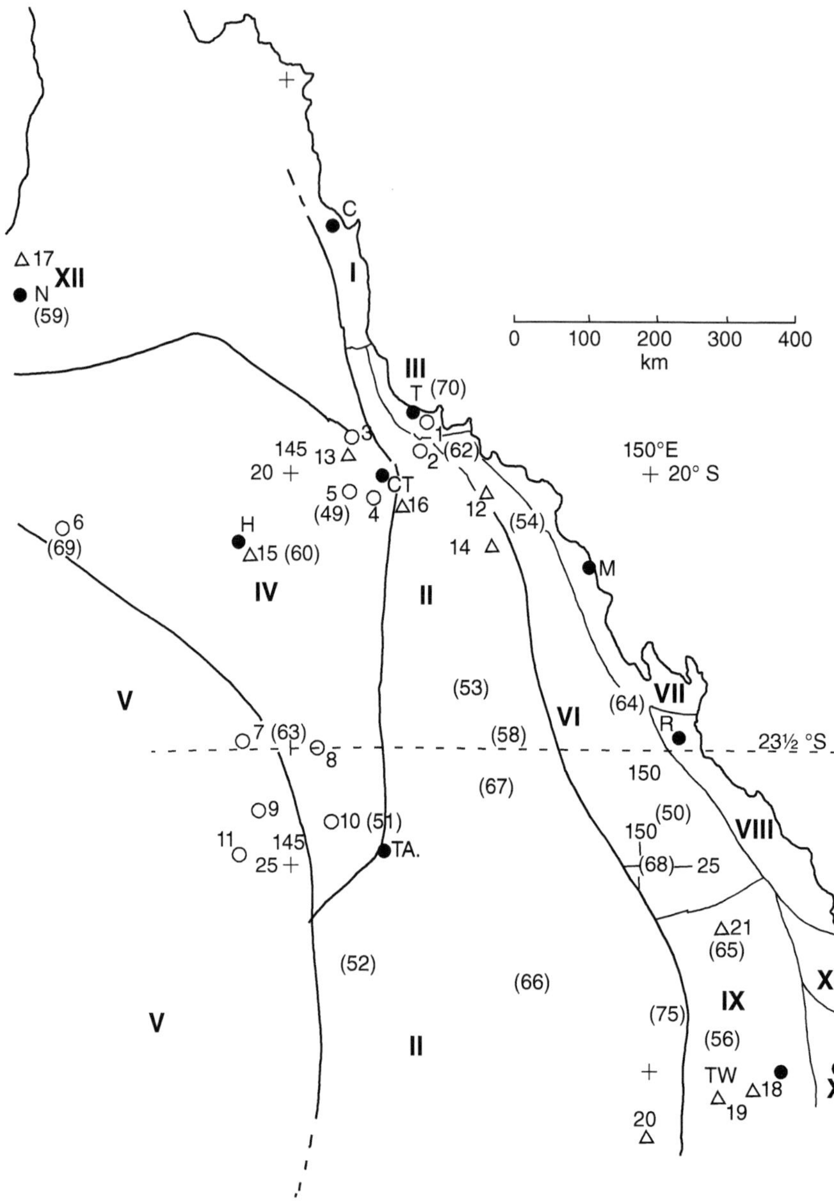

○: Sites 1 to 5 used by one of the authors in the lead-up to the work at sites 6 to 11.
△: Other evaluation sites mentioned in the text

1. Alligator Creek;	8. Barcaldine;	15. Hughenden;
2. Lansdowne;	9. Isisford;	16. Trafalgar;
3. Hillgrove;	10. Blackall;	17. Normanton;
4. Cardigan;	11. Yaraka;	18. Pittsworth;
5. Balfe's Creek;	12. Mguna;	19. Kindon;
6. Toorak, Julia Creek;	13. Bluff Downs;	20. Toobeah;
7. Rosebank, Longreach;	14. Havilah;	21. Naragen;

●: Towns.
B, Brisbane; C, Cairns; CT, Charters Towers; H, Hughenden; N, Normanton; R, Rockhampton; T, Townsville; TA, Tambo; TW, Toowoomba

Fig. 9.1. Evaluation site geographical locations and associated climate types.
Climatic types are numbered I to IX as nominated by Papadakis (1966) and are shown for Northern Australia by Reid et al. (1976). For further details see Appendix 9.3. Numbers in brackets are those used for place names in the climatic analysis (Chapter 7, this volume, Table 7.2 and Fig. 7.1).

Table 9.1. Testing site locations and environments.

Latitude °S	Nearest town	Property name	Site description	Mean annual rainfall (mm)	Mean rainy days/year	Temperature °C	
						Mean maximum	Mean minimum
21° 39′	Julia Creek	Toorak	Mitchell Grass Downs	463	33	33	18
23° 26′	Longreach	Rosebank	Mitchell Grass Downs	443	40	31	16
23° 30′	Barcaldine	Blendon	Cleared clay Gidgee	494	48	30	16
24° 15′	Isisford	Taranaway	Cleared pebbly clay Gidgee	446	41	31	15
24° 20′	Blackall	Woodbine	Cleared clay loam Gidgee	521	50	30	15
24° 50′	Yaraka	Mt Marlow	Open woodland clay loam	360	27	n.a.	n.a.

n.a., not available.

Standard accessions were included in the trial (accessions 1 to 10, Table 9.2). These include an *Alysicarpus* accession (1) that is native to the clay soils of the region, *Arachis* (2 and 3) representing the only leguminous genus found on a clay soil in dry tropical Brazil (Burt *et al.*, 1979), *Vigna trilobata* from a similar situation in India that had shown promise before succumbing to insect attack, *Macroptilium atropurpureum* (8 and 9), and *Chamaecrista rotundifolia* (4) and *Clitoria ternatea* (5 and 6) as species often recommended for clay soils. The details are presented in Appendix 9.1.

The second type of experiment utilizing small swards is the type of experiment often recommended at this stage of evaluation. Inoculated seed was hand broadcast onto plots measuring 4 m × 5 m. There were three replicates.

The preferred type of experiment was only logistically possible at the two QDPI field stations Toorak and Rosebank, and then only in a more simplified form than previously undertaken. It utilized transplants with some initial irrigation to improve establishment (based on the methodology employed earlier by Burt, Chapter 8, this volume).

9.2.3 Evaluation methods

The results obtained herein were obtained from 13 experiments comprised of three types: two were GRAZNET experiments (Experiments 1 and 2), eight were small sward experiments (3 to 10), and three were transplant experiments (11 to 13) (Table 9.3).

As noted in Chapters 6–8 (this volume), the GRAZNET experiments were part of a state-wide evaluation programme involving a wide range of genera and species. The methodology, of necessity, had to be simple and sparing with seed. At each site 0.5 g of seed previously inoculated with a relevant strain of *Bradyrhizobia* was hand broadcast along a 4-m row. There were two replicates only. More information on all three types of experiment is presented in Appendix 9.2.

9.2.4 Climate conditions and measurements taken

Six experiments were sown at Toorak and Rosebank in 1988, two GRAZNET (1 and 2), two small swards (3 and 5) and two transplants (11 and 13). It proved to be a drought year (Figs 9.2 and 9.3). In the GRAZNET experiments seedling establishment was poor to non-existent and thus both experiments were terminated in the first growing season. Establishment in the small swards was also poor but some useful information was recorded (Table 9.3 and the Results section).

The rainfall received in 1989, 1990 and at the start of the 1991 growing season was much higher, to the extent that flooding caused serious problems with access. The dates that data were recorded are given in Table 9.3, the

Table 9.2. Plant accessions used in the experiments and seasons planted (A, B, C: Seasons accessions planted: first, second and third growing seasons).

Plant no. in experiment	Genus and species[a]	Accession number[b]	Season	Type of experiments and site sown[c]: GRAZENET	Small swards	Transplants
Non-*Stylosanthes*/*Desmanthus*						
1	*Alysicarpus* sp.	TQ 1		1, 2	3 to 10	11, 12, 13
2	*Arachis* sp.	LD 1		n.a.	n.a.	
3		LD 2				
4	*Chamaecrista rotundifolia*	cv. Wynn		1, 2	3 to 10	
5	*Clitoria ternatea*	47187				
6		48337				
7	*Centrosema pascuorum*	55697				
8	*Macroptilium atropurpureum*	cv. Siratro				
9		61232				
10	*Vigna trilobata*	13671				
Stylosanthes						
11	*S. fruticosa*	60354		1, 2	3 to 10	11, 12, 13
12	*S. hamata*	70522				
13		70525				
14		72850				
15		cv. Verano				
16		TQ 2				
17		TQ 3				
18	*S. scabra*	cv. Seca				
19		55870				
20	*S. sympodiales*	65960				
Desmanthus						
21	*D. bicornutus*	81337	A	1, 2	3 to 10	11, 12, 13
22		84508	B	1, 2		
23		90857	A	1, 2		
24		91162	C	1, 2		
25	*D. covillei*	90311	B	n.a		
26		90877	C	n.a.	3 to 6	
27	*D. fruticosa*	84960	C	1, 2	8 to 10	11, 12 only
28	*D. leptophyllus*	37143	A	1, 2	3 to 10	11, 12, 13
29		38351	A	1, 2		
30		55719A	A	1, 2		
31		76053B	B	1, 2		
32		92655	n.d.	1, 2		

Continued

Table 9.2. Continued.

Plant no. in experiment	Genus and species[a]	Accession number[b]	Season	Type of experiments and site sown[c]: GRAZENET	Small swards	Transplants
33		92746	C	1, 2		
34		92809	B	1, 2		
35		LD 3	A	1, 2		
36		LD 4	B	1, 2		
37		LD 5	B	1, 2		
38	*D. pernambucanus*	40071	A	1, 2		
39		49728	A	1, 2		
40		83565	B	1, 2		
41	*D. pubescens*	92800	C	n.a.		
42		92802	A	1, 2		
43		cv. Uman	A	1, 2	3 to 10	
44		92804	C	n.a.		
45	*D. tatuhyensis*	37538	B	1, 2	3 and 5 only	
46		90362	C	1, 2		
47	*D. virgatus*	57960	A	1, 2	3 and 5 only	11, 12, 13
48		67643	A			
49		78372	A			
50		cv. Marc	A			
51		78382	A			
52		79653	A			
53		85173	C			
54		85178	A			
55		85182	B			
56		90751	A			
57		91181	C			
58	*D.* sp.	33426	B		5, 6, 8 only	
59		70338	B		5 to 10 only	
60		83563	B			
61		83570	C		n.a.	

[a]*Desmanthus* identities usually as given in Gardiner *et al.* (2004); [b]accession numbers are CPI (Commonwealth Plant Introduction) unless otherwise noted. [c]site sown, see Table 9.3; TQ, Townsville, Queensland; LD, collections from an evaluation site at Lansdown near Townsville; n.a., not available for sowing; n.d., missing data.

types of data for each trial are in Tables 9.4 and 9.5 and the relationship of the observations to monthly rainfalls are presented in Figs 9.2 and 9.3.

Those small sward experiments sown in the 1988 growing season (numbers 3 at Toorak and 5 at Rosebank) were affected by drought and after poor establishment were terminated in the first growing season. Experiment 2 similarly provided little information after poor establishment.

Experiments 6 to 10 were established at the beginning of the 1989 growing season and this season, together with that of 1990, experienced well above average rainfalls. Record taking was inhibited at Experiments 8, 9 and 10 due to access

Table 9.3. Experiment sites and dates of establishment and sampling.

Experiment no. and trial type[a]	Site[b]	Sowing/ transplant date	Sampling dates and growing seasons		
			First season (A)	Second season (B)	Third season (C)
GRAZNET					
1	Toorak	30/1/88	20/3/88		
2	Rosebank	2/2/88	10/5/88, 7/9/88		
Small sward					
3	Toorak 1	30/1/88	3/5/88		
4	Toorak 2	26/11/88	14/4/89, 6/6/89		
5	Rosebank 1	2/2/88	10/5/88, 8/9/88		
6	Rosebank 2	29/11/88	6/3/89, 8/9/89	12/12/90	
7	Woodbine	30/11/88	9/3/89, 8/6/89	13/11/89, 20/7/89	13/12/90
8	Taranaway	11/12/89	21/7/90	12/12/90	
9	Blendon	14/12/89	9/7/90	1/12/90	
10	Mt Marlow	12/12/89	21/7/90	3/12/90	
Transplants					
11	Toorak 1	29/1/88	23/5/88, 15/7/88	23/11/88, 9/1/89, 3/6/89	4/5/90
12	Toorak 2	24/11/88	13/4/89, 3/6/89	24/11/89	4/5/90
13	Rosebank	3/2/88	13/3/88, 10/5/88, 7/9/88	28/11/88, 5/6/89	26/11/89, 22/7/90, 10/12/90

[a]Type: see Evaluation Methods section in text; [b]site: given as property names – see Table 9.1 for locations.

being limited by flooding. Favourable rainfall at the beginning of the next season in December allowed the final set of records to be taken.

The data recorded are presented in Tables 9.4 and 9.5 and their timing relative to rainfall events shown on Figs 9.2 and 9.3. Visits to aid record-taking were not taken at regular intervals but were rather timed to fit into important events at each station such as the start and end of the growing seasons, flowering and seeding, and regrowth after flooding. Because of logistical problems and the large variations in plant form and height, regular dry matter harvesting was not deemed to be favourable or warranted for the present purpose, survival being the paramount character (e.g. Burt and Ebersohn, 1969).

The transplant experiments (11 to 13) at the two QDPI field stations all provided information over three growing seasons. Because the growth and development of the original transplants could be more accurately monitored, the data are much more reliable and interpretable.

Further information from some of these experiments was obtained in 1999 and 2003 by Gardiner *et al.* (2004).

9.3 Results and preliminary discussion

9.3.1 Preliminary considerations

The experiments conducted generated a very large volume of data, too bulky for complete presentation and ease of interpretation. Much information will therefore be presented in appendices and more detailed information on the experiments themselves can be obtained from ACS and on more botanical aspects from RLB.

To ease presentation and interpretation the results have been presented in three sets. The first comprises information from Experiments 1 to 6 (Table 9.3). These were surface sown and short-lived, usually lasting for only one growing season. From these trials useful information about the abilities of the various plants to establish from seed, and to flower and seed in harsh environmental conditions, can be extracted.

The second set is derived from Experiments 7 to 10 (Table 9.3), also surface sown, but experiencing more favourable environmental conditions. Some plants survived until the experiments were curtailed and were surviving some

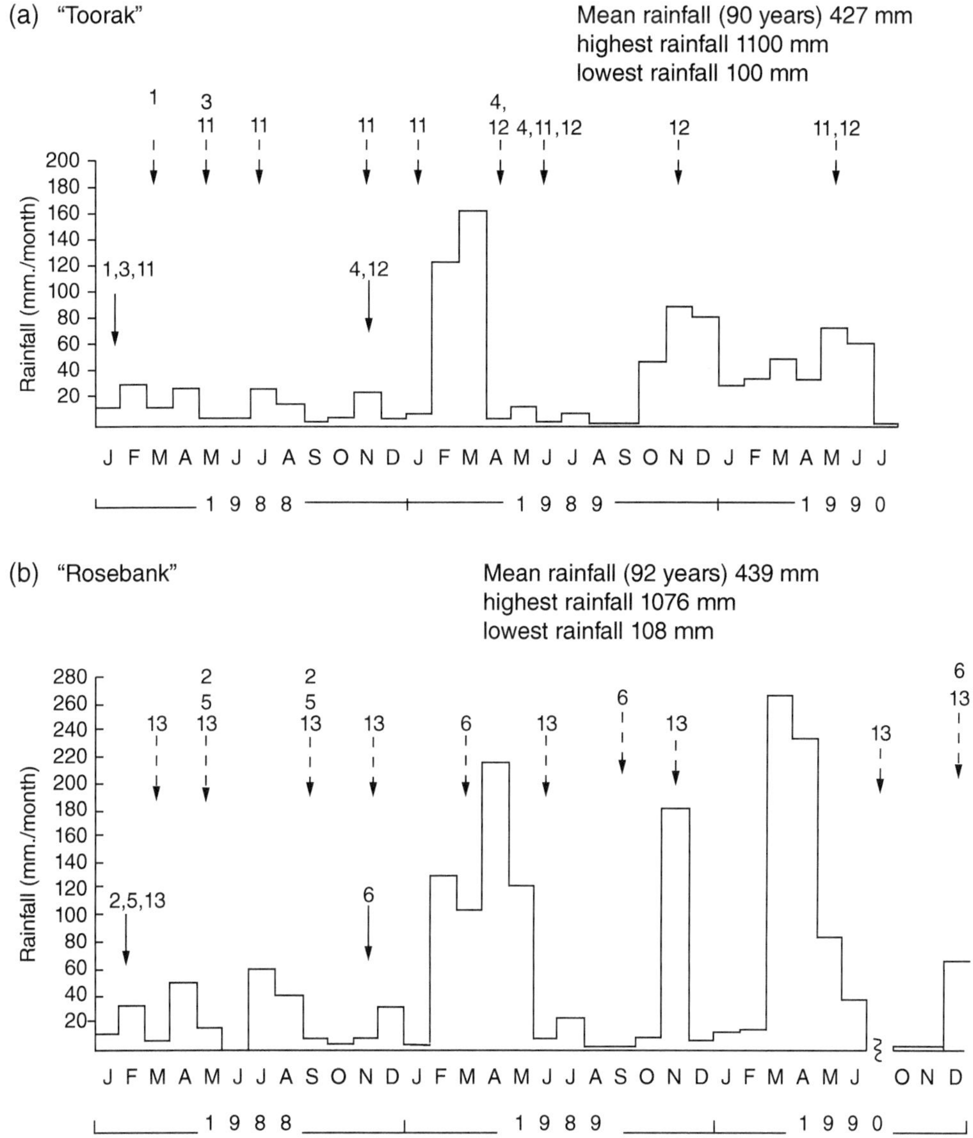

Fig. 9.2. Monthly rainfall over the experimental periods at Toorak and Rosebank. Solid arrows indicate the times at which the experiments numbered as in Table 9.3 were sown, and dashed arrows the recording times (see Table 9.3).

10 years later (Gardiner *et al.*, 2004). Here differences in persistence, vigour and ability to flower and set seed can readily be discerned. Such results can be compared with perceived differences in persistence and production some 10 years later (Gardiner *et al.*, 2004).

The final set presents data from the transplant experiments (11 to 13, Table 9.3). Here much more precise ecological information about the various plants, their persistence, ability to flower and set seed, ability to regrow after frost, fire and flood, can be presented. Graphical representations are used to illustrate the behaviour of the most promising accessions at the most amenable site, Rosebank. This consolidates the analysis. A full analysis is presented in Appendices 9.4 and 9.5.

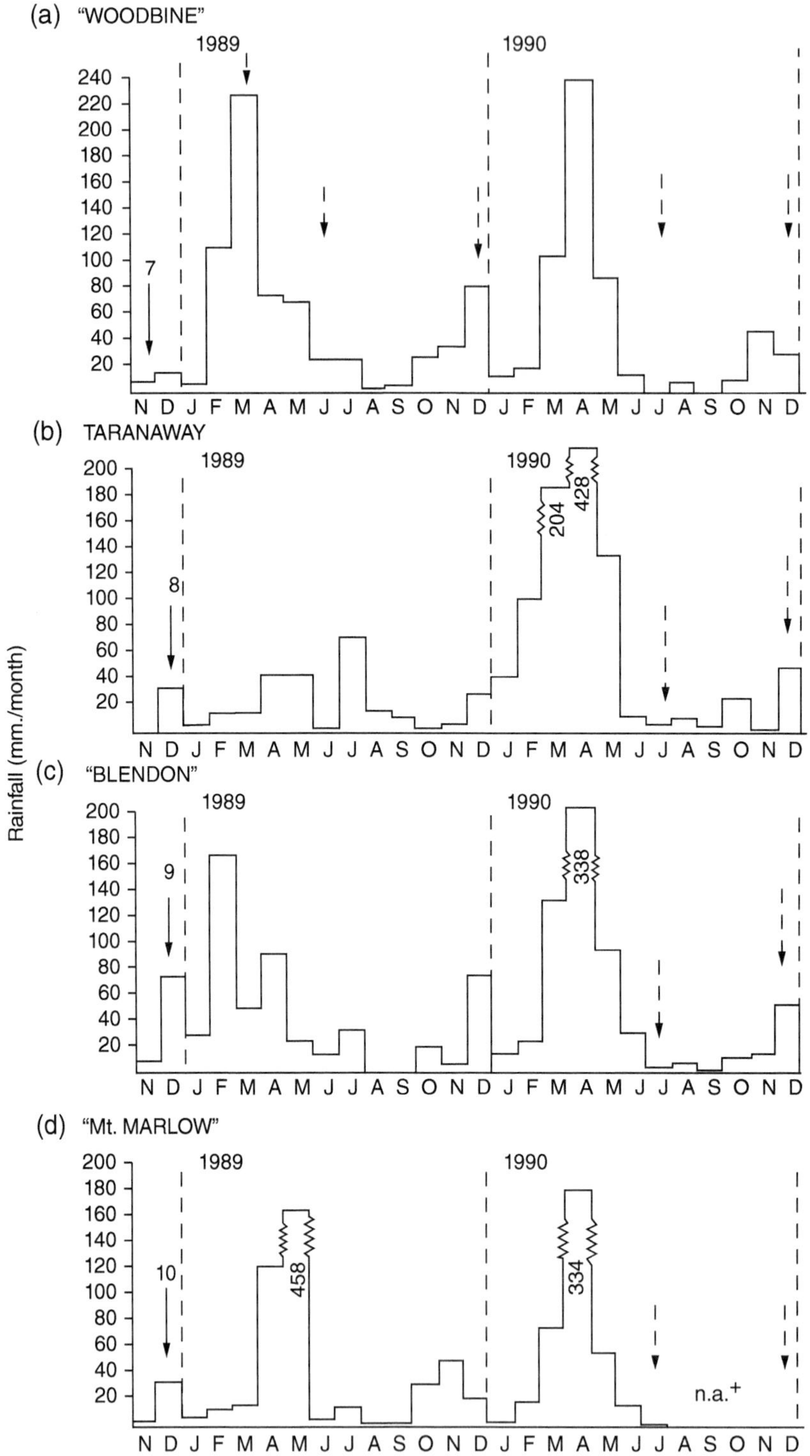

Fig. 9.3. Monthly rainfall over the experimental periods at Woodbine, Taranaway, Blendon and Mt Marlow. Notation as in Fig. 9.2; n.a., not available.

Table 9.4. Observations taken in the small sward experiments.

Trial			
Number[a]	Location	Date	Observations
4	Toorak	14/4/89	Number of plants/plot
			Presence or absence of flowers or seeds
		6/6/89	As above
5	Rosebank	*First sowing*	
		10/5/88	Number of plants/plot
		8/9/88	As above
6	Rosebank	*Second sowing*	
		6/3/89	Number of plants/plot
		8/9/89	Number of plants/plot
			Presence or absence of flowers or seeds
			Plant growth and vigour rated 0–10 (0 = dead)
		12/12/90	Number of plants/plot
			Plant growth and vigour rated 0–10 (0 = dead)
7	Woodbine	9/3/89	Number of plants/plot
		8/6/89	Number of plants/m^2
			Legume dry matter yield (g/m^2) cf. grass dry matter yield of 216 g/m^2
			Plant regrowth and vigour rated 1–10 (0 = dead)
		13/12/89	Plant growth and vigour (after recovery from frost in winter)
		20/7/90	Plants regrowing after flooding in previous months
			Number of plants/m^2
			Plant regrowth and vigour rated 1–10 (0 = dead)
		13/12/90	Number of plants/m^2
			Plant growth and vigour rated 0–10 (0 = dead)
			Presence or absence of seeds
8	Taranaway	21/7/90	Number of plants/plot
			Plant growth and vigour rated 0–10 (0 = dead)
			Presence or absence of seeds
		2/12/90	Same as for 21/7/90
9	Blendon	19/7/90	As for Taranaway
		1/12/90	As for Taranaway
10	Mt Marlow	21/7/90	As for Taranaway
		3/12/90	As for Taranaway

[a]Experiments numbered as in Table 9.3. Relevant rainfall records and sampling dates shown in Figs 9.2 and 9.3.

Performance of non-Desmanthus accessions

The performance of these accessions (details presented in Appendix 9.4) was, in every case, disappointing. They usually died in the first growing season and even in the tropical experiment (12), at the most amenable site (Rosebank), only three accessions showed any survival up to the end of the third growing season (July 1990, Fig. 9.2b), and three showed no growth at all at the start of the next season. The first of these, a native *Alysicarpus* species, had not been sighted previously and it may have survived in the soil seed bank. The second was an annual *Arachis* species that had performed poorly at all sites and times (and is unlikely to be economically useful because of its subterranean seed production and requirement for seedbed preparation at sowing). The third, *Clitoria ternatea*, had failed elsewhere. All were collected originally from clay soils in semi-arid tropical environments.

Although all non-*Desmanthus* accessions ultimately failed and none were to re-establish at a later date (Gardiner *et al.*, 2004), some accessions of *Stylosanthes* and *Clitoria ternatea* showed some signs of adaptation. Both accessions of

Table 9.5. Observations taken in the transplant experiments.

Trial			
Number[a]	Location	Date	Observations
11	Toorak	3/5/88	Number of transplants surviving/plot Presence or absence of flowers or seeds
		3/17/88	As for 3/5/88
		23/11/88	Number of transplants surviving/plot Plant growth and vigour rated 0–10 (0 = dead)
		19/1/89	Number of transplants surviving/plot Presence or absence of flowers or seeds Number of seedlings per plot
		13/6/89	As for 19/1/89
		4/5/90	Number of transplants surviving/plot Presence or absence of flowers or seeds Number of seedlings/plot Plant growth and vigour rated 0–10 (0 = dead)
12	Toorak	13/4/89	Number of transplants surviving/plot Plant growth and vigour rated 0–10 (0 = dead)
		3/6/89	Number of transplants surviving/plot Presence or absence of seeds Plant growth and vigour rated 0–10 (0 = dead)
		24/11/89	Number of transplants surviving/plot Presence or absence of seeds Plant growth and vigour rated 0–10 (0 = dead) Number of seedlings/plot
		4/5/90	Number of transplants surviving/plot Presence or absence of flowers or seeds Number of seedlings/plot
13	Rosebank	13/3/88	Number of transplants surviving/plot Plant growth and vigour rated 0–10 (0 = dead)
		10/5/88	Number of transplants surviving/plot Plant growth and vigour rated 0–10 (0 = dead) Presence or absence of flowers or seeds
		7/9/88	As for 10/5/88
		28/11/88	As for 10/5/88
		5/6/89	As for 10/5/88
		26/11/89	As for 10/5/88
		22/7/90	Number of transplants surviving/plot Plant growth and vigour rated 0–10 (0 = dead) Number of seedlings per plot Vigour and growth of seedlings, rated 1–10
		12/12/90	Number of transplants surviving/plot Plant growth and vigour rated 0–10 (0 = dead) Presence or absence of seeds

[a]Experiments numbered as in Table 9.3. Relevant rainfall records and sampling dates shown in Figs 9.2 and 9.3.

C. ternatea had persisted and seeded to the beginning of the third season (November 1990, Fig. 9.2b). The same was true for *S. fruticosa*, *S. hamata* and *S. scabra*, four accessions of which were noted to have produced seedlings on 22/7/90, at the end of the third growing season (Fig. 9.2b). This same material was among the best adapted on a clay soil in dry tropical conditions (Chapter 8, this volume). However, their ultimate failure, and the success of *Desmanthus*, anticipates the findings of Gardiner *et al.* (2004).

Performance of Desmanthus accessions

Many *Desmanthus* accessions survived until the closure of the experiments and have been noted to survive for a much longer period (Gardiner *et al.*, 2004). Here the behaviour of some of the better adapted accessions in three of the present experiments is used to illustrate the adaptation of the genus, in which, as was just noted, accessions of all other genera failed.

EXPERIMENT 7. This small sward experiment was situated at Woodbine, where the site was accessible for record taking despite above average rainfalls (Fig. 9.3). *D. leptophyllus* showed the highest plant establishment and dry matter yields in the first season (1989) and had the highest plant vigour ratings at the start of the second (Fig. 9.4). The ratings declined at the end of the second season (20/7/90) and start of the third (12/12/90). *D. virgatus* produced poorly in the first season but its vigour ratings improved thereafter and its plant numbers increased markedly at the start of the third season when, unlike the other species, most of its accessions were flowering, *D. bicornutus* also showed some slight improvement in plant numbers at this time.

Plant numbers changed markedly throughout the course of this experiment (Table 9.6) and it is notable that in some accessions such as *D. pubescens*, *D. tatuhyensis* and *D.* sp. 70338, plants could be absent in one sampling period only to reappear at the next. Gardiner noted similar findings in this experiment and at Taranaway (Experiment 8) (Gardiner *et al.*, 2004). A similar pattern of behaviour has been noted to occur in the nearby Charters Towers region (Brownson, 1976, personal communication), and in semi-arid areas in north-east Brazil. In the latter situation seed germination has been said to be promoted by fire, and this has been noted to occur in *Desmanthus* (Rangel and Gardiner, 1996; Rangel, 2005). Certainly some form of seed scarification is needed to promote germination in some species of *Desmanthus* and the plants could be from the seed sown originally or from that produced by plants in the experiment itself. It is also possible that some regeneration arises from the woody taproots present in all of the species, a situation much more readily observable in the transplant experiments.

A summary of the performance of the various accessions at the start of the third growing season (12/12/1990, Fig. 9.4) is presented in Fig. 9.5. Performance is given as plant number × plant vigour ratings and the species concerned, and the origins of the various accessions are shown.

The outstanding accessions are almost entirely composed of *D. virgatus* accessions from Argentina, north-east Mexico and Guatemala. Accessions of the species from other areas, from Cuba, Costa Rica and even other parts of Mexico,

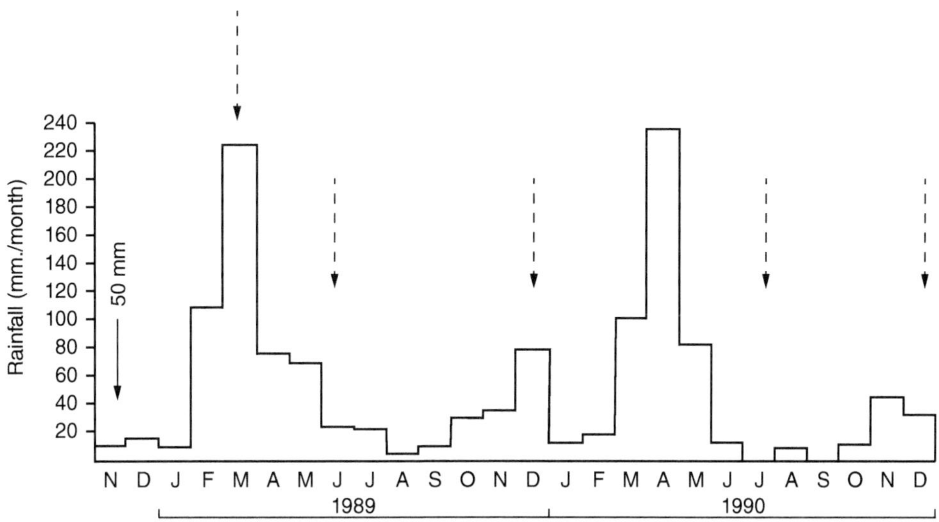

Fig. 9.3. Monthly rainfall over the experimental periods at Woodbine.

Table 9.6. Performance characteristics of *Desmanthus* in Experiment 7: mean data for the various species.

	Date observed								
	9/3/89	8/6/89		13/12/89	20/7/90		12/12/90		
Desmanthus species	Plants/plot[a] (no.)	Plants/plot[a] (no.)	Dry matter[b] (g)	Vigour[c] (0–10)	Plants/plot[a] (no.)	Seed[d] (%)	Plants/plot[a] (no.)	Vigour[c] (0–10)	Seed[d] (%)
D. bicornutus	17	142	323	4.75	38	25	59	39	0
mean, 4	0–45	50–349	212–817	3.5–6.0	10–60		2–107	2.5–4.7	
D. leptophyllus	58	281	388	6.7	60	40	98	4.7	20
mean, 10	0–209	174–543	121–838	3.7–7.8	20–201		0–322	2.0–7.0	
D. pernambucanus	88	194	216	3.7	71	33	56	3.1	33
mean, 3	57–130	140–261	80–308	2.2–5.0	47–101		47–87	2.0–6.3	
D. pubescens	0	99	223	2.9	26	20	26	6.3	20
mean, 4	(0–0)								
D. tatuhyensis	0	67	34	1.7	0		0	–	–
mean, 1									
D. virgatus	84	198	215	6.4	58	73	174	6.1	64
mean, 11	0–177	60–637	54–637	5.3–7.7	13–121		54–302	2.5–8.0	
D. sp. 70338	121	460	220	3.5	0		20	7.0	–
D. sp. 83563	71	194	134	4.8	2.0		101	5.0	–

[a]Mean number per 20 m^2 plots; [b]mean dry matter production of the legumes, g/20 m^2 plot; [c]vigour rating of 0 to 10; [d]mean percentage of accessions bearing seed.

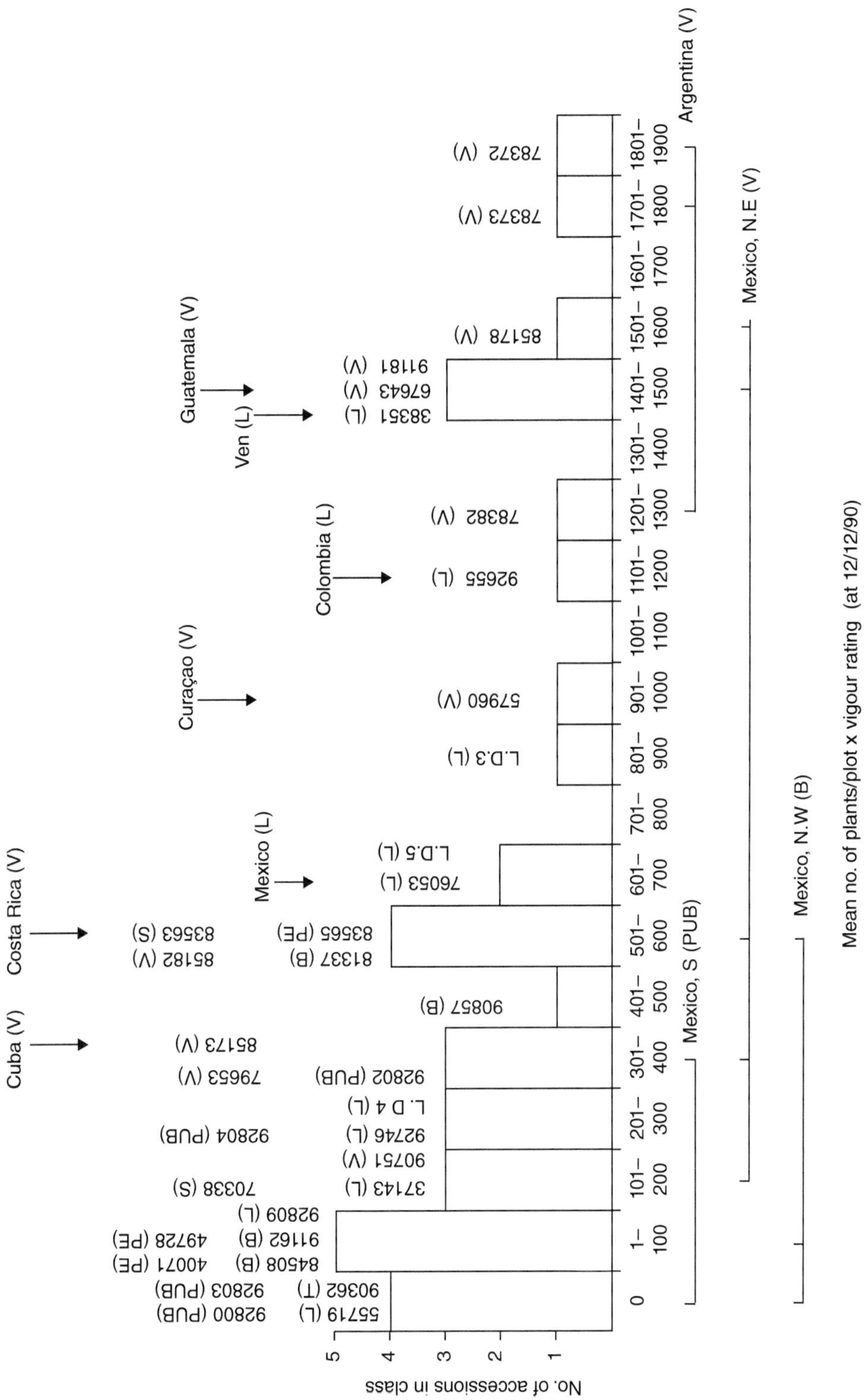

Fig. 9.5. Classes of vigour (plant number × vigour rating) in Experiment 7; third growing season; importance of species and origin.

performed much more poorly. Some species such as *D. pubescens* were poorly adapted.

Experiment 11

This transplant experiment was established at Toorak during a drought year and soon after sowing experienced a locust attack that killed most of the accessions. Details of those that survived are presented in Table 9.7.

All accessions were well eaten and the only accessions capable of survival and regrowth were from the genus *Desmanthus*. Some, *D. covillei*, *D. pernambucanus*, *D. pubescens* and two accessions of those seven surviving accessions of *D. virgatus*, had only a single surviving plant and these died subsequently.

Most of the *D. virgatus* accessions that survived the locust attack had persisted well at Rosebank (cf. Table 9.7). With the sole exception of accessions 67645 and 83563, from Guatemala and a research station in Costa Rica respectively, all were from dry regions (440–680 mm M.A.R.) in Argentina and Mexico. Where the relevant provenance records are available, all were collected on clay or clay-loam soils. Those that were best adapted in terms of their ability to persist, flower, seed and to produce seedlings, accessions 78376, 78373 (cv. Marc), 75182, 81181 and 83563 (the latter labelled *D.* sp. but almost certainly *D. virgatus*) are clearly worthy of further study. To these must be added *D. bicornutus* 81337, one of only four accessions of that species to be tested, and the sole remaining accession to survive drought. It appears not to have been studied elsewhere in Australia, but has shown promise in the USA (Muir *et al.*, 2005), along with the other accessions included herein.

EXPERIMENT 13. This transplant experiment at Rosebank, unlike Experiment 11 at Toorak described above, received better than average growing conditions throughout. The behaviour of some of the accessions that have been able to survive the very harsh conditions is illustrated in Fig. 9.6. Half of the transplants of *D. bicornutus* 81337 present in the early part of the first growing season (1988) survived up to the onset of the fourth season (December 1990). They showed higher ratings for vigour after the first season and in the third season had flowered, seeded and produced seedlings. They had also flowered early at the onset of the fourth growing season.

D. virgatus cv. Marc showed higher levels of persistence than *D. bicornutus* 81337 and the other accession of *D. virgatus* illustrated (91181). Both *D. virgatus* accessions flowered and seeded throughout the year and both produced established seedlings by the end of the third season.

9.3.2 Summary and preliminary discussion

For this climate and these soil types *Desmanthus* has proved to be by far the most promising genus for use as a legume in long-term pastures. Data presented at a later date confirm the persistence of accessions of *Desmanthus* and the failure of other genera at these locations (Gardiner *et al.*, 2004). In later studies by the same author (Gardiner, 2003), it is shown that *Stylosanthes seabrana*, which is suitable for clay soils in subtropical conditions, failed to persist in the more tropical, arid climate at Hughenden.

Although it is tempting to consider several accessions for commercial release, for instance *D. virgatus* 85178 (a Mexican form), or *D. virgatus* 78382 (from Argentina) (see the appendices for more detailed information), there are reasons why this may not be entirely appropriate. First the genus as such has been poorly tested. Only four accessions of *D. bicornutus* have been examined, and none of the other potentially useful species (see the main Discussion). Existing collections may thus contain valuable germplasm, for example, as yet undefined morphological/agronomic types of *D. virgatus*, some from areas likely to yield adapted material (such as Mexico).

Second, important differences between species and accessions have not been explored. *D. bicornutus* for instance differs from *D. virgatus* in that it is able to retain seed in the pod for two or three months after dehiscence (Luckow, 1993). Such seed could be an extremely valuable source of quality animal feed in the dry season. In addition, there are large differences in the nutritional quality of the seeds (Schlink and Burt, 1993), response of seed banks to fire (Rangel and Gardiner, 1996), ability of feed to stimulate wool growth (Rangel and Gardiner, 2009), and passage of the seeds through and possibly

Table 9.7. Performance characteristics of surviving *Desmanthus* accessions in Experiment 11.

			Survival seedling no.[b]		Dead (D), flowering (F) and seeding (S)						No. seedlings (mean/row)[e]			Plant vigour[e]		
					1988[c]			1989[c]	1990		1989		1990	seedlings	transplants	
Plant no.[a]	Species[a]	CPI accession no.	3/5/88	4/5/90	13/5	13/7	23/11	19/11	4/5[c]	4/5[d]	19/11	3/6	4/5	4/5/90[d]	23/11/88	4/5/90
21	*D. bicornutus*	81337	0.7	1.0	–	–	–	–	S	–	–	–	–	–	5.0	4.0
25	*D. covillei*	90311	0.3	–	D	–	–	–	–	–	–	–	–	–	–	–
40	*D. pernambucanus*	83565	0.3	–	–	–	–	D	–	–	–	–	–	–	3.5	–
44	*D. pubescens*	92804	0.3	–	–	–	S	D	–	–	–	–	–	–	4.0	–
48	*D. virgatus*	67643	0.3	–	F	F	S	D	–	–	–	–	–	–	4.0	–
49		78372	2.7	0.7	FS	FS	S	S	S	S	133.0	83.0	10.0	7.5	7.0	3.0
50		cv. Marc	2.7	0.3	FS	FS	S	S	–	S	5.0	7.0	1.7	6.0	7.5	6.5
51		78382	1.3	1.3	–	–	S	S	–	–	–	–	–	–	6.0	8.0
54		85178	0.3	–	–	–	S	D	–	–	–	–	–	–	4.0	–
55		85182	0.7	0.3	–	–	S	–	S	–	–	–	0.3	4.0	4.0	6.0
57		91181	0.0	0.7	–	–	S	S	S	S	–	18.3	1.0	4.0	5.0	6.0
60	*D.* sp.	83563	3.0	0.3	FS	FS	S	S	–	S	1.0	8.0	3.0	6.0	7.0	3.0

[a]See Table 9.2; [b]survival of original transplants: mean number/row; [c]original transplants; [d]new seedlings; [e]vigour of plants: rated 1–10.

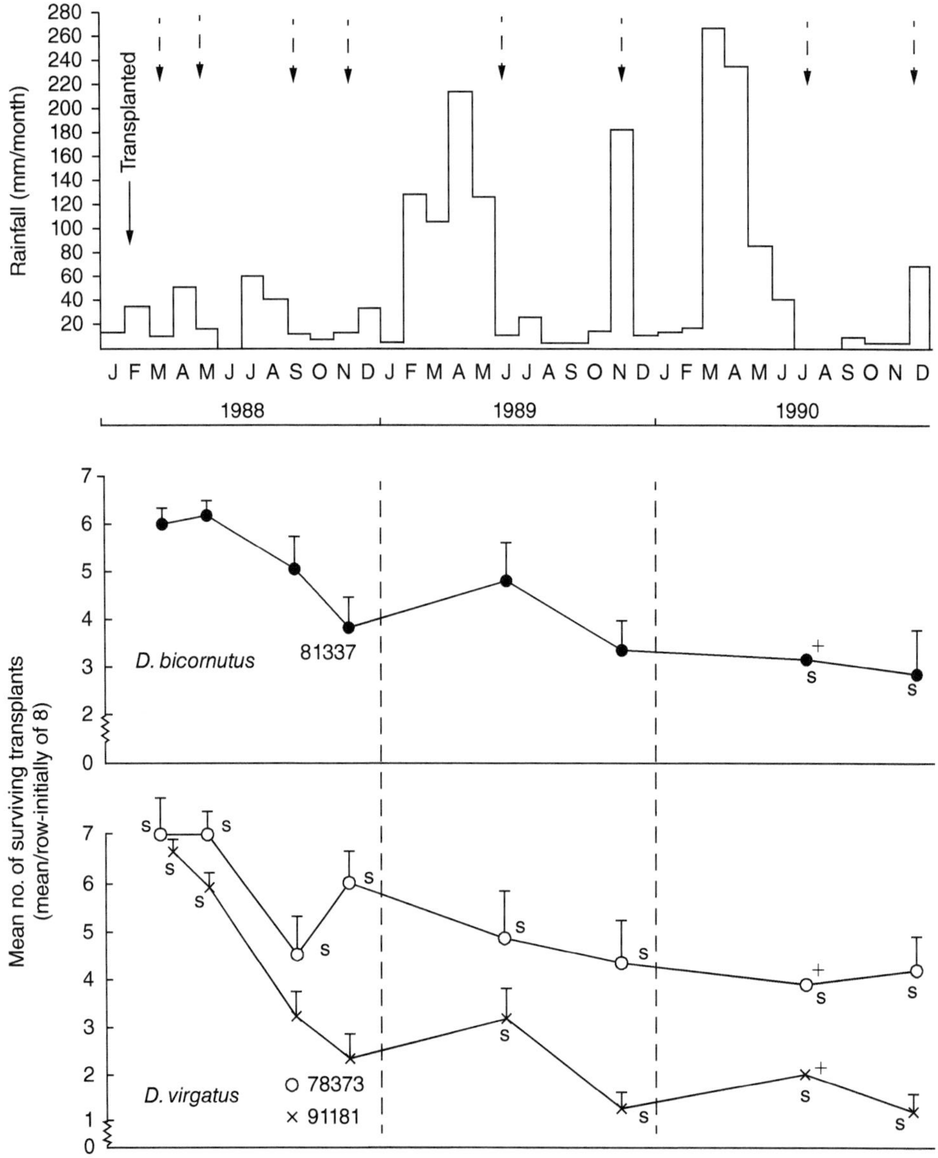

Dashed arrows - Sampling times;
Barred vertical lines on graphs - Vigour ratings shown at half the scale of that for plant number
S – Flowering and/or seeding
T – Seedlings present

Fig. 9.6. Performance characteristics of *Desmanthus* in Experiment 13: data for two accessions.

dissemination by the grazing animal (Lubanga, 1993). *D. virgatus* CPI 91181, from Mexico, Figures highly in some of these respects and could be useful, possibly as part of a composite variety for these environmentally harsh areas. The existing composite 'Jabiru' contains two species, both of which have performed poorly in these environments. Further strategic work undertaken on some of these aspects before cultivar release should be of considerable value.

9.4 Discussion

9.4.1 Preliminary considerations

In 1976 it was considered that there were some 50 leguminous genera, containing about 2000 species that warranted evaluation as tropical pasture legumes (Williams *et al.*, 1976). From the work described herein and in the detailed overview of Cook *et al.* (2005) it is now clear that two genera, *Stylosanthes* and *Desmanthus*, are likely to be the prime source of pasture legumes for use in dry tropical regions on clay soils. Serendipity may occur when an adapted plant is found from an unlikely source, as has happened with *Stylosanthes* (Chapter 8, this volume), but this is becoming increasingly unlikely as tropical material becomes better known and communications between relevant authorities improve.

The concerted development of tropical pasture legumes is a relatively recent phenomenon and two of the most commonly used new species *Stylosanthes hamata* and *S. scabra* were not domesticated until 1973 and 1977 respectively. *Stylosanthes seabrana* was made commercially available in 1996, and was not delineated as a separate species until 2002. The state of knowledge about, and the genetic resource development in temperate and tropical pasture legumes can be conveniently summarized by reference to, first, *Advances in Legume Science* (Summerfield and Bunting, 1980), a report from an authoritative international conference held at the Royal Botanic Gardens, Kew. References made to the two most important temperate genera, *Trifolium* and *Medicago* – the first for more acidic soils, and *Stylosanthes* and *Desmanthus*, are listed in Table 9.8.

At the time of this conference a great deal was known about temperate legume genera, especially the perennials *T. repens* and *M. sativa*. Interest had awakened in *Stylosanthes*, especially in *S. guianensis* and the annual *S. humilis*, and the two new species suited to acid soils, *S. scabra* and *S. hamata*, were mentioned. *Desmanthus* featured twice, once as *D. virgatus* being used as a pasture plant and once as 'a promiscuous and effective nodulator'. It now seems likely, since the taxonomic revision of the genus (Luckow, 1993), that an earlier suggestion concerning the identity of *D. virgatus* was correct (Burt, 1993a); it was probably *D. pernambucanus* (Cook *et al.*, 2005). Today *Desmanthus* species are usually deemed to be 'highly specific' in their rhizobial requirements (Cook *et al.*, 2005).

The importance of *Stylosanthes* as a provider of tropical pasture legumes on clay soils and the adequacy of genetic resource collections for this purpose have been discussed previously (Chapter 3, this volume). Here the principal concern with *Desmanthus* is with the information gained herein to 'further develop a body of theory and general principles' (Harlan, 1983) with which to aid the effective development and utilization of genetic resource collections.

Table 9.8. References to *Trifolium*, *Medicago*, *Stylosanthes* and *Desmanthus*.[a]

	Temperate		Tropical[b]	
	Trifolium	*Medicago*	*Stylosanthes*	*Desmanthus*
Total number	84	63	52	2
No. of chapters with references	18	18	11	2
No. species listed	18	18	11	1
Species per genus				
No. of references	*T. repens* 21 *T. subterraneum* 12	*M. sativa* 20 *M. trunculata* 6	*S. guianensis* 12 *S. hamata* 4 *S. humilis* 12 *S. scabra* 3 *S. seabrana* 0	*D. virgatus* 1

[a]In the combined index of Subjects and Authors in *Advances in Legume Science* (Summerfield and Bunting, 1980).
[b]The genera discussed in the text of this report.

9.4.2 The demonstrated potential of *Desmanthus*

Desmanthus plants have long been valued as pasture plants in their native environments. The names 'anil de bode' (food of the goat) and 'fijolillo de burro' (donkey bean) refer to its ability to support the growth of valued animals in dry tropical environments. *Desmanthus* was introduced into India many years ago and was recorded by Hooker (1879) in the *Flora of British India*. It probably arrived on a shipment as part of the 'East Indies–West Indies connection' and found a use as 'hedge lucerne' in alley cropping systems and as 'protein banks' in local farming systems. In Australia it is currently being developed for use in ley farming systems in subtropical Australia (Pengelly and Conway, 2000).

In the more southerly regions of subtropical Australia, more temperate species such as lucerne (*Medicago* spp.) are commonly used in ley farming systems (Williams and Burt, 1982; Jones and Rees, 1997). The latter authors sought more tropical legumes to complement them and after sowing on fully prepared seed beds evaluated some 48 accessions representing 16 genera and 27 species (at sites 18–20, Fig. 9.1). Few survived and the most promising legume was *Desmanthus virgatus* 78382 (plant 51, Table 9.2; see Table 9.19), one of the only two accessions of *Desmanthus* on trial.

In more northerly latitudes (Naragen, site 21, Fig. 9.1), eight accessions were evaluated as potential pasture species under grazing on clay soils. They were again sown into fully prepared seedbeds and rolled in to assist establishment. Although then listed as *D. virgatus* (plants 50, 51, 52 and 54 in Table 9.18), they also contained accessions *D. leptophyllus* (plants 29 and 30), *D. pernambucanus* (plant 38) and *D. pubescens* (plant 43). The best adapted plants, *D. virgatus* accessions 78382 and 78373 cv. Marc, were both Argentinean with the former much outperforming the latter. Variation within the species was very evident with accession 85178 from Mexico also persisting while 79653 died (Jones and Brandon, 1998).

Clem and Hall (1994) sought to find pasture legumes for clay soils in tropical and subtropical areas. They selected 37 accessions from genera that occur on clay soils in low to moderate rainfall environments in the tropics and evaluated them at sites 12, 13 and 14 (Fig. 9.1, Table 9.18) using transplant experiments. Four accessions of *Desmanthus*, one of *D. leptophyllus* and three of *D. virgatus* (including plants 47 and 50, the latter cv. Marc) were included. Only the tree legume *Leucaena leucocephala* and *Clitoria ternatea* were rated as being better than *Desmanthus*. The authors noted that 'its improvement in productivity and regeneration in the fourth growing season at Havilah, the site where the grass competition was the greatest was in contrast to that of the other legumes which generally declined in number and vigour ... there is merit to widening the range of genetic material tested in future evaluations'. In this evaluation study *D. virgatus* cv. Marc was the best adapted accession of those studied.

In these studies and in the work described therein, *Desmanthus* has been proved to be well adapted to clay soils in both tropical and subtropical regions. However, few accessions have been tested and evaluation methodologies have differed. As shown herein, surface-sown experiments in particular can produce contradictory results. These considerations rendered sensible comparisons between plant accessions difficult and there is still much confusion about the genus itself.

9.4.3 Characterizing *Desmanthus*

The agronomic development of the genus *Desmanthus* was, for many years, hindered by the lack of a taxonomic revision (Burt 1993a,b). Many accessions were simply placed in the *D. virgatus* 'hold-all' even up to the mid- to late 1990s. Confusion between agronomists, collectors and genetic resource managers followed. Although the taxonomic confusion has been alleviated with the publication of a review (Luckow, 1993), considerable uncertainty still remains. In addition, while the genus is primarily subtropical (Burt, 1993a), of the 24 species recognized by Luckow (1993), germplasm collections such as that held in the Australian Tropical Forages Genetic Resource Centre (ATFGRC) are heavily biased toward four tropical species, *D. virgatus*, *D. pubescens*, *D. pernambucanus* and *D. leptophyllus*. There are even misconceptions about its general habit. *D. virgatus* is sometimes regarded as a tree legume to be evaluated as such, while in other studies it is deemed to be a plant '1.5 to 3 m tall'. Luckow refers to it as being 'prostrate, decumbent or erect herbaceous perennials to 1.5 m tall, much branched from the base' and 'In extreme sites it can be a very small prostrate herb ... in favourable sites it may be decumbent or erect'.

In reality *D. virgatus* and other species such as *D. leptophyllus* are quite unlike other pasture legumes with which they may be compared in evaluation studies. They possess a highly lignified structure between the stem and the root that has variously been described as 'a thickened tap root', 'crown' or 'caudex' which acts as a 'lignotuber'. If the plant is defoliated by cutting, grazing, fire, frost or flood, it produces new adventitious shoots and the lignotuber increases in diameter to form a 'crown'. The productive structure changes to one suited to grazing systems and part of the lignotuber remains buried, protected from frost or fire (Gardiner and Burt, 1995). In many cases, unlike shrubby Stylo (*S. scabra*) and Leucaena (*L. leucocephala*), the stems are eaten and have proved to have animal feeding value. The seed can have protein values higher than that of the soya bean (Schlink and Burt, 1993) and is a valued source of dry season animal feed. There are no annual species. *Desmanthus* is a 'new' and different type of pasture legume and this fact needs to be taken into consideration in the evaluation process.

Other species in the ATFGRC collection are less likely to be of agronomic utility. *D. covillei* and *D. fruticosa* from arid and desert conditions in north-west Mexico are poorly persistent drought deciduous trees. *D. pernambucanus*, although occasionally found in wetter niche situations (as was the case for *D. pernambucanus* 55718, collected in a water run-on situation in Brazil), is wet tropical, has brittle, pithy stems that render it susceptible to grazing, and is regarded as invasive in some Pacific islands. Although useful as 'hedge lucerne' in India , where its high tannin levels provide bypass protein for ruminant digestion, it is known that the incorporation of tannins into the soil can restrict nitrogen release to plants and thus detract from its use in ley systems. *D. pubescens* is again wet tropical (Luckow, 1993), and had shown little agronomic variation in its native area (Lazier, 1978, personal communication), and collections show limited genetic variation (Pengelly and Liu, 2001). It has been found to be poorly persistent under grazing (Jones and Brandon, 1998) or in dry environments (see Table 9.18) and the cultivar 'Uman' has found little use in Australia (Cook *et al.*, 2005).

The ATFGRC germplasm collection also contains small numbers of accessions, which were not selected for use in the current experiments because of their late arrival and subtropical/temperate origins. They have also not featured in the Australian evaluation work described earlier in the discussion. However, they are of interest for a variety of reasons. *D. obtusus* can be found in parts of Texas with climates very similar to that of Bourke, NSW (using the Papadakis system of climatic classification). Several species, including *D. leptolobus* and *D. illinoensis*, occur around Fort Worth on clay soils in cooler climates (see Fig. 6.1, Chapter 6, this volume). These plants are close taxonomically and *D. illinoensis* with high levels of protein rich seeds (higher than soya bean) is currently being considered for use as a dryland pulse crop.

D. acuminatus, very similar to *D. virgatus* and at one time considered to be a taxonomic variety of it, can be found both in Texas and in north-west Argentina. In the latter country it is found in the same region as the very promising accessions of *D. virgatus* (plants 49, 50, 51). *D. cooleyi* (with most records stemming from the USA, but also found in Mexico) is reportedly 'good browse for antelope' and is a 'common component of short grass prairies'. It grows over a wide range of climates and geographical locations. The accessions of this species in the ATFGRC collection are from the southern part of the range in very arid conditions in Arizona. Based on the use of homologous grass floras, which have been shown to be valid for plant introduction, Hartley (as reassessed by Williams and Burt, 1982), suggested Arizona as a likely source of adapted plant materials for NSW, Australia. Several of these species are being evaluated as summer forage legumes for use for grassland reconstitution, especially the Blackland Prairies in Texas, USA (e.g. Muir *et al.*, 2005). The similarity between these American regions and the clay soil areas in Australia is emphasized by the native legumes selected for use. Many of these, *Neptunia*, *Rhynchosia*, *Indigofera*, *Desmodium* and *Acacia*, are typically found on the clay soils in Australia (see Chapter 6, this volume).

The main species of interest in Australia at the present time are *D. virgatus*, *D. leptophyllus* and *D. bicornutus*. *D. bicornutus* is poorly represented in the ATFGRC collection; Pengelly and Liu (2001) list only six accessions, while one of the authors (RLB) has tentatively identified eight. The species is found in a wide range of geographical locations from 18°N to 32°N and altitudes from 0 m to 1400 m (Luckow, 1993). Prior to the revision, *D. bicornutus* was recognized by an eminent taxonomist as three separate species (Turner, 1950, 1959). The ATFGRC collection lacks accessions from the southernmost part, situated in the Rio Balsas Depression (zone 13, Fig. 9.7).

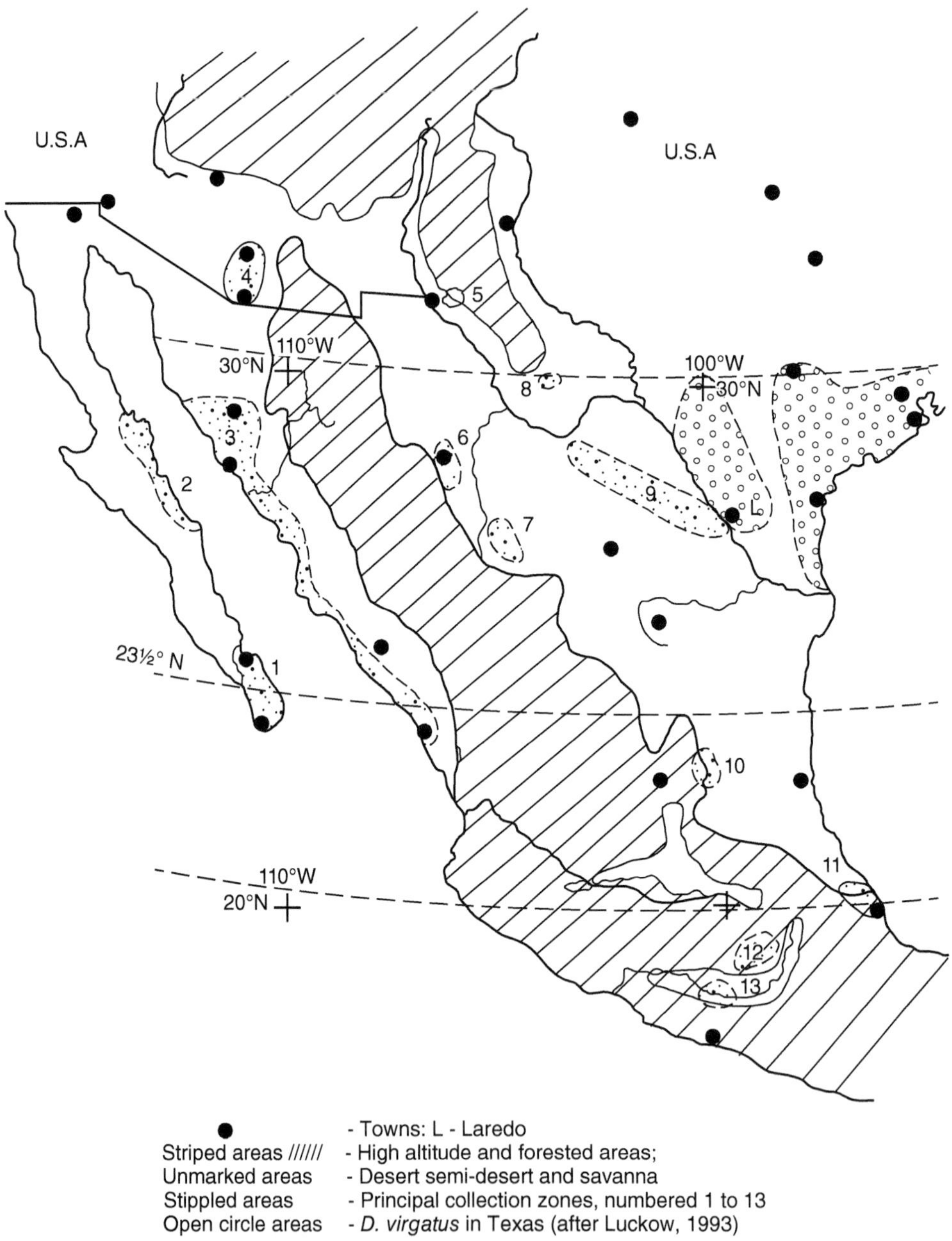

Fig. 9.7. Australian Tropical Forages Genetic Resources Centre *Desmanthus* collection zones in Northern Mexico.

This area has been designated as a 'special terrestrial ecosystem'. It is dry tropical forest, well separated from similar regions elsewhere, and with very different soils. It is a major centre of diversity with many endemic species and is 'also the centre of diversification for the family *Fabaceae* ... for example the endemic *Macroptilium esculenta* ... is only found in this habitat'. The environment is over 180 km long with altitudes ranging from 200 to 1000 m and has an 8-month dry season (www.Worldwildlife.org, text 2001, retrieved 2007). *D. bicornutus* is found in this area 'in grassland or cut-over thorn scrub in agricultural areas ... in sandy clayey or rocky soils ... 0–1,400 m', in a climate and vegetation type similar to those that had yielded cultivars in *Stylosanthes* that have been successful in dry tropical environments. Further collections are warranted.

The other species of major interest in Australia are *D. leptophyllus* and *D. virgatus*. These have a much wider range of distribution than *D. bicornutus*. *D. virgatus* extends from 30°N to 30°S, and in particular can be found in a very wide range of environments. In the USA, for instance, *D. virgatus* has been noted to occur near Fort Worth, a clay soil area with a cold winter 'steppe' climate (Papadakis, 1966 and see Fig. 9.1; Chapter 6, this volume). Not surprisingly, *D. virgatus* contains an enormous amount of morphological and agronomic variation (Burt and Beattie, unpublished data; Pengelly and Liu, 2001) and this is coupled with considerable taxonomic confusion because it is regarded as a 'complex', and is seen as grading into, for instance, *D. acuminatus*. Examples of variation within the species are provided in the text (e.g. Appendix 9.2).

9.4.4 Development and utilization of genetic resource collections in *Desmanthus*

For some species such as *D. bicornutus*, development and utilization of their genetic resources is conceptually straightforward. The potential utility of *D. bicornutus* has been documented and thus further collection missions are clearly warranted. It has a well-defined distributional range and more detailed provenance records could be used to describe the variation present in the species in relation to its environment, including disturbance (Burt *et al.*, 1983). Given that the species will almost certainly be considered as a component of grass pastures in drier areas, subsequent evaluation could easily be carried out in an appropriate fashion. This is not the case for other species and discussion on this aspect follows.

Commercial development of cultivars using our existing knowledge

Our knowledge of *Desmanthus* is very limited, so much so that the *Desmanthus* cultivar 'Jabiru' is a composite of three species (*D. virgatus* cv. Marc, *D. pubescens* cv. Uman and *D. leptophyllus* cv. Bayamo) 'used in the absence of precise information to provide a broader genetic base to accommodate differences in soil, climate and management' (Jones and Clem, 1997). Pengelly and Conway (2000) have noted that there are 'weaknesses in the components of "Jabiru"' and some of these have been summarized in Cook *et al.* (2005), presented in results of a grazing experiment (Jones and Brandon, 1998). Two of the three species in the composite cv. Jabiru come from wetter climates and have been unsuccessful in drier regions.

Despite a piecemeal approach to evaluation, research results also show that despite only a small proportion of the ATFGRC collection being tested, some accessions of *D. virgatus* can outperform existing cultivars. *D. bicornutus*, which has been little tested in Australia, has shown much promise in the work reported herein. All four accessions tested here have been made commercially available for grazing and regeneration in the USA (Ocumpaugh *et al.*, 2004). Moreover the successful accessions of *D. virgatus* come from many areas (Guatemala, Argentina, Mexico and Cuba) and offer a wide range of types from which to choose. A suitable composite of *D. virgatus* and *D. bicornutus* might well be considered for use in pasture fortification or regeneration in drier regions on clay soils.

Any such development should carry with it some form of risk/benefit assessment (e.g. Bennett and Virtue, 2004). With *D. virgatus* such risk assessments at first appear to yield conflicting results. In Hawaii the species is rated as being likely to be invasive (PIER, Pacific Islands

Ecosystems at Risk Assessment, accessed in 2007 from the update on 10/2/2008), while in tropical Australia, where it has been present for many years it is not so rated (Batianoff and Butler, 2002). The first assessment is based purely on published work. As is noted earlier, the name *D. virgatus* was widely used as a hold-all for many *Desmanthus* species, including for instance, the very weedy *D. pernambucanus* (also naturalized on Hawaii). Even the true *D. virgatus* is a very variable species 'prostrate, decumbent or erect herbaceous perennials to 1.5 m tall' (Luckow, 1993). It is taken to be a 'nitrogen fixing shrub 1.5 to 3 m tall' that will 'grow well' (in soils) with pH from about 5 to 8. In contrast, Cook *et al.* (2005) notes that it has 'only limited ability to grow on even moderately acid soils, and to have very specific rhizobial requirements'. Nor do 'many forms persist under intensive grazing' or defoliation (cf. Burt, 1993a). The author of the PIER report notes that plant ratings are subject to change as new information becomes available, and this would seem to be the case for at least some forms of *D. virgatus*.

An ecological approach to plant collection and evaluation might well provide the required information, and this has been discussed elsewhere (Chapter 5, this volume). Sufficient to note that some of the more promising types collected by one of the authors (RLB) were to be found as individual plants, interspersed between grass. In evaluation studies (see herein) they have behaved in the same way, closely resembling the native *Neptunia* with which they are morphologically very similar. This is a pattern of behaviour very unlike aggressive ruderals or the enveloping climbers. As they have specific rhizobial requirements that limit them to clay soils, their distribution is likely to be even further limited.

Use of the existing ATFGRC germplasm collection

Pengelly and Conway (2000) highlight 'an urgent need to identify suitable cultivars' and this need is exacerbated by the threat of global warming, droughts and the need to use less water for irrigation to provide animal feed. Reference has already been made to the development of a composite cultivar for drier areas.

As noted in Jones and Clem (1997), many of the accessions of even the promising species *D. virgatus* remain untested and the same is true for very similar species such as *D. acuminatus*, *D. paspalaceus* and *D. tatuhyensis* (Luckow, 1993), which are to be found in those climatic types that have yielded accessions adapted for Australian environments.

For ley legumes, where long-term persistence is not a requirement, accessions less well adapted to the target area may be of value. Some of those with potential on clay soils in subtropical Australia were introduced from more tropical, often wetter environments, such as those found in Belize (Cook *et al.*, 2005). These plants are less likely to be invasive in the better environments in which ley farming is practised, and those that flower late and produce less seed are less likely to show unwanted regeneration (Pengelly and Conway, 2000). Knowledge of such plants is limited however and 'there is an urgent need to identify suitable cultivars and to develop appropriate guidelines for both their establishment and grazing management' (Pengelly and Conway, 2000).

9.4.5 Strategic research

Desmanthus is different from most pasture legumes. It is always perennial, has a woody lignotuber that aids perenniation and regrowth after frost, fire or flood, it can produce large yields of very nutritious seed, and it thrives on clay soils. Some areas of research that could greatly aid in its development and suitable use are briefly summarized below.

Efficient classification

The key to unlocking the secrets of a genetic resource collection is in efficient classification (Krull and Borlaug, 1970), a subject discussed further for *Stylosanthes* (Burt, 1983). In the latter genus the collection was split into 'morphological/agronomic' groups, representatives of which could be evaluated over a wide range of environments. Once a type adapted to a particular environment had been nominated, other members could be tested, and in many instances suitable areas for further collection specified.

This has only been undertaken for a small proportion of the *Desmanthus* collection (Chapter 5, this volume).

Feeding value of the seed

Desmanthus can produce high seed yields and Cook *et al.* (2005) quote yields of over 1000 kg/ha. Long-term, anecdotal information from areas in which the genus is native strongly suggests that this is a very valuable source of animal feed, particularly at times of feed shortage or when the nutritional value of the associated grass is low. Such seed availability could be particularly useful in rangeland conditions and the early detection of accessions that confer this ability is important.

Luckow (1993) notes that some species, notably *D. bicornutus*, retain seed in the pod for some weeks after dehiscence. One species, *D. oligospermus* is indehiscent. Such characteristics may favour presentation of the seed to animals and at the same time prevent the deterioration implicit when it falls to the soil. Total indehiscence, which led to the domestication of beans and peas (Smartt, 1990), could be used to produce a valuable new pulse/grazing plant.

The nutritive value of *Desmanthus* seeds is higher than soya bean, and efforts are currently underway to domesticate *D. illinoensis* as a perennial pulse crop. There are large differences between the chemical composition of seeds of various *Desmanthus* accessions and it is likely that these will result in differences in animal feeding quality (Schlink and Burt, 1993). There are also differences in the digestibility of the seeds and in the germination of those that remain undigested (Lubanga, 1993). The documentation of such differences before accessions are selected for commercial release would greatly aid the selection process and provide insights into the potential utility of the material present.

Feeding value of the vegetative growth

There are large differences in the feeding value of *Desmanthus* accessions and these are likely to be of importance (Rangel and Gardiner, 1996). Accession 78382, from Argentina, has shown much promise as a legume for long-term pastures (herein, and Jones and Brandon, 1998). Its leaves are much more readily digested than those of another *D. virgatus* accession 79654 from Cuba, but its whole seed, much less so. The latter would favour passage through the animal and regeneration in the pasture. The latter accession has higher nitrogen levels in all of its above-ground components, has much higher levels of sulphur and stimulates wool production more than the former (Rangel *et al.*, 1996). When mixed with Mitchell grass (*Astrebla* sp.), a major component of the clay soil plains of western Queensland, it increases the efficiency of wool production from 0.83 g/day/cm^2/kg of food intake for Mitchell grass alone to 1.03 g/day/cm^2/kg of food intake, thus a given level of wool production may be achieved at lower grazing pressures.

D. virgatus is a highly variable species and this extends to its animal feeding characteristics. There are differences in tannin content (see information from Singh *et al.*, 2000, quoted in Cook *et al.*, 2005) and tannin content is also known to affect nitrogen release from material incorporated into the soil (possibly an important feature in ley farming).

Further studies on the different types of *Desmanthus* are clearly warranted.

Ecological understanding

In reviewing the various species of *Desmanthus*, Cook *et al.* (2005) draw attention to the problems associated with its ability to establish on heavy clay soils and to its quite specific rhizobial requirements. Both aspects are very important in the long-term success of the plants in permanent pastures. Yet, the material concerned, in the results presented, has often proved to be capable of regeneration from uninoculated seed when grown on clay soils. Some prior information about these abilities would be of considerable utility and aid in their sensible evaluation and appreciation. To quickly and efficiently screen a range of soils for the ability of legume genotypes to nodulate, Norris's simple approach can be used (referred to by Burt, Chapter 8, this volume). It involves incorporating high carbon organic material into the soils in question and planting seeds of the legumes to be tested. Legume survival indicates the ability to utilize the indigenous *Bradyrhizobia*.

Other relevant information is provided by Rangel and Gardiner (1996) and Rangel (2005).

Differences within *D. virgatus* will again be used for illustrative purposes. They noted that accession CPI 83563, from Costa Rica, had failed to establish seedlings in the four years prior to a grass fire. Seedling recruitment then occurred. After measuring soil temperatures under such fires, they showed that dry heat treatments of 80°C raised germination percentages from 17.5% at room temperature to 80.0% at 80°C. With cv. Marc from Argentina, the equivalent Figures are 8% and 32%, for accession CPI 78382 another Argentinean accession known to sometimes outperform cv. Marc, they are 4% and 68%. A promising accession from Guatemala CPI 67643 had its hard seed broken at a lower temperature of 60°C. It has been suggested (Burrows and Porter, 1993, quoted in Jones and Brandon, 1998) that the half-life of the breakdown of hard seeds under dense cover is 60–70 months. But this is specifically for accession CPI 78382 and would probably be far less for CPI 83563, which had germination rates of 22% after exposure to temperatures of only 40°C, and for most such accessions, because there are major shifts in germination after grass fires.

In field conditions, especially in rangelands, where grass fires occur, hard seededness may not be a problem. Indeed, as in the Caatinga region of Brazil (Da Lima, 1976, personal communication), response to fire may constitute a useful management tool. Simple screening experiments similar to those devised by Rangel and Gardiner (1996) would provide information of value in utilizing genetic resource collections.

9.4.6 The future development of *Desmanthus* collections

There is a clear need to develop legume cultivars for use on clay soils, both for leys and for long-term pasture plants (e.g. Pengelly and Conway, 2000 for ley legumes in Australia and Ramesh *et al.*, 2004 for the vast clay soil areas in India). The work described herein has shown that *Desmanthus* is almost certainly likely to be the major source of such legumes.

The need for additional collections has been described elsewhere (Chapter 5, this volume) and only the major points are summarized here.

The existing ATFGRC *Desmanthus* collection numbers about 300, some 30 or 40 of which are reintroductions. It contains about 100 accessions of *D. virgatus*, probably the most economically important species. By contrast the equivalent collection of the tetraploid form of *S. hamata*, a relatively uniform species from a small coastal area of northern Venezuela and Colombia, numbers over 200 (Edye and Maass, 1997). *D. virgatus* on the other hand occurs naturally from 34°S to 34°N, and in the Caribbean archipelago alone, over a range of 25° of latitude. From this area the AFTGRC collection contains only nine *Desmanthus* accessions, mostly provided by an IDRC/ University of the West Indies project based in Antigua. The wide genetic variation within these has been illustrated by Pengelly and Liu (2001). Its agronomic utility across the archipelago was noted by Stehle (1956) and this has been reinforced by Keoghan (1980). Collections of genetic resource collections of temperate species are even more remarkable. The USDA collection of *Medicago* species contains over 7500 accessions.

The genus *Desmanthus* contains over 24 species (Luckow, 1993) and these, unlike *Stylosanthes*, are primarily from North and Central America, rather than from South America. Several are subtropical and temperate. Although some species are probably not of direct agronomic value, such as the woody, drought-deciduous species *D. fruticosa* and *D. covillei*, several species warrant collection and evaluation. Some of those being proposed for use in the revegetation of Blackland Prairie in Texas (Muir *et al.*, 2005) are to be found in areas likely to yield adapted material for, among other areas, Australia (Hartley, 1963; Williams and Burt, 1982). Of especial importance is *D. oligospermus*, a species with non-dehiscent pods and of great potential importance in future plant breeding programmes. It is restricted to a small geographical area in Baja California, where it is uncommon, and it should be considered a 'threatened or endangered' species (Luckow, 1993). Global warming in recent years exacerbates this threat. Given the availability of information in many of the areas to which *Desmanthus* is native, studies of the type described by Maxted and Guarino (1997) could well be of value in the collection and evaluation of such material.

The AFTGRC collection contains samples of some of the major species of interest from most, but not all, of the collecting areas specified in earlier studies (Burt and Williams, 1979a;

Pengelly and Liu, 2001). However, in tropical areas species occurrence and distribution may be poorly documented. Indeed, now geographical systems are needed to suggest where new target species may be found (Jones and Clem, 1997). Much of the earlier collection work has been exploratory, relying on road transects, frequent stops and in many cases roadside collections. In *D. bicornutus*, for example, the accession CPI 91162 is the sole entrant from the Rio Balsas depression in Mexico. It has shown much promise herein. Luckow (1993) records that, in the depression, the species occurs over a wide range of ecological conditions, soil types and altitudes. It has been shown (Chapter 5, this volume) that plants collected from grazed conditions may be more suited for use in pastures than those collected in roadsides. A 'follow-up' collection aimed at finding representatives of the species on different ecological and environmental conditions is warranted.

Most tropical pasture legume cultivars are plants that have been introduced from the wild, subjected to suitable evaluation programmes and made commercially available (e.g. information contained in Cook *et al.*, 2005). Even in *Stylosanthes*, arguably the best-developed tropical genus, the main impact of plant breeding has been to improve suitable wild 'types', especially for disease resistance (Chakraborty, 2004). The delineation of these types, which even modern chemotaxonomic methods fails to accomplish (as pointed out in Pengelly and Liu, 2001), followed by meaningful evaluation is a practical necessity, and without such information it is difficult to sensibly specify the species of key interest and the geographical zones that most warrant collection.

Appendix 9.1

Selecting *Desmanthus* accessions for use in the experiment

At the time this programme commenced there was little international interest in the development of cultivars of *Desmanthus*. In 1979, after notable successes with *Stylosanthes*, the entire collection of *Desmanthus* in Australia numbered only 79 accessions (Burt, 1993a). In the absence of a taxonomic review of the genus (not available until 1993), such information as was available was enshrined in publications that were often difficult to access, and virtually impossible to collate. A given species could have many different names in the various countries to which it was native or to which it had been introduced. One species name *D. virgatus* was used as a 'hold-all' for what can now be identified as many different species (Luckow, 1993). As the size of the collection increased and evaluation became paramount, three separate morphological/agronomic classifications (*sensu* Burt *et al.*, 1971; Edye *et al.*, 1974) were carried out to aid with the selection of material for evaluation. However, each covered only the part of the collection that was available when the work was undertaken. Meantime the size of the collection has grown to over 300 accessions for *D. virgatus* alone.

The *Desmanthus* suite (Table 9.2,) comprised 41 accessions and what is now known to represent nine named species. It contains three sets of material labelled A to C in Table 9.2. Set A was chosen to represent all of the various morphological/agronomic (MA) groupings as designated by the aforesaid classifications and so to represent the genus. This methodology had been used previously with *Stylosanthes* (Burt *et al.*, 1974) in GRAZNET, a multi-generic legume evaluation programme carried out by the Queensland Department of Primary Industries over a wide range of environments in Queensland.

Set B was comprised of plants that had shown promise on Lansdown clay. To these were added additional plants of some of the MA types that had shown promise, together with newly arrived material of species of interest. The species were named using a composite key described in Burt (1993b).

The third set, C, again contained several 'new' species, many accessions of which were unavailable in the earlier experiments. *D. fruticosa* (plant 27) and *D. tatuhyensis* (plant 46, previously ascribed to *D. brevipes*) were from zones A and B respectively in semi-arid Mexico (Fig. 9.8), *D. bicornutus* (previously ascribed to *D. subulatus*) was from northerly and southerly sections of zone A and from the most southerly extension of this species (in zone Z). All were collected on clay soils. To accessions from clay

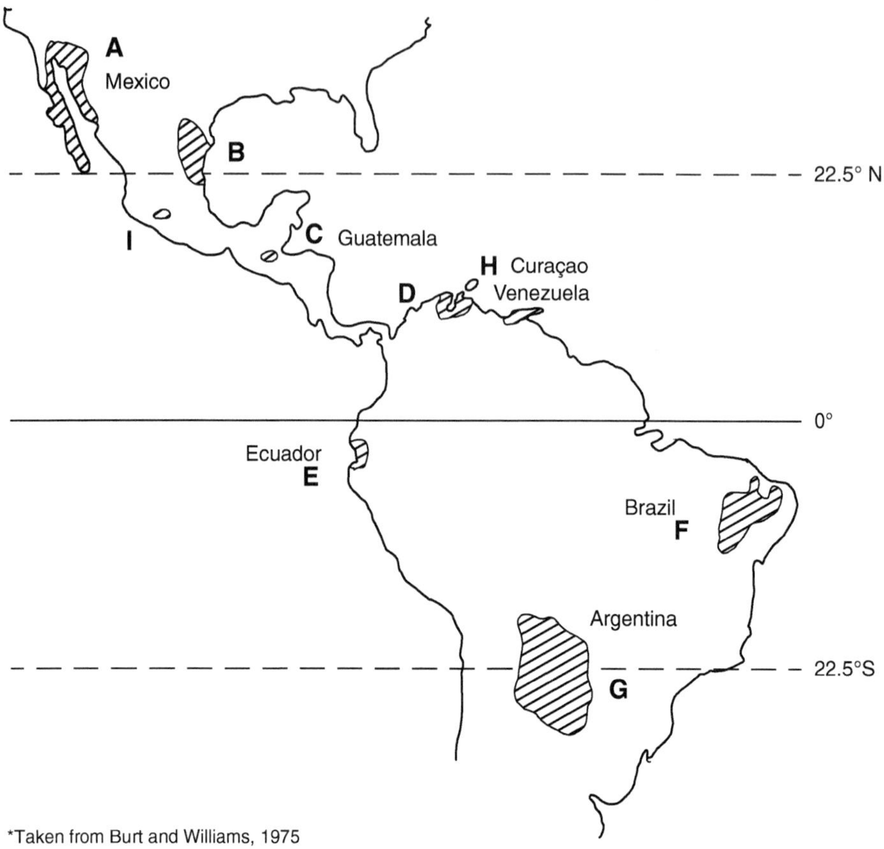

*Taken from Burt and Williams, 1975

	Country	Climate+		Country	Climate+		Country	Climate+
A	Mexico	III, XI	D	Venezuela	III	G	Argentina	IV,VIII
B	Mexico	IV	E	Ecuador	III	H	Curaçao	III
C	Guatemala	III	F	Brazil	III	I	Mexico++	

+ Climate type III semi-arid tropical; IV hot semi-tropical; VII semi-monsoon, subtropical; XI hot tropical or subtropical desert. After Papadakis, (1966)

++ Not located in original study but collected by Reid – semi-thorn scrub vegetation climate types (Papadakis, 1966)

Fig. 9.8. Areas of Central and South America deemed to be most likely to provide pasture legumes for the dry tropical and subtropical portions of Australia (taken from Burt and Williams, 1975).

soils in zones C, E and G and the Caribbean Islands were added plant 53, from a black, cracking clay soil in zone B. Plant 26 (cf. *D. covillei*) was from a particularly dry part of zone B, with a mean annual rainfall of 250 mm, and plants 33 (cf. *D. leptophyllus*) and 61 (*Desmanthus* spp.) represented newly defined MA types from wetter conditions and sandy-textured soils.

With the benefit of later studies (Pengelly and Liu, 2001; Luckow, 1993) it is likely that species such as *D. paspalaceus* would merit inclusion. It is found on clay soils and in pastures in that part of Argentina (G in Fig. 9.8) which has yielded accessions adapted to the Lansdown clay. For more subtropical zones in particular, species from some of the southern stages of the USA should merit attention (Muir *et al.*, 2005). Hartley's Agrostological Index suggests that such areas should contain material adapted to southern Queensland and northern NSW (Williams and Burt, 1982). But for the purpose on hand, and to explore the potential of the genus in western Queensland, the collection assembled was more than adequate.

Appendix 9.2

Details of experimental methods

GRAZNET experiments

These experiments were carried out at Toorak and Rosebank and the dates on which all experiments were sown are listed in Table 9.3 in the main body of this chapter. At Toorak the experiment was established in an area previously used for hay production. The grass was first removed and the area harrowed to break the soil surface. Shallow furrows were then made and 4 m rows of inoculated seed were sown by hand. The first sowing failed and the trial was resown the following year (Experiment 1, Table 9.3). At Rosebank the grass on the area to be sown was removed by grazing with sheep and shallow furrows were again cultivated and sown (Experiment 2). No fertilizer was applied at either site.

Small sward experiments

These were carried out at sites where only minimal plot preparation was feasible and access by agronomists might sometimes be difficult. At Toorak only, the grass was first removed from the experimental area and the ground harrowed to break the soil surface. At the remaining sites grass was removed by grazing with sheep, and soil cultivation was not possible. Inoculated seed was hand sown onto plots measuring 4 × 5 m at the rate of 5 g/plot. There were three replicates. No fertilizer was applied and all plots were fenced to prevent grazing.

Transplant experiments

These were carried out at Toorak and Rosebank. Individual seedlings of the various accessions were raised in small peat pots in the shade house at the CSIRO Davies Laboratory, Townsville and inoculated with the relevant *Bradyrhizobia*. After 4 weeks the plants were trimmed and placed under shade cloth outside the shade house. Watering was reduced and the plants allowed to 'harden off'. These plants were transplanted into the field, again unfertilized, and irrigation applied to facilitate plant survival. The accessions were allocated at random to plots measuring 5 m × 1 m arranged in three blocks, each plot contained eight plants spaced at 20 cm.

Before the first planting at Toorak (Experiment 11), the experimental area was ploughed and banks were raised to permit flood irrigation. This continued until early April in the first growing season (Fig. 9.2, main body of the text). For the second sowing at this site (Experiment 12) the area was first ploughed and levelled and the transplants watered using Hardi-Pope 'biwall' drip feed. At Rosebank the same methods were used as in Experiment 12 and irrigation was applied for 2 months after transplanting (Fig. 9.2, main body of the text). Unlike the previous transplanted experiments as reported for the Lansdown clay soil site (Chapter 8, this volume) and as used for instance at the Balfe's Creek site (site 5, Fig. 9.1), it was logistically possible to leave uncultivated areas between accession rows.

Appendix 9.3

Climates at points of collection and evaluation

Although *Desmanthus* is frequently referred to in the literature as being a tropical genus, this is not entirely correct. It is true that the species most usually cited, *D. virgatus*, is commonly found in tropical regions, but other members of the genus can be found at latitudes of up to 30° on both sides of the tropics. One species, *D. illinoensis*, extends almost up to the Canadian border. Most species are subtropical (Burt, 1993a).

Potentially therefore, the genus could provide legumes for a very wide range of climatic conditions. This would involve the use of different species and different morphological types within the species. It seems that current genetic resource collections do not adequately represent some species. *D. covillei* for instance is a species that provides good browse for antelope (quoted in Luckow, 1993) and is capable of growth in a wide range of more temperate environments, is poorly represented in Australian collections and appears to have not been included in evaluation programmes in more subtropical areas. In the development and use of this significant genetic resource it is important therefore that recognition be taken of the types of climate from which the accessions originate and those in which they are evaluated.

The locations of some of the main testing sites in the Australian tropics have been presented in Fig. 9.1. The various types of climate encountered are numbered I to XII *sensu* Papadakis (1966). These climatic types are defined by their ability to sustain the growth of different suites of crop species and the system is generally useful for plant introduction and evaluation purposes (Reid *et al.*, 1976). Because there is an increasing awareness of the need to develop ley legumes for use in cropping systems (Pengelly and Conway, 2000), it is particularly appropriate here. The characteristics of the climate types I to XII are described fully elsewhere (Papadakis, 1966) and a brief resume is presented in Table 9.8. Here we simply note as an example that cold-requiring crops such as wheat cannot be grown in climate IV but are feasible in the winter seasons in 4.21 and 4.24 if there is sufficient moisture. Rainfall, particularly the winter component, increases from climate VI to IX.

However, climatic boundaries for crop and pasture plants rarely coincide. In climate II, for instance, the tropical genus *Stylosanthes* provides the two main pasture species *S. hamata* and *S. scabra* only in more northerly latitudes. At its southern end the Mediterranean legumes dominate, *Medicago truncatula* in dry zones and *M. sativa* and *T. repens* in the wetter areas. *Stylosanthes* does not appear. It is simply not possible to produce a climatic classification of the Papadakis type for the tropics using pasture species because unlike crop plants the cardinal temperature and moisture requirements of tropical pasture species are not known. Indeed, in *Desmanthus* it is likely that 'new' species or species forms will come into use.

When faced with a similar problem in developing the *Stylosanthes* gene pool, the approach that proved to be most useful was to first display the variation patterns in the climates from which the various accessions had been collected. This was achieved by the use of principal coordinate analysis (as in Fig. 6.1, Chapter 6, this volume). Onto this Figure was superimposed the distribution of the various species and 'morphological/agronomic' (MA) types (Burt, 1975a,b; Burt *et al.*, 1976). After evaluating representatives of the various MA types over a wide range of environmental types (Reid *et al.*, 1976), it was possible to interrelate the types of climate that had yielded the various species and the MA types with the climate zones in which they were most likely to be useful. It was then possible to select which MA types merited detailed examination for a given environment.

This type of information could be useful in ecogeographical studies (Maxted *et al.*, 2000; Maass and Sawkins, 2004), and in predicting possible invasive behaviour (Daehler and Carino, 2000). Incorporation of other relevant data such as plot soil type, disturbance level, run on/run off should greatly increase the utility of this methodology (see Discussion section herein).

Appendix 9.4

Performance of non-*Desmanthus* accessions in western Queensland

In the seed-sown experiments, germination and establishment were erratic and usually poor. In Experiments 1 and 3, at Toorak (Table 9.3) no germination occurred. In Experiments 2, 4, 5 and 6 survival was so poor that the experiments were curtailed after one growing season (Table 9.9). Even when flowering and seeding took place (notably in Experiment 6, especially with *Stylosanthes* accessions), no seedlings were produced at the onset of the second growing season.

In Experiments 7 to 10 (Fig. 9.2, Table 9.10) there were much better rainfalls in the establishment phases and as a result germination, and seedling vigour and establishment were often excellent (e.g. data for *Stylosanthes* in Experiment 7). However, there was no persistence or seedling production at the start of the second growing season.

In the transplant experiments establishment was ensured by the use of supplementary irrigation in the early phases of the experiments. In the first season at Toorak (Experiment 11), plants were stripped of their leaves by locusts and none subsequently survived. When replanted the following growing season (Experiment 12), good establishment was achieved because of favourable conditions, and good growing conditions continued in subsequent seasons since they had above average rainfall totals (Fig. 9.2). Both accessions of *Clitoria ternatea* proved to be well adapted in having reasonable persistence and

Table 9.8. Climates of evaluation sites and selected collection sites.[a]

Climate	Country of collection	Location/species
Tropical		
1.2 Tropical moist monsoon		
1–3 dry months		
Rain grown sugar cane, bananas, rice. Too warm for cryophilous crops	Australia	Cairns, Townsville (70)
	Cuba	*D. leptophyllus* cv. Bayamo
1.3 Dry semi-hot tropical, dry monsoon		
4–6 dry months		
Maize, cotton, irrigated rice, sugar cane. Too warm for cryophilous crops	Brazil	*Stylosanthes scabra* cvs Seca, Fitzroy
	Ecuador	*Desmanthus virgatus* CPI 65947, *S. sympodiales* CPI 66705
	Curaçao	*D. virgatus* CPI 57960
1.48 Hot tropical, moist monsoon		
4 humid months		
Sorghum, millet, groundnuts, irrigated sugar cane and rice. Too warm for cryophilous crops	Australia	Darwin, Normanton XII, 17 (59)
	Mexico	*D. virgatus* cv. Uman
1.5 Semi-arid tropical		
1.53 Hot equatorial or hot tropical		
1–3 humid months	Australia	Kununurra, Wyndham
Drought-resistant crops in wetter types, irrigated rice, sugar cane, cotton. Too warm for cryophilous crops	Venezuela	Maracaibo; Paraguana Peninsula (18)
		S. hamata cvs Verano, Amiga, CPI 61670, *S. humilis* cv. Khon, *D. virgatus* CPI 55719
	Guatemala	Zacapa (12)
		D. virgatus CPI 67643
	Brazil	Juanzeiro (17)
		Arachis pusilla, *Macroptilium martii*, *S. humilis* CPI 55797, *Centrosema pascuorum* CPI 55697
1.54 As for 1.53		
No month is humid	Australia	Daly Waters
	India	Bellary (43)
		Vigna trilobata, *Heylandia laterbrosa*

Continued

Table 9.8. Continued.

Climate		Country of collection	Location/species
3.2 Hot subtropical desert			
	Cotton, rice. Winter cool enough for citrus		
3.22 Monsoon desert			
	Winters too warm for cryophilous crops	Australia	Cloncurry
3.24 Mediterranean desert			
	Winters cool enough for wheat, citrus	USA, Arizona	*D. cooleyi*
3.25 Isohygrous desert			
	Winters cool enough for wheat, citrus	Australia	Thargomindah
3.26 Monsoon desert			
	Winters cooler than 3.22	Australia	Longreach V, 7 (61)
Subtropical			
4.2 Monsoon subtropical			
	Rainfed sorghum, millet, groundnuts in wetter types, cotton on moisture-holding soils (clays), irrigated cotton and rice, wheat in winter		
4.21 Hot subtropical, dry monsoon			
	Summer/winter grains and oilseeds	Australia	Biloela VI (50); Theodore VI (68)
		Mexico	near Monterrey, *D. virgatus* CPI 85178
4.22 Hot subtropical, semi-arid monsoon			
	Summer/winter grain, oil crops, sheep, cotton	Australia	Roma II (66); Emerald II (58); Springsure II (67); Clermont II (53)
	Grazing in Queensland Australia		Charleville II (52)
	Summer/winter grain, oil crops; pigs		Dalby IX (56)
4.24 Continental subtropical			
		Australia	Miles IX; Narayen IX (65)
4.3 Hot semi-tropical			
	Sorghum, groundnut, cotton in wetter types, rice with irrigation. Winter usually too warm for winter crops, grazing		
4.31 Hot semi-arid monsoon			
	No crops without irrigation	Australia	Charters Towers; Hughenden; Tambo (all IV)
			Balfe's Creek (49); Hughenden (60); Julia Creek nr Toorak (69); Blackall (61) (all IV except Julia Creek, which borders on V)
			Evaluation sites: 3, 4 5, 6, 8, 13, 15

4.36 Moist monsoon		Argentina	Corrientes, *D. acuminatus*
2.2 Low tierra fria	Cool enough for cryophilous crops, too frosty for coffee		
2.21 Dry monsoon and 2.26 Humid 5 or more dry months		Argentina	Salta (19), San Salvador de Jujuy, yielded *D. virgatus* cv. Marc (400 m alt.) and CPI 78382 (1000 m alt.)
2.25 Low tierra fria	Wheat maize, oilseeds, fruit sorghum, cotton and soya beans	Australia	Toowoomba, placed in IX by Reid *et al.* (1976), on fertile black volcanic clays of the Darling Downs
5. Pampean	Wide range of cooler climates, e.g. ranging in NSW, Australia from 5.7 Semi-arid peri-Pampean to 5.4 Marine Pampean	Australia	Bourke, Coonabarabran Armidale
		S. America, southern USA	*D. reticulatus*; *D. cooleyi*; *D. acuminatus*; *D. tatuhyensis* var. *brevipes* and var. *tatuhyensis*; *D. virgatus*; *D. velutinus*; *D. paspalaceus*

[a]As defined by Papadakis (1966); Numerals: evaluation sites as shown on Fig. 9.1; Latin numerals: climatic types as shown in Fig. 9.1; Numbers in round brackets: as shown on Fig. 8.1.

Table 9.9. Performance of non-*Desmanthus* accessions in Experiments 1 to 6.

	Location		Rosebank		Toorak		Rosebank					
	Experiment no.		2[a]		4[b]		5[c]		6[d]			
	Year		1988		1989		88	89	1989			1990
	Date (day/month)		10/5	7/9	14/4	6/6	10/5	8/9	6/3	8/9		12/12
Plant no.		Accession no.									Vigour[e]	
1	*Alysicarpus* sp.	TQ 1	58.0	00	9.3	14.0 fs	0.0	0.0	188.0	107.0s	6.7	0.0
4	*Chamaecrista rotundifolia*	cv. Wynn	2.5	0.5	0.0	0.0.	0.3	0.0	77.3	11.3	3.7	0.0
5	*Clitoria ternatea*	47187	3.0	0.5	2.0	3.7	0.0	0.0	0.0	2.3	3.7	0.0
6		48337	1.0	0.0	18.0	12.7 f	0.0	0.0	6.3	8.0fs	5.0	0.0
7	*Centrosema pascuorum*	55697	15.0	0.0	0.0	0.0	0.3	0.0	0.0	0.0	0.0	0.0
8	*Macroptilium atropurpureum*	cv. Siratro	28.0	0.0	0.0	0.0	0.0	0.0	0.0	0.0	0.0	0.0
9		61232	29.0	1.5	0.0	0.0	0.0	0.0	0.0	0.0	0.0	0.0
10	*Vigna trilobata*	113671	0.0	0.5	0.0	0.0	0.0	0.0	1.3	6.0	4.0	0.0
11	*Stylosanthes fruticosa*	60354	0.0	0.0	n.a.	n.a.	n.a.	n.a	n.a.	n.a.	n.a.	n.a.
12	*S. hamata*	70552	0.0	0.0	14.3	6.7	0.0	0.0	3.3	16.7f	2.0	0.0
13		70525	0.5	0.0	0.0	0.3	0.0	0.0	68.7	21.0f	3.7	0.0
14		72850	6.5	0.0	69.0	59.0 f	n.a.	n.a.	137.7	44.3f	4.0	0.0.
15		cv. Verano	20.0	0.0	37.0	20.3 fs	0.0	0.0	71.7	83.0f	4.7	0.0
16		TQ 2	0.0	0.0	3.7	6.7 fs	n.a.	n.a	162.0	102.3fs	5.3	0.0
17		TQ 3	12.0	6.0	4.0	2.7f	n.a.	n.a	107.0	91.7f	5.3	0.0.
18	*S. scabra*	cv. Seca	1.5	0.0	0.0	0.0	0.0	0.3	31.0	5.7	2.0	0.0
19		55870	0.0	0.0	40.3	36.0	0.0	0.0	67.2	34.0f	2.7	0.0
20	*S. sympodiales*	65960	2.0	0.5	20.7	17.7 fs	0.0	0.0	51.3	68.0	6.7	0.0.

As listed in Table 9.3 Experiments 1 and 3 sown at Toorak in a drought year (Fig. 9.2) failed to germinate and no data were recorded. Plant no. as listed in Table 9.2. n.a., not available for sowing; f, flowering; s, seeding at that date; [a]Sown 2/2/88: mean number of plants per 4 m row; [b]Sown 26/11/88: mean number of plants per 20 m^2 plot; [c]Sown 2/2/88: mean number of plants per 20 m^2 plot; [d]Sown 29/11/88: mean number of plants per 20 m^2 plot; [e]Mean vigour rated 0 to 10.

Table 9.10. Performance of non-*Desmanthus* accessions in experiments 7 to 10.

	Location		Woodbine			Taranaway		Blendon		Mt Marlow	
	Experiment no.		7[a]			8[b]		9[b]		10[b]	
	Year		1989			1990		1990		1990	
	Date (day/month)		9/3	8/6		9/7		9/7		9/7	
Plant no.		Accession no.	no.	no.	g/plot	no.	vigour	no.	vigour	no.	vigour
1	*Alysicarpus* sp.	TQ 1	314.7	13.3	6.7	1.3	1.0	3.0	2.7	1.3	2.0
4	*Chamaecrista rotundifolia*	cv. Wynn	48.0	5.7	0.7	0.0	0.0	0.7	2.0	0.0	0.0
5	*Clitoria ternatea*	47187	1.3	2.0	1.0	0.3	2.0	2.0	4.0	0.3	2.0
6		48337	18.3	3.7	3.3	0.0	0.0	1.0	2.0	0.0	0.0
7	*Centrosema pascuorum*	55697	12.3	1.3	0.7	0.0	0.0	4.7	4.7	0.0	0.0
8	*Macroptilium atropurpureum*	cv. Siratro	17.7	1.3	2.7	3.0	3.3	1.0	2.0	3.0	3.3
9		61232	33.3	4.3	2.7	5.3	4.0	1.7	3.5	5.3	4.0
10	*Vigna trilobata*	113671	129.6	6.0	10.0	18.0	4.3	9.7	3.3	18.0	4.3
11	*Stylosanthes fruticosa*	60354	n.a.	n.a.	n.a.	1.0	2.0	0.0	0.0	1.0	2.0
12	*S. hamata*	70552	186.7	16.0	1.7	10.7	3.0	11.0	4.0	10.7	3.0
13		70525	100.0	13.0	7.0	0.3	2.0	1.0	2.0	0.3	2.0
14		72850	257.0	0.0	0.0	4.7	2.7	1.7	2.0	4.7	2.7
15		cv. Verano	162.7	11.3	7.0	1.3	1.5	5.7	2.0	1.3	2.0
16		TQ 2	169.0	26.0	8.7	0.7	2.5	2.7	2.5	1.0	2.5
17		TQ 3	76.3	23.3	4.3	0.3	2.0	0.3	2.0	1.0	2.0
18	*S. scabra*	cv. Seca	208.0	23.0	1.3	6.7	4.0	9.0	2.7	6.7	4.0
19		55870	261.3	21.7	2.7	7.7	2.0	16.0	3.3	7.7	2.0
20	*S. sympodiales*	65960	214.7	18.0	4.3	0.0	0.0	0.0	0.0	0.0	0.0

Experiment numbers: As listed in Tables 9.3 (Exp. 7) and 9.2 (Exp. 8 to 10); Accession number: CPI, or other specified accession number. n.a., Not available for sowing. [a]Sown 30/11/88: mean number of plants/20 m^2 , with legume yields (g/plot); grass yields at that time were 216 g/plot. No survival recorded at 13/1/89. [b]Sown 11–14/12/89: mean number of plants/20 m^2 with ratings (0–10) for vigour. No survival in December 1990.

vigour, but despite flowering and seeding, no seedlings were subsequently found (Table 9.11). *S. hamata* and *S. scabra* behaved similarly.

In the third transplant experiment at Rosebank near Longreach (Experiment 13), plant establishment was reasonable but heavy flooding occurred in the second and third growing seasons (Fig. 9.2). *C. ternatea* and *Stylosanthes* were again the only accessions to persist, flower and set seed (Table 9.12) but they failed to survive at the start of the fourth growing season. Two annual species *Arachis* from Brazil and a native *Alysicarpus* were often apparently absent for periods during the experiment, but temporarily reappeared as seedlings at the end of the third growing season.

None of the non-*Desmanthus* accessions survived in these environments, as confirmed some 10 years later (Gardiner *et al.*, 2004), but *Clitoria* and *Stylosanthes* were the most promising. This reflects the results obtained in a less-stressful clay soil situated in the dry tropics (Chapter 8, this volume), where these same accessions persisted.

Appendix 9.5

Performance characteristics of *Desmanthus* accessions in western Queensland

Seed sown experiments (1–10, Table 9.3)

In these experiments germination and establishment were very often poor with zero germination in Experiments 1 and 4. Such germination as was noted was inconsistent in the remaining trials. In Experiment 2 at Rosebank, *D. leptophyllus* showed better establishment in a drought year than it did at the same site in a better year (Experiment 6, Fig. 9.2, Table 9.13), while the opposite was observed with *D. virgatus*.

In some of the later sown experiments (7 to 10), establishment conditions were more favourable. Data for Experiment 7 at Woodbine are presented in Table 9.14. Plant numbers often increased with time even up to the start of the third growing season (1990), and plant vigour ratings were often high (e.g. plants 21 (*D. bicornutus*), 29 (*D. leptophyllus*) and 48, 50 and 54 (*D. virgatus*)). Many had produced seeds at the end of the second growing season (20/7/90) and early in the third (12/12/90). Accessions of *D. virgatus* were outstanding, particularly some from Argentina (plants 49 to 51) and Mexico (plants 54 and 57). The sole accession from Guatemala (plant 48) had shown high vigour, dry matter yield and plant numbers. Accessions of *D. bicornutus*, initially slow to establish (on 9/3/89), improved with time. Plants at the site were subjected to severe flooding in 1990. In subsequent years they experienced drought (rainfall of only 260 mm in 2003), and severe overgrazing. The data published by Gardiner *et al.* (2004) showed that many persisted up to at least 2003 and the most productive at that time were the accessions of *D. virgatus* noted above.

Persistence records of Experiment 8 at Taranaway (Table 9.15) were taken in 1990, and later in 1999 and 2003 at Taranaway near Isisford (Gardiner *et al.*, 2004). This is in a drier environment than Woodbine (Table 9.1) and is on a light clay cleared to plant Buffel grass (*Cenchrus ciliaris*) that formed a mixture with the native *Astrebla* sp. (Mitchell grass). It was severely flooded in 1990 and access and record taking were very restricted. Here, unlike the situation at Woodbine, the best adapted species in the first growing season were *D. bicornutus* and *D. leptophyllus*.

At the start of the next growing season (12/12/90) only two accessions of *D. virgatus* were visible, one of these being the sole introduction from Guatemala (plant 48). Some of the best adapted accessions at Woodbine, plants 49–51 (from Argentina) and 54 and 57 (from Mexico), were not apparent. At this time, using plant vigour × plant number as a criterion, the best accessions in the surviving species were plants 23 (*D. cornutus*), 29 (*D. leptophyllus*), 38 (*D. pernambucanus*), 44 (*D. pubescens*), 56 (*D. virgatus*) and 60 (*D.* sp. probably *D. virgatus*).

In the following years the experimental area was subjected to severe grazing and drought. In 1995 the annual rainfall was only 210 mm but in 1999 the area received 860 mm and the plots were re-examined (Gardiner *et al.*, 2004). All of the above accessions just mentioned were sighted with *D. bicornutus*, plant 23 showing the highest dry matter production of all of the accessions on trial (Table 9.15). Plant 48, *D. virgatus* from Guatemala was the second highest yielding. Many of the accessions that were not

Table 9.11. Performance of non-*Desmanthus* accessions in Experiment 12 at Toorak.[a]

			Dates of Observation								
			13/4/89			24/11/89			4/5/90		
Plant no.[b]	Species	Accession no.[c]	no.[d]	seed[e]	vigour[f]	no.[d]	seed[e]	vigour[f]	no.[d]	flow/seed[e]	vigour[f]
1	*Alysicarpus* sp.	TQ 1	6.3	s	5.0	2.7	s	4.5	0.0	–	–
2	*Arachis* sp.	91423	2.7	–	5.0	0.0	–	–	–	–	–
3		58114	1.3	–	3.3	0.0	–	–	–	–	–
4	*Chamaecrista rotundifolia*	cv. Wynn	1.7	–	2.5	0.0	–	–	–	–	–
5	*Clitoria ternatea*	47187	7.7	s	4.0	6.7	s	3.7	3.3	fs	5.6
6		48337	6.7	s	4.7	6.7	s	2.3	4.3	–	5.3
7	*Centrosema pascuorum*	55697	5.7	s	2.7	0.0	–	–	–	–	–
8	*Macroptilium atropurpureum*	cv. Siratro	5.7	–	2.3	2.7	–	2.3	0.0	–	–
9		61232	7.0	–	3.0	1.0	–	2.0	0.0	–	–
10	*Vigna trilobata*	113671	2.0	s	1.0	0.0	–	–	–	–	–
11	*Stylosanthes fruticosa*	60354	6.7	s	2.7	3.3	s	1.7	0.0	–	–
12	*S. hamata*	70552	7.0	s	3.0	6.7	s	1.3	1.7	f	5.0
13		70525	5.7	s	3.0	5.7	s	1.0	0.3	–	6.0
14		72850	6.3	s	3.3	1.0	s	1.0	0.3	–	4.0
15		cv. Verano	8.0	s	4.0	7.0	s est.[g]	1.0	0.0	–	–
16		TQ 2	7.7	s	3.7	7.7	s	1.0	0.7	–	4.0
17		TQ 3	8.0	s	4.0	3.3	s	1.0	1.7	– est.[g]	6.0
18	*S. scabra*	cv. Seca	7.3	s	4.7	7.3	s	4.7	1.3	–	4.0
19		55870	7.3	s	4.7	7.3	s	3.7	2.0	–	3.7
20	*S. sympodiales*	65960	6.7	s	3.0	6.7	s	3.0	0.0	–	–

[a]Transplanted 24/11/88; [b]number assigned to plants in trial; [c]accession number: CPI number unless otherwise noted; [d]mean number of surviving transplants/row (out of 8); [e]seeding and/or flowering; [f]vigour: mean rating of plant growth rated 1 to 10. – means dead; [g]est., seedlings established.

Table 9.12. Performance of non-*Desmanthus* accessions in Experiment 13 at Rosebank.[a]

			Dates of observation						
			13/3/88		28/11/88		26/11/89		22/7/90
Plant no.[b]	Species	Accession no.[c]	no.[d]	vigour[e]	no.[d]	vigour[e]	no.[d]	vigour[e]	no.[d]
1	*Alysicarpus* sp.	TQ 1	0.0	–	0.0	–	0.0	–	0.0
2	*Arachis* sp.	91423	5.7	2.0	3.3	2.0	0.0	–	0.3 s
3		58114	0.0	–	1.3	4.0	0.0	–	–
4	*Chamaecrista rotundifolia*	cv. Wynn	0.0	–	0.0	–	0.0	–	–
5	*Clitoria ternatea*	47187	7.7	4.3	5.0 f	2.0	0.7s	3.0	0.7 s
6		48337	6.3	3.0	0.7	2.0	0.6 s	2.5	–
7	*Centrosema pascuorum*	55697	1.3 s	2.0	0.0	–	0.0	–	–
8	*Macroptilium atropurpureum*	cv. Siratro	4.7	1.7	0.3	4.0	0.0	–	–
9		61232	4.7	2.0	0.0	–	0.0	–	–
10	*Vigna trilobata*	113671	3.6 s	3.0	3.7	2.7	0.0	–	–
11	*Stylosanthes fruticosa*	60354	4.7 s	2.7	3.0	3.0	1.3 s	1.5	–
12	*S. hamata*	70552	5.7 s	3.0	5.7 f	3.7	1.7 s	3.0	–
13		70525	7.0 s	3.7	5.3	3.0	0.7 s	1.0	–
14		72850	4.3 s	2.3	4.3	2.0	0.0	–	–
15		cv. Verano	7.3 s	4.0	7.3	4.3	4.3 s	2.0	–
16		TQ 2	7.3 s	4.3	8.0 f	5.0	5.3 s	2.0	–
17		TQ 3	8.0 s	5.3	8.0 f	4.7	4.0 s	2.0	–
18	*S. scabra*	cv. Seca	4.7	2.7	5.3	4.0	2.0 s	2.0	–
19		55870	6.7 s	3.3	5.7 f	4.3	3.7 s	2.0	–
20	*S. sympodiales*	65960	4.3 s	4.0	3.7	2.7	0.0	–	–

[a]Transplanted 3/2/88; [b]number assigned to plants in trial; [c]accession number: CPI number unless otherwise noted; [d]mean number of surviving transplants/row (out of 8); [e]vigour: mean rating of plant growth rated 1 to 10. – Means dead. s, seeding; f, flowering.

Table 9.13. Performance of *Desmanthus* accessions in Rosebank Experiments 2, 5 and 6.

			Experiments, observation dates						
			2[c]		5[d]		6[e]		
			1988		1988	1989	1989		
			20/5	7/9	10/5	8/9	6/3	8/9	
Plant no.[a]	Species	Accession no.[b]	no.	no.	no.	no.	no.	no.	Vigour[f]
21	*D. bicornutus*	81337	n.a.	n.a.	n.a.	n.a.	7.7	0.0	
22		84508	0.0	0.0	1.7	3.0	0.0	0.0	
23		90857	3.5	0.5	0.3	0.0	9.0	0.3	2.0
24		91162	10.0	4.5	0.7	0.7	0.0	7.3	3.0
25	*D. covillei*	90311	n.a.	n.a.	0.0	0.0	0.0	0.0	
26		90877	0.0	0.0	n.a.	n.a.	0.0	0.0	
27	*D. fruticosa*	84960	5.5	2.0	0.3	0.0	n.a.	n.a.	
28	*D. leptophyllus*	37143	27.0	1.0	0.0	0.0	19.0	1.7	2.0
29		38351	30.5	3.0	1.0	11.3	19.0	0.0	
30		55179	8.5	2.0	0.7	0.7	19.7	0.3	2.0
31		76053	7.5	1.5	0.0	0.3	8.7	0.0	
32		92655	24.0	4.0	0.0	0.0	20.0	1.0	1.5
33		92746	8.5	8.0	1.3	3.7	0.0	0.0	
34		92809	2.0	0.0	n.a.	n.a.	0.0	0.7	2.0
35		L.D. 3	10.0	1.5	n.a.	n.a.	0.3	2.0	1.0
36		L.D. 4	3.5	0.5	n.a.	n.a.	0.0	0.0	
37		L.D. 5	17.5	1.5	n.a.	n.a.	0.0	0.0	
38	*D. pernambucanus*	40071	4.0	1.5	1.0	1.0	13.7	0.0	
39		49728	2.0	0.0	0.3	0.3	15.7	1.7	2.0
40		83565	7.0	1.0	0.7	0.7	7.0	0.3	1.0
41	*D. pubescens*	92800	n.a.	n.a.	n.a.	n.a.	0.0	0.0	
42		92802	3.0	2.0	n.a.	n.a.	0.0	0.0	
43		92803 cv. Uman	1.0	0.5	1.7	7.3	0.0	0.0	
44		92804	n.a.	n.a.	n.a.	n.a.	n.a.	n.a.	
45	*D. tatuhyensis*	37538	0.0	0.0	n.a.	n.a.	n.a.	n.a.	
46		90362	3.5	0.5	n.a.	n.a.	n.a.	n.a.	
47	*D. virgatus*	57960	1.0	0.5	n.a.	n.a.	28.7	1.0	3.3
48		67643	1.5	0.0	n.a.	n.a.	18.7	3.0 fs	6.0
49		78372	8.5	2.0	n.a.	n.a.	21.0	0.3	2.0
50		78373 cv. Marc	5.0	2.0	1.0	14.3	23.3	3.0 s	2.0
51		78382	0.0	0.0	0.3	1.3	20.7	0.0	
52		79653	2.0	0.0	n.a.	n.a.	38.7	0.0	
53		85173	1.0	1.0	n.a.	n.a.	20.0	5.3	4.5
54		85178	0.50	0.0	0.3	0.7	6.0	0.0	
55		85182	0.5	0.0	0.7	0.0	1.0	0.0	
56		90751	0.0	0.5	0.7	5.7	10.7	0.0	
57		91181	3.0	0.5	n.a.	n.a.	n.a.	n.a.	
58	*D.* spp.	33426	0.0	0.0	n.a	n.a	33.0	1.0	6.0
59		70338	1.5	1.0	n.a	n.a	4.0	0.0	
60		83563	4.5	0.0	0.7	3.7	9.7	0.0	
61		83570	n.a.	n.a	n.a	n.a	n.a	n.a.	

Experiments as listed in Table 9.3. Seeds sown in Experiments 1 and 3 at Toorak in a drought year (Fig. 9.2) failed to germinate and no data were recorded. Seeds in Experiment 4 at Toorak failed to germinate as well. [a]Plant number used in trials; [b]accession number, CPI unless otherwise noted; [c]sown 2/2/88, mean number of plants per 4 m row; [d]sown 26/11/88, mean number of plants per 20 m^2 plot; [e]sown 29/11/88, mean number of plants per 20 m^2 plot; [f]mean vigour, rated 0 to 10. Flowering (f) and seeding (s) recorded at the last observation. n.a., not available for sowing.

Table 9.14. Performance of *Desmanthus* accessions in Experiment 7, Woodbine, near Blackall.

			Observation dates								
			1989				1990				
			9/3	8/6		13/12	20/7	12/12		1999[f]	2003[f]
Plant no.[a]	Species	Accession no.[b]	no.[c]	no.[c]	Dry wt.[d]	vigour[e]	no.[c]	no.[c]	vigour[e]	no.	no.
21	*D. bicornutus*	81337	45	80	220	5.5	27	106	5.3	T	T
22		84508	0	180	133	2.5	13 s	7	4.0	–	–
23		90857	16	87	120	3.8	53	93	5.0	–	–
24		91162	0	347	814	6.0	27	33	2.5	–	–
25	*D. covillei*	90311	n.a.								
26		90877	n.a.								
27	*D. fruticosa*	84960	n.a.								
28	*D. leptophyllus*	37143	129	187	760	7.0	200 s	33	5.0	–	–
29		38351	113	253	140	6.5	87 s	320 s	4.5	–	–
30		55179	63	233	360	7.3	87 s	0	–	–	–
31		76053	62	173	320	7.8	27 s	87	7.0	T	1.0
32		92655	209	313	354	7.3	68 s	107	7.0	T	T
33		92746	0	140	120	5.0	33	40	5.0	T	0.3
34		92809	0	307	260	4.0	27	20	2.0	–	–
35		L.D. 3	0	273	454	7.2	60 s	167	5.3	–	T
36		L.D. 4	0	5540	834	6.7	33 s	107	5.0	–	–
37		L.D. 5	0	195	254	7.8	43	40	6.3	–	–
38	*D. pernambucanus*	40071	73	180	327	3.3	47 s	11	1.0	–	0.5
39		49728	130	140	80	4.0	33 s	47	2.0	0.4	0.6
40		83565	71	167	133	4.8	20	100	5.0	T	T
41	*D. pubescens*	92800	0	80	67	3.0	7	40	6.0	–	–
42		92802	13	127	453	3.5	0	53	7.0	T	T
43		cv Uman	0	107	147	4.0	40	0	–	–	–
44		92804	0	80	227	3.2	20 s	47 s	5.5	T	–
45	*D. tatuhyensis*	37538	n.a.								
46		90362	0	67	33	1.2	0	0	–	–	–
47	*D. virgatus*	57960	73	253	474	5.8	40	120 s	8.0	1.0	1.5
48		67643	163	567	634	7.2	127 s	200	7.0	1.3	1.5
49		78372	96	220	280	6.5	40 s	300 s	6.3	1.0	4.0
50		cv Marc	61	193	300	7.3	80 s	314 s	7.3	3.0	9.0
51		78382	64	127	187	7.7	80 s	153 s	8.0	0.8	2.0
52		79653	44	87	54	4.8	33	54	7.0	1.0	2.0
53		85173	173	180	160	6.5	7	53 s	7.0	1.0	1.0
54		85178	105	120	47	6.5	13 s	240 s	5.3	2.5	9.0
55		85182	0	253	193	5.0	73 s	107	2.5	1.0	0.7
56		90751	50	120	60	5.3	60	80	2.0	2.7	5.1
57		91181	87	167	113	5.3	47 s	213 s	6.7	2.0	3.5
58	*D.* sp.	33426	n.a.								
59		70338	120	230	220	3.5	0	7	7.0	T	0.6
60		83563	57	260	307	5.0	100	87 s	6.3	–	–
61		83570	n.a.								

[a]Plant number used in trials as listed in Table 9.3; [b]accession number, CPI unless otherwise noted; [c]number of plants per 20 m^2 plot; [d]dry weight (g) per 20 m^2 plot; [e]vigour: mean, rated 1 to 10; [f]From Gardiner *et al.* (2004), mean number of plants/m^2 or T trace (< 0.3 plants/m^2). n.a., not available for sowing; s, seeding.

Table 9.15. Performance of *Desmanthus* accessions in Experiment 8 at Taranaway.

			Observation dates						
			1990						
			21/7		12/12			1999[e]	2003[e]
Plant no.[a]	Species	Accession no.[b]	no./20 m[2c]	vigour[d]	no./m[2c]	vigour[d]	no. × vig.	no./m²	no./m²
21	*D. bicornutus*	81337	2.3	4.5	0.3	4.0	1.2	T	–
22		84508	2.3	4.0	–	–	–	0.3	–
23		90857	0.7	3.0	1.3	7.0	9.1	1.4	–
24		91162	5.7	4.3	–	–	–	T	–
25	*D. covillei*	90311	0.3	4.0	–	–	–	T	–
26		90877	n.a.		n.a.		n.a.		n.a.
27	*D. fruticosa*	84960	–	–	–	–	–	–	–
28	*D. leptophyllus*	37143	27.0	4.3	2.3	4.0	9.2	–	–
29		38351	10.3	4.0	3.3	3.5	11.6	T	–
30		55719	8.7	4.0	–	–	–	–	–
31		76053	6.7	4.7	–	–	–	–	–
32		92655	7.7	4.3	0.3	3.0	0.9	–	–
33		92746	7.3	4.3	0.1	8.0	8.0	T	–
34		92809	5.7	3.7	–	–	–	–	–
35		L.D. 3	0.7	4.0	–	–	–	–	–
36		L.D. 4	3.0	4.5	–	–	–	–	–
37		L.D. 5	4.0	4.0	2.0	5.5	11.0	–	–
38	*D. pernambucanus*	40071	3.0	4.5	0.7	6.0	4.2	T	–
39		49728	0.3	4.0	–	–	–	T	–
40		83565	0.3	4.0	–	–	–	T	–
41	*D. pubescens*	92800	–	–	–	–	–	–	–
42		92802	2.0	4.7	0.3	2.3	0.7	T	–
43		cv. Uman	3.7	5.0	0.3	5.0	1.5	–	–
44		92804	10.7	4.0	1.0	2.0	2.0	T	–
45	*D. tatuhyensis*	37538	–	–	–	–	–	–	–
46		90362	1.3	3.0	–	–	–	–	–
47	*D. virgatus*	57960	2.0	3.3	–	–	–	0.5	–
48		67643	–	–	0.3	9.0	2.7	1.0	–
49		78372	6.0	4.0	–	–	–	0.3	–
50		cv. Marc	0.3	2.0	–	–	–	0.4	–
51		78382	–	–	–	–	–	–	–
52		79653	0.3	4.0	–	–	–	–	–
53		85173	1.3	4.0	–	–	–	–	–
54		85178	–	–	–	–	–	0.3	–
55		85182	1.7	3.0	–	–	–	0.6	–
56		90751	0.3	2.0	0.7	6.0	6.0	0.5	–
57		91181	0.3	4.0	–	–	–	0.4	–
58	*D.* sp.	33426	–	–	–	–	–	0.3	–
59		70338	5.7	4.6	–	–	–	T	–
60		83563	13.0	4.0	4.3	6.3	27.1	T	–
61		83570	n.a.		n.a.		n.a.	n.a.	n.a.

[a]Plant number used in trials as listed in Table 9.3; [b]Accession number, CPI unless otherwise noted; [c]number of plants per 20 m² plot; [d]vigour: mean, rated 1 to 10; [e]from Gardiner *et al.* (2004), mean number of plants/m² or T trace (< 0.3 plants/m²); n.a., not available for sowing; s, seeding.

found on 12/12/90, especially *D. virgatus*, had measurable yields (unlike those of *D. pubescens*, *D. leptophyllus* and *D. pernambucanus*). In 2003, with only 284 mm rainfall (after a drought year of 164 mm in 2002), no *Desmanthus* growth was apparent.

These observations confirm that accessions of two species, *D. bicornutus* and *D. virgatus*, can be established by surface sowing and persist for many years in grass pastures in these difficult environments. The most promising plants, at both Woodbine and Taranaway, were becoming evident in the small sward experiments. It has long been accepted that long-term persistence in these difficult environments is the proximate requirement (e.g. Burt and Ebersohn, 1969). Simple measurements taken at strategic intervals, rather than logistically difficult dry weight harvests at regular time intervals, meet the initial experimental aims.

The remaining small sward experiments (Experiments 9 and 10, Table 9.16) were carried out at Blendon and Mt Marlow. Both sites experienced severe flooding in the first two growing seasons (Fig. 9.3) and access and record taking were again very limited. By the onset of the third growing season (3/12/90) only 11 accessions were visible at Blandon and four at Mt Marlow. *D. leptophyllus* was again most promising initially but two accessions of *D. pubescens*, atypically for these experiments, had survived and shown good vigour. This is a species from, and adapted to, wetter environments (as in Central America; Lazier, 1995 onwards, personal communication) and according to Luckow (1993) occurs in 'marshy coastal areas or near lakes'. Its good performance in this experiment may be the result of two very wet growing seasons. Similarly where good provenance records are available for the two surviving accessions of *D. leptophyllus*, one (plant 32) is from a relatively high rainfall area in Colombia. The results obtained from these experiments, carried out over a short time period, almost certainly reflect the highly unusual consecutive years of flooding.

Transplant experiments (numbers 11, 12 and 13, Table 9.3)

These experiments were carried out at Toorak (Experiments 11 and 12) and Rosebank (Experiment 13), the latter being more subtropical (Fig. 9.1). Results from the former experiments have been presented previously in the Results section of the main text and a summary of all of these experiments showing long-term persistence, vigour and ability to produce seedlings is presented in Table 9.17.

In the first sowing at Toorak (Experiment 11), few accessions survived the locust attack and the drought encountered in the first growing season. *D. bicornutus*, plant 22, and several accessions of *D. virgatus*, with the Argentinean (plants 49 to 51), Mexican (plants 54 and 57), Guatemalan (plant 48) and *D.* sp. (plant 60, probably *D. virgatus*) plants again being the most persistent and vigorous. In the second sowing, with much better growing conditions, more plants survived, including accessions from *D. leptophyllus* and more accessions of *D. pubescens* and *D. pernambucanus*. The greatest vigour was shown by Argentinean and Guatemalan accessions of *D. virgatus* (plants 49 to 51 and plant 48), by *D. bicornutus* and by *D.* sp. (plant 60). The Argentinean accessions again flowered and set seed more than the other plants studied. At the southerly site Rosebank (Experiment 13), this same situation was repeated with *D. leptophyllus* performing more poorly. The behaviour of representatives of eight of the species studied at this site is examined in more detail in Fig. 9.9a–h.

D. virgatus was the most successful species investigated in these experiments (Fig. 9.9d–f). Supplementary irrigation was terminated in April 1988 and this was followed by the death of some transplants in all accessions. Plant 47, *D. virgatus* from Curaçao, died out completely within 3 months (Fig. 9.9d). Transplant death in the Argentinean accessions (plants 50 and 51, 9e) was relatively slight in the first growing season, and plant numbers were very stable thereafter. In one accession (78382, plant 51), they possibly increased at the start of the fourth growing season (December 1990). They showed high vigour throughout the experiment and flowered and seeded throughout the growing season, producing established seedlings.

The Mexican accessions (plants 54 and 57, 85178 and 91181, Fig. 9.9f) declined sharply in number in the first growing season and continued to decline thereafter. They were often

Table 9.16. Performance of *Desmanthus* accessions in Experiments 9 and 10 at Blendon and Mt Marlow.

			Experiment							
			Exp. 9 at Blendon				Exp. 10 at Mt Marlow			
			1990				1990			
			19/7		1/12		21/7		3/12	
Plant no.[a]	Species	Accession no.[b]	no.[c]	vigour[d]	no.[c]	vigour[d]	no.[c]	vigour[d]	no.[c]	vigour[d]
21	*D. bicornutus*	81337	0.3	2.0	–	–	0.3	4.0	–	–
22		84508	1.3	3.0	–	–	2.7	3.5	–	–
23		90857	2.0	4.7	–	–	0.7	2.0	–	–
24		91162	0.3	4.0	–	–	3.0	3.3	–	–
25	*D. covillei*	90311	0.3	4.0	0.3	1.0	0.3	4.0	–	–
26		90877	n.a							
27	*D. fruticosa*	84960	0.0	–	–	–	0.0	–	–	–
28	*D. leptophyllus*	37143	6.7	4.5	1.0	7.0	4.3	3.7	0.3	7.0
29		38351	3.0	2.3	0.7	6.0	4.0	4.0	–	–
30		55179	2.0	3.0	–	–	0.7	2.0	–	–
31		76053	3.7	4.0	–	–	7.0	4.0	–	–
32		92655	8.3	5.7	5.3	1.3	3.3	3.0	0.3	2.0
33		92746	19.0	5.0	0.3	2.0	3.0	3.0	–	–
34		92809	20.3	3.3	2.7	1.0	3.0	3.3	–	–
35		L.D. 3	4.3	3.7	1.0	4.0	8.7	4.3	–	–
36		L.D. 4	27.0	3.0	–	–	13.0	2.0	–	–
37		L.D. 5	5.0	3.0	–	–	1.0	4.0	–	–
38	*D. pernambucanus*	40071	3.7	4.7	0.7	8.0 s	3.7	2.5	–	–
39		49728	1.3	2.7	0.7	2.7	0.3	2.0	–	–
40		83565	0.7	6.0	–	–	0.0	–	–	–
41	*D. pubescens*	92800	0.7	2.0	–	–	0.0	–	–	–
42		92802	2.0	3.3	–	–	6.7	4.0	0.3	2.0
43		92803 cv. Uman	5.3	3.7	–	–	5.3	3.3	–	–
44		92804	5.0	5.0	–	–	9.3	4.0	0.7	9.0
45	*D. tatuhyensis*	37538	n.a.							
46		90362	0.3	4.0	–	–	0.0	–	–	–
47	*D. virgatus*	57960	1.0	4.0	–	–	7.3	3.0	–	–
48		67643	1.0	2.0	–	–	0.0	–	–	–
49		78372	5.3	5.0	–	–	0.7	3.3	–	–
50		78373 cv. Marc	0.0	–	–	–	0.0	–	–	–
51		78382	0.7	2.0	–	–	0.3	2.0	–	–
52		79653	0.0	–	–	–	0.3	4.0	–	–
53		85173	2.3	7.0	–	–	0.0	–	–	–
54		85178	1.0	2.0	–	–	0.3	2.0	–	–
55		85182	1.7	4.0	0.3	4.0	3.3	2.0	–	–
56		90751	0.7	2.0	–	–	0.0	–	–	–
57		91181	0.0	–	0.3	2.0	0.0	–	–	–
58	*D.* spp.	33426	n.a.							
59		70338	12.0	3.0	–	–	5.3	4.0	–	–
60		83563	2.0	4.0	–	–	2.3	2.5	–	–
61		83570	n.a.							

[a]Plant number used in trials as listed in Table 9.3; [b]accession number, CPI unless otherwise noted; [c]mean number of plants per 20 m^2 plot; [d]vigour: mean, rated 1 to 10. n.a., not available for sowing; s, seeding.

Table 9.17. Performance of transplanted *Desmanthus* accessions at Toorak (Experiment 11) and at Rosebank (Experiments 12 and 13).

			Experiments and dates of observation								
			11			12			13		
			13/5/1988		1990	1989		1990	1990		
			13/5		4/5	24/11		4/5	12/12	22/7	12/12
Plant no.[a]	Species	Accession no.[b]	no.[c]	vigour[d]	vigour[d]	no.[c]	vigour[d]	vigour[d]	no.[c]	seedlings	vigour[d]
21	*D. bicornutus*	81337	0.7	5.0	4.0	3.0	1.5	5.0	2.7	–	6.0
22		84508	–			–	–		–	–	–
23		90857	–			–	–		0.7	–	7.0
24		91162	–			3.7	1.5	7.0	0.7	–	8.0
25	*D. covillei*	90311	0.3	–	–	1.0	2.0	5.0	1.3	–	1.0
26		90877	–	–	–	1.3*	5.0	6.0 s	0.3	–	1.0
27	*D. fruticosa*	84960	–	–	–	–	–		–	–	–
28	*D. leptophyllus*	37143	–	–	–	0.3	2.0		–	–	–
29		38351	–	–	–	4.3	3.3 s	4.0 s	–	–	–
30		55179	–	–	–	0.3	2.0		–	–	–
31		76053	–	–	–	–	–		0.7	–	6.0
32		92655	–	–	–	1.3*	1.7 s	6.0s	–	–	–
33		92746	–	–	–	–	–		–	–	–
34		92809	–	–	–	4.0	2.0		–	–	–
35		L.D. 3	–	–	–	0.7	2.0		1.3	2	4.0
36		L.D. 4	–	–	–	0.3*	1.0	4.0	–	–	–
37		L.D. 5	–	–	–	0.3*	4.0	4.0	–	–	–
38	*D. pernambucanus*	40071	–	–	–	1.3	3.0		1.0	–	7.0
39		49728	–	–	–	1.0*	2.0	6.0	–	–	–
40		83565	–	3.5	–	–	–		0.3	–	5.0
41	*D. pubescens*	92800	–	–	–	2.0	2.0		–	–	–
42		92802	–	–	–	0.3	2.0		–	–	–
43		92803 cv. Uman	–	–	–	1.7*	3.0 s	4.0	–	–	–
44		92804	0.3	5.0		2.0*	2.5	8.0	–	–	–
45	*D. tatuhyensis*	37538	–	–	–	–	–		–	–	–
46		90362	–	–	–	–	–		0.3	1	5.0
47	*D. virgatus*	57960	–	–	–	–	–		–		–
48		67643	0.3	1.3	4.0	1.0*	3.0	8.0	–		–
49		78372	2.7*	6.5	3.0	1.7*	2.7 s	6.6 s	4.0	80	6.0
50		78373 cv. Marc	2.7*	7.5	6.5	1.38	3.5 s	6.0 s	3.3	1	5.0
51		78382	1.3*	7.0	8.0	1.38	2.0 s	8.0 s	2.0	–	4.0
52		79653	–	–	–	0.3	2.0	–	0.7	2	7.0
53		85173	–	–	–	–	–	–	1.3	–	4.0
54		85178	0.3	4.0	–	–	–	–	0.7	–	2.0
55		85182	0.7*	4.0	6.0	–	–	–	0.3	–	3.0
56		90751	–	–	–	1.0	1.3	–	1.3	–	6.0
57		91181	*	–	–	0.9	2.0	–	1.0	–	3.5
58	*D.* spp.	33426	–	–	–	–	–	–	0.3	1	8.0
59		70338	–	–	–	–	–	–	–	–	–
60		83563	3.0	2.3	2.0	2.7*	2.5	8.5	0.7	–	10.0
61		83570	–	–	–	–	–	–	–	–	–

[a]Plant number used in trials as listed in Table 9.3; [b]accession number, CPI unless otherwise noted; [c]number of surviving transplants, mean/row; *indicate survival on 4/5/90; [d]vigour: mean, rated 1 to 10. n.a., not available for sowing; s, seeding.

less vigorous than the Argentinean material and flowered and seeded less frequently. They too were able to establish seedlings. The Guatemalan accession, plant 48 (67643, Fig. 9.9d) showed poor vigour, had neither flowers nor seeds and died out in the third growing season. Plant 60, *D.* sp. 83563, from a research station in Costa Rica, is probably *D. virgatus* and is conveniently regarded as such here. Although weakly perennial (Fig. 9.9c), it flowered and seeded throughout, produced established seedlings and showed high vigour, particularly in the later stages of the experiment.

After *D. virgatus*, the next most successful species was *D. bicornutus* (Fig. 9.9a). After the first growing season the surviving plants proved to be very persistent and vigorous. They seeded less heavily than *D. virgatus* but produced seed early in the final growing season (December 1990) and produced established seedlings.

Data for the other species may be similarly viewed and compared. The well-known colonizer *D. pernambucanus*, for instance (Cook *et al.*, 2005), is a very vigorous free-seeding plant that establishes seedlings in much drier environments than those with which it is normally associated. One of only two accessions of *D. tatuhyensis*, plant 44, from semi-arid Mexico (90362, Fig. 9.9b), persistently showed increasing vigour in the later stages of the experiment, flowered, seeded and produced established seedlings. Formerly known as *D. brevipes*, the species is found in grassland areas in Texas, and may be useful as a pasture species.

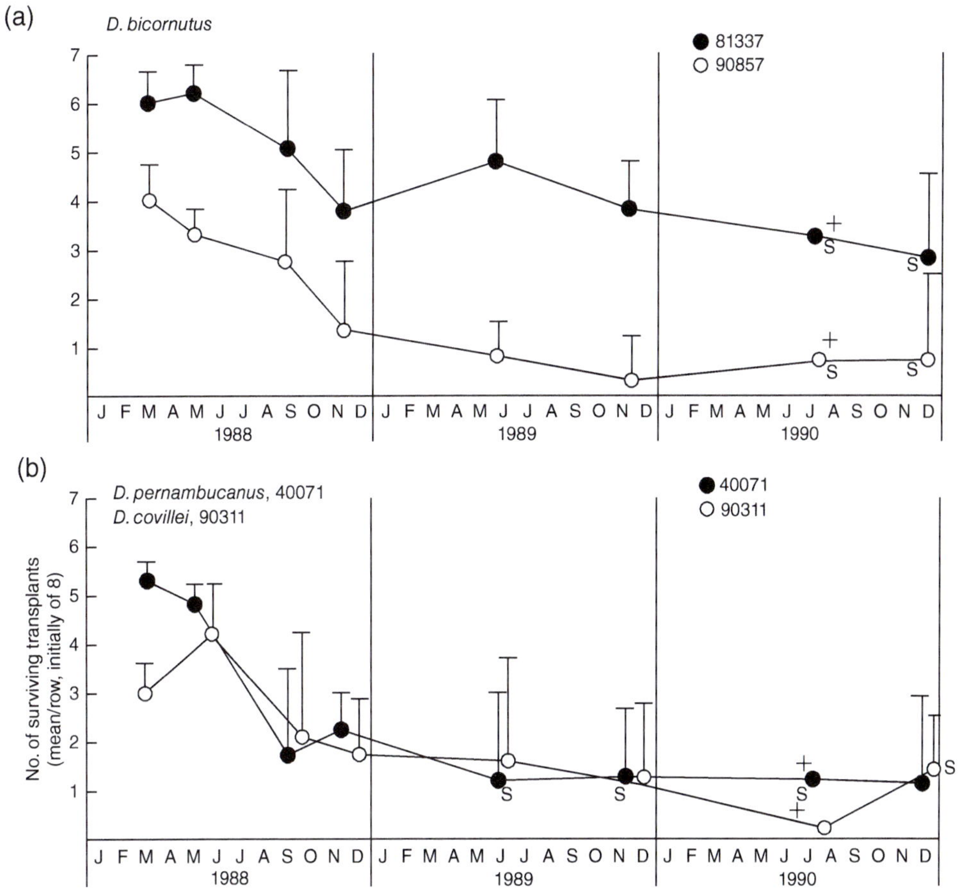

Fig. 9.9. Experiment 13, number of surviving transplants at Rosebank. (a) *D. bicornutus*; (b) *D. pernambucanus*, *D. covillei*; (c) *D.* sp., *D. pubescens*; (d) *D. virgatus* (Guatemala and the Caribbean); (e) *D. virgatus* (Argentina); (f) *D. virgatus*, *D. tatuhyensis* (Mexico); (g) and (h) *D. leptophyllus*.

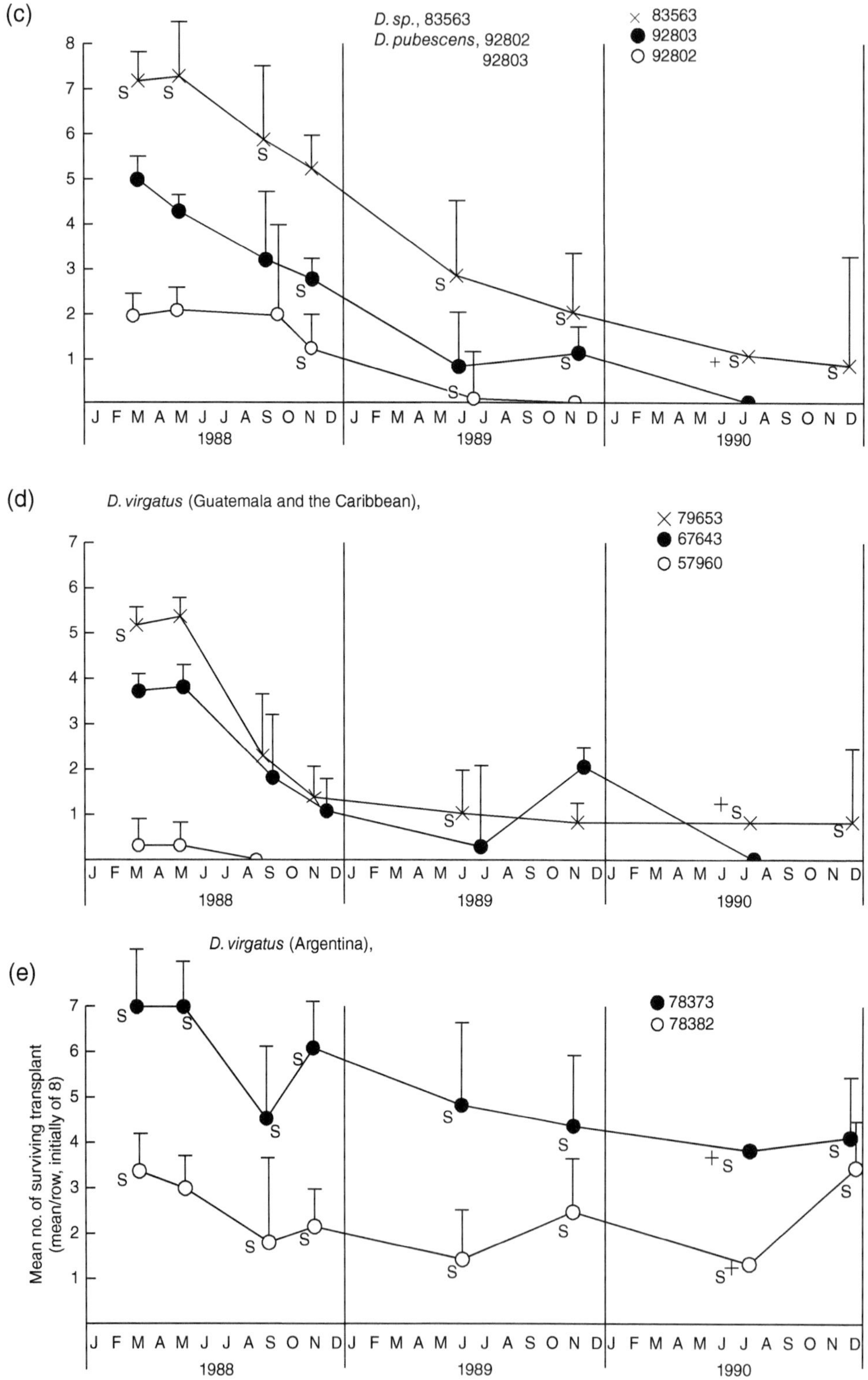
(c)
D. sp., 83563
D. pubescens, 92802
92803
83563
92803
92802
(d)
D. virgatus (Guatemala and the Caribbean),
79653
67643
57960
(e)
D. virgatus (Argentina),
78373
78382
Mean no. of surviving transplant (mean/row, initially of 8)
J F M A M J J A S O N D J F M A M J J A S O N D J F M A M J J A S O N D
1988
1989
1990

Fig. 9.9.

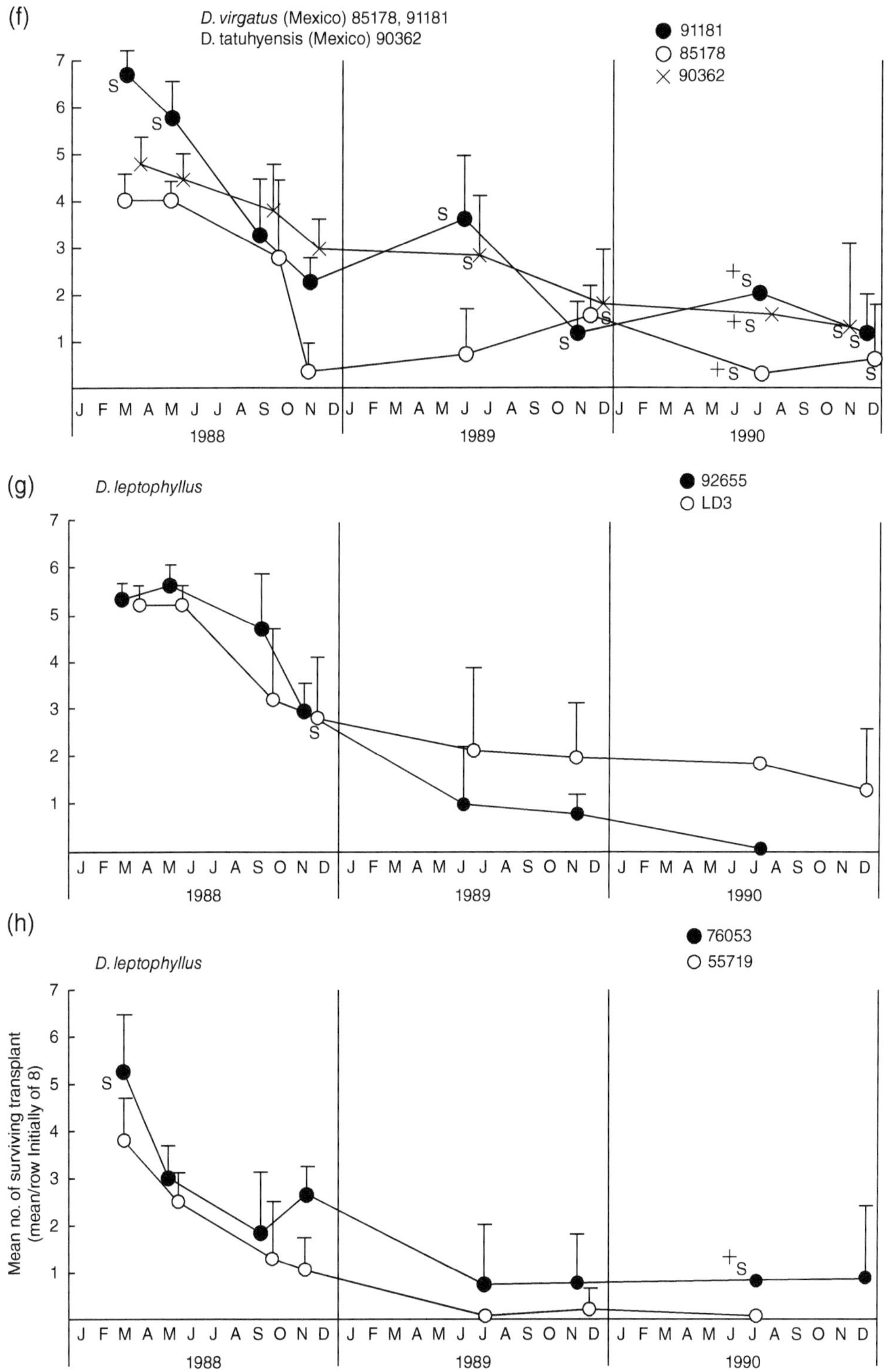
(f)
D. virgatus (Mexico) 85178, 91181
D. tatuhyensis (Mexico) 90362
91181
85178
90362
(g)
D. leptophyllus
92655
LD3
(h)
D. leptophyllus
76053
55719
Mean no. of surviving transplant (mean/row Initially of 8)
J F M A M J J A S O N D
1988
1989
1990

Fig. 9.9.

Table 9.18. Summary of available evaluation studies on *Desmanthus* species.

Desmodium species and accession numbers[a]	No.[b]	Study																	
		I					II			III						IV	V	VI	VII
		AC		LD			T		R	W		Ta		B	M				
		Sa	Cl	Sa	Cl[1]	Cl[2]	11	12	13	7	G	8	G	9	10				
D. bicornutus																			
81337	21						S	S	S	S	T	S	T	–	–				P
84508	22						–	–	–	S	–	–	0.3	–	–				P
90857	23						–	–	S	S	–	S	1.4	–	–				P
91162	24						–	S	S	S	–	–	T	–	–				P
D. leptophyllus																			
38351	29	S	S	S	S	S**	–	–	–	1440	–	S	T	S	–			5	P
55719	30	S	S	S	S	S	–	–	–	–	–	–	–	–	–			6D	
76053	31	S	S	S	S	S*	–	–	S	S	1.0	–	–	–	–				
37143	28						–	–	–	S	–	S	T	–	–				
92655	32						–	S	–	749	T	S	–	S	S				
92746	33						–	–	–	S	0.3	S	T	S	–				
93809	34						–	–	–	S	–	–	–	S	–				
LD 3	35						–	–	S	885	T	–	–	S	–				
LD 4	36						–	S	–	749	–	–	–	–	–				P
LD 5	37						–	S	–	S	–	S	–	–	–				P
D. pernambucanus																			
40071	38						–	–	S	S	0.5	S	T	S	–			3	P
49728	39	S	S	S	S	S	–	S	–	S	0.6	–	T	S	–				P
83565	40						–	–	S	S	T	–	–	–	–				
D. pubescens																			
76057		S	S	S	S	S	N	N	N	N	N	N	N	N	N				
92800	41						–	–	–	–	–	–	–	–	–				
92802	42						–	–	–	S	T	S	T	–	S				P
92803	43						–	S	–	–	–	S	–	–	–			8D	P

92804	44						–	S	–	S	–	S	T	–	S				P
D. virgatus																			
67642		N	N	N	N	N	N	N	N	N	N	N	N	N	N	2			
57960	47						–	–	–	9.0	0.5	–	0.5	–	–	3			
67643	48	S	S	S	S	S**	S	S	–	1400	1.5	S	1.0	–	–				
78372	49						S	S	S	1890	4.0	–	0.3	–	–				
65847		–	S	–	S	S	N	N	N	N	N	N	N	N	N	4			
68369	+	–	S	S	S	S	N	N	N	N	N	N	N	N	N				
78373	50	S	S	S	S	S*	S	S	S	2292	9.0	–	0.4	–	–	1		2	P
78380		–	–	–	S	S	N	N	N	N	N	N	N	N	N				
78382	51	–	S	S	S	S**	S	S	S	1224	2.0	–	–	–	–		1	1	
79653	52	–	–	–	S	S**	–	–	S	S	2.0	–	–	–	–			7D	P
85173	53						–	–	–	S	1.0	–	–	–	–		2		P
85178	54						–	–	–	1272	9.0	–	0.3	S	–			4	
85182	55						S	S	–	S	0.7	–	0.6	–	–				
90751	56						–	–	–	S	5.0	S	0.5	S	–				
91181	57						S	S	–	1427	3.5	–	0.4	–	–				
D. sp.																			
33426	58	–	S	S	S	S*	–	–	–	N	N	–	–	N	–				
70338	59						–	–	–	N	0.6	–	T	–	–				
83563	60						S	S	S	–	–	S	T	–	–				
83570	61						–	–	–	N	N	N	N	N	N				

[a]Accession numbers are CPI numbers unless otherwise noted; [b]numbers assigned *Desmanthus* accessions in the trials numbered 1 to 13 reported herein; S, survival; N, not sown, – died. Study I: At Alligator Creek (AC) and Lansdown (LD) with Sa (sandy textured) and Cl (clay) soils. S, survival at 1984 (Cl[1]) and 1985 (Cl[2]). **Best adapted accessions; * moderately well adapted (Chapter 5, this volume). Study II: Results taken from Experiments 11 to 13 at Toorak (T) and Rosebank (R) described herein. S, survival at 4/5/90; N, not sown; – died by 4/5/90. Study III: Results taken from Experiments 7 to 10. W, Woodbine; Ta, Taranaway; B, Blendon; M, Mt Marlow. For Woodbine: S, survival at 12/12/90. Numerical data are the highest scores for mean plant number/20 m^2 × plant vigour rating. G, plant number/m^2 in 2003 (Gardiner *et al.*, 2004). T, trace. Study IV: From Clem and Hall (1994). Accession 65947 given as *D. virgatus* is possibly *D. leptophyllus*. Study V: From Jones and Rees (1997). Ranked 1 to 2 in order of yield and adaptation. Study VI: From Jones and Brandon (1998). Rated 1 to 8 as above. D, died before end of trial. Study VII: Noted as promising in Cook *et al.* (2005) for global environments, not just for the dry tropics. Other accessions so noted but not included herein are: *D. glandulosus:* CPI 90319A; *D. leptophyllus:* CPI nos. 38820, 63473, 92806, 92809, 92818; *D. pernambucanus*: CPI nos. 52401, 30205, 33201, 55718; *D. pubescens:* CPI nos. 92801, 92807, 92811, 92812, 92813; *D. virgatus:* Q 9153, CPI 91326; + *D. virgatus* CPI 78369. This accession from N.W. Argentina is of doubtful taxonomic provenance and may be elsewhere referred to as *D. acuminatus*. It is difficult to distinguish *D. acuminatus* from *D. virgatus* in plants from this region. The species also grades morphologically into *D. tatuhyensis*. *D. acuminatus* was originally considered to be a taxonomic variety of *D. virgatus*.

Table 9.19. Provenance data, where available, for the accessions used in the various experiments and for accessions identified as promising by Cook *et al.* (2005).

Desmanthus species and accession numbers[a]	Plant no.[b]	Country	Latitude (°)	Altitude (m)	Mean annual rainfall (mm)	Soil	Notes
D. bicornutus							
81337	21	Mexico	25 N	10	658	clay	
84508	22		23 N	10	849	clay?	
90857	23		28 N	300	350		
91162	24		18 N	340	800		
D. glandulosus							
90319A		Mexico	28 N	400	370		
D. leptophyllus							
38351	29	Venezuela	9 N	2240	not dry		
55719	30		11 N	80	600	sand	roadside
76053	31	Mexico	20 N	50	1100	sand	
37143	28		19 N	150	not dry		
92655	32	Colombia	4 N	1900	1000		
92746	33		10 N	200	1000		
93809	34	Belize	18 N	10	1400	clay loam	
LD 3	35	?					
LD 4	36	?					
LD 5	37	?					
38820		Venezuela					
63453		Brazil	5 S				
92806		Belize	18 N		>1200		
92809			18 N		>1200		
92818			18 N		>1200		
D. pernambucanus							
49728	39	Brazil	8 S				
40071	38		8 S				
83565	40		9 S				
30205		ex. India[c]					
33201		Guadeloupe	16 N		1800		
52401		ex. Mauritius[d]	20 S		1750		
55718		Brazil	11 S	390	586	sand, acid	roadside
D. pubescens							
76057		Belize	18 N	10	1400	clay	
92800	41		18 N	10	1400	clay	
92802	42	Mexico	19 N	10	1250	clay	
92803	43		18 N	50	1250	clay	
92804	44		19 N	100	1250	sand	
92801			19 N	10	1250	sand	
92807		Belize	18 N	50	>1300	sand	
92811			18 N	50	>1300	clay loam	
92812			18 N	75	>1300		
92813			18 N	50	>1300		
D. virgatus							
67642		Guatemala	15 N	240	550	sand	roadside collection
67643	48		15 N	185	500	clay, pH 8.0	in grass sward
65947	*	Ecuador	2 S	100	500		
78369	**	Argentina	24 S	1000	700		
78373	50		23 S	300	650		cv. Marc
78380			25 S	1330	690		
78382	51		24 S	1000	700	clay loam	

Continued

Table 9.19. Continued.

Desmanthus species and accession numbers[a]	Plant no.[b]	Country	Latitude (°)	Altitude (m)	Mean annual rainfall (mm)	Soil	Notes
79653	52	Cuba	20 N	5	700		
57960	47	Curaçao	12 N	50		alkaline	
78372	49	Argentina	23 N	400	680	clay	
85173	53	Mexico	28 N	400	400	clay	cracking clay soil
85178	54		27 N	220	400	clay pH 8.5	
85182	55		27 N	174	440	clay loam pH 7.5	
90751	56		27 N	1350	320		
91181	57		26 N	280	500		
Q 9153		Peru	4 S				
91326		Mexico	16 N	20	1600	heavy dark clay	
D. sp.							
33426	58	Guyana	7 N		>1200		probably clay
70338	59	Antigua	17 N	5	980	clay	alkaline
83563	60	Costa Rica					experimental station
83570	61	Brazil					Mato Grosso

[a]CPI number (Commonwealth Plant Introduction), unless otherwise noted; [b]see Table 9.2 in text; [c]not native to India; [d]not native to Mauritius; *Given as *D. virgatus* in Clem and Hall (1994), but probably *D. leptophyllus*; **Given as *D. virgatus* but possibly *D. acuminatus*.

10 Recent Development and Commercial Adoption of Legumes for Heavy Clay Soils in Queensland

K.G. Cox[1]
Queensland Department of Agriculture and Fisheries, Mareeba, Australia

Abstract
In previous chapters of this volume, various authors describe the development of herbaceous legumes for pastures on clay soils in Queensland until about the 1980s. Emphasis is on the collection and evaluation of the genus *Desmanthus*, given its relatively recent addition to agriculture and considerable potential for providing useful pasture legumes for clay soils, particularly in the seasonally dry areas of northern Australia. Other genera are also discussed, including early assessments of herbaceous legumes that were later developed for clay soils (*Clitoria, Macroptilium* and *Stylosanthes*). This chapter provides a summary of the development of herbaceous legumes for clay soils in Queensland from these earlier assessments until present.

Beef cattle farming is the principal agricultural enterprise in seasonally dry areas of northern Australia, including large areas of clay soils in Queensland. Sown and naturally occurring grasses provide the key feed resource, and the inclusion of sown legumes can significantly improve live-weight gain and reproductive performance per unit area. Queensland has been the centre of development for legumes for clay soils in tropical and subtropical areas of Australia, mostly through assessing and developing plants held in the Australian Tropical Forages Genetic Resource Collection (ATFGRC) (now a component of the Australia Pastures Genebank (APG)).

The systematic appraisal of genetic material for clay soils was a focus of well-resourced government research up to the early to mid-1990s, but declined thereafter as sown pasture research teams were dismantled and funding to maintain the ATFGRC declined. Cultivar development is now conducted by small government, private enterprise and university research teams that collaborate where possible. In recent studies the use of experienced researcher knowledge and old plant evaluation sites has been particularly valuable for identifying potentially useful material.

Cultivars for long- and short-term pastures on clay soils have been developed to the level of commercial seed production for *Desmanthus* (five cultivars from four species with two cultivars (one composite) in current use), *Clitoria ternatea* (one cultivar), *Macroptilium bracteatum* (two) and *Stylosanthes seabrana* (two). Other potential cultivars of these species are currently in various stages of development. Each species has different production niches depending on climate, clay soil type and grazing strategy. Adoption of these cultivars is occurring but has variously been impeded by limited promotion, mismatch of seed supply and demand, and difficulty establishing legumes in pastures of some key grass species. Recent renewed investment by the Australian Beef Industry has seen revived government research into pasture legumes in Queensland and rejuvenation of the APG.

[1] E-mail: Kendrick.Cox@daf.qld.gov.au

10.1 Introduction

The beef cattle industry is Australia's most valuable livestock industry and the principal agricultural sector in Queensland. In 2008, the beef industry comprised some 27.3 million cattle nationally (44% of which were in Queensland), covered 332 million ha (43% of the Australian landmass) and was valued at AUD 8000 million for meat plus AUD 646 million for cattle exported live (MLA, 2010). By 2011, the beef industry in Queensland was valued at AUD 3280 million, over 30% of the total value of primary industries in the State, with dairy and sheep production valued at AUD 230 million and 197 million, respectively (Queensland Government, 2011).

Sown introduced grasses and legumes are the primary feedbase for beef growing and dairy enterprises in Queensland. Sown pastures were originally used in moderate to high rainfall (700 to 1200+ mm average annual rainfall [aar]) areas, including infertile to fertile soils in the coastal strip between Cairns (17°S) and Brisbane (27°S) and fertile soils in the sub-coastal areas, particularly in central and southern Queensland (Walker and Weston, 1990). Natural native grasslands, utilized *in situ* or oversown with legumes, are most commonly used for beef breeding enterprises and sheep production. These occur mostly in moderate to low rainfall areas on less fertile soils. However, some extensive areas of naturally occurring grasses on heavy clays in low rainfall (500–700 mm aar) environments are an important resource for beef growing enterprises, for example the Mitchell Grass (*Astrebla* spp.) Downs in central and northern inland areas.

The approach of sowing introduced tropical and subtropical grasses and legumes to increase live-weight gain, stocking rate and conception rates has been highly successful in northern Australia, particularly in Queensland where most of the development has occurred. By 1997, it was estimated some 5 million ha had been sown to 'improved' grasses and/or legumes. The net present value of sown pastures to the beef industry was estimated in 1997 to be AUD 712 million, with an annual gross benefit of AUD 80 million (Walker *et al.*, 1997). Based on projected rates of pasture sowing and tree-clearing regulation, Peck *et al.* (2011) recently estimated the current area sown to pastures at 12 million ha, approximately half of the area considered suitable for sown pastures in 1990 (Walker and Weston, 1990). Investment in sown pasture research, mostly conducted by federal (CSIRO) and state (particularly Queensland) research and development organizations, has been positive overall: conservative cost–benefit ratios from economic reviews conducted in the mid- to late 1990s ranged from 1.4 for all sown pasture research over 40 years to 21.8 for more recent sown pasture development in crop/graze systems in southern and central Queensland (Chudleigh and Bramwell, 1996; Scattini and Bartholomew, 1998).

Alkaline clay soils in Queensland, broadly represented by the Mitchell Grass Downs (*Astrebla* spp.; ~23.9 million ha), Brigalow (*Acacia harpophylla*; ~8.5 million ha) and Gidgee (*Acacia cambagei*; ~2.6 million ha) vegetation communities are a key resource for the beef grazing industry in sub-coastal and inland Queensland south of Townsville (19°S). The alkaline, low-fertility grey clays of the Balbirini land system (Normanton grey clays) (~4.9 million ha) in extensive coastal and sub-coastal floodplains in north Queensland (18–20°S) have a lesser, but still significant, role in beef production (Tothill and Gillies, 1992) (Fig. 10. 1).

The development of sown legume pastures on clay soils in Queensland (and northern Australia generally) is less advanced than for other soil types, particularly in coastal and sub-coastal areas. Sown grasses, principally *Cenchrus ciliaris* (syn. *Pennisetum ciliaris*), have been used extensively on some clay soils (Brigalow, Gidgee), while others mostly retain their naturalized grasses (Mitchell and Balbirini grey clays) (Walker *et al.*, 1997). Whereas legume cultivars have been developed for coastal areas (*Aeschynomene*, *Arachis*, *Centrosema*, *Desmodium*, *Macroptilium*, *Neonotonia*, *Stylosanthes* and *Vigna*) and non-clay soils in drier, inland environments (*Centrosema*, *Chamaecrista*, *Stylosanthes*), none (with the exception of the tree legume *Leucaena leucocephala*) had been released for clay soils before 1990 (Walker *et al.*, 1997).

Decreased productivity of grasslands under grazing is an emerging and economically significant issue in Queensland. A reduction in native legume populations under grazing along with invasion of weeds are two proposed causes (Tothill and Gillies, 1992). Pastures based on sown grasses (without substantial amounts of legumes) have also been affected by 'pasture rundown'. This is principally due to a reduction in the

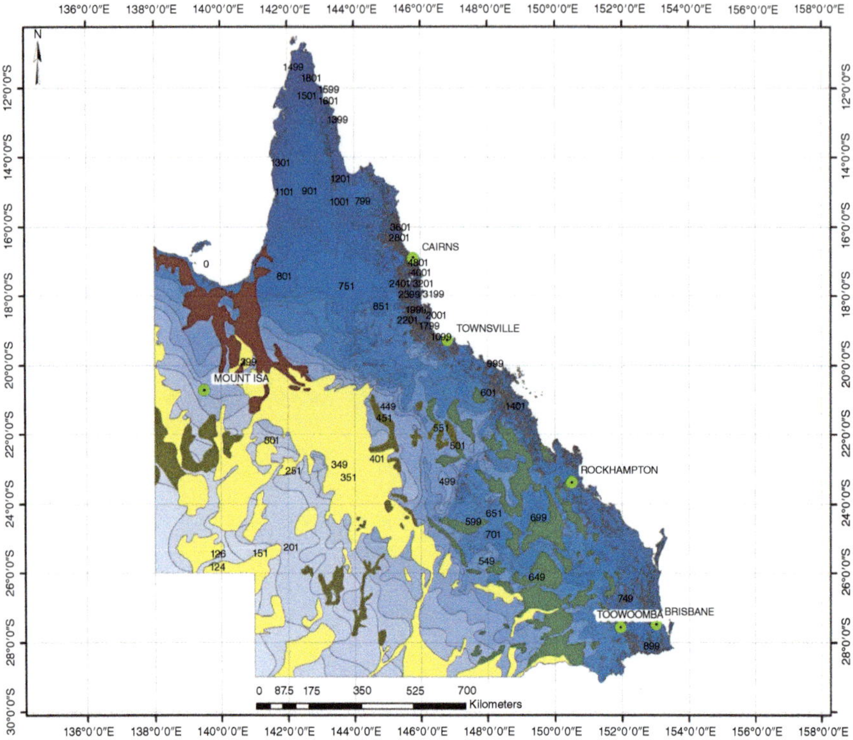

Fig. 10.1. Key vegetation communities based on clay soil areas in Queensland. ▭, Mitchell grassland (*Astrebla* spp.); ▬, Bluegrass (*Dichanthium* sp.)/Browntop (*Eulalia* sp.) grassland; ▬, Gidgee (*Acacia cambagei*) and ▬, Brigalow (*Acacia harpophylla*). Numbers represent mean annual rainfall isohyets based on 1920–1969 records. Image composed by N. Gobius.

amount of plant-available nitrogen, due to incorporation within older ungrazed components of the grass plant (roots and stems). This particularly applies to large, strategically important areas of clay soils in central and southern Queensland that are now dominated by buffel grass. In these areas stocking rates have declined by half over 10–20 years and graziers believe rates are continuing to decline (Peck *et al.*, 2011). A range of mitigation (ripping to mineralize nitrogen, use of fertilizers) and renovation (fallows to mineralize bound nitrogen) strategies were investigated, but the reviewers concluded 'The only long-term solution that provides good economic returns for the beef industry and individual graziers is to establish a range of adapted legumes into the existing grass-only pastures. Establishing legumes into a grass pasture can reclaim 30–50% of the lost production from pasture rundown and improve economic returns' (Peck *et al.*, 2011).

The strategic importance of beef production on clay soils in Queensland, and their decline in productivity, has recently seen renewed investment in the development of legumes for clay soils by the beef grazing industry, through levied funds, and the Federal and State Governments (particularly Queensland). This follows a period of slowed cultivar development when federal and state government sown pasture researchers declined to approximately 10% of their former numbers, in response to which small research teams in universities and private companies began to develop new legume cultivars (Cox, 2013).

This chapter provides an overview of legume development for clay soils in the seasonally dry tropics and subtropical areas of Queensland since about 1990. It focuses on the development of herbaceous (or shrub) legumes for beef production used for permanent pastures, emphasizing the development of *Desmanthus* due to the particular potential of the genus, although other key legumes

(*Clitoria ternatea*, *Stylosanthes seabrana* and others) are also discussed because they also have important roles for pastures on clay soils.

Two other species of commercial significance in Queensland are not covered here; *Leucaena leucocephala* and *Medicago sativa*. It should be noted the browse tree *Leucaena leucocephala*, although relatively expensive to establish, seasonally prone to damage by the psyllid *Heteropsylla cubana* in coastal areas and requiring careful management of stock to degrade mimosine (Cook *et al.*, 2005), is now a robust and economically significant production resource for graziers on suitable clay soils in southern and central Queensland. Shelton and Dalzell (2007) estimated 150,000 ha of a potential 13 million ha had been sown to *L. leucocephala* in central and southern Queensland, including substantial areas of cracking and non-cracking clays. Similarly, the use of *Medicago* spp. is not discussed, although the use of sown and naturalized medics and lucerne are important on clay soils in some areas of southern Queensland.

10.2 Preamble: potential legume germplasm for clay soils based on early research

The accumulation of genetic resources, approaches to describing and prioritizing accessions, and evaluation at clay soil sites in north, central and southern Queensland before the early 1990s are described in Chapters 6–9 (this volume). A consistent theme is the relative dearth of genetic material suitable for permanent pastures on alkaline clay soils held in the Australian Tropical Forages Genetic Resources Collection (ATFGRC) compared with genetic material used to develop cultivars for lighter textured soils (often with an acid reaction). This is a legacy of the past focus on targeted roles for legumes on lighter acid soils, and therefore when the majority of the ATFGRC material was being compiled, collectors targeted legumes on these soils for development (yielding from Brazil, for example, *Stylosanthes*, *Centrosema* and *Macroptilium*). Despite this limitation to cultivar development, many of the genera collected contain species that occur on clay soils and could therefore yield useful cultivars; there are too many to list here, but species of particular relevance to later research occur within the genera *Alysicarpus*, *Clitoria*, *Indigofera* and *Macroptilium* (see below).

In early testing for adaptation on clay soils, a number of useful species were identified from 'best bet' genera (Chapters 6 and 7, this volume) and these were tested on clay soils in a range of environments (Chapters 8 and 9, this volume). Some, however, appeared to have a greater affinity with heavy clay soils, notably *Desmanthus* and *Leucaena*. Interestingly, diploid forms of *Stylosanthes hamata* (for example, from Guatemala) were commonly collected on clay soils, whereas tetraploid types used to develop 'Verano' and 'Amiga' were not. *Stylosanthes seabrana* was also highlighted for use on clay soils, possibly based on later research, and is also discussed below.

Floristic studies (Chapter 7, this volume) highlighted the potential breadth of *Desmanthus* spp. suited to clay soils, and the potential for many to form effective symbiotic relationships with *Bradyrhizobia* associated with *Neptunia*. *Teramnus* and some *S. hamata* were also noted as having a high affinity with clay soils. The authors emphasized the value of collecting trips in countries such as India and Sudan that have areas that better match the vegetation found on clay soils in Australia (*Neptunia*, for example), rather than the areas where the majority of collecting had previously been done. Areas of South America and the mesoamerican region (including the southern USA) were also recommended.

Field assessment of 'promising' material for clay soils was initially conducted during the early 1980s at three sites (coastal and sub-coastal) at approximately 17.5°S in the seasonally dry tropics (Chapter 8, this volume). The genera included *Alysicarpus* (four spp.), *Centrosema* (two), *Clitoria* (one) (including a component of 'Milgarra'), *Desmanthus* (two spp., mostly what was classed as *D. virgatus* including what was to become cv. 'Marc'), *Dolichos* (one), *Haydonia* (one), *Heylandia* (one), *Macroptilium* (one, *M. atropurpureum*), *Strophostyles* (one), *Stylosanthes* (four) and Vigna (one, *V. trilobata*). These were assessed on clay and non-clay soils as spaced plants in small fertilized plots and emphasis was placed on persistence over above-ground biomass production. *Alysicarpus* and *Desmanthus* showed a high affinity with clay soils, as did some *Stylosanthes*, including diploid and tetraploid *S. hamata* types. *Clitoria ternatea*, *Macroptilium atropurpureum* and *Centrosema pubescens* also performed well under light grazing. Interestingly, *Desmanthus* and diploid *S. hamata* showed a capacity to compete with invading *B. pertusa* and *Desmanthus* showed a superior capacity to spread from the original rows.

A broader-scale assessment was conducted between 1988 and 1991, including sites in arid, western areas between 21.4°S and 24.5°S (Chapter 9, this volume). Other sites were included for the purposes of analysis. Non-*Desmanthus* spp. performed poorly, overall, with the exception of *Clitoria ternatea* (including CPI 47187, a future component of 'Milgarra') (see below) and accessions of *Stylosanthes fruticosa*, *S. hamata* and *S. scabra*. The *Desmanthus* accessions, however, generally persisted well. With two exceptions (CPI 67645 and 83563), the best *D. virgatus* accessions were those collected from dry regions in Argentina and Mexico. The following were deemed worthy of further study: *D. virgatus* CPI 78376, 78373 ('Marc'), 75182, 81181 and 83563; *D. bicornutus* CPI 81337.

The authors also noted persistence of *Desmanthus* on clay soils in other tropical and subtropical areas, and the excellent performance of *Clitoria ternatea* and *Leucaena leucocephala* (Clem and Hall, 1994). In the 4-year study on heavy, cracking-clay (vertisol) soils (between 1981 and 1984), the best yielding legumes were: *C. ternatea* CPI 47187, 49664, 49963 and 74821; *D. virgatus* CPI 57960 and (later) 'Marc'; *L. leucocephala* CPI 61227, and *S. scabra* CPI 55868. *Indigofera schimperi* produced the highest first-year biomass, but was not palatable and was removed from the experiment.

10.3 Development of legumes for clay soils during the 1990s

The previous studies indicated that while *Desmanthus* spp. (sub-shrub, *Mimosaceae*) such as *D. virgatus* (Fig. 10.2) provide significant opportunities for developing robust, perennial and palatable legumes for heavy clay soils across a range of environments in Queensland, there are also opportunities within some other species. With the exception of *Leucaena leucocephala* (small tree, *Mimosaceae*), these (and later legumes for clay soils) were mostly from the *Fabaceae*. *Clitoria ternatea* (herbaceous twining, see Fig. 10.3), for example, showed promise as a pasture plant on clay soils. In a review of plant evaluation in northern Australia, Hall (1985) concluded accessions of *C. ternatea* were persistent and productive under grazing on a range of clay soils in north-west Queensland. These included low fertility grey, cracking clay soils (740 mm aar) where it had persisted for 14 years (at the time of reporting), a fertile black-cracking clay under light grazing (5 years, 700 mm aar), clay-loams along river frontages (~ 900 mm aar), and a high-fertility brown cracking clay (400 mm aar, not grazed). Persistence was more variable at higher latitudes, because of succumbing to frost. It has been reported to nodulate well with *Bradyrhizobium* hosted by *Sesbania*, which is frequent in clay soils in Queensland, and is highly palatable to cattle and sheep. Initial perceived potential roles for *C. ternatea* were as a pasture legume to provide early–dry season protein, a high-quality hay or a cover crop (Anon, 1992), but more recently it has been considered useful as a cover crop or grazing ley (Cook *et al.*, 2005).

10.3.1 The first cultivars of legumes for clay soils

Three cultivars of *Desmanthus* and one cultivar of *Clitoria ternatea* were registered as pasture plants for use on clay soils in Queensland in 1991 (Graham *et al.*, 1991a,b; Anon, 1992). The release process was structured, involving peer review of performance data and the development of a 'cultivar release case', and was coordinated with the Queensland Government seed increase programme in North Queensland. The characteristics of the *Desmanthus* cultivars were:

- *Desmanthus virgatus* cv. Marc (CPI 78373), to 50 cm, early-season flowering, collected in Argentina, 300 m asl and 650 mm aar.
- *Desmanthus leptophyllus* cv. Bayamo (CPI 82285), to 150 cm, mid-season flowering, collected in Cuba, 75 m asl, 1200 mm aar.
- *Desmanthus pubescens* cv. Uman (CPI 92803), to 250 cm, late-season flowering, collected in Mexico, 50 m asl, 1250 mm aar.

The three cultivars were licensed to a company under Plant Breeders' Rights (PBR) protection and marketed as a composite line named 'Jaribu' to, theoretically, instil a breadth of adaptation (Cook *et al.*, 1993). Although commercial seed

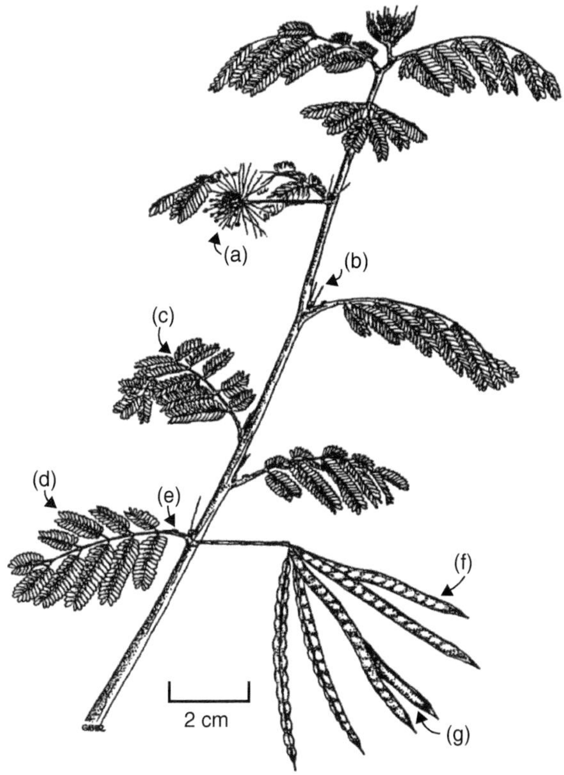

Fig. 10.2. Morphological features of a *Desmanthus* species*: (a) flower, (b) stipules, (c) bipinnate leaves, (d) pinna, (e) gland, (f) pod and (g) dehisced pod. (From Gardiner and Burt, 1995; permission granted by Tropical Grasslands.) * Published as *D. virgatus* before the review of *Desmanthus* by Luckow (1993), the plant depicted is probably *D. leptophyllus* (Chris Gardiner, 2015, personal communication).

production was undertaken for all three cultivars, concerns over potential poor persistence of the later flowering 'Uman' and 'Bayamo', and potential weediness of 'Uman', resulted in 'Marc' being the sole variety available in Queensland by 2000. In 2015, 'Marc' was being grown in north Queensland to service key perceived markets in central and southern Queensland (Fig. 10.4).

Clitoria ternatea cv. Milgarra is a composite line comprising two key components (CPI 47187 and Q 7006) plus 18 other lines included as minor components (Anon, 1992). The composite was grown for three seasons by the Queensland Government in north Queensland to produce 'Milgarra', which is recognized as Q 17476. The composite nature of the cultivar is reflected in variable flower colour (blue and white types), but the cultivar is otherwise quite uniform and believed to be diploid. 'Milgarra' was released as a public cultivar (no plant breeders' protection), and promoted by the Queensland Government in collaboration with seed companies.

Seed increase for cultivar release was mostly conducted at the Queensland Government seed production facility at Walkamin (17°S) in north Queensland, and seed was provided to the seed industry for commercial adoption. The research station is located centrally within the key seed production area for tropical pasture plants in northern Australia, and is characterized by an upland tropical climate with fertile basalt-derived clay soils. 'Marc' was first grown at Walkamin in 1989 and 1990, along with 'Uman', CPI 38351, 79653 and 78382 (also grown earlier in 1986). First seed crops were excellent overall (~900 kg/ha), but seed production was poor in the second year (~200 kg/ha). 'Marc' was initially grown commercially in southern Queensland, but poor production led to moving production to the north Queensland seed production area (10 km from

Fig. 10.3. Morphological features of *Clitoria ternatea*: (From 't Mannetje and Jones, 1992 as re-produced in Humphreys and Partridge, 1995; permission granted by Backhuys Biological Books, backhuys@ euronet.nl.)

Fig. 10.4. A commercial seed crop of 'Marc' *Desmanthus virgatus* grown on the Atherton Tablelands in north Queensland, 2015. Dense seed crops such as these are usually chemically desiccated prior to harvest and harvested with a conventional header fitted with a cut-down front to handle the bulk of wet material.

Walkamin) in 1993. A further seed crop was produced at Walkamin (747 kg/ha direct headed, plus 321 kg/ha vacuum harvested), further demonstrating the potential of these cultivars to produce high yields of seed. The locally produced commercial crop was also successful. 'Uman' seed production was up-scaled in 1992 and 1993, and yields between 590 and 1055 kg/ha were achieved at the Walkamin facility. Commercial production was again undertaken locally, resulting in 1100 kg/ha harvested by direct heading. The first commercial 'Bayamo' seed crops were undertaken about the same time, but further south in the Burdekin district. The above information was sourced from various Queensland Government annual technical reports completed by B.H. English and J.M. Hopkinson prior to 1998 and M.A. Keating and K.G. Cox thereafter.

To provide affordable seeds to graziers, research was undertaken between 1994 and 1998 into optimizing seed crop agronomy of the *Desmanthus* cultivars (Cox, 1998). Key areas of investigation included measuring the development of seed yield and determining plant population densities and pre-harvest techniques to maximize presentation yield for harvest, and the recovery of seed during harvesting. Effective control of weeds in Queensland seed crops was also investigated. The tolerance of 'Marc' *Desmanthus* to a total of 55 pre- and post-emergence applied selective herbicides was tested, initially in pot-based screenings followed by combinations of promising herbicides in field studies (Cox, 2005). Suitable pre- and post-emergence applied herbicides were identified to control most grass (trifluralin, imazethapyr and haloxyfop) and broadleaved (imazethapyr and bentazone) weeds commonly experienced within Queensland seed crops, the notable exception being leguminous weeds.

Seed increase of the composite *Clitoria ternatea* cv. Milgarra was conducted at Walkamin during the mid-1980s. Some seed yields were comparatively low (122–153 kg/ha) due to weed pressure and it was proposed the crop be grown as an annual seed crop. Later production of a range of *C. ternatea* accessions at Walkamin was, however, highly successful (Fig. 10.5). Much of the commercial production was conducted in central Queensland and direct-headed seed yields of 700 kg/ha were readily achieved, enabling a reliable supply of inexpensive seed.

Fig. 10.5. *Clitoria ternatea* accessions grown at Walkamin, north Queensland, prior to field evaluation as ley pastures.

10.4 Refining the types and management of legumes for clay soils

Plant evaluation research concurrent with and subsequent to the release of the first cultivars for clay soils saw a transition towards grazing studies in central and southern Queensland, often with legumes sown in combination with grasses. In a small-plot study (1987 to 1995) on clay soils used for cropping and grazing in southern Queensland, 48 annual, biennial and perennial legumes, including *Alysicarpus rugosus*, *Clitoria ternatea*, *Desmanthus virgatus*, *Lablab purpureus*, *Indigofera schimperi*, *Macroptilium atropurpureum*, *M. bracteatum* and five *Vigna* species, were compared for establishment density and biomass production with, or without companion grasses under light and irregular grazing (Jones and Rees, 1997). *Desmanthus virgatus* CPI 78382, *Indigofera schimperi* CPI 52621 and 69495 and *Glycine latifolia* CQ 3368 (one site) were the best performing perennial legumes.

Importantly, these legumes co-existed with grasses over the 5–7-year period of the experiments, indicating good potential as long-term pastures. Promising legumes for short-term pastures were also identified, including representatives from three *Macroptilium* spp. (notably *Macroptilium bracteatum*), *Macrotyloma uniflorum* cv. Leichhardt and four *Vigna* spp.

The capacity for eight *Desmanthus* accessions, including 'Marc' and 'Uman' sown in companion with *Chloris gayana* and *Panicum maximum* var. *trichoglume* to tolerate grazing under five stocking rates (based on presentation yield), was assessed over 7 years at one Brigalow clay soil site (25.4°S, 710 mm aar) over three dry years (Jones and Brandon, 1998). The best varieties were 'Marc' and CPI 78382, similar plants from Argentina. They comprised ~10% of total pasture biomass over most of the study (but up to 20%), and were believed to persist through early flowering, thereby seeding during grazing (late flowering lines steadily decreased in number). The authors therefore proposed that the mid- and late-flowering components of the 'Jaribu' mixture would slowly decline under a similar grazing regime, leaving cv. Marc as the dominant cultivar.

10.5 The Legumes for Clay Soils project

A substantial multi-agency (CSIRO, University of Queensland and Queensland Government) project was co-funded by the beef industry in 1992 to further the development of legumes for clay soils, including identifying new types, developing robust methods for pasture establishment, measuring live-weight gain and generally promoting the adoption of legumes for clay soils (Clem and Jones, 1996). This complemented other well-resourced, industry-funded collaborative projects used to identify promising genetic material held within the ATFGRC (Pengelly and Staples, 1996), and legumes for use on lighter-textured soils should *Stylosanthes* spp. be decimated by a new strain of anthracnose (*Colletotrichum gloeosporioides*) (Bishop and Hilder, 2005).

The research and promotion activities were principally focused on supplying legumes for beef finishing operations in grazing and crop/grazing systems on clay soils between Roma (23°S) and Middlemount (27°S) and included 16 research sites. Emphasis was placed on testing and progressing cultivars (where present), and there was a strong focus on *Desmanthus* cultivars and promising accessions in early grazing experiments. However, promising accessions of *Clitoria ternatea*, *Glycine latifolia*, *Indigofera schimperi*, *Lablab purpureus* and *Stylosanthes seabrana*, among others, were also evaluated in on-property assessments (under grazing) following good performance in multiple species assessments conducted at the onset of the project (see below). The assessments occurred over a 'run' of drought years, providing a rigorous 'test' of the capacity of legumes to establish and persist in pastures.

Animal live-weight gain was evaluated in a replicated grazing experiment at the Brian Pastures Research Station near Gayndah (Clem and Jones, 1996). *Desmanthus virgatus* cv. Marc was oversown into pasture originally sown to CPI 78382 and *Setaria incrassata*, but *Desmanthus* establishment was poor. Marginal increases in live-weight gain due to the presence of *Desmanthus* were initially attributed to this. However, the *Desmanthus* persisted and seeded under extremely dry conditions, even when grazed all year and down to the crowns. An extension to the project (1996–1998) saw excellent recruitment of new plants from fallen seed and a benefit to live-weight gain

(0.47 kg/head/day) compared with grass-only pasture (0.38 kg/head/day) even though *Desmanthus* still only comprised 10% of total pasture yield. Analysis of carbon isotope contents in dung confirmed *Desmanthus* was a key component of the 'non-grass' portion of the diet. At three other clay soils sites improvements in live-weight gain due to the presence of *Desmanthus* over those sown only to grass, including paddocks sown to *Cenchrus ciliaris*, were not measured. Growth rates at these sites by 1998 in the grass-only pastures were, however, relatively high (0.43–0.63 kg/head/year), indicating they were yet to 'run-down'. *Desmanthus* content was also low (6–18%), but there was also strong evidence of seeding (through counts of seeds in soil cores) and recruitment of new plants indicating the *Desmanthus* accessions sown would persist in pastures under grazing, thereby potentially preventing or reducing the impact of pasture run-down over time.

The effectiveness of nitrogen fixation by *Desmanthus* was also assessed in collaboration with University of Queensland researchers. Nitrogen isotope studies on samples collected from the grazing studies showed the nitrogen in *Desmanthus* was initially sourced from mineralization (presumably from renovation), but a few years after establishment was derived mostly from nitrogen fixation. Further studies revealed *Desmanthus* could form effective symbiotic relationships with native *Bradyrhizobia*, but not on all soil types, so it was considered best practice to inoculate seed with strain CB 3126 when sowing (Clem and Jones, 1996). Survival of the strain was shown to be significantly reduced beyond 14 days if stored at temperatures above 35°C and beyond 3–7 days at temperatures above 45°C. Such temperatures commonly occur under normal seed-bed conditions for summer-sown *Desmanthus* in Queensland. This, in combination with sporadic rainfall, presents a challenge for achieving effective nodulation on soil types where effective native *Bradyrhizobia* are not present.

Nutritional studies were also conducted to investigate causes of *Desmanthus* 'yellowing', which had been observed in pastures and seed crops during the mid-1980s (Clem and Jones, 1996). The application of sulphur was shown to improve plant growth and colour initially in pot-based experiments and later in field experiments. Fertilizer application was recommended in soils with less than 5 mg/kg extractable sulphur. Applications in the order of 30–60 kg/ha were found to give maximal responses. Molybdenum was also shown to benefit plant growth in the pot-scale study, and studies at Walkamin revealed responses to phosphorous and liming in seed crops (high biomass) grown on a slightly acid basalt-derived clay with an extractable P of 40 mg/kg.

To date, there appear to be no serious pests or diseases associated with *Desmanthus* grown in pastures in Queensland. Occasional defoliation caused by a native psyllid (*Acizzia* sp. (*Psyllidae*)), which are commonly found on native or naturalized species including *Acacia*, *Dodonaea*, *Apophyllum*, *Hakea*, *Eremocitrus* and *Leptospermum* (Carver *et al.*, 1991), has been observed in southern Queensland, but has not been a problem elsewhere. This was first noticed in sub-coastal areas of south-east Queensland during the 1980s (Loch, 1996, personal communication, as cited by Cox, 1998) and during the Legumes for Clay Soils project near Mundubbera during the early 1990s (Clem and Jones, 1996). In other areas of south-east Queensland (Kilkivan and Gayndah), the psyllid caused extensive damage to growing tips (young leaves) on 'Marc', 'Bayamo' and 'Uman' grown in seed production experiments (Cox, 1998). Control using dimethoate was required between March and June. To date, serious damage seems limited to mono-cultures, such as those used for seed crops, rather than a mixed grass/legume pasture. Observations by Clem and Jones (1996) and Cox (1998) indicate there may be some genetic basis to susceptibility to the psyllids (some accessions and species appeared more badly affected than others), but this is yet to be tested.

Overall, *Desmanthus* had been confirmed as a key genus for developing legumes for clay soils, albeit with a limited range of genetic material. Accessions, particularly within *D. virgatus*, had persisted in grazing experiments and were known to produce high-quality fodder. Seed crop agronomy studies had been undertaken and the transition made to commercial seed production in north Queensland by specialist seed growers. These looked particularly well adapted to central and southern Queensland. One variety of *Clitoria ternatea* had also been released, with a potential role as a pasture plant in central and northern Queensland, and as a ley legume or 'fodder bank'. However, some other legume genera remained of interest: *Alysicarpus*, *Indigofera*, *Macroptilium* and *Stylosanthes*. Of these, *Stylosanthes seabrana* was to prove to have considerable potential as a pasture plant on clay soils, particularly in central and southern Queensland.

10.6 *Stylosanthes seabrana*, an emerging legume for clay soils

Stylosanthes (sub-shrub, *Fabaceae*) has long been considered a key genus for developing legumes for pasture systems in Queensland (Burt and Miller, 1975). Cultivars currently traded in Queensland include those of *S. scabra* ('Seca' and 'Siran' – perennial pasture plants suited to the seasonally dry tropics and subtropics); *S. hamata* ('Verano' and 'Amiga' – annual or biennial pasture legumes also suited to the seasonally dry tropics and subtropics); *S. guianensis* var. *intermedia* ('Oxley' – a perennial pasture plant suited to the subtropics in sub-coastal areas) and *S. guianensis* var. *guianensis* (ATF 3308 and 3309 traded under various trademarks – anthracnose-resistant long- or short-term pasture legumes for tropical moderate to high rainfall environments). The above are all best adapted to light-textured soils with a neutral or acid reaction, and none suited to heavy alkaline clay soils, particularly the black-cracking clays (Cook *et al.*, 2005).

In 1988 (although there are unofficial reports of detection some 10 years earlier), an unusual perennial *Stylosanthes* was found contaminating a *S. scabra* accession grown at Lansdowne, near Townsville. It was intermediate in morphology and growth between *S. hamata* and *S. scabra* and was classified *S.* aff. *scabra* (as cited by Jansen and Edye, 1996). Seed of similar material was imported and plant morphology and growth compared with a range of *S. hamata*, *S. scabra* and *S. humilis* accessions in a spaced-plant experiment on a fertile alluvial soil at Lansdowne (Jansen and Edye, 1996). Separations were based mainly on growth habit, perenniality, seed colour, the presence of leaf bristles, stipule horn lateral bristles, stem viscidity and inflorescence bristle viscidity (Fig. 10.6). Five groups, broadly homogenous by species, were identified, although more than one species was represented in four of them. *Stylosanthes* aff. *scabra* lines were described as having thick, relatively woody stems, elongate glabrous (or near to) leaves, straight bristles (usually viscid) on the margins and back of the

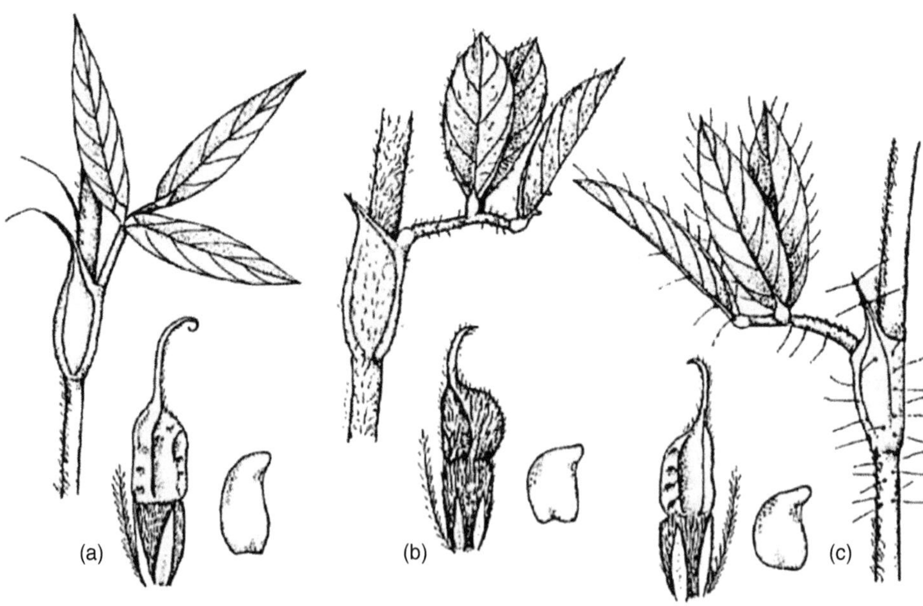

Fig. 10.6. Morphological features of the stem, node, leaf, seed pod and seed of (a) *Stylosanthes hamata* cv. Verano (tetraploid), (b) *S. scabra* cv. Seca and (c) *S. seabrana* (syn. *S.* sp. aff. *scabra*) CPI 115993). (From Jansen and Edye, 1996, permission granted by CSIRO (copyright owners) and CSIRO Publishing (original publishers) http://www.publish.csiro.au/nid/40/paper/AR9960985.htm.)

leaf, the stipule, the node (sometimes on the internode) and inflorescence. The stipule horn had no terminal bristle and strong lateral bristles. All originated from Brazil. Two key groups were identified: 'a prostrate anthracnose-susceptible group and an erect, high-yielding anthracnose-resistant group'. The authors suggested that cultivars would be developed from the latter: members of this group included CPI 92476, 92838B, 105546B, 110361, 115994 and 11995.

Eleven *S. seabrana* accessions were assessed on heavy clay soils at five sites in subtropical Queensland (24.5°S–27.2°S) between 1991 and 1996 and compared with 'Seca' *S. scabra* as a benchmark (Edye *et al.*, 1998). The accessions included those from groups 2 and 3 in the phenology study, and all originated from Brazil (Jansen and Edye, 1996). Two series of experiments were completed: in the first, small plots of scarified and inoculated (CB 2126 and 1650) seeds were established without a companion grass (existing types allowed to regenerate); the second included *Cenchrus ciliaris* sown with the legume seeds inoculated with *Bradyrhizobium* strains CB 2126 and 3053. Plant populations and biomass production were assessed under dry-season grazing during a run of 'dry' years. *Stylosanthes seabrana* biomass yields up to 3510 kg/ha were recorded, and plant densities were considerably higher than for *S. scabra* 'Seca'. Plant biomass of *Stylosanthes seabrana* accessions CPI 115994, 115995, 110361 and 115994 were almost twice that of 'Seca' when not grown with buffel (various sites). When grown with *Cenchrus ciliaris* the best accessions based on establishment, plant populations and biomass yield over 5 years were CPI 92838B, 110370C, 115994, 110361 and 104710 (first to fifth), and all outperformed *S. scabra* despite not having effective nodulation in all instances. Accessions were short-listed for commercial release beginning with performance across high- and low-yielding sites (CPI 92838B, 110370C, 115994 and 110361), reported resistance to anthracnose (CPI 92838B, 110361 and 115994) and seed yield (CPI 110361 and 92838B). New *Bradyrhizobium* strains were isolated from material sourced in 1994 to overcome deficiencies of previous strains (Date *et al.*, 1996).

Accessions CPI 110361 and 92838B (both from phenology group 3) were released under plant breeders' rights in 1996 as cultivars 'Unica' and 'Primar', respectively (Anon, 1996). The varieties were selected to provide pasture options for clay soils in subtropical ('Primar' – cold tolerant and earlier flowering) and more tropical ('Unica' – later flowering) environments (Cook *et al.*, 2005). Seed production was undertaken at Walkamin in 1995 and 1996 to support commercial adoption (English and Hopkinson, internal seed production reports). Seed yields were excellent, ranging from 400 to 883 kg/ha for 'Primar' and 504 to 1390 kg/ha for 'Unica'. The first commercial seed crops were undertaken under licence in north Queensland (Fig. 10.7).

10.7 Other legumes for clay soils

Although the *Desmanthus* spp., *Stylosanthes seabrana* and, to a lesser degree, *Clitoria ternatea* were arguably the best suited for use as permanent pastures on clay soils, some other legumes have been developed as potential options for pastures on clay soils. These were assessed under grazing, along with the previously mentioned *Desmanthus* and *Stylosanthes seabrana* accessions and cultivars, as a component of the Legumes for Clay Soils project (Clem and Jones, 1996). The legume species assessed included *Aeschynomene* (one species), *Alysicarpus* (two), *Arachis* (two), *Atylosia* (one), *Cajanus* (one), *Centrosema* (two), *Clitoria* (one), *Desmanthus* (three, mostly *D. virgatus*), *Desmodium* (three), *Galactia* (one, possibly three), *Glycine* (one), *Heylandia* (one), *Indigofera* (one), *Lablab* (one), *Macroptilium* (six), *Macrotyloma* (three), *Medicago* (one), *Neptunia* (four, possibly five), *Psoralea* (two), *Rhynchosia* (three), *Stylosanthes* (three, mostly *S. seabrana*), *Teramnus* (one) and *Vigna* (six). Assessments were completed at three neutral to alkaline heavy clay sites located between 23.3°S and 25.4°S over a run of dry years (400–600 mm per annum). Selected groups of legumes were sown in replicated small plots with and without a grass, and the remainder sown with a grass adapted to the local area (either *Cenchrus ciliaris* or *Dichanthium sericeum*). The best available *Bradyrhizobium* inoculum was used.

Establishment was variable and overall smaller seeded species established better than large-seeded ones. Most *Desmanthus* accessions, *Glycine latifolia*, *Indigofera schimperi* and *Stylosanthes seabrana* were the most persistent overall, a requisite for permanent pastures. *Indigofera*

Fig. 10.7. A dense seed crop of *Stylosanthes seabrana* 'Primar' grown near Mareeba in north Queensland, June 2015. (Photo: Kendrick Cox.)

schimperi biomass yields were highest overall (1000–2000 kg DM/ha), although *S. seabrana* and some early-flowering *Desmanthus* accessions (TQ 90 and CPI 40071, along with 'Marc') also performed well. Competition from companion grasses appeared to suppress the yields of *Glycine latifolia* and *Clitoria ternatea* and their plant populations declined at some sites, although they showed the potential to recruit from seed. Overall, these species were considered the best suited to permanent pastures. However, concerns about the acceptability of *I. schimperi* to cattle resulted in its removal from experimental plots, and the species was later removed from cultivar pre-release and targeted for eradication from plant evaluation sites (Bishop, 2003). It has since proven to be a persistent contaminant of pastures (Cox, 2014).

A number of potentially useful legumes for ley pastures were also identified, including *Macroptilium bracteatum* CPI 27404, which produced up to 2400 kg DM/ha in the second year after sowing. *Macroptilium bracteatum* CPI 55755 also performed well at one site and 'Milgarra' *Clitoria ternatea* performed well despite low initial plant density.

The best-performing 'advanced lines' were included in on-farm assessments under grazing at 12 clay-soil sites between 22.3°S and 26.4°S (Clem and Jones, 1996, 1998). The sites were sown in either 1994 or 1996, the latter sowing including 23 accessions across 23 species (including the three *Desmanthus* cultivars). The on-farm assessments confirmed the performance of the legumes from the previous comparisons. Legumes with high first-year biomass included *Lablab purpureus*, *Vigna trilobata* CPI 13671, 'Uman' *Desmanthus pubescens*, *Macroptilium bracteatum* CPI 27404, 'Unica' *Stylosanthes seabrana* and either 'Aztec' or 'Siratro' *Macroptilium atropurpureum*. Of the palatable herbaceous legumes, the highest yielding in the second year (and potentially suited to ley systems) were 'Siratro' *M. atropurpureum*, *M. bracteatum* CPI 27404, 'Milgarra' *C. ternatea*, the three *Desmanthus* cultivars and 'Unica' *S. seabrana*. By the third year the best legumes were 'Milgarra' *C. ternatea* and the *Desmanthus* and *S. seabrana* cultivars. Importantly, the broad range of sites enabled some trends in adaptation to be identified. 'Milgarra' *C. ternatea* was found to be best suited to lighter textured 'Downs' soils where grass competition was less, whereas 'Marc' *Desmanthus* performed best on the heavier Brigalow soils. 'Unica' *S. seabrana* tended to yield more on Downs soils. 'Marc' *Desmanthus* was more persistent than the later flowering cultivars 'Bayamo' and 'Uman' because it set more seed and recruited more seedlings. *Stylosanthes seabrana* also set seed well and recruited new plants from seed.

Leaf and stem samples of the best performing legumes were collected from four of the evaluation sites and analysed for nutritive quality as it relates to ruminants. The percentage of leaf was consistently lowest in the *Stylosanthes* spp. Leaf tip percentage nitrogen varied widely, with 'Milgarra' *C. ternatea* the best (4.5%; separate tests found some as high as 6% N for fertilized plants; T. Hall, personal communication), followed by the three *Desmanthus* cultivars, 'Unica' *S. seabrana* and 'Aztec' *M. atropurpureum* (3.5–4%). Values for *M. bracteatum* tended to be lower (3%). Nitrogen content of stems was considerably lower for all legumes, ranging from 1.5% to 2.2%. Acid digestible fibre values of leaves or stems were similar for the best performing legumes.

Activities to refine seed production systems were undertaken by the Queensland Government in order to accelerate the adoption of some of these promising legumes. Twenty-eight post-emergence herbicides were tested on the promising legumes *Clitoria ternatea*, *Glycine latifolia*, *Macroptilium bracteatum*, *Indigofera schimperi* (since rejected) and *Vigna trilobata* to identify options to control weeds in established seed crops and to control the legumes should they become a 'plant out of place' (Loch and Harvey, 1997). 'Safe' options were identified to potentially control a wide range of grass (e.g. fluazifop, haloxyfop, quizalofop and sethoxydim) and broadleaved weeds (e.g. imazethapyr and pyridate) in seed crops when applied to young (1–3 true leaves) plants, and some were found to provide useful selective control between different legumes (e.g. acifluorfen, bentazone, imazaquin).

10.8 Ley pastures on clay soils in Queensland

Although ley farming (rotating cropping with pastures) had long been investigated in northern Australia, increased concern about declining soil fertility (particularly nitrogen) in the 1990s in northern Australian cropping lands saw calls by industry representatives to undertake research into the development of ley legumes, particularly in central (Sykes, 1997) and southern inland Queensland (Weston *et al.*, 2000). The results of the plant evaluation programmes presented above highlighted the potential roles for *Lablab purpureus*, *Clitoria ternatea*, *Desmanthus* spp. and *Stylosanthes seabrana*, among others, in fodder production and farming systems ranging from short-term leys or manures through to longer term pastures.

By 2000, approximately 40,000 ha of annual cultivars of *Lablab purpureus* had been established to provide high yields (up to 4000 kg/ha) of high-quality, short-term biomass over summer on clay 'cropping' soils (Pengelly and Conway, 2000). Attempts to commercialize a perennial lablab ('Endurance') have not been successful. Of the longer term grazing leys, approximately 6000 ha of 'Primar' or 'Unica' *S. seabrana* had been established in the clay-cropping zone of Queensland (based on seed sales), principally within permanent pastures in southern Queensland, and 12,000–20,000 ha of 'Milgarra' *C. ternatea* in central Queensland where it is mostly used as a ley legume. This compared to about 3000 ha for the three *Desmanthus* cultivars that were sown principally as permanent pastures in the clay cropping zone (Pengelly and Conway, 2000).

Macroptilium bracteatum (herbaceous twining, *Fabaceae*), a previously relatively undeveloped species (Fig. 10.8), also showed promise as a short-term ley pasture on clay soils. In an evaluation of potential ley legumes on clay soils in central and southern Queensland that included *Alysicarpus*, *Centrosema*, *Clitoria*, *Desmanthus*, *Lablab*, *Macroptilium*, *Macrotyloma* and *Vigna* spp., *M. bracteatum* was the outstanding species, particularly CPI 55769 (Pengelly and Conway, 1998). Other accessions included CPI 27404, 37608 and 68892. *Desmanthus* and 'Milgarra' *C. ternatea* also performed well. *Desmanthus* accession CPI 27404 was shown in glasshouse experiments to be more persistent and regrow better after repeated cutting than 'Highworth' *Lablab purpureus* (an annual form) and *Macrotyloma daltonii* CPI 60303, legumes considered at the time to also have potential as short-term ley legumes (Dalzell *et al.*, 1997).

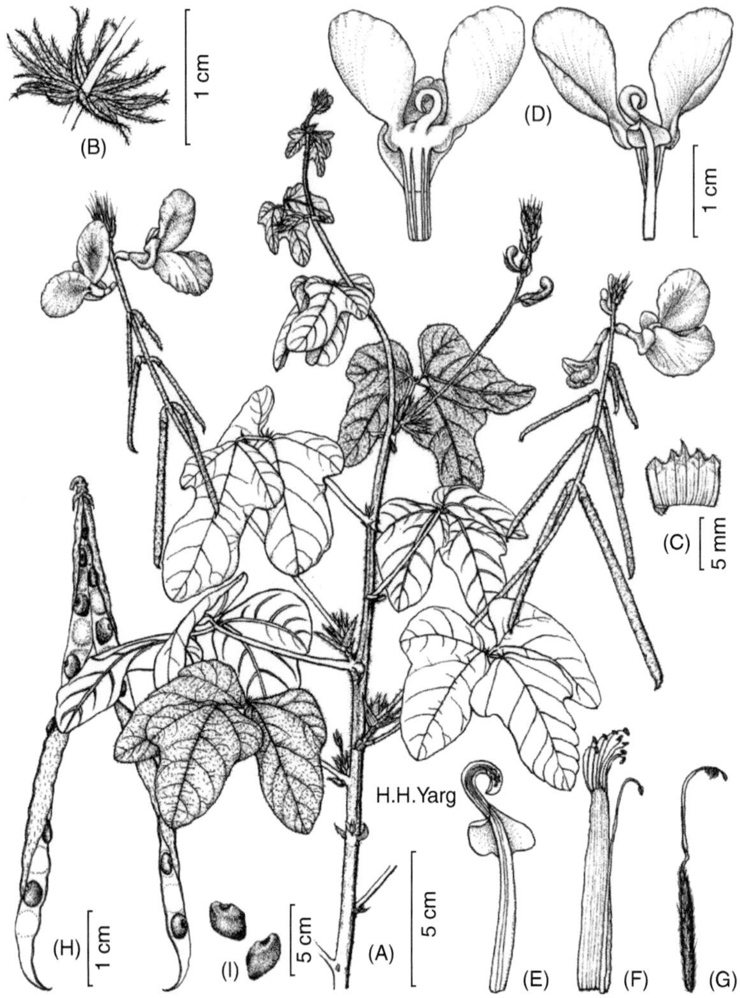

Fig. 10.8. Morphological features of *Macroptilium bracteatum*: (A) habit, (B) bract, (C) calyx, (D) corolla, (E) keel with stamens, (F) stamens, (G) pistil, (H) pod and (I) seeds. (From Wang and Chen, 2012.)

Two *M. bracteatum* cultivars, both originating from Brazil, were released in Queensland under plant breeders' rights protection in 2000: 'Cadarga' (CPI 55769), an erect form evaluated across a range of sites in the subtropics, but partially susceptible to bean mosaic virus, and 'Juanita' (CPI 68892), a decumbent form less affected by bean mosaic virus and generally more persistent than 'Cadarga' (Cook *et al.*, 2005). Pre-commercial seed crops were successfully grown by the Queensland Government at Walkamin in 1998, with 'Cadarga' and 'Juanita' yielding 941 and 1142 kg cleaned seed/ha respectively using a combination of direct heading and vacuum harvesting (recovering seed from the soil) (Keating and Cox, internal seed production reports). Commercial seed production has since been conducted successfully in north Queensland (Fig. 10.9).

10.9 Legumes for clay soils beyond 2000

By 2000, varieties of *Desmanthus virgatus* ('Marc'), *D. pubescens* ('Uman') and *D. leptophyllus* ('Bayamo') had been released, commercial seed production begun and adoption by graziers undertaken. These cultivars were principally used for permanent

Fig. 10.9. A vigorous seed crop of *Macroptilium bracteatum* 'Juanita' grown at Walkamin in north Queensland, June 2015. (Photo: Kendrick Cox.)

pastures on heavy clay soils in central and southern Queensland. 'Primar' and 'Unica' *Stylosanthes seabrana* were also available for the development of permanent pastures in central and southern Queensland, with a complementary role given to the *Desmanthus* cultivars as they had also performed well on lighter clay soils. 'Milgarra' *Clitoria ternatea* and 'Cadarga' and 'Juanita' *Macroptilium bracteatum* were available, principally for use as short-term pastures in crop/graze systems or feed banks, but they could potentially be used in permanent pastures if stock was managed carefully to avoid being grazed out. 'Milgarra' also had a potential role on the grey clay soils in northern Queensland.

At this time, Pengelly and Conway (2000) identified key areas for future legume development in the clay cropping zone that included developing cold-tolerant cultivars of *Clitoria ternatea* (and *Leucaena leucocephala*), establishing legumes without excessively mineralizing nitrogen (during cultivation), refining grass/legume management under grazing and improving the survival of rhizobia for the available cultivars of *Desmanthus* and *Stylosanthes seabrana*. Jones and Clem (1997) suggested one approach to overcoming the latter problem (and other deficiencies in current cultivars) is to ensure the supply of adequate genetic resources with which to develop new cultivars.

Although a number of promising *Desmanthus* accessions held within the ATFGRC had previously been identified as potentially useful legumes for clay (and other) soils in Queensland (e.g. *D. virgatus* CPI 85178 and Q 9153, *D. leptophyllus* TQ 90 and CPI 38351) (Cook *et al.*, 2005), no further cultivars were advanced to commercial production over the next decade. An exception was the 'Marc-like' *D. virgatus* CPI 78382, which was briefly commercially grown for seed, but not developed further by the seed company. Similarly, no new varieties of *C. ternatea*, *M. bracteatum* or *S. seabrana* were advanced to commercial production.

The decline in new cultivar development occurred as state and federal governments decreased spending on sown pasture research, and universities and seed companies, which had previously not been at the forefront of cultivar development, were yet to begin to fill the vacuum. The dismantling of government sown pasture research teams arose with changes to the 'models' of research funding whereupon there was a greater emphasis on sourcing revenue from industry or private enterprise. Industry investment in sown pastures declined over this time and projected returns from royalties of PBR cultivars were generally not realized. The PBR model, which was expected to provide the funding for future development of cultivars by government research agencies, saw the opposite occur as seed companies had little capacity to undertake sown pasture research and there was insufficient market to justify investment in the research (Illing, 2009). In addition, availability of PBR cultivars was occasionally inadequate (e.g. *Desmanthus*) as companies struggled to match supply with demand, or seed was contaminated by off-types (e.g. *Stylosanthes seabrana*). These factors, combined with a decline in government pasture extension staff, impeded adoption of the legumes by graziers.

This period also saw stagnation in the maintenance and use of the ATFGRC as governments and levied research and development corporations struggled to fund the operation of the collection. The collection was reduced from ~26,000 accessions to ~10,000 (614 species) of warm season grasses and ~2700 accessions (255 species) of legumes (Cox, 2013). Regeneration of the ailing collection was only conducted through extraordinary funding opportunities, whereupon ~1000 accessions were regenerated and described in north Queensland (Cox *et al.*, 2009). *Tropical Grasslands*, Australia's premier journal for publication of research on sown pastures, was disbanded as numbers of contributors and editorial staff declined, and by 2010 there were fewer than 20 sown pastures research and technical staff employed by the Queensland Government loosely allied to a few researchers in CSIRO (the premier federal research organization) and two universities. In response to declining capacity (globally) in sown pasture research in the tropics and subtropics, the *'Tropical Forages'* website (Cook *et al.*, 2005) was developed to capture knowledge before it was lost and provide a research tool to better enable future sown pasture research and extension. An Australian version followed shortly afterwards that targeted extension and adoption (Brown *et al.*, 2008).

Some momentum in sown pasture research and cultivar development was maintained by the small research teams (Cox, 2013). 'Elite' grass and legume cultivars were imported, seed increase was undertaken by the Queensland Government and the cultivars were commercialized as public and PBR-protected cultivars. However, most of them were not primarily suited to heavy clay soils but were better suited to other growing environments more similar to areas where initial selection and development had occurred (mainly in South America). Researcher (often retired) experience was also used by government agencies and seed companies to identify potentially useful accessions from the ATFGRC, including *Desmanthus*, *Lablab* and *Macroptilium* spp. Seed increase was undertaken in readiness for plant evaluation and commercial production if deemed appropriate. These included some of the *Desmanthus* accessions previously identified as having good potential as permanent pasture plants in Queensland, for example Q 9153 and TQ 90.

Old plant evaluation sites proved a particularly rich source of legume germplasm as research teams sought efficient ways to identify genuinely useful plants for regions where no, or very few, productive and persistent legumes were available. Two key examples relate to legume development for seasonally dry regions of Queensland. One, conducted during the early to mid-2000s, principally targeted clay soils from north to southern Queensland, and the other, conducted from approximately 2010, involved *Stylosanthes* as well and targeted light-textured soils in southern inland Queensland. The former work championed by university staff is particularly important within the context of this chapter because it relates to the persistence of *Desmanthus* on clay soils in semi-arid north Queensland (Mitchell Grass Bioregion) (Gardiner *et al.*, 2004) and the comparative testing of *Desmanthus* spp. with *C. ternatea*, *M. bracteatum* and *S. seabrana* on 'Gidgee' clay soils in north Queensland (Gardiner, 2003). The former work is discussed in a separate case study in Chapter 11 (this volume); the latter is briefly discussed here to illustrate the potential benefit of capitalizing on previous evaluation work.

The 'reselection' of legumes from plant evaluation sites in southern Queensland provides an excellent example of the value of building on previous evaluation work and maintaining a research record to guide subsequent generations of researchers in the development of potentially useful cultivars. In this instance, Queensland government researchers surveyed legume populations at 33 old plant evaluation sites some 10–30 years after they were originally sown (Peck *et al.*, 2012). The sites included clay soil sites from inland central and southern Queensland, and loam and sandy soil sites in southern inland Queensland. This followed a 'run' of dry years and variable grazing pressure, providing the opportunity to identify lines that were persistent under typical grazing management. *Desmanthus* lines were found to persist well on clay soils, but poorly on sandy soils and basalt-derived clay soils in central Queensland. *Stylosanthes seabrana* performed well across a range of soil types and regions, including basalt-derived clay soils, whereas *Clitoria ternatea* persisted best on basalt-derived vertisols in central Queensland and appeared to benefit more from periodic grazing than the others. *Macroptilium bracteatum* did not perform well as a permanent pasture. As an extension to the survey, seed was collected from sites with light-textured soils in frost-prone areas (where there were no commercially available options for graziers) from approximately 30 groups of phenotypically similar plants of *Stylosanthes scabra* and *S. seabrana*, and a few of *Desmanthus*. Seed increase for plant evaluation was undertaken in north Queensland, enabling assessment of seed yield potential in a typical seed production environment and susceptibility to anthracnose. Evaluation of selected lines in southern Queensland is planned beyond 2015.

Testing new grazing systems using the recently developed ley legumes for clay soils was also assessed during this period. Whitbread *et al.* (2005) assessed the biomass production and nitrogen contribution of recently released cultivars of *Lablab purpureus* ('Endurance'), *Macroptilium bracteatum* ('Cadarga') and *Clitoria ternatea* ('Milgarra') over 3 years at three clay soil sites in southern inland Queensland. Short-period grazing was used. 'Endurance' yielded highest in the first two years (to 6000–7000 kg DM/ha) but declined as plants died in year three. 'Cadarga' and 'Milgarra', however, produced higher yields in years two and three (in the order of 4000–5000 kg/ha). 'Milgarra' plants showed the highest level of persistence, overall. After three seasons, the system showed the potential to contribute in the order of 100–190 kg nitrate N/ha to a depth of 1.5 m (depending on site management).

Collaborative international research between Australian researchers and partner countries through Australian government-funded programmes served to further research into legumes for clay soils, particularly within farming systems. A pertinent example was the 'Tropical Legumes for Sustainable Farming Systems in Southern Africa and Australia' project, undertaken by federal and state researchers with co-funding from the Australian Centre for International Agricultural Research (Whitbread and Pengelly, 2004). For Queensland, this included plant evaluation of a wide range of legumes on moderately fertile clay soils in sub-coastal southern Queensland (Clem and Cook, 2004). The work confirmed the role of *Macroptilium bracteatum* as a short-term, moderately persistent grazing ley and demonstrated the potential of a range of *Desmanthus* lines, including *D. glandulosus* CPI 90319A, *D. virgatus* CPI 91491, 91326 and ATF 3009, plus a range of *D. leptophyllus* including some that out-yielded 'Bayamo' (although results were not consistent across the two sites). *Stylosanthes seabrana* 'Primar' and 'Unica' were also confirmed to be productive at one of the sites.

The programme also provided the opportunity to record animal live-weight gain data on ley pastures on clay soils (Clem, 2004). On a brown to dark self-mulching clay (basaltic origin) in sub-coastal southern Queensland, 'Milgarra' *Clitoria ternatea* and *Macroptilium bracteatum* CPI 27404 grown as mono-cultures were shown to produce live-weight gains in the order of 0.40–0.65 kg/ha/day at moderate stocking rates and the inclusion of *S. seabrana* and *C. ternatea* in grass pasture increased production by 30–70 kg/ha over 5 years compared with grass-only pasture.

Success of the various legumes, particularly *Desmanthus*, in central and southern Queensland saw interest in sowing legumes in clay soils move further south into crop/graze areas in northern New South Wales (summer rainfall, but winter frosts). State Government researchers undertook two experiments on vertisols near Moree (29.4°S) (Boschma *et al.*, 2012). These were sown during 1998 (a legume adaptation experiment) and 2001 (performance of a smaller range of accessions under grazing). In the 1998 experiment a range of *Desmanthus* accessions (eight *D. leptophyllus*, two *D. pernambucanus*, five *D. pubescens* and 13 *D. virgatus* were compared with two of *Glycine latifolia* and two of *Medicago sativa* on a fertilized site that was slashed annually. *Glycine latifolia* performed best overall for plant frequency and biomass, but *D. virgatus* CPI 78372 and 67643, *D. leptophyllus* 38820 and 38351, *D. pernambucanus* 83565 and 40071 and *D. pubescens* 92805 also performed well. There was considerable variation in leaf–stem ratio within *Desmanthus* and stemmier types tended to have higher biomass. The entries selected for the 2001 grazing study (*D. bicornutus* CPI 91162, *D. glandulosus* 90319A, *D. leptophyllus* 38351, 38820 and 63453 and *D. virgatus* 78372, Q 9153 and cv. Jaribu) were all well grazed. Biomass ranking altered over time as populations changed due to the success or failure of recruiting new plants from seeds (although the large *D. bicornutus* plants maintained high biomass without recruiting new plants). Some early flowering types (particularly Q 9153) were particularly successful recruiting new plants. The experiments indicated that there is a role for *Desmanthus* on vertisols in northern New South Wales.

Overall, the 2000s saw continued, but limited, gains in legume development for clay soils (and sown pastures in general) at a time when government research resources were limited. There were, however, legumes in the 'pipeline' and progress was made towards testing the cultivars released during the 1990s and developing new pasture/graze systems, particularly in central and southern Queensland. For *Desmanthus*, there was further evaluation of the available genetic material, and testing was extended into north Queensland and northern New South Wales. During this period, some smaller Queensland seed companies were purchased by larger multi-national companies and these began their own development programmes, including the development of new *Macroptilium bracteatum* lines. Towards the end of the 2000s, a commercial company was also established to develop and market new *Desmanthus* cultivars developed within a university programme that had seed increase activities in north Queensland at the Queensland Government facility (Fig. 10.10; also see the case study in Chapter 11, this volume).

It would be remiss not to mention a more than 10-year breeding programme undertaken

Fig. 10.10. A 'foundation' seed crop of one of the components of the 'Progardes' *Desmanthus* composite variety grown at Walkamin in north Queensland, June 2015. This early generation seed crop is grown on weed mat to maximize seed yield through vacuum harvesting fallen seeds and minimize time spent on weeding. The crop layout allows easy routine application of insecticides using a tractor and boom. (Photo: Kendrick Cox.)

by the University of Queensland, in which *Leucaena leucocephala* × *L. pallida* hybrids were developed to produce psyllid-resistant (*Heteropsylla cubana*) *Leucaena* options for clay soils in coastal and sub-coastal areas of Queensland (Shelton, 2014). The development of these hybrids follows a significant research effort championed over decades by long-term researchers from the Universities of Hawaii and Queensland. This ultimately led to the field testing of genetic material of simple species and hybrids for psyllid resistance, dry-matter production and cold tolerance in south-east Queensland (27.3°S) during the early to mid-1990s (Castillo *et al.*, 1997), and further development of hybrids by University of Queensland research teams. The selected lines were under final evaluation during 2014/15 and a cultivar is due for release in 2015/16. If successful, they will add substantially to the already significant performance of *Leucaena* on clay soils, particularly in coastal and sub-coastal (psyllid-prone) areas of central Queensland and potentially in northern Queensland.

10.10 Recent research activity

Since about 2010, there has been renewed interest in sown herbaceous legumes, particularly for the seasonally dry areas of Queensland (and northern New South Wales). Significantly, the Australian beef industry contributed grower levies (with matching federal government contributions) towards sown pasture research, with a view to long-term investment in sown pastures. This followed the basic tenet that the growth of beef livestock is primarily based on feed intake and digestibility of feed (McLennan *et al.*, 1998), and nitrogen availability was limiting the amount and quality of feed provided by grass pastures (other factors such as rainfall being non-limiting). The use of pasture legumes was identified as the best option to overcome 'pasture rundown' (Peck *et al.*, 2011). Funding principally targeted the seasonally dry areas of central and southern Queensland because these areas carry the greater number of cattle. However, research in seasonally dry areas of north Queensland was also supported in recognition of significant potential gains to reproductive and grower performance in these inherently less-productive systems. Clay soils were key components of the programmes because they represent economically significant beef grazing resources and there were few (if any) suitable legumes (or legume systems) available for graziers.

Government research activities targeting uptake of 'known' types (when this book went to press) included: appraisal of renovation methods to establish small-seeded legumes in grass (particularly buffel) pastures in central and southern Queensland; comparative assessment of recently developed legumes (and grasses) under grazing across a range of soil types (including grey clays and vertisols) and rainfall environments in northern and central Queensland; and the appraisal of methods to prolong the viability of specific *Bradyrhizobia* strains on legume seed (particularly for *S. seabrana*) sown in areas of unreliable rainfall and typically hot (summer) soil conditions. Levied funds were also used to review all previous plant evaluation data and prioritize future plant evaluation efforts. Government and university researchers have contributed data sets for the compilation of a 'master set' and digital platform to be used to identify broadly adapted genera and species. *Desmanthus* will likely be identified as a key genus for continued investment.

There was also success achieving funding to conserve and develop the Australian Tropical Forages Genetic Resources Collection. Following transfer from Queensland to South Australia in 2014, the tropical forages collection is now a component of the Australian Pastures Genebank (APG), a national collection comprising temperate and tropical forages. Levied grazing (particularly the beef industry) and cropping industries now co-fund the management of the collection, including servicing requests for research and cultivar development, viability testing and long-term storage. A public-access database, using a platform similar to GRIN™, is planned to include collection and plant characterization information. The regeneration and characterization of 'priority' germplasm with small amounts of viable seeds began at Walkamin in late 2014, and is scheduled to continue for 5 years in the first instance. First-year accessions included *Clitoria*, *Desmanthus* and *Macroptilium* (K. Cox, personal communication).

10.11 Genetic resources

The key themes of this chapter are the considerable genetic scope, particularly within *Desmanthus*, which can be utilized to develop useful plants for clay soils in northern Australia, and the relative dearth of genetic material of species of potential held by the Australian Tropical Forages Collection. Successful pasture legume cultivars have often been developed from genetic material collected from areas of the world (principally Brazil) with similar soils and climate to northern Australia, and Queensland in particular (see Chapter 7, this volume). There are considered to be good opportunities to collect legumes for clay soils from other parts of the world that have better floristic analogues of Australian clay soil areas than many of the areas where legumes have been collected previously. These include clay soil areas of India and Africa and certain locations within the mesoamerican region where there have been only relatively minor collecting efforts.

To date, useful herbaceous pasture cultivars (or cultivar components) for clay soils in Queensland have been developed from *Clitoria* (*C. ternatea*), *Desmanthus* (*D. bicornutus*, *D. leptophyllus*, *D. pubescens* and *D. virgatus*), *Macroptilium* (*M. bracteatum* and, to a lesser extent, *M. atropurpureum*) and *Stylosanthes* (*S. seabrana*). However, there is no current commercial production of *D. leptophyllus* or *D. pubescens* cultivars (the others are all available). These species that have contributed cultivars continue to be developed, albeit on a smaller scale than during the 1970–1990 period, when much of the material was assessed. There has, however, been relatively little development of *Stylosanthes hamata* for use on clay soils despite previous researchers demonstrating potentially useful adaptation and breadth of genetic material in the APG (Chapter 2, this volume). This also applies to some other genera such as *Alysicarpus*.

The status of the APG collection in May 2015 for all *Desmanthus* spp. is presented in Table 10.1 and the other four key species for clay soils (*C. ternatea*, *S. bracteatum*, *S. hamata*, *S. seabrana*) in Table 10.2. The tables include the countries from which the accessions were sourced, in most cases these being their countries of collection. It should be noted that seed stocks may be limited or of low viability where regeneration has yet to be completed.

The APG contains some 440 *Desmanthus* accessions across (at least) 14 species; 154 of these are *D. virgatus*, of which ~80% originate from Mexico, Argentina and Paraguay. *Desmanthus leptophyllus* is the next most represented species (70 accessions), with most accessions originating from the Mexican peninsula and Brazil. The next best represented species are *D. tatuhyensis*, *D. pernambucanus*, *D. covillei*, *D. bicornutus*, *D. acuminatus* and *D. paspalaceus*. *Desmanthus oligospermus* and *D. interior*, recommended for collection (Chapter 2, this volume), are not represented.

There have been two key *Desmanthus* collecting 'zones': the Mexican peninsula (Southern California and the Baja Peninsula), and especially mainland Mexico and Central America (~10–30°N), where 37% of accessions were collected; and 'central latitudes' of South America (~20+°S) at 34%. Almost 10% originated from Brazil. Very few have been collected from the Caribbean or South American countries between the equator and 10°N, and, not surprisingly due to its American centre of origin (Cook *et al.*, 2005), few have originated from other parts of the world. Only two accessions originate from India. Based on the above, and the reasoning presented in this text, it would appear there remains strong merit in extending the breadth of the APG collection of *Desmanthus* through undertaking carefully targeted collections in future.

There has been no recent development of *Stylosanthes hamata* for clay soils in Queensland. In 2015, the APG contained some 476 accessions, over half of which were sourced from Colombia and Venezuela. In contrast to the other legumes, however, a considerable number (66) originate between 10°N and 30°N in Mexico, Central America and the Caribbean. There is considerable variability within *S. hamata* and the *S. hamata sensu stricto* group appears to be particularly promising for clay soils. Field testing of this material, and perhaps further collection, appear warranted, particularly if there are deficiencies in the other legumes.

10.12 Conclusions

The development of well-adapted, persistent and palatable herbaceous legumes for clay soils in Queensland, and on similar soils in the tropics and subtropics, has been an evolutionary process whereby research teams have built on research data and resources compiled in well-resourced government evaluation programmes conducted during the 1970s and 1980s. The decline in government

Table 10.1. Source countries and the number of *Desmanthus* accessions, by species, held within the Australian Pasture Genebank (APG), June 2015.

Location	*D. acuminatus*	*D. bicornutus*	*D. covillei*	*D. fruticosus*	*D. glandulosus*	*D. illinoensis*	*D. leptophyllus*	*D. painteri*	*D. paspalaceus*	*D. pernambucanus*	*D. pubescens*	*D. pumilus*	*D. tatuhyensis*	*D. virgatus*	*Desmanthus* sp.	Total
United States (25°–35°N)		1				4				2			1	1	2	11
Mexico, Central America (10°–30°N)																
Belize							6				13			1	9	29
Costa Rica														1		1
Guatemala														1		1
Mexico		15	22	16	2		22	1			3	1		41	8	131
Panama							1									1
Caribbean (10°–25°N)																
Cuba							5							2		7
Guadeloupe										1						1
Guyana															1	1
Puerto Rico										1						1
St Kitts and Nevis														1		1
US Virgin Islands		1												2		3

Continued

Table 10.1. Continued.

Location	*D. acuminatus*	*D. bicornutus*	*D. covillei*	*D. fruticosus*	*D. glandulosus*	*D. illinoensis*	*D. leptophyllus*	*D. painteri*	*D. paspalaceus*	*D. pernambucanus*	*D. pubescens*	*D. pumilus*	*D. tatuhyensis*	*D. virgatus*	*Desmanthus* sp.	Total
Northern South America (1) (10°–0°N)																
Colombia							4							2		6
Venezuela							8									8
Northern South America (2) (0°–20°S)																
Brazil		1					17		2	13			1	3	3	40
Bolivia														2		2
Ecuador							1									1
Peru														3	1	4
Central South America (20°+S)																
Argentina	13						1		6	1			7	40	11	79
Paraguay									7	1			11	45	1	65
Uruguay	3								1				1			5
Africa																
Ethiopia										1					1	2

Mauritius										2				1		3
Seychelles														1		1
Zimbabwe															1	1
Asia																
India										2						2
Malaysia										1						1
Philippines										1						1
Pacific																
Fiji										1						1
New Caledonia														1		1
Australia														1		1
Europe																
Germany														5		5
Unknown		1				2	5			1			11		3	23
Total accessions	16	19	22	16	2	6	70	1	16	28	16	1	32	154	41	440

Data provided by the curator of the APG. Most values represent points of collection, but exceptions may apply where a donor institute provided the seeds. Clearly as *C. ternatea* is native to Africa, and the other species in the tables to the Americas, accessions sourced from outside of a species' normal range are either duplicates or have been sourced from primary sources in other geographical areas (e.g. Ethiopia has sourced material directly from Belize research activities). *C. ternatea*, however, has been distributed widely for many years and sources from beyond its native range may exhibit genetic variation.

Table 10.2. Source countries and the number of *Clitoria ternatea*, *Macroptilium bracteatum*, *Stylosanthes hamata*, *S. seabrana* and total *Desmanthus* spp. accessions held within the Australian Pasture Genebank (APG), June 2015.

Location	Latitude	*Clitoria ternatea*	*Macroptilium bracteatum*	*Stylosanthes hamata*	*Stylosanthes seabrana*	Total *Desmanthus*	Total
United States (25°–35°N)				34		11	45
Mexico, Central America (10°–30°N)							
Belize		2		1		29	32
Costa Rica				1		1	2
Guatemala		1		5		1	7
Honduras		3					3
Mexico		8		2		131	141
Panama						1	1
Caribbean							
Antigua and Barbuda				13			13
Barbados				3			3
Bahamas				2			2
Cuba		3		5		7	15
Dominican Republic		1		3			4
Guadeloupe				2		1	3
Guyana						1	1
Monserrat		1					1
Panama		3					3

Puerto Rico				4		1	5
St Kitts and Nevis		1		17		1	19
St Lucia				4			4
US Virgin Islands		1	1			3	5
Northern South America (1) (10°–0°N)							
Colombia				82		6	88
Venezuela		6	1	191		8	206
Northern South America (2) (10°–0°N)							
Brazil		23	40	11	45	40	159
Bolivia			12			2	14
Ecuador						1	1
Peru						4	4
Central South America (20°+S)							
Argentina			11			79	90
Peru		1					1
Paraguay			13			65	78
Uruguay						5	5

Continued

Table 10.2. Continued.

Location	Latitude	*Clitoria ternatea*	*Macroptilium bracteatum*	*Stylosanthes hamata*	*Stylosanthes seabrana*	Total *Desmanthus*	Total
Africa							
Côte d'Ivoire		1					11
Ethiopia		1		36		2	39
Gambia				1			1
Ghana		2					2
Kenya		10					10
Madagascar		2					2
Mauritius						3	3
Senegal		10					10
Seychelles						1	1
Somalia		1					1
Sudan		4					4
Tanzania		2					2
Zambia		1	1				2
Zimbabwe						1	1
Asia							
Hong Kong		2					2
India		5				2	7

Indonesia		5					5
Malaysia		2				1	3
Papua New Guinea		4					4
Philippines		1				1	2
Sri Lanka		3					3
Thailand		2					2
Pacific							
Fiji		1				1	2
New Caledonia						1	1
Australia		7	3			1	8
Europe							
Belgium		1					1
Germany						5	5
Unknown		21		59	5	23	108
Total		142	82	476	50	440	1190

See Note to Table 10.1.

research capacity during the 1990s saw small research teams within private companies, universities and federal and state government departments assess available genetic material and develop a limited number of new cultivars. Restricted funding for research meant that much of this would not have been possible without the knowledge of older (often retired) researchers, the prior development of the Australian Tropical Forages Genetic Resource Collection (ATFGRC) and the existence of a broad network of (mostly) decommissioned pasture evaluation sites across Queensland to source promising material. The continued operation of a Queensland Government seed production service to supply seeds for evaluation and commercial adoption has contributed significantly to cultivar development. There has, however, been a deficiency of comparative evaluation of new genetic material and cultivars that is only recently being addressed.

Much of the focus of research and development has been on *Desmanthus*, but useful plants have been identified for clay soils from a range of legume genera. Commercially available varieties have been produced from *D. bicornutus*, *D. leptophyllus*, *D. pubescens* and *D. virgatus*, although by 2015 only *D. bicornutus* and *D. virgatus* cultivars were commercially available. Other species for clay soils represented by commercially available varieties include *Clitoria ternatea*, *Macroptilium bracteatum* and *Stylosanthes seabrana*. These all have slightly differing roles in terms of grazing management (long- versus short-term pastures) and adaptation to climate and type of clay soil, although various evaluation experiments indicate there is often the capacity for a legume to perform well in a range of environments and management regimes.

Although a range of varieties are available for use on clay soils, adoption is currently well below projected estimates of potential use, and many significant areas of clay soils under grazing do not utilize herbaceous legumes in pastures. Adoption is particularly low in the seasonally dry, and semi-arid areas of north-west Queensland (*Astrebla* and *Dichanthium*/*Eulalia* vegetation communities), which historically have a lower emphasis on pasture development than the brigalow (*Acacia harpophylla*) and gidgee (*A. cambagei*) areas of central and southern Queensland. A decrease in government extension capacity has, no-doubt, contributed to lower uptake by beef producers, as have cyclical deficiencies in seed supply compared with demand. Unreliable establishment of pasture legumes, particularly in dense grass pastures such as *Cenchrus ciliaris* pastures in central and southern Queensland, will also have impeded adoption as graziers examine the costs of renovating pastures compared with returns. Difficulty maintaining *Bradyrhizobium* viability under harsh environments has also been an issue for some legumes.

Renewed interest in herbaceous pasture legumes for clay soils in recent years has, however, seen re-investment by the beef industry into federal research and state government research and extension teams. This, along with investment by a few seed companies and universities, has maintained some momentum in the development and adoption of legumes for clay soils.

The ATFGRC, now the Australia Pastures Genebank (APG), remains the key source for genetic material for developing legumes for clay soils, and there is evidence that it contains further promising material which has not been commercialized to date. Unreleased lines of *D. leptophyllus* (for example TQ 90 and CPI 38351) and *D. virgatus* (CPI 85178 and Q 9153) have performed well in previous plant evaluation research but are yet to be commercially adopted. There is also potential for clay pasture legumes within other *Desmanthus* species. One species that appears to have been neglected despite a broad range of material held within the APG is *Stylosanthes hamata* (*sensu stricto*). Recent investment into regenerating APG genetic seed stocks, and prioritization towards developing legumes for areas of the seasonally dry tropics of strategic importance for the beef industry, should enable enhanced future evaluation activities on clay soils in Queensland.

While the APG already contains potentially useful legume germplasm for clay soils in Queensland, many of the authors referred to in this chapter have called for more targeted collection of legumes from environments better matched to the soils, climate and vegetation of Queensland clay soils than many of the areas from which the bulk of the collection originated. Whereas, current research efforts seek primarily to encourage the adoption of known genetic material and cultivars, there may be significant gains in seeking new plants from clay soils in regions such as the Caribbean, India or North Africa.

11 Developing and Commercializing New Pasture Legumes for Clay Soils in the Semi-arid Rangelands of Northern Australia: The New *Desmanthus* Cultivars JCU 1–5 and the Progardes Story

C.P. Gardiner

Abstract

Vast areas of semi-arid clay soil rangeland regions of northern Australia, such as the Mitchell Grass Downs Bioregion, have (until recently) no commercially available or adapted sown pasture legume. Other regions and land types with clay soils too, such as the Brigalow region, have had a very limited range of sown pasture legume species available. Low livestock productivity in northern Australia is largely due to low protein content and low digestibility of the diet during the dry season. An adapted pasture legume could ameliorate this problem and enhance the productivity and sustainability of the region's grazing sector. Of all the legume species tested to date in this region, a number of *Desmanthus* species are the preeminent survivors and are proving to be well adapted. After extensive surveys of old abandoned trial sites, selection and further evaluation, and seed increase, a number of *Desmanthus* accessions have been developed and released commercially for semi-arid tropical/subtropical clay soil regions of northern Australia. These new selections come from *D. bicornutus*, *D. leptophyllus* and *D. virgatus* genotypes, and the new cultivars are named JCU 1, JCU 2, JCU 3, JCU 4 and JCU 5 and Plant Breeders' Rights has been granted on these five cultivars. A commercial composite blend of these five cultivars has been named Progardes™ and commercial seed production and cultivation is well advanced, with some 20,000 ha being sown to date, mostly in Queensland.

11.1 Introduction and background: setting the scene for legume introduction into tropical/subtropical semi-arid clay soil areas of northern Australia

11.1.1 The Mitchell Grass Downs

Significant numbers of cattle and to a lesser extent sheep are grazed in the vast semi-arid central-western and north-western rangelands of Queensland, such as in the Mitchell Grass Downs Bioregion (some 29.5 million ha) and its associated land types, for instance cleared Gidgee (*Acacia cambagei*) (Fig. 11.1). In these regions annual rainfall is very variable and typically less than 500 mm/annum (250–500 mm per annum across the region), and the soils are characteristically vertisols (brown and grey cracking clays) (Orr, 1975).

The region has a short growing season (summer rainfall), a long dry season (typically more than 6 months), and has been in drought (2012–2015) (Fig. 11.2). The predominant pasture grasses in the region are the native 3P (Perennial, Palatable and Productive) *Astrebla* species

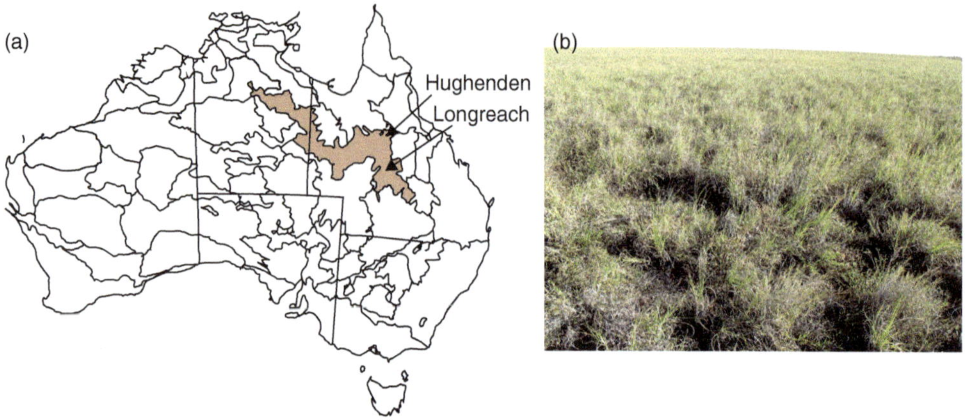

Fig. 11.1. (a) The location of the Mitchell Grass Downs Bioregion and (b) an image of the Mitchell Grass Downs (near Julia Creek) in good condition prior to drought in August 2012. (Map adapted from: http://www.environment.gov.au/system/files/resources/a8015c25-4aa2-4833-ad9c-e98d09e2ab52/files/bioregion-mitchell-grass-downs.pdf Accessed January 2014.)

(Mitchell grasses) and the annual *Iseilema* species (Flinders grass). Both grasses are highly regarded by farmers/graziers because of their adaptation to the harsh climate and difficult edaphic environment. Both grasses respond and grow rapidly in the short summer wet season, but their nutritive quality quickly deteriorates as they mature into the long annual dry season. Buffel grass (*Cenchrus ciliaris*) has been introduced to the region, particularly in areas cleared of the native leguminous tree Gidgee (*Acacia cambagei*) that occurs on some 4.8 million ha (Weston and Turner, 1987). During the dry season livestock are typically fed protein and energy supplements for maintenance.

11.1.2 The Brigalow Region

To the east of the Mitchell Grass Downs Bioregion is the Brigalow region of some 8.7 million ha in central Queensland with an average annual rainfall of 590 mm. This region is predominantly also clay soils, and the native leguminous Brigalow tree (*Acacia harpophylla*) has been cleared mainly for buffel grass monoculture pasture and cropping. Like Gidgee, the Brigalow regions after clearing are initially very productive but yields decline over time as the available nitrogen becomes less available to plants (Weston and Turner, 1987). Low livestock productivity in northern Australia is largely due to low protein content and low digestibility of the diet during the dry season (Poppi and McLennan, 2010). Calving/lambing percentage, weaning and cow rebreed rates are low across the region, with many northern beef producers achieving weaning rates below 50% (McCosker *et al.*, 2010), and for decades a lambing rate of only 48% (Hegarty, 2012). Nutrition is often stated as a major factor contributing to the region's poor animal production (McCosker *et al.*, 2010; Hegarty, 2012). High sheep weaner mortality in Queensland has been linked to the extensive system of management and the poorer and more variable nutrition available in rangeland environments (Chapman *et al.*, 1973). MLA (2014) says that the causes of poor reproductive performance in northern Australian beef herds are multifactorial and lists seasonal pasture quality, body condition score and country type among the factors related to pastures and nutrition issues.

These tropical and subtropical rangelands are difficult environments for plant growth because they have low and variable rainfall and hot and humid environmental conditions. The mean long-term rainfall for Hughenden and Longreach and the 2013 rainfall records (Table 11.1) illustrate the current drought and the variability of rainfall in the region. In addition, the vertisol soils of the Mitchell Grass Downs (Fig. 11.3) are hostile environments for many pasture plants because they tend to have an alkaline pH, crack deeply when dry, swell and have poor water penetration when wet, and are often saline at depth (ACIAR,

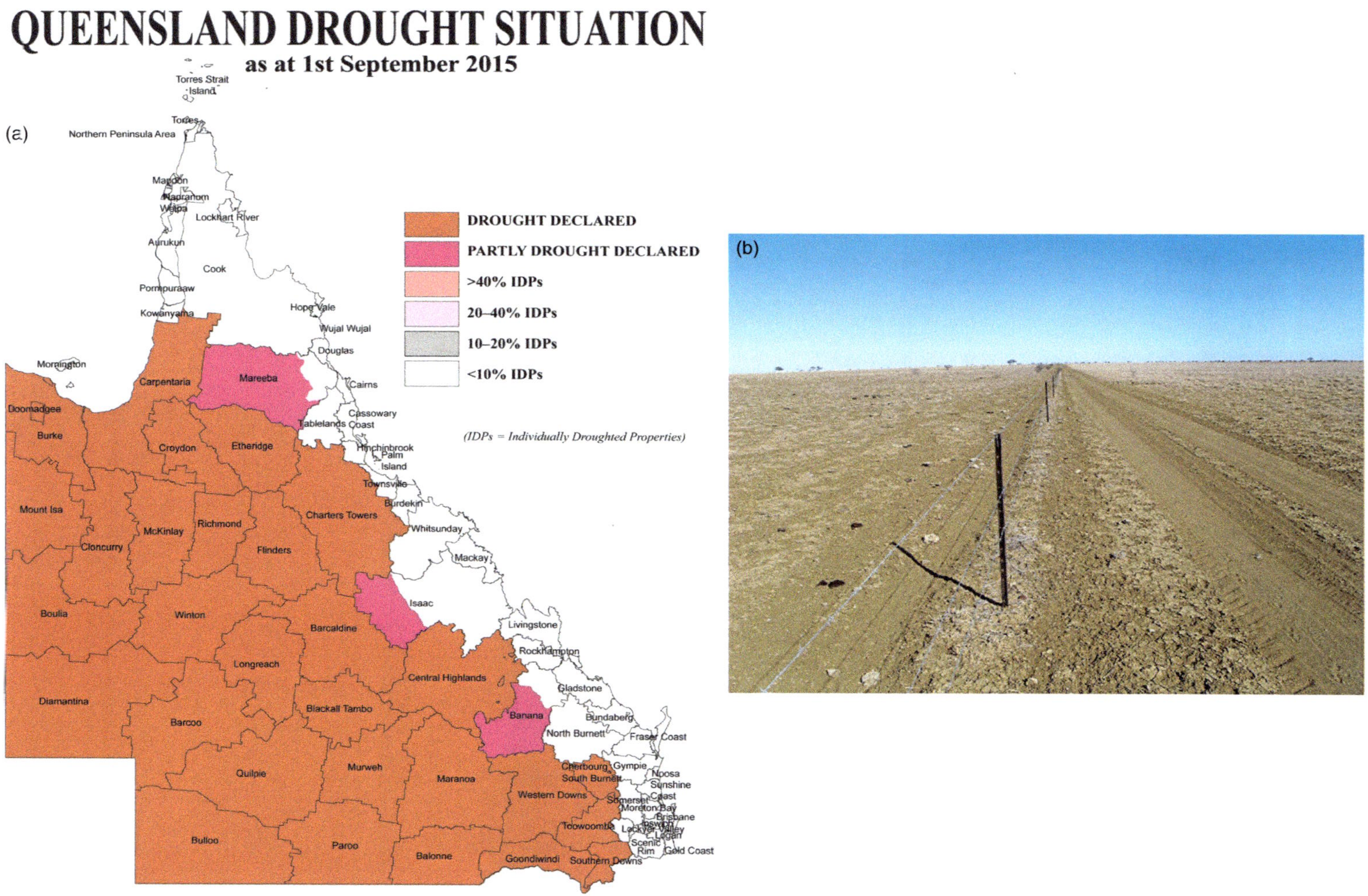

Fig. 11.2. (a) Queensland 2015 drought map and (b) drought-affected paddocks near Hughenden (March 2013). (Map source: https://www.longpaddock.qld.gov.au/queenslanddroughtmonitor/queenslanddroughtreport/2015/jul1.pdf Accessed July 2015.)

Table 11.1. Long-term mean and 2013 rainfall in mm for Hughenden and Longreach sites.

Period	Month												Total
	J	F	M	A	M	J	J	A	S	O	N	D	
Hughenden – Redcliffe Station													
Mean	114.5	98.0	58.2	26.0	17.8	18.6	11.6	7.9	9.0	22.3	36.0	71.1	492.4
2013	31.4	25.2	0	22.2	19.4	0.2	0	0	0.4	0.2	25.8	4.8	129.6
Longreach – Colanya Station													
Mean	65.4	82.3	56.7	26.2	22.3	17.7	14.4	7.4	8.3	19.1	25.5	46.3	390.8
2013	0	33	0	0	3	0	0	0	0	0	0	37	73

Source: http://www.bom.gov.au/climate/averages/tables/ca_qld_names.shtml. Accessed January 2014.

Fig. 11.3. A vertisol soil profile on the Mitchell Grass Downs near Hughenden.

1984). All of these factors severely limit the range of pasture plants adapted to such semi-arid clay soil environments.

Table 11.2 illustrates two examples of the soils from the region. The current (Fig. 11.2) and previous extensive droughts in the region illustrate the harshness of the climate and the resilience needed in pasture plants and livestock alike to survive during drought and revive after. The Mitchell grasses demonstrate this resilience because they are long-lived native perennials with deep specialized roots that can penetrate and survive the dry climate and cracking clay soils (Orr, 1975).

One of the major ways that livestock production can be improved is to increase the legume content within the pasture. McMeniman (1986), for example, suggested that for optimum animal performance in the Mitchell Grass Downs region, pastures should be managed so as to maintain a high proportion of forbs (broadleaf herbaceous plants that includes legumes). In general, when legumes are present in a pasture, livestock have, compared with grass-only pasture, a higher quality diet, because legumes have higher digestibility, higher intakes, are used more efficiently than grasses and their nutritive value tends to remain

Table 11.2. Top soil properties from Hughenden (Redcliffe site), a pebbly 'Gidgee' soil, and a Mitchell Grass Downs (MGD) soil type from the Longreach region (Colanya site).

	Texture	pH	C Organic	N Nitrate	S (MCP)	P (Colwell)	K (Amm.Ac.)	Electrical conductivity	Na (Amm.Ac.)	CEC
Soil type	% clay	1:5 water	% C	mg/kg	mg/kg	mg/kg	meq/100g	dS/m	meq/100g	meq/100g
Gidgee	22	8.5	1.2	7.4	9	19	1.54	0.13	0.09	44.42
MGD	44	7.8	0.3	3.6	4	8	0.73	0.06	0.30	27.2

higher as plants mature. Grazing livestock always increase their dry matter intake when legumes are present, because legumes leave the rumen faster (Bogdan, 1977; Minson, 1977; Teitzel and Middleton, 1983; Leng, 1986). These various attributes can all translate into increased animal production. Having legumes in a pasture also promotes a healthier soil via nitrogen fixation and improved organic matter that can help arrest declining soil fertility (Rotar, 1983; Bowen *et al.*, 2015).

In addition, legumes have tap roots that can give them an advantage over grasses in extracting soil moisture and thus providing some resistance to drought (Friedel *et al.*, 1990), and in extracting nutrients from deep in the soil profile; thus they can stay green and productive longer than associated grasses in semi-arid environments. Rao *et al.* (2015) discuss many of the social, economic and environmental benefits of improved forage-based systems including the use of legumes. They state that the sowing of improved forages can lead to sustainable intensification of tropical forage-based systems that can, for example, improve productivity, restore degraded land, enhance system resistance to drought and waterlogging associated with climate change, accumulate carbon in the soil, fix N (legumes) and reduce greenhouse gas emissions per unit of livestock product.

Native legumes, such as *Crotalaria* spp., *Cullen* spp., *Glycine* spp., *Indigofera* spp., *Rhynchosia* spp., *Sesbania* spp. and *Vigna* spp., to name some, are present, but are often described as being of no grazing significance, toxic and/or not palatable, and those that are palatable tend to disappear under normal commercial grazing utilization rates (Hacker, 1990; Orr and Phelps, 2013; Chapter 6, this volume). Native herbaceous and shrub legumes have, however, been assessed by O'Keeffe *et al.* (2015) as an alternative feed source for livestock, particularly as some appear to fix nitrogen under low P soils conditions. Long-term continuous grazing by sheep in this region has depleted some of the palatable interstitial species from grazing lands (Orr and Phelps, 2013), including valuable herbaceous legumes (Leslie *et al.*, 1987). In other regions of Queensland botanical changes and land degradation due to overgrazing of cattle have also been noted (Gardener *et al.*, 1990). Losses or change in botanical composition of palatable forbs and especially the legumes component of these may contribute to production issues including 'pasture run-down' and thus to the demise of the sheep and wool industry in north-western Queensland. Stewart-Moore (2015, personal communication) stated that the productivity of his Merino flock on the Mitchell Grass Downs near Hughenden has been in decline for decades. Reinstating well-adapted productive and persistent legume species into these grazed regions would provide beneficial ecosystem services and revive and even enhance productivity. Rangel and Gardiner (2009) have demonstrated the benefit of providing a legume to sheep fed a diet of Mitchell grass hay in pen studies as manifested in higher feed intake, reduced weight loss and improved wool growth.

Finding legumes for northern Australian clay soils (vertisols) has been seen as a major need (Staples *et al.*, 1985). An adapted pasture legume could ameliorate this problem and enhance the productivity and sustainability of the region's important grazing sector as diet quality is a major driving force behind productivity (Coates, 2000). It has long been a goal of graziers and researchers to seek adapted pasture legumes for such rangelands because the major limitation to increased livestock production is nutritional (Eyles and Cameron, 1985), and the best option for increasing productivity is using legumes (Thomas and Grof, 1986).

11.1.3 Importance of plant persistence

Persistence is critical for the success of a pasture legume in tropical and subtropical rangeland environments and if a legume is persistent it implies (Whiteman, 1980) that the plant has the following characteristics:

- tolerant of grazing;
- can survive water stress or waterlogging;
- can withstand low and high temperatures;
- can withstand fire/burning;
- longevity/permanence;
- disease and insect resistance;
- a suitable flowering and seed maturity time to cope with the sometimes short and variable wet seasons;
- can regenerate from stolons, rhizomes or seed;
- adequate seed production and hard seededness to establish good soil seed reserves.

For semi-arid areas, pasture yield has been considered as being a secondary consideration to longevity (Burt and Ebersohn, 1969), and plant survival as a critical criteria followed by nutritive value (Otsyina *et al.*, 1999). The ability to regenerate from seed is important to ensure longer term persistence (Boschma *et al.*, 2012) and maintaining a viable (hard seed) seed bank of tropical pasture legumes is important for ensuring legume persistence (Lawrence *et al.*, 2012). Jones (2014), in discussing the rise and fall of the utilization of Siratro (*Macroptilium atropurpureum*), suggests that the major limitation of Siratro was that soil seed banks were generally inadequate to ensure persistence, especially through dry years. Persistence is complex however and difficult to define and a total of 29 causes of poor legume persistence can be found in a summary of a trilateral workshop on persistence of forage legumes (Marten, 1989). Biotic and abiotic aspects of persistence, including plant morphology such as root depth, as well as factors such as grazing and the interactions with microbes are also important factors to consider (Ludemann *et al.*, 2015). A successful sown legume therefore needs to mimic the adaptability and environmental tolerances of native plants, but is also required to be productive, palatable, grazing tolerant, nontoxic and of minimal weed risk.

11.1.4 Acceptability to graziers

The rangeland grazing communities of northern Australia appreciate the value that an adapted pasture legume would provide and are keen to improve the productivity of their livestock and sustainability of both their native and introduced grass pastures in a manner similar to what the *Stylosanthes* spp. have delivered to the light-textured soil regions of northern Australia. The very extensive nature of the grazing properties (as well as the previously mentioned climate and edaphic difficulties) across the area poses problems for sowing and establishing a new pasture species. Other difficulties include a lack of a sown pasture culture, the scarcity in the region of broad-acre farming equipment and the potential expense of sowing such immense rangelands (Gardiner *et al.*, 2012).

11.1.5 Legumes tested

It has long been recognized that an adapted pasture legume would greatly enhance the productivity of the pastures on vertisol soils in semi-arid tropical regions, but unlike southern Australia, few if any sown pasture legumes have been successfully incorporated up until now. In fact, there is a real dearth of pasture legumes adapted to these tropical/subtropical clay soils (Burt, 1993a; Jones and Clem, 1997; Pengelly and Conway, 2000; Chapter 9, this volume). Indeed, immense areas of northern Australia's semi-arid clay soil regions and beyond have no sown pasture legume with proven adaptation and persistence through their long annual dry seasons (Gardiner and Swan, 2008).

In northern Australia, the success of incorporating the *Stylosanthes* legumes, particularly the cultivars Seca (*S. scabra*) and Verano (*S. hamata*), into native grass pastures such as black spear grass (*Heteropogon contortus*) on light-textured soils is well known to be associated with benefits in cattle live-weight gains in the order of 35–65 kg/head/year and improved stocking rates and herd fertility (Coates *et al.*, 1997). However, in semi-arid northern regions with heavy textured soils (vertisols), the stylos are not usually well adapted and few other sown legume species have been shown to persist in such environments.

Desmanthus is one of the very few legumes that have been consistently observed to persist under heavy grazing on clay soils in its native environment (Pengelly and Conway, 2000). As discussed elsewhere in this volume, numerous accessions of *Desmanthus* were collected and introduced into Australia by various institutions (notably CSIRO and QDPI) over the past 50 years (Reid, 1983; Pengelly and Liu, 2001) as potential legumes for clay soils. After years of multi-site field evaluation of *Desmanthus* and other species, in 1991 QDPI released three *Desmanthus* cultivars (cv. Marc (*D. virgatus*), cv. Bayamo (*D. leptophyllus*) and cv. Uman (*D. pubescens*)) that were marketed as a blend named 'Jaribu' (Cook *et al.*, 1993). Of those three, only cv. Marc is currently available commercially, mainly promoted in southern subtropical Queensland markets. Pengelly and Conway (2000) state, however, that due to Marc's low dry matter production, its contribution to animal diet and soil N is limited. Nevertheless, in a review of evaluation sites Gardiner *et al.* (2004) found that of the entire range of accessions of legume species tested to date, a number of *Desmanthus* genotypes are the best survivors in the above-mentioned region. More recently, Peck *et al.* (2012) found *Desmanthus* and *Stylosanthes* survivors in southern Queensland.

11.2 The history and development of 'Progardes *Desmanthus*' (cultivars JCU 1–5)

There are numerous old discontinued pasture evaluation sites across northern Australia in many different environments and land types; as illustrated here, such sites, including their surviving plants, grazing history, soils data and climate records, are potentially very valuable resources. The plants that survive and persist over long periods of time from such abandoned trial sites are valuable in the development and release of promising new varieties and for the retrospective evaluation of species. Unfortunately, many early evaluation experiments were not conducted long enough, often only 2–3 years, and thus ignored critical plant attributes such as persistence (Jones, 2014). This assessment is supported by the good plant densities of *Stylosanthes* that were recorded in the early years of western Queensland trials (Chapter 9, this volume), but some years later no evidence was found of any *Stylosanthes* plants (Gardiner *et al.*, 2004). This evidence highlights the value of long-term trials and of revisiting old discontinued trial sites. Revisiting such sites can also draw attention to plants that are less desirable, for example unpalatable potentially weedy species such as *Aeschynomene paniculata* and *Acacia angustissima* (Cox *et al.*, 2007).

In the 1990s, Chris Gardiner at James Cook University (JCU), Townsville started to survey legumes in a number of old abandoned trial sites across remote northern and central-western Queensland's semi-arid clay soil regions (<500 mm annual rainfall). Work began in 1996 on sourcing information and locating these. With the assistance of CSIRO and QDAF staff (R. Burt, R. Coventry, W. Messer and B. Walker), a list of 74 old sites across northern Queensland was compiled, which was reduced by a later focus on those in semi-arid clay soil regions of central-western, north and north-western Queensland, particularly, but not only, the six sites (>1000 plots) in the Mitchell Grass Downs Bioregion described by Burt (Chapter 9, this volume) at Blackall, Barcaldine, Longreach, Julia Creek, Isisford and Yaraka (see Fig. 11.4). These were critical to the eventual selection, re-evaluation and release of some of the components of Progardes. In addition to these, two other key sites were selected at Kelso near Townsville and Wrotham Park Station near Chillagoe (see Fig. 11.4).

Other old abandoned trial sites in the north and north-west that have also been surveyed for surviving legumes include: Redlands Station (planted 1976) near Charters Towers; Hillgrove Station (Chapter 9, this volume, planted 1983/4), which is also near Charters Towers (Gardiner and Swan, 2008); Monavale Station (planted in the 1970s) near Hughenden (R. Coventry and W. Messer, personal communication); plus other sites in the Townsville region. New selections have been made from these that are currently in trials across Queensland. These may eventuate in additional *Desmanthus* releases. Likewise in southern Queensland, Peck *et al.* (2012) have reported recently revisiting old abandoned trial sites seeking persistent survivors, particularly *Stylosanthes* and *Desmanthus*, seeds of which are being collected for evaluation.

Since many years had passed since the trial sites had been abandoned, locating them was somewhat challenging, as was identifying their individual plots, and in some cases exactly what accessions were planted and where. In most cases fences around the original trials sites had

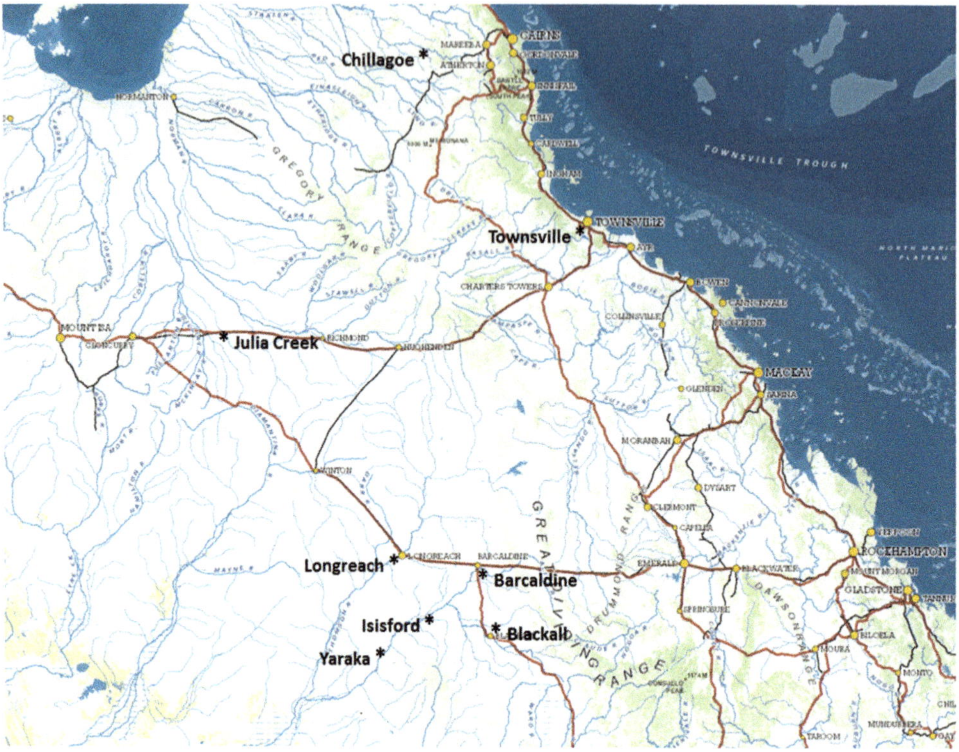

Fig. 11.4. Towns in Queensland nearest key abandoned legume evaluation sites relating to the development and release of Progardes. (Overdrawn on Google Earth image.)

been removed or realigned, no GPS waypoints were available, properties had been sold and new owners were unaware of the location of the trial sites, plot pegs had disappeared, rotted away or had been knocked out of the ground by livestock. For some sites no plot layouts could be located, or plot layouts had no key features/reference points to correctly orient the trial plans. At two sites plots had been disturbed by a bulldozer that had been clearing trees nearby and at another site strips had been cultivated through the old trial and a new trial planted over the old. At each of the six key sites mentioned above and shown in Fig. 11.4, seven species and 37 accessions of *Desmanthus* plus accessions of *Alysicarpus*, *Arachis*, *Atriplex* (non-legume), *Centrosema*, *Chamaecrista*, *Clitoria*, *Desmanthus*, *Macroptilium*, *Stylosanthes* and *Vigna* were originally planted in replicated trials. At some sites additional experiments or observation plots were also planted adjacent to these trials.

The six key sites were surveyed several times over a number of seasons and years, recording population locations, plant density, plant habit and characteristics, soil seed banks and, when available, seed and pressed specimens were collected. In some cases, if no seed was available (e.g. due to heavy grazing), whole plants were collected, potted and transported to Townsville where they were cultivated to produce seed. Leaf samples were taken for molecular studies and soil samples were collected from each site. At JCU, Townsville, a spaced plant trial of the collected survivors was planted along with a set of seed (sourced from CSIRO, Davies Laboratory, Townsville) of the accessions that were originally planted at these sites (see Chapter 9, this volume) to compare the morphological aspects of the survivors with their putative parents.

Of all the genotypes of the genera sown, only some *Desmanthus* accessions survived and thrived. These persistent accessions had at that time survived a decade or more (by now more than two decades), and under harsh conditions that included grazing, floods, fires, frost, droughts and insect attacks (locusts and army worms), and had therefore withstood the full gambit of biotic and abiotic field tests (Gardiner *et al.*, 2004; Gardiner and Swan, 2008; Gardiner *et al.*, 2013). Identifying the many *Desmanthus* species and multiple accessions was challenging because the taxonomy of *Desmanthus* has long been described as being complex and confused (Burt, 1993a; Luckow, 1993; Pengelly and Liu, 2001).

Later, when small amounts of seed of these survivors was produced and available, a considerable number of small plot field trials were established and evaluated across several regions of northern and north-western Queensland on various soil types (e.g. Gardiner, 2003). Accessions that were well ranked, particularly for persistence, in these semi-arid clay soil environments were sent to DAF Walkamin for seed increase and later planted again in further evaluation trials in north and north-western Queensland, and later again seed was increased. Five accessions from this work were eventually selected to be commercialized and to be put forward for PBR (Plant Breeders Rights). The PBR evaluation process included DUS trials that ensure the new cultivars have distinctness, uniformity and stability compared with other known *Desmanthus* cultivars such as Marc and Bayamo and the new cultivars' putative parents. The selection and breeding of the surviving plants from these and other similar abandoned trial sites and their subsequent evaluation in new field trials have led to the development of new cultivars of *Desmanthus* for northern Australia and similar environments.

A blend of the five selected new *Desmanthus* cultivars namely JCU 1 (*D. leptophyllus*), JCU 2 (*D. virgatus*), JCU 3 (*D. virgatus*), JCU 4 (*D. bicornutus*) and JCU 5 (*D. virgatus*) has been commercialized under the name 'Progardes' (www.progardes.com.au). A description of the five cultivars is available in the *Plant Varieties Journal* (Loch, 2015), and PBR has been granted by IP Australia. Agrimix Pty Ltd, the partner of James Cook University in the commercialization of this blend, coined the name; 'Pro' refers to protein, 'gar' to Gardiner and 'des' to *Desmanthus*. The blend of species and genotypes in Progardes gives it a broad range of plant types: early to late maturity, herbaceous to suffruticose habit and broad edaphic and climatic tolerances. In the field, a simple practical trait that helps discern the three species in the Progardes blend is their nyctinastic leaf movements (night leaf movements), as illustrated in Fig. 11.5 and described by Luckow (1993). *D. bicornutus* (JCU 4) has nyctinastic leaf movements in which both the pinnae and the petiole move downward at night; with *D. leptophyllus* (JCU 1), the rachis assumes an arched position and the pinnae fold forward and downward at night and with *D. virgatus* (JCU 2, 3 and 5) both the pinnae and petiole move upward at night.

11.2.1 Cultivar JCU 1 (*D. leptophyllus*)

JCU 1 is a *D. leptophyllus* genotype that was observed flourishing as adventive plants at Kelso near Townsville on a duplex soil (contrasting texture between soil horizons). It was noted to be a very vigorous, very leafy multiple-branched plant with a small shrubby habit and prolific seed production (see Fig. 11.6). It had persisted through the 1990s, which included a number of drought

Fig. 11.5. Nyctinastic leaf positions (left to right) of *D. bicornutus* (cultivar JCU 4), *D. leptophyllus* (JCU 1) and *D. virgatus* (JCU 2, 3 and 5).

Fig. 11.6. Cultivar JCU 1 (*D. leptophyllus*) (a) on weed mat and (b) a seed crop in the Burdekin irrigation area south of Townsville.

and average (1150 mm) rainfall years. Seed was collected and initially grown on weed mat in a spaced plant arrangement at James Cook University, Townsville with many other *Desmanthus* accessions and particularly the accessions listed by Burt (Chapter 9, this volume) for comparative morphological studies and seed increase. It was later planted on a larger scale at the Queensland Department of Agriculture and Fisheries Walkamin Research Station. Morphologically, its close parent was deemed to be CPI 76053. Seed was later used in evaluation trials with other legumes and other *Desmanthus* accessions at several sites, including Fletcherview and Trafalgar Stations near Charters Towers, both on Gidgee clay soils, and also at Redcliffe station, Dunluce Station, Como Station and Barragunda Stations near Hughenden on Mitchell Grass Downs and/or associated pebbly Gidgee clay soils, similar to those outlined in Table 11.2. In these trials JCU 1 regularly ranked highly, particularly for persistence, which was deemed to be a critical trait in these semi-arid clay soil environments (Gardiner, 2003). As a result of these trials, seed was again increased at Walkamin Research Station, and later for commercial seed production on the Atherton Tablelands and in the Burdekin Irrigation area, where it has consistently produced bulky crops and good seed yields. Currently JCU 1 is proving to be productive and well regarded across a number of clay soil regions, including the central Queensland Brigalow belt.

An important attribute of JCU 1 and some other *Desmanthus* genotypes is that under grazing or cutting they have the ability to produce a low crown and soboliferous-like shoots, as seen in Fig. 11.7, which provides the plants with resistance to heavy grazing, frost and fire.

11.2.2 Cultivar JCU 2 (*D. virgatus*)

JCU 2 is a *D. virgatus* genotype; morphological and molecular studies indicate that its putative parent is accession CPI 91335. JCU 2 was collected by Bob Walker (DAF) for Chris Gardiner (JCU) from outside an old abandoned legume evaluation site at Wrotham Park Station near Chillagoe in far north Queensland. The trial site had been established some years previously on a dark clay soil and these well-established, long-term survivor plants were observed as being of possible interest and value. In the same manner as for JCU 1, a very small amount of seed (a few grams) was sown initially at JCU for morphological studies. Seed was then increased before it was included in field trials, where it showed excellent attributes, including rapid growth and seed set even in the short, late wet seasons of western Queensland (see Fig. 11.8). JCU 2 is quite distinctive compared with the other *Desmanthus* cultivars and Progardes constituents because its leaves are bluey-green (compared with green to pale green in many other *D. virgatus* types such as cv. Marc), and it has clusters of pods that produce copious amounts of seeds. JCU 2 is performing well across a number of environments with buffel grass and native grasses in northern and central Queensland, including Mitchell Grass Downs, Gidgee soils, Brigalow and Blackwood

Fig. 11.7. Low crown and soboliferous – like shoots on CPI 38351 (a) and also JCU 1 (*D. leptophyllus* (b).

Fig. 11.8. Cultivar JCU 2 (*D. virgatus*) (a) on weed mat and (b) in the field.

(*Acacia agyrodendron*) land. Figure 11.8 illustrates typical JCU 2 plants. It is also more fully described by Loch (2015).

11.2.3 Cultivar JCU 3 (*D. virgatus*)

A genotype of *D. virgatus*, JCU 3, was selected from the abandoned Taranaway Station trial site near Isisford in central-western Queensland. It was one of only a few *Desmanthus* accessions that had survived at the abandoned site for many years. Many other accessions had failed in this particularly harsh stony clay environment. Seed and living plants were collected and multiplied at JCU and later at DAF Walkamin, and seed was sown in a number of trials as detailed above for JCU 1 and 2. In those trials JCU 3 ranked well, especially at Hughenden in north-western Queensland in a Gidgee land type (see Fig. 11.9). Seed from JCU 3 plots at a Hughenden (492 mm) trial site were selected and multiplied for inclusion in the PBR process.

11.2.4 Cultivar JCU 4 (*D. bicornutus*)

JCU 4 (*D. bicornutus*) is a robust early maturing suffruticose plant (see Fig. 11.10). Like JCU 3, it was found to be surviving in the abandoned

Fig. 11.9. Cultivar JCU 3 (*D. virgatus*).

Fig. 11.10. Cultivar JCU 4 (*D. bicornutus*) (a) on weed mat and (b) as a seed crop in the Burdekin district.

Taranaway trial in a harsh pebbly Gidgee semi-arid clay soil environment (451 mm average annual rainfall) with a good number of plants present some 22 years after originally being sown. It is still present in the abandoned plots. As with JCU 2 and JCU 3, seed and living plants were collected, multiplied at JCU for seed and morphological studies in a spaced plant trial with many of the other accessions that were in the trials described by Burt (Chapter 9, this volume). Seed was later sown in a number of trials as described above, and was later multiplied again at DAF Walkamin for inclusion in Progardes. See Loch (2015) for a more detailed morphological description.

11.2.5 Cultivar JCU 5 (*D. virgatus*)

The putative parent of JCU 5 (*D. virgatus*) is CPI 67643, which was one of the many *Desmanthus* accessions planted in the original six CSIRO western Queensland trials (the abandoned trials; Fig. 11.4). Seed and plants were collected from the Woodbine (Blackall) abandoned site and planted in other trials where it consistently ranked highly. It is a prostrate to decumbent fine-multi-stemmed plant (see Fig. 11.11). See Loch (2015) for additional botanical details.

11.3 Agronomic and production experiences with Progardes

11.3.1 Establishment

Progardes has been sown in numerous trials and demonstration areas and more recently in some 20,000 ha of commercial paddocks in a wide range of environments, including: open Downs, cleared Gidgee/Boree, Brigalow and Blackwood land types across Queensland, and into the Northern Territory and northern New South Wales. Numerous planting techniques have been used, including: aerial seeding, particularly in rangeland environments at the end of the dry season or following a fire; i.e. aerial seeded into an ash seed bed; seeded via or following a blade plough or stick raking; broadcast onto cultivated seedbeds in a buffel grass renovation process; broadcast onto unprepared native grass pasture with or without herbicide treatments to control grass competition; into fully cultivated seed beds; faecal seeding. Plant establishment success across all these planting techniques and environments has varied from excellent to poor. Gramshaw and McKeon (1985) and Cook *et al.* (1993) state that pasture establishment results are often variable and are generally specific to the location, to the pattern and nature of rainfall events at germination and establishment, and to competition factors. As with most crops and pastures, the key to successful establishment is the use of good agronomic practices, timing, controlling or minimizing competition, and having adequate stored soil moisture and follow-up rainfall. With Progardes cultivation and good seedbed preparation, a shallow planting depth and weed control have been shown to be very beneficial particularly in central and southern areas of Queensland. The timing of planting in northern areas has generally been at the end of the dry season/start of the wet season, which gives the establishing plants a good opportunity to grow with the sown or native grasses and enables seed set and some grazing to occur late in the summer/autumn period. However, false starts to the wet season are major risks because the annual rainfall of Australian rangelands is very variable (McKeon, 2006). This risk can be mitigated to some degree by planting 3 kg of Progardes seed/ha as a mixture of hard and soft (scarified) seed, usually in a 50:50 ratio. The ratio can be altered according to expected conditions, e.g. with a higher proportion of scarified seed used in more favourable planting environments.

There is, however, some debate relating to the best planting time because of the vagaries and variability of the start, end and duration of the northern Australia wet season. Bishop (no date) outlined the benefits and risks associated with early and late planting. False starts and weed competition are risks associated with early planting, but benefits include a full growing season allowing good plant establishment, seed set and grazing. With late planting there may be poor and slow plant development, and the possibilities of

Fig. 11.11. Cultivar JCU 5 (*D. virgatus*) (a) on weed mat and (b) as a seed crop on the Atherton Tablelands.

no seed set and limited grazing until a second season of growth has occurred. Benefits of late sowing include less weed competition, and a better soil moisture profile for plant establishment. For best perennial legume establishment in southern and central Queensland, Mace *et al.* (2015) advocate a fallow period of 1–4 months or more followed by a late summer/early autumn planting period. Clem *et al.* (1994) discuss many aspects of pasture establishment in the Brigalow region, including the merits of using hard seed if planting early in the season.

Scarification of Progardes seed by hot water, or for large commercial batches of seed the use of a mechanical abrasive disc typically enhances germination from about 10% (untreated) to 70–80% (with scarification), which is similar to results reported elsewhere (see Hopkinson and English, 2004). Experience has shown that two to three wet seasons may be required for Progardes to become well established in the harsher, drier environments such as on the Mitchell Grass Downs. This slow but progressive establishment over several seasons may be related to the time needed to break down and soften the hard seed coat. Legume hard seededness is caused by the inability of the seed coat to transmit water and oxygen freely to the inner structures of the seed for germination to occur. Hard seededness is of ecological importance as it lengthens the life span of the seed and allows for progressive germination over time, thus increasing opportunities for species survival (Argel and Paton, 1999). Seed dormancy (in this case due to hard seededness) therefore provides an important mechanism for plant species to calibrate germination with environmental conditions in a way that will maximize the probability of recruitment (Baskin and Baskin, 1998).

In a study of soil seed banks and dormancy-releasing mechanisms of nine *Desmanthus* genotypes, Rangel *et al.* (2015) found persistent seed banks under all nine genotypes. It was also found that an increase in the percentage of softened seeds correlated with higher temperatures in two patterns. Some accessions were not significantly affected by temperatures below 80°C, but others became soft when temperatures rose to 60°C. The authors suggested that high seed production and dependence on elevated temperatures to soften *Desmanthus* seeds is an important strategy for survival in dry regions. In a study examining the effect of fire on soil temperature and heat on *Desmanthus* germination using thermocouples buried from 0 to 3 cm deep in the soil, Rangel and Gardiner (1996) found the average soil temperatures under a grass fire to be: 242°C on the soil surface, 76°C at 0.5 cm soil depth, 62°C at 1 cm and 53°C at 3 cm. Seedcoat dormancy of *Desmanthus* seed may well be affected by the near ideal range of heat in the 0.5–1 cm zone. Fire in savanna landscapes may thus assist *Desmanthus* seedling recruitment and pasture sustainability. Jones and Carter (1989) found that the higher the soil temperature and the longer the dry season, the greater the amount of hard seededness breakdown and with *Stylosanthes* they found this was correlated with the number of days during which maximum soil temperatures exceeded 50–55°C. Van Klinken *et al.* (2006) discuss the role of wet heat, dry heat, drought, microclimates and fire on seed dormancy and hard seededness of legumes. It has also been noted that mature *Desmanthus* plants are capable of developing new stems from the base of the plant after fire.

During the current drought (2012–2015), a study was undertaken of a Progardes legume trial plot near Hughenden on a Mitchell Grass Downs site to determine the Progardes' soil seed bank. The 2.5 ha site was sown in 2010 and established to become a good stand of the legume; although grazed, a considerable amount of seed was produced and seed naturally entered the soil seed bank over time prior to the drought. The 2013 rainfall at Hughenden totalled only 129.6 mm for the year, while the long-term mean annual rainfall is 492 mm (Table 11.1). In February 2014 the plants were considerably stressed by drought, but had a population of 3.7 plants/m^2 with a frequency of 66%. Towards the end of 2013, 16 soil cores were taken (using a cylindrical steel ring 7 cm in diameter × 5 cm deep) at random across the 2.5 ha Progardes trial plot and the resulting soil cores were sieved to recover any Progardes seeds. The result showed that there are 6349 seeds/m^2 across the plot in the top 5 cm of the soil seed bank. In a subsequent germination test of these recovered soil seed bank seeds, 43% germinated (Gardiner, 2014a).

The results from this Hughenden study confirms that in normal seasons Progardes *Desmanthus* produces abundant seeds, that a large soil seed bank can accumulate over time and that a good proportion of those seeds are available to potentially germinate, while also a good proportion remain hard in the soil seed bank for germinating at a later date. The real test of the value of hard seededness of Progardes and its soil seed bank in terms of germination and establishment at this and other sites will be seen when the current drought terminates. Lawrence *et al.* (2012) state that *Desmanthus* seeds may persist in the soil for many years and Murat (a Progardes and *Desmanthus* commercial seed grower) (2010, personal communication) observed mass *Desmanthus* recruitment in a paddock some 8 years after a single *Desmanthus* crop was planted and harvested. Other soil seed bank studies reported in the literature have shown *Desmanthus* to have from 87 seeds/m^2 to several thousand seeds/m^2 (Burrows and Porter, 1993; Rangel and Gardiner, 1996; Jones and Brandon, 1998; Gardiner *et al.*, 2004), and even at the Isisford site (Gardiner *et al.*, 2004) with just 87 seeds/m^2, a population of *Desmanthus* still persists today, a decade later. Jones (2014), however, suggests that the demise of Siratro among a number of factors is related to its lack of an adequate seed bank, which averaged about 200 seeds/m^2.

As mentioned above, the very extensive nature of the grazing properties across northern Australia poses problems for sowing and establishing a new pasture species such as Progardes *Desmanthus*. Such problems include a lack of a sown pasture culture, the scarcity of broad-acre farming equipment in the region and the potential expense of sowing such immense rangelands (Gardiner *et al.*, 2012). The methods selected or used to sow and establish pastures can depend on many factors, including: economics and available finance, topography, rainfall and the soil moisture status, the degree of tree cover, the amount of timber or rocks on the ground, the degree of vegetation and ground cover, available time, access to equipment, experience and personal preferences for a particular planting method. One approach to address some of these issues and in some circumstances is the potential use of livestock to faecal seed Progardes across the vast semi-arid tropical clay soil regions where it is adapted. Due to the long annual dry season in the region, supplementary feeding of cattle with, for example, molasses and/or dry licks is common practice and adding seed to these supplements is considered a simple practical way of introducing seed.

Seeds of leguminous species with hard seeds are known to survive digestion and be dispersed in faeces by ruminants (endozoochory) (Simao Neto and Jones, 1987; Russi *et al.*, 1992). There is evidence, particularly with the stylos in northern Australia's light-textured soil regions, that grazing stylo with mature seed pods or incorporating stylo seed into dry season supplements is a means of spreading this valuable legume through paddocks. It is known that some stockfeed companies regularly get requests from their clients to include stylo seed into their various supplements for cattle (Coleman, 2009, personal communication). There are also examples of other pasture legumes being successfully spread by cattle; for example, Bishop (1988) was very successful in using cattle to graze (harvest) and disperse Glen Joint Vetch (*Aeschynomene americana*) seed from a 2.5 ha 'protein bank' to an adjacent 120 ha paddock over a 2-year period with a resulting legume frequency of up to 91%, while conventional sown stylos had only a 21% frequency. Using Switch grass (*Panicum virgatum*), Ocumpaugh *et al.* (1996) demonstrated significant advantages of faecal seeding over conventional broadcast seeding in terms of seedling emergence, establishment and growth. Armke and Scott (1999) found that 75% of seeds of *D. illinoensis* (Illinois bundleflower) fed to cattle were recovered in faeces and had a high germination rate. They suggest that *D. illinoensis* is a potential candidate for faecal seeding.

At JCU, Townsville Progardes seed fed in molasses to a mixed herd of cows as part of a regular dry season supplementation programme was noted to be abundant in cow dung 2 days after feeding. Hundreds of seeds were found per cow pat, and the seed recovered from these cow pats had 72% germination (see Fig. 11.12). This suggests that *Desmanthus* is readily consumed in supplements and that a considerable amount of seed passes through the cattle in a viable state. As Progardes *Desmanthus* has a hard seed coat, it not only survives the passage through the digestive tract, but will also survive sitting in the cow pat for extended dry periods before the wet season arrives. Dung beetles were noted to be active in these cow pats, but with no apparent adverse effect on seed viability or germination in the field Gardiner (2014b). Indeed, dung beetles may be beneficial in

Fig. 11.12. (a) Cows licking Progardes seed in dry season molasses supplement at JCU, Townsville. (b) Progardes seeds germinating in dung in a paddock at Charters Towers, North Queensland after being fed with molasses as a dry season supplement.

breaking down cow pats and creating a more friable seed bed. Using sheep fed cultivar JCU 2, Gardiner *et al.* (2012) found most seeds passed through the animals in 48 hours, but only 9% of the seed fed was recovered, of which about 60% remained viable.

Various authors, including Cook *et al.* (1993), Ocumpaugh *et al.* (1996), Armke and Scott (1999) and Gokbulak and Call (2004), list and discuss the benefits and drawbacks of faecal seeding for the seed and establishing plant in dung pats. These are summarized below.

The dung pat characteristics that enhance plant establishment include:

- high nutrient status;
- the pat or microsite has high organic matter;
- good moisture and moisture retention;
- short-term reduction in competition with pre-existing vegetation;
- short-term grazing avoidance of the dung pat site.

The drawbacks may include:

- possible poor distribution of the seeds across the whole paddock (seed may end up mostly near feed troughs or watering points);
- some seeds may be digested;
- some seed may not germinate from within a deep/thick cow pat;
- the dung pat may form a crust and restrict germination;
- it may be a slow method of establishing a new pasture;
- not all seeds are suitable for faecal seeding.

Some of the possible drawbacks with faecal seeding can be managed by widely separating the feed and water troughs in the target paddock, and by regularly moving the position of the feed troughs to different parts of the paddock. Agrimix Pty Ltd (Kempe, 2015, personal communication) recommend adding only small amounts of Progardes seed in supplements at any one time, but doing this continually for the duration of the dry season with the aim of distributing the seed widely and over many dung pats over time. Faecal seeding could also be combined with feeding phosphorus supplements in the wet season when cattle graze further from feed troughs. The grazing system used will also influence potential seed distribution, e.g. in smaller paddocks (cell grazing) cattle may utilize the whole paddock, resulting in more even distribution of dung and seed.

In 2010, a 250-ha paddock of 'run-down' buffel grass was ploughed and Progardes seed was broadcast onto the ploughed soil surface near Springsure in central Queensland's Brigalow region. A Progardes plant density of 7 plants/m^2 with a frequency of 75% was achieved at the end of the first growing season. The current drought has thoroughly tested the resilience and persistence of Progardes at this and other sites. The 2012/13, 2013/14 and 2014/15 wet seasons failed across northern Australia, but where rain has occurred, Progardes has proven that it is well adapted and has re-sprouted from the crowns of previously established plants and/or from its hard seed in the soil seed bank. Friedel *et al.* (1990) state that plants that allocate considerable resources to large root structures are more tolerant of drought and of low levels of nutrients. Across the range of *Desmanthus* species there are a variety of root morphologies (Luckow 1993), ranging from deep tap roots as seen on a Progardes plant extracted from a field site on a vertisol soil in the Mitchell Grass Downs near Hughenden to napiform (turnip shaped) roots on *D. tatuhyensis*, as shown in Fig. 11.13.

11.3.2 Persistence in buffel grass stands

The Progardes/buffel paddock near Springsure has had a regular commercial beef cattle grazing regime since its establishment in 2010 and a botanical survey in 2015 showed that it remains in good condition having come through the latest dry period and forms a good pasture in association with buffel grass, with 7.25 Progardes plants/m^2 with a frequency of 74%. This result is particularly good because in central Queensland buffel run-down is problematic (Robertson *et al.*, 1997; Peck *et al.*, 2011) and the grazing community are fully aware of this and are seeking solutions to the problem. Robertson *et al.* (1997), however, notes that few legumes appear to be compatible with buffel, but Reid (1983) noted when collecting *Desmanthus* in Mexico that in certain regions on clay soils it often occurred naturally with the introduced buffel grass. There are a number of examples of *Desmanthus* persisting and thriving with buffel in Queensland for a decade or more, such as at the Woodbine Station site near Blackall (Chapter 9, this volume), now 26 years since planting and at Redcliffe Station near Hughenden, now 13 years since planting. Progardes appears to be one of the very few legumes available and is proving to be adapted and persistent on such run-down clay soils with buffel grass.

11.3.3 Inoculation

The industry recommendation is to inoculate *Desmanthus* seed at sowing with *Bradyrhizobia* strain CB3126 because *Desmanthus* is said to be highly specific in its rhizobia requirements (NSW Ag, 2004; Cook *et al.*, 2005). However, Date (1991) suggests that *Desmanthus* does not have a highly specific requirement for rhizobia for effective N fixation. Our lack of success with inoculation of Progardes *Desmanthus* in the

Fig. 11.13. Examples of the rooting systems of *Desmanthus* that aid their ability to tolerate drought. (a) Deep Progardes taproot; (b) napiform tap roots of *D. tatuhyensis*.

seasonally dry tropics is most probably due to the commercial rhizobia not surviving in the hot dry soil conditions that often prevail at planting time. This seems to be the case whether the inoculums are peat based, clay based or even water injected. For these reasons, to date most commercially sown Progardes seed has not been inoculated with CB3126, yet in northern Australia nodulation is often noted in the field.

It has long been suspected that indigenous rhizobia in neutral to alkaline clay soils associated with the native legume *Neptunia* species effectively nodulates *Desmanthus* (Brandon *et al.*, 1997). To verify this, this author collected nodules from *Neptunia gracillis*, *N. major* and *N. monosperma* from north and western Queensland plus nodules from *Desmanthus leptophyllus* plants that had not been inoculated with CB3126. Seed of *Desmanthus* (JCU 2) seed and the nodules were provided to the Australian Inoculants Research Group in New South Wales so they could evaluate rhizobial isolates for their suitability as inoculants for *Desmanthus*. The nodules provided 35 presumptive rhizobial isolates and these were screened in a basic plant growth test for nodulation and growth response to nitrogen fixation by visual score. Seven isolates scored higher for plant response to inoculation than the commercial strain CB3126. The results confirmed that rhizobia isolates from *Neptunia* spp. from north and western Queensland effectively nodulate Progardes *Desmanthus virgatus* (variety JCU 2). This research suggests that these seven isolates have the potential to improve nitrogen fixation and subsequent *Desmanthus* yields (Hartley and Gemell, 2015). Interestingly, CB3126 originates from Mexico and was sourced from *Leucaena* plants on an acidic soil (Mullen *et al.*, 1998; Fornasero *et al.*, 2014). As most Progardes, and for that matter *Leucaena*, in Queensland is sown on neutral to alkaline soils, these new rhizobium isolates may potentially improve the performance of both of these legumes. Fornasero *et al.* (2014) in Argentina also found that inoculation with local rhizobium produced higher *Desmanthus paspalaceus* aerial plant dry weights compared with CB3126 inoculated plants and concluded that the local isolates have remarkable tolerances to several abiotic stresses including acidity, salt and temperature, while exhibiting prominent N fixation, thus indicating their suitability as candidates for inoculation of *D. paspalaceus*. They also state that, from a practical point of view, the temperature tolerance displayed by some of the Argentinean isolates would be expected to constitute a positive trait, likely to favour rhizobial survival both in soil and while on the surface of inoculated seeds.

These finding suggest we should test the isolates from *Neptunia* species and undertake more extensive studies to collect and test rhizobial isolates from different regions, and edaphic and climatic situations where Progardes is adapted, as local native rhizobia may overcome the deficiencies noted above of current commercial strains of rhizobia, particularly if local isolates survive in the soil and on the seed coat in high temperatures, as suggested by Fornasero *et al.* (2014).

11.3.4 Animal production

It is expected that even small amounts of Progardes will make improvements to animal production in the targeted dry tropical environments due to what Bell (2009) terms the marginal value of feed that such nutritious plants offer. Kretschmer and Pitman (2001) state that as little as 10% of a legume in poor quality grass pasture will make significant impacts on animal intake and therefore production. Only a limited number of grazing and feeding trials have been undertaken to date with the new cultivars JCU 1–5 individually, or as the blend Progardes. Steers grazing a buffel grass and Progardes mixed pasture near Springsure in central Queensland during a cool dry winter in 2011 gained an additional 40 kg live weight over a 90-day study period compared with steers on an adjacent buffel grass-only pasture (Gardiner and Parker, 2012). More recently (May 2015), a 250-head mob of steers in the same Progardes/buffel paddock had an average daily gain of 800 g/head/day. This paddock is illustrated in Fig. 11.14. Results from other grazing and/or feeding trials with cattle grazing or fed *Desmanthus* include Hill *et al.* (2004), who found average daily gains for steers grazing *Desmanthus* pasture to be in the range 400–600 g/head/day; Sukkasame and Phaikaew (1998) reported live-weight gains of 890 g/head/day. Rangel and Gardiner (2009) demonstrated that Merino sheep on a Mitchell grass diet supplemented with *Desmanthus* produced more wool

Fig. 11.14. An excellent 250-ha paddock of Progardes/buffel grass pasture near Springsure in the Brigalow region of central Queensland.

and lost less live weight than sheep on a grass-only diet. A trial feeding sheep cv. JCU 1 and Flinders grass vs Flinders grass alone had a positive N balance and significantly improved intake and weight gains when sheep had access to the Progardes in the diet (Ngo, 2012). In informal monthly observations, including regular live weight weighing of kids at James Cook University, Townsville, Aoetpah (2015, personal communication) observed that 29 female growing kids with an average live weight of 15 kg offered cv. JCU 1 hay and Rhodes grass (*Chloris gayana*) hay plus some soybean meal had a very strong preference for the *Desmanthus* hay and gained on average 103 g/head/day for a 46-day period. Although an informal study (it is to be repeated in a formal trial in the future), the results compare well with those of, for example, Kanani *et al.* (2006), who evaluated lucerne, Lab, *Leucaena* and *Desmanthus bicornutus* with growing goats, the average daily gain of which was 78.4 g/day (range: 60.9–93.9 g/day).

11.3.5 Nutritive value and seeds

Crude protein levels of Progardes and *Desmanthus* leaf, stem and pods with seeds have been recorded as being variously in the order of 20%, 11%, and 17% respectively, and metabolizable energy levels of the whole plant at some 10.3 MJ/kg dry matter (DM) (Agrimix Pty Ltd, no date; Gardiner, 1992; Adjei and Pitman, 1993; Gardiner and Rangel, 1996; Jones *et al.*, 2000; Cook *et al.*, 2005; Kanani *et al.*, 2006; Rangel and Gardiner, 2009). The mature seeds of *Desmanthus* also have good levels of crude protein of about 29%, which is similar to lupins and peas (Schlink and Burt, 1993; Gardiner and Rangel, 1996), high concentrations of S, and adequate levels of other macro and micro minerals (Schlink and Burt, 1993). As discussed elsewhere in this chapter, *Desmanthus* has a high degree of hard seededness, as illustrated by, for example, up to 75% seed recovery from faeces after being fed to cattle (Armke and Scott, 1999). However, there was only about 9% recovery when fed to sheep (Gardiner *et al.*, 2012), suggesting that in this case a considerable amount of seed was digested. Legume seed can be an important source of protein as well as macro and micro minerals, particularly during the dry season, for grazing livestock when other feed is of poor nutritive value. This can be especially valuable if the seed is held in the pods on the plant and accessible to livestock for some time into the dry season (Schlink and Burt, 1993). A range of pod dehiscence has been observed in Progardes and other *Desmanthus* genotypes with, for example, some accessions of *D. bicornutus* retaining seed in the pod late into the dry season in north Queensland.

11.3.6 Other uses

Anthelmintic

Other developments with Progardes and *Desmanthus* genotypes include interest in them as anthelmintics, hay, green manure/break crops, for revegetation purposes, and for their anti-methanogenic potential. In a study investigating the anthelmintic effects of tropical shrub legumes in ruminant animals, Cresswell (2007) notes that a number of forage species containing high concentrations of tannins have been shown to have anthelmintic properties. *Calliandra calothyrsus*, *Leucaena leucocephala* and *Desmanthus virgatus* were identified as having possible anthelmintic activity against nematode eggs and larva and *Desmanthus* has the most promise because egg hatch rates of *Haemonchus contortus* (barbers pole worm) were below 50% and in *Trichostrongylus colubriformis* (black scour worm) larvae exposed to the *Desmanthus* extract, over 35% degenerated. She goes on to suggest that the structure and concentration of the condensed tannins in *Desmanthus* may help determine plant anthelmintic properties.

Hay

Interest in producing hay from well-adapted tropical legume species in northern Australia for dry season supplementation, particularly for weaner calves is growing, especially where irrigation is available. The dairy and beef finishing industries in north Queensland require a year-round supply of affordable, high-quality (digestible protein and carbohydrates) feed to achieve meat and milk production at levels suitable for maintaining profitable enterprises (Cox *et al.*, 2012). However, poor winter/dry season growth of pastures results in a seasonal feed shortage, which is overcome through the use of expensive supplements. Perennial herbaceous legumes grown for hay could provide an affordable alternative. Lucerne (*Medicago sativa*) is a highly productive legume hay in the ssubtropics (Cook *et al.*, 2005), but (anecdotal) producer experience suggested lucerne has poor tolerance for the northern tropical wet summer climate (Cox *et al.*, 2012). Alternative promising legumes for hay include *Stylosanthes guianensis* varieties and a range of *Arachis* spp. (Cox *et al.*, 2012). Experience with *Desmanthus* as a hay crop is very limited and the bipinnate leaf is considered fragile. Several Progardes crops have been baled however and interest in and inquiry for such a clay soil-adapted potential fodder crop in north-western Queensland, which is remote from existing hay-producing regions, is growing and field trials have commenced.

In western Queensland near Hughenden, Progardes/Rhodes grass hay has been cut and baled successfully several times (Reid, 2014, personal communication). A crop of cv. JCU 1 was grown in the Burdekin Irrigation Area just south of Townsville; the 5 ha paddock was cut and baled once and yielded 276 round bales (Fig. 11.15), an equivalent of 6072 kg DM/ha. Some Progardes and other *Desmanthus* accessions such as JCU 5 may hold leaf and have finer stems, or indeed may be much higher yielding, which may make them more suitable for hay production. However, as with any new crop, considerable testing is required to learn, for example, exactly how and when to cut and bale to maximize yield and leaf retention.

Antimethanogenic potential

In Australia, enteric emissions from the livestock sector contribute some 10% of Australia's greenhouse gases, and the northern tropical beef industry is said to be a major contributor to those emissions. The dominant pastures in the north are composed of C4 grasses, which generally have lower nutritional value than temperate grasses.

Fig. 11.15, Examples of Progardes hay crops. (a) 276 round bales of JCU 1 yielded > 6 t DM/ha; (b) an experimental line of *D. bicornutus* (JCU 6) with massive growth yielding > 20 t DM/ha in one cut at JCU, Townsville.

With these and other factors, methane emissions from the northern Australian beef herd are considered to be higher than for more intensive systems (Charmley *et al.*, 2008). In a review of enteric methane abatement, Beauchemin *et al.* (2007) state that legumes hold promise for CH_4 mitigation, but farmers are unlikely to adopt such measures unless there are positive economic impacts on animal production. Improving forage quality can both improve animal performance and reduce CH_4 production, but it can also improve efficiency by reducing CH_4 emissions per unit of animal product (Eckard *et al.*, 2010). As Progardes *Desmanthus* is proving to be well adapted to much of the north's very extensive semi-arid clay soil areas where few other legumes have persisted, it may thus prove valuable not only directly to livestock production but also to the environment as an antimethanogenic feed resource.

Ramírez-Restrepo *et al.* (2011) and MLA (2012) have reported on the *in vitro* antimethanogenic potential and nutritive value of a number of tropical legumes, including *Calliandra*, *Clitoria*, *Desmanthus* (including cultivars JCU 1, 2 and 4), as well as cv. Marc, *Leucaena*, *Stylosanthes* and others, compared with the temperate forage lucerne/alfalfa (*Medicago sativa*). Of the tropical forages tested there was great variability in the antimethanogenic potential in the rumen. Some of the forages had diet quality parameters similar to lucerne but with significantly lower methane production. Compared with the methane production of lucerne, *Calliandra calothyrsus*, *Leucaena leucocephala* and *Desmanthus* spp. produced lower methane (mL/g/DM) by 56%, 36.5% and 32% respectively. The variation among the *Desmanthus* accessions tested (Progardes (JCU 1, 2 and 4) and cv. Marc) was considerable. Of the legume species evaluated in the MLA report, those in the *Mimosaceae* subfamily may have superior antimethanogenic properties to those in the *Fabaceae*, and the report concludes that the results indicate that there is an opportunity to use some plants for methane mitigation in grazing ruminants in northern Queensland. Vandermeulen *et al.* (2015) evaluated the *in vitro* mitigation potential of varieties of Progardes (*D. bicornutus*, *D. leptophyllus* and *D. virgatus*). Samples of the legumes and the control Rhodes grass (*Chloris gayana*) were incubated *in vitro* using ruminal fluid from grazing Brahman (*Bos indicus*) steers. Overall, the *in vitro* organic matter degradability (OMD) and methane production between *Desmanthus* species differed ($p < 0.001$). It is suggested that the impact of these attributes may benefit farmers, expand productivity, improve livelihoods and meet the need for growing food demand.

Condensed tannins (CT) have been shown to reduce CH_4 production by 13–16% (DM intake basis) (Waghorn *et al.*, 2002, cited by Eckard *et al.*, 2010). Eckard *et al.* (2010) say that many plants contain CT and that they may be used to introduce CT into the diet of animals when daily supplementation is not practical or economic. As discussed above in relation to CH_4 mitigation, the three species showing potential as antimethanogens, *Calliandra*, *Leucaena* and *Desmanthus*, all contain CT to varying degrees (Ahn *et al.*, 1989; Adjei and

Pitman, 1993; Otsyina *et al.*, 1999; Cook *et al.*, 2005; Cresswell, 2007).

Green manure

There is increasingly interest in Progardes and yet-to-be-commercialized *Desmanthus* genotypes as potential green manure crops (Fig. 11.15). At the International Rice Research Institute, rice yields were significantly increased by the incorporation of *Desmanthus* as a green manure and were the same as those given 90 kg N/ha. The *Desmanthus* in those trials yielded 13.9 t DM/ha in 7 months (Carangal *et al.*, 1994). As an irrigated green manure/break crop in the Burdekin region of north Queensland, Progardes can accumulate 15 t/ha of above-ground dry biomass (Ockerby, 2013). It was estimated that about one-third of the biomass is carbon, so 5 t/ha of carbon in the form of mostly woody organic matter can be incorporated into the soil. About one-third of that carbon can be converted to humus, which results in an increase of about 0.15% soil carbon per crop over the long term. The amount of carbon in stable organic matter generated by a Progardes break-crop is thus small, but any incremental increase in soil organic carbon is an important change to the health of sugarcane farming systems. Progardes thus offers farmers an alternative break-crop to those such as soybean, with the advantage that Progardes has more residues which are much woodier, and therefore has much greater capacity to store and slowly release nutrients.

Ecological potential

Research is also ongoing in other areas, such as the use of Progardes and other tropical legumes and grasses in revegetation projects along the Burdekin River in north Queensland in an attempt to reduce erosion, sediment and nutrients reaching the Great Barrier Reef from grazed lands in the river catchments. Hall (2014) outlines the 'Reef Rescue Project' and its treatments, and in 2015 (Hall, 2015, personal communication) found excellent plant density and seedling recruitment of Progardes on several of the clay soils in the revegetation trial plots.

11.4 Conclusions

The use of adapted and persistent legumes such as the *Desmanthus* cultivars and particularly Progardes *Desmanthus* in the semi-arid clay soil regions of northern Australia, where few if any other herbaceous sown pasture legumes are currently adapted, will give farmers and graziers new and exciting opportunities to improve pastures and increase livestock productivity. The new cultivars JCU 1–5, either as stand-alone cultivars or as the blend Progardes, have the potential to make significant impacts on animal production and ecosystem function, whether in live-weight gains, turn off, improved wool production, reproductive performance, as an antimethanogenic feed resource, a green manure, a possible anthelmintic or providing N to sustain productive and healthy grasslands.

12 The Cattle Industry of Belize: A Brief History of Research and Development to the Mid-1970s

J.R. Lazier*[1]
*Formerly International Livestock Centre for Africa

Abstract
Belize, one of the sites of the collaborative IDRC/UWI–CSIRO research reported in this volume, is a small Central American country whose economy is little known compared with that of Australia, thus background information on its livestock industry is presented here to set the scene for the chapters that follow. With a small population and extensive areas of grassland, it has been seen as having potential as a source of protein for the countries of CARICOM. Though cattle had long been used as draught animals in logging, a cattle industry for beef has been considered seriously only since the 1930s. Despite the industry having since been given high priority in Government-development plans, research has been sporadic, mainly dependent on special programmes funded from abroad and on temporary personnel. Considerable progress was made in developing the industry, but by the mid-1970s there were major problems including lack of infrastructure, uncertain markets and lack of management skills. Some adapted, improved pasture species were identified, but more work was required, particularly on forage legumes. Cattle raising had been encouraged with some success as a step in the development of sedentary farming from shifting agriculture. In 1970 the cattle population was 38,000 head and 92% of the 1322 herds in the country had 50 animals or less.

12.1 Introduction

As the site of much of the field activities of the collaborative CSIRO–IDRC/University of the West Indies Forage Legume Research Programme, reported in this volume, Belize's livestock industry has been little reported compared with that of Australia. It is reviewed briefly here to provide background for the following six chapters. Additional background to the research is provided by a description of the native pastures of central Belize in the next chapter (Chapter 13, this volume). Formerly known as British Honduras, Belize is situated in Central America on the east coast of the Yucatan Peninsula (Fig. 12.1). A small country, about half the size of Jamaica, it has a land area of about 23,000 km^2. The presence of soils suitable for agriculture, a moderate rainfall (1500–4000 mm) and a low population have resulted in Belize being viewed as a potential source of food for the more densely populated West Indian islands. A member of CARICOM, the Caribbean Common Market, its main agricultural export potential is to the other members of that organization.

12.2 Early settlement years

In the early years of settlement, the late 17th and early 18th centuries, the area that is now

[1] E-mail: john.lazier@sympatico.ca

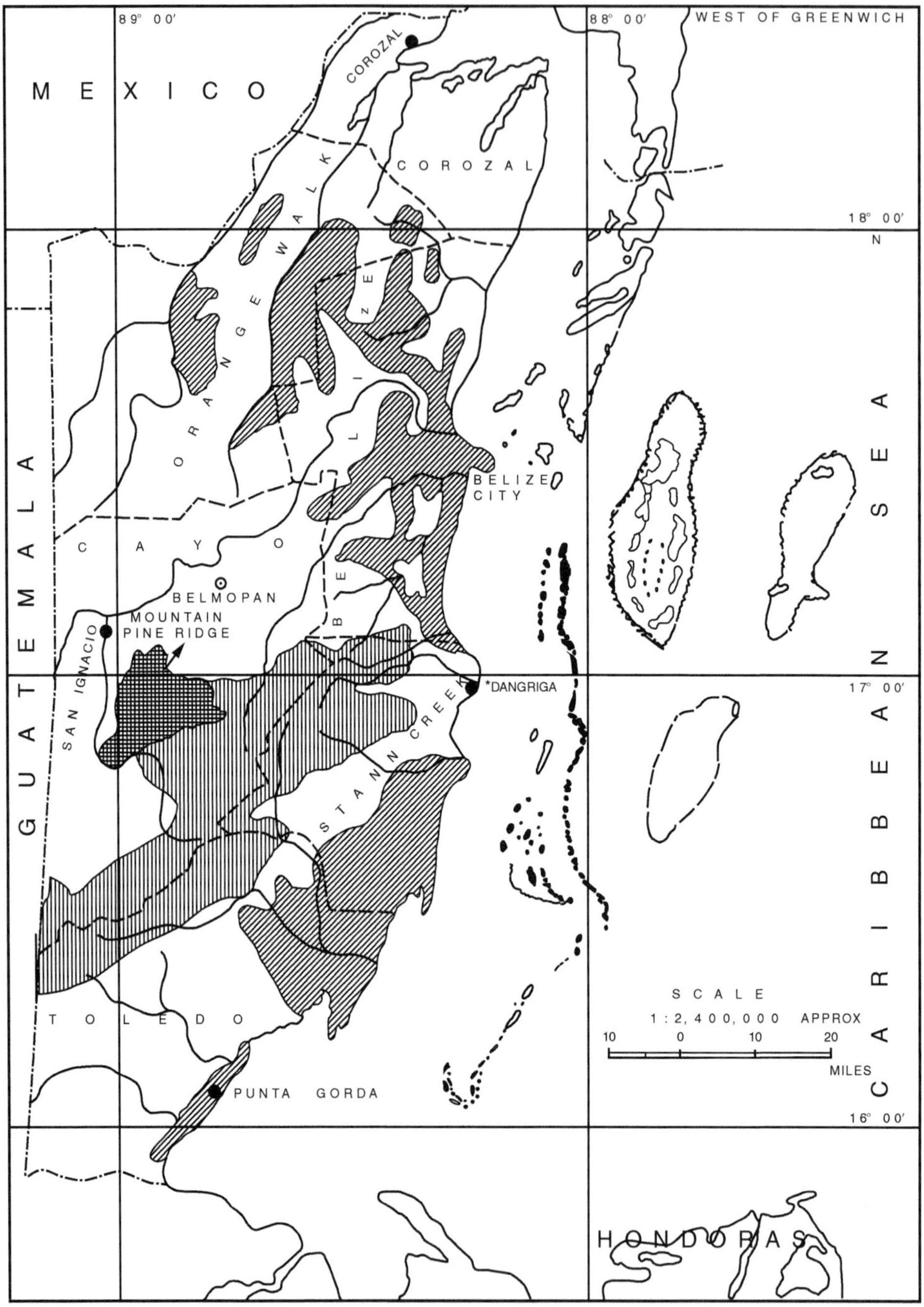

- - - District boundaries; ⊙ Belmopan- Capital; ● District center; Vertical lines: mountains 600–1000 m; Slanted lines: Low pine savanna; Hatched areas: High pine savanna;

Fig. 12.1. Belize, physical and administrative.

Belize was ostensibly under Spanish control, but British interests dominated the colony. During this period the entire economy and thus the entire population was concerned with logging and no attention was paid to agriculture. This attitude was encouraged by the Spanish, whose inspection teams, enforcing treaty terms, destroyed any crops found. After Spanish influence waned, agriculture continued to be ignored because of the shortage of labour, the larger profits made in logging and the attitude of the traders, major land owners, who made substantial profits importing food.

Cattle had been raised in ranches throughout this period mainly for use as draught animals in logging operations; meat production for the mahogany concessions and local settlements was only of minor importance. As other agricultural industries such as bananas developed, they too used bullocks for hauling produce. All pastures were based on native grasses until the beginning of the 20th century when Guinea (*Panicum maximum*, Jacq.), Para (*Brachiaria mutica*, Stapf.) and Bermuda (*Cynodon dactylon* (L.) Pers.) grasses were introduced. However, they were little used (Dobson, 1973).

12.2.1 The 1930s

By the 1930s, sufficient interest in agriculture had developed that agricultural stations were established, one in each of the four areas of the country (Corozal, El Cayo, Stann Creek and Punta Gorda). Professional staff in the mid-1930s numbered three. Though there was little interest in cattle raising generally, they were still being produced for the traditional markets and beef was assuming somewhat more importance than draught as tractors had been replacing oxen in logging since the 1920s.

As late as 1935, the emphasis on cattle raising in the Agriculture Department was not on beef but on draught (DOA, 1935) for the aim was to encourage a switch from the universally used 'shifting agriculture' (milpa farming) based on maize to a sedentary agriculture with cultivation of the main tropical crops. Cattle were being promoted as a cheap source of farm energy.

Because cattle raising was slowly increasing (Table 12.1), some attention was given to its problems. Due in part to poor communications there was little export of meat from each district. The meat had to be sold in one day because of a lack of refrigeration facilities, thus only small animals that had very low dressing-out weights were being butchered. Belize City was a particular problem, for as the largest consumer of meat in the country it had very poor transportation to the districts, no pastures for holding animals until sold and no refrigeration facilities to hold the animals once slaughtered. Mainly as a consequence of poor communications, the beef needs of Belize City were satisfied by imports from Honduras (Table 12.1).

On the ranches there was a scarcity of improved forage plants and lack of stored feed for the 3-month dry season. The lack of stored feed caused semi-starvation of herds and considerable weight loss or death of large numbers of animals, depending on the length and severity of the dry seasons. The main cattle breed was Criollo, which is derived from Spanish stock and is hardy, but the breed was unimproved and slow to gain weight. Department research of the 1930s included the establishment at Corozal and Stann Creek of trial plots of grasses and legumes for grazing, and work on stall feeding, reserve feed for the dry season and green manure. Small dairy herds were established at the two sites and progeny were sold to small farmers to help upgrade their herds; rotational grazing was used at the Stann Creek station and pit silage was successfully made at Corozal. 'Bush' silos were tried at both stations, but no report was made on the result.

12.2.2 The 1940s

In the 1940s, as the Department of Agriculture began to appreciate the value of the beef industry, an assessment was made of limitations to its growth (DOA, 1948). Problems cited included poor productivity of native pastures, absence of fencing and virtually no improved management practices other than branding by 90% of owners. There were no improved pastures, no storage of feed for the dry seasons, no rotation of herds on pastures and much praedial larceny (DOA, 1948).

Generally the incidence of disease was low, but with no Veterinary Officer in the country the statement could not be made with any confidence. By 1947 a Veterinary Officer had been recruited

Table 12.1. Statistics of the Belizean cattle industry 1930–1974.

Year	Total slaughter (head)	Imports for slaughter (head)	Average dressed weight	Exports (head)	Imports breeding (head)	Reference
1930	–	576	–	0	0	B 1933
1931	–	640	–	0	0	B 1933
1932	–	398	–	0	0	B 1933
1933	–	600	–	0	0	B 1933
1936	–	65	–	14	0	A 1935
1938	–	960	–	0	96	A 1938
1939	–	812	–	–	–	A 1943
1940	–	151	–	0	0	A 1943
1941	–	650	–	0	189	A 1943
1942	–	755	–	0	94	A 1943
1943	–	1628	–	0	42	A 1943
1944	–	1362	–	0	10	A 1944
1945	–	1017	–	0	24	A 1945
1946	–	1263	–	0	0	A 1946
1947	–	622	–	0	2	A 1947
1948	2742	509	91	–	25	A 1948
1949	2974	–	–	–	–	A 1949
1951	3226	120	–	–	10	A 1951
1952	3074	213	106	–	17	A 1952
1953	3250	–	98	–	35	A 1953
1954	3036	–	103	–	54	A 1954
1955	3651	13	–	–	40	A 1955
1956	3728	4	102	–	18	A 1956
1957	–	–	–	–	58	A 1957
1958	–	27	–	–	0	A 1958
1959	–	–	–	–	30	A 1959
1960	3554	–	–	–	46	A 1960
1961	3315	–	102	5	1	A 1961
1962	3411	–	116–136	–	–	A 1962
1963	3444	–	122	4	–	A 1963
1964	4195	–	112	46	–	A 1964
1965	4231[a]	–	113	653	–	A 1965
1966	4578	–	114	967	–	A 1967
1967	4804	–	112	950	–	A 1967
1968	4945	–	120	391[b]	153[c]	A 1968
1969	3081	–	143	–	–	A 1970
1969	3114	–	–	–	–	C 1969
1970	3310	–	144	–	–	A 1970
1971	4390	–	155	26	–	A 1971
1972	5004	–	153	–	–	A 1972
1973	6910	–	148	228,400[d]	–	A 1973
1974	5670	–	153	72,400[d]	–	A 1974

[a]Incomplete; [b]ceased November 30; [c]unspecified imports; [d]kg; A, DOA (1933–1974); B, Government of British Honduras (1934); C, Government of British Honduras (1969).

and his services were offered free for the first year (DOA, 1947). Work continued at Stann Creek and Corozal stations on pasture grasses and a third milk herd was established in Cayo District. Guinea and Para grasses were found to do well. Forage legumes were not mentioned again, so presumably either the plants or enthusiasm for the trial failed.

In hopes of encouraging the abandonment of shifting agriculture, a ley cropping and grazing system was established on all three stations to demonstrate that once the land was cleared, given proper management, it could continue to be cropped. Grazing was introduced into the cropping system at Cayo out of necessity because continued crop production had resulted in weeds becoming unmanageable (DOA, 1942).

Feeding trials at Corozal were started using cut grass and at Stann Creek using grass plus local supplements (coconut meal and rice bran). However, there were insufficient concentrates and the trial was abandoned (DOA, 1943). Holstein-Zebu bulls were imported for upgrading local bulls, and a campaign was started in 1944 to encourage fencing of pastures (DOA, 1944). A war and post-war shortage of barbed wire gave the programme a slow start, so the Department bought barbed wire in bulk for resale to the farmers (DOA, 1946). The response to the programme was good and fences, rotational grazing and improved pastures for reserve feed appeared, especially in Cayo District.

The first farm demonstrator's course in cattle management had to be taught by farmers (DOA, 1945).

An agricultural policy committee reported (DOA, 1948) that declining timber production and lack of minerals meant that the country was destined to rely on agriculture as its main source of revenue. It noted that farming systems were still largely based on shifting agriculture and what livestock there were, were of poor condition, were on unimproved pastures and made uneconomic weight losses in the dry season. Research was recommended in the cultivation, management and utilization of adapted grasses for each major ecological zone, plus identification of practical methods for conserving fodder in the dry season.

The Veterinary Officer wrote (DOA, 1948) that it had not yet been demonstrated that a profit could be made at raising beef, and stock raising was mainly a side-line to chicle or mahogany as a source of income. Many cattle raisers were on Crown lands and had no incentive to improve land not their own. He also criticized the general lack of farmer expertise and the lack of government services for the industry.

Land was purchased in the Cayo District for a 550 ha research farm (Central Farm) and the first crops were planted in 1949. The location of the farm emphasized the importance given to the livestock industry because Cayo District had over one half of the country's pasturage and herd.

A survey and analysis of the beef industry (Arnold, 1950) noted existing drawbacks to development; it recommended a larger effort be made to improve the national herd and that studies be made on pasture grasses and legumes, including fertilizer trials on the Pine Ridge savanna soils. Suggestions for greater productivity included leguminous feeds, leguminous browse species for fodder in the dry season and a rotational grazing system utilizing 70% Guinea grass pastures and 30% native grasses. The soft bones of Belizean stock indicated that mineral supplements were essential.

Cattle censuses in 1943 (10,131 animals) and 1947 (16,658 animals) showed a 64% increase in the herd (Table 12.2); however, a severe drought in 1949 resulted in the loss of roughly 3500 animals. At the end of the decade there was one dairy herd (300 animals) producing mainly cheese because the population preferred imported condensed milk.

12.2.3 The 1950s

Work of the Department of Agriculture was based on a series of development plans derived from earlier studies.

The one for the period 1952–1955 (Government of British Honduras, no date a) was carried out with the aid of Commonwealth Development and Welfare funds. Additionally a Land Use Survey

Table 12.2. Census figures for the cattle population of Belize 1943–1974.

Year	Number of cattle	Reference
1943	10,131	A 1944
1947	16,658	A 1948
1951	15,644	A 1952
1960	22,100	B 1964–1970
1961	23,098	A 1962
1964	28,829	A 1965
1968	29,600	A 1968
1968	30,274	A 1969
1970	38,399	A 1971
1974	45,000 (est.)	A 1974

A, DOA (1944–1974); B, Government of British Honduras (no date b).

team, which was in the country from 1951 to 1954, produced a report (Wright *et al.*, 1959) and 1:40,000 and 1:250,000 scale maps that have been very valuable tools for further development planning. Extra staff members were procured for the country (a chemist, an experimental officer and a livestock officer); limited funds were also provided for research.

The Development Plan Part III, 1955–1960 (Government of British Honduras, 1957) saw a scheme started to boost livestock production that had been proposed in the earlier Plan. Commencing in 1956 and run through two phases until the end of 1960, it involved expenditure in the form of subsidies for pasture (about £119,000) and improvement of breeding stock including a 'revolving bull' (sic.) fund (£70,000). More funds were produced for research and one member of the Land Use team remained as chemist and, along with the experimental officer, continued trials laid down in the second phase; new trials were also initiated. The experimental officer resigned in 1959 and the chemist in 1960, ending a most fruitful period of rapid expansion of knowledge of Belize and its potential.

Many grass and legume species and varieties had been introduced, observed and selected, perhaps with the aid of advice by Stehle (1956), who published recommendations of forage and grass species for Belizean pastures. The more successful introductions were tried on different soils with varying rates of fertilizer (Wright *et al.*, 1959) and from these came Pangola (*Digitaria decumbens* Stent.) and Jaragua (*Hyparrhenia rufa* Nees.) grasses as successful pasture plants for Belize. Demonstration plots were set up in various parts of the country and cuttings of the better grasses provided to farmers. 'Scatter plot' trials demonstrated the need for phosphate for successful growth on most Belizean soils, particularly those of the Low Pine Ridge. Three valuable studies on Belizean pastures (native, Jaragua and Guinea) were published (Romney, 1960, 1961; Kemp *et al.*, 1961) and these were the first results of Belizean pasture research to be published internationally

Among other work of the period was an attempt at making pit silage at Central Farm, for which no results were reported, and soil surveys of small areas, which were carried out for a few years after the departure of the Land Use Survey team. Soil analyses and fertilizer and crop recommendations for farmers were initiated by the chemist.

12.2.4 1960s

On the advice of the *Report of the Committee on Agriculture* (Government of British Honduras, 1960) and of a United Nations Technical Assistance Mission, the Development Plan for 1964–1970 (Government of British Honduras, no date b) aimed to have the cattle industry become an important part of the country's economy. It recommended that pasture and livestock management research be continued and that the areas of the country with the best promise should receive most attention. Specifically, Cayo and Orange Walk Districts were being referred to, for the former had more than half the national herd and the latter was thought to have much potential for beef. The 1960 Report encouraged further research into the Pine Ridge savanna lands and pointed out that personnel whose jobs were exclusively research were needed and that continuity was required if productive work was to be done. The study estimated that land resources were available for about 180,000 head of cattle.

Pastures by this time, though improving, were still far from being generally developed because even in 1969 it was estimated that only 10% of 160,000 grazed hectares were in improved grasses (Cowlishaw, 1969). Pastures were becoming over-stocked because pasture expansion was not keeping up with the rate of increase in stock, and the high cost of clearing land and low market prices for beef were suggested as constraints to investment in the industry. Stock losses in the dry season and malnutrition generally were cited repeatedly in Annual Reports as the main animal health problems.

Exports gradually crept up during the 1960s to about 1000 head in 1968, due mainly to higher prices being offered in neighbouring countries. The 1968 cattle census (Table 12.2) showed that 2000 females had been slaughtered. These mainly consisted of young breeding age animals because of the market's preference for small carcasses. In addition, instead of an estimated 34,000 animals in the country, there were only 29,600. In order to increase the national herd, exports were banned and slaughter of breeding females was forbidden. As a result, cattle numbers grew rapidly thereafter.

Hurricanes Abby in 1960 and Anna and Hattie in 1961 discouraged investment in tree crops and reinforced the Government's interest in cattle production. Trials were conducted with grasses and legumes at Central Farm, on the Low and Mountain Pine Ridges, and at the District Agricultural Stations. Greater emphasis was put on forage legumes after the arrival of an FAO livestock advisor in 1965 for 2 years. He stimulated the pasture programme, many new species were tested and replicated fertilizer trials were established. Unfortunately little of the work was published (Snook, 1968). Somewhat more sophisticated trials were attempted in this period, including an arable ley rotation trial with a 6-year cycle and a rotational grazing trial on Pangola grass.

From 1968 to 1971, a team of experts from the British Land Resources Division, Ministry of Overseas Development, worked in Belize on the agricultural potential of the Belize River Valley (Jenkin *et al.*, 1976). Important climatic, edaphic and economic studies were done and information was obtained on the productivities and nutrient values of the alluvial soils and of the under-utilized savanna soils (Rose-Innis, 1973).

12.2.5 1970s

Conditions described above for the 1960s continued through the 1970s, with most animals being raised on native pastures (Fig. 12.2).

Department of Agriculture research in the early 1970s was concerned mainly with testing introduced grass species and varieties in various edaphic and climatic environments. Yields were obtained of a wide range of grass species in introduction plots. Replicated trials demonstrated that high fresh weight yields can be obtained from Kazangula Setaria (*Setaria sphacelata* on low Pine Ridge soils, and Jaragua and Nunbank buffel (*Cenchrus ciliaris*) on alluvial soils (DOA, 1973). However, due to the dramatic worldwide rise in fertilizer prices in the early 1970s, the development of more intensive livestock production based on fertilized African grass pastures became uneconomic. The costs of establishment of such popular African grasses as Guinea and Jaragua were too high considering their low productivity and likelihood of failure when unfertilized.

In 1973 a pasture research programme funded by IDRC (The International Development and Research Centre of Canada) and executed by the University of the West Indies (UWI) was established to identify forage legumes for Belizean pastures. The programme was undertaken with the collaboration of CSIRO, Australia, and the UWI field activities were undertaken in Belize and Antigua. It was envisioned that the addition of a forage legume to Caribbean pastures of African grasses would eliminate the need for nitrogenous fertilizers and thus make cattle-raising more economic. The Belizean research programme continued until 1985. It collected native legume germplasm from the gene pools in the region (Belize, Guatemala and the Yucatan) and screened accessions of over 50 genera and 110 species of legumes, including exotic material in a series of increasingly complex trials established in a variety of edaphic and climatic environments. Legumes were identified that will grow successfully on such difficult soils as ultisols, vertisols and oxisols, thus potentially bringing into production large areas of unutilized land. The following chapters in this volume provide some of the details of the work undertaken and the results achieved, while the previous

Fig. 12.2. (a) Low-productivity native pasture, central Belize River Valley; (b) better pasture Sibun River.

chapters have been concerned with the research in Australia that utilized germplasm and knowledge gained by the programme. The more productive and persistent lines identified belonged to such species as *Macroptilium atropurpureum*, (DC.) Urban, *Codariocalyx gyroides* DC., *Desmodium cinereum* Poepp., *Desmanthus virgatus* Willed., *Centrosema schottii* (Millsp.) K. Schuw., *Leucaena leucocephala* (Lam.) de Wit. and *Stylosanthes guianensis* (Alub.) Sw.

The Department tested the productivity of progeny of breeds of cattle: Brahman, Costa Rica-derived Brahman, Indo-Brazil, Angus, Brown Swiss, Red Poll, Hereford and Charolais. The Costa Rica-derived Brahman–Charolais crosses were usually superior (DOA, 1973, 1974). Though the unimproved native Criollo breed still formed the majority of the national herd, private and Department imports of stock had greatly increased the variety of breeds represented and the number of quality breeding animals, including a private breeding herd of Santa Gertrudis cattle.

In 1972 a new livestock station was initiated at Richmond Hill near Orange Walk Town, and was developed over the following 4 years to enhance the long-established Government policies of providing breeding stock for farmers and encouraging expansion of the beef industry in the north. One small private dairy was established in the Mennonite colony at Spanish Lookout, which supplied part of the fresh milk requirements of Cayo and Belize Districts. Some good-quality cheese was made as well. Unfortunately the community was not interested in expanding the operation because dairy cows had been only a peripheral part of their normal farm operations.

Plans were made and concessions granted in the mid-1960s for the construction of a meat packing plant that could meet United States Department of Agriculture standards and thus export surplus national stock. The plant was not in operation with USDA certification until 1973.

The implications of the 1970 survey of Belizean cattle (38,400 animals) was discussed by Jenkin and colleagues in their 1976 report. Most farmers had less than 50 head of cattle (Table 12.3) and were mainly *milperos* (farmers who utilized shifting cultivation). They initiated permanent pastures by grazing their cattle on newly cleared land from which a crop of maize had just been harvested, and among which a grass, normally Jaragua, was planted about the same time as the maize. Their herds expanded as the land was cleared. A major problem for these *milperos* was woody regrowth, which demands strenuous control efforts for at least 10 years. Fencing and other capital improvements were very difficult for small farmers to finance.

Table 12.3. Size of cattle herds on Belizean farms in 1970.

Size of herd (head)	Number of farms
1 to 10	780
11 to 25	308
26 to 50	128
51 to 100	65
101 to 500	34
501 to 1000	5
1000+	2
	Total 1322

Farmers with 50 to 100 animals were usually solely dependent on cattle. Development was leading them to a semi-intensive system for, as their herds built up and as less land was available for expansion, higher production could come only through establishment of more intensive management. Thus a major problem for this group was lack of land for expansion; however, better management would solve their other problems which included mediocre pasture, bad drainage, poor farm layout and unproductive stock. The report pointed out, as many previous reports had, that better pastures, reserve feed or silage for the dry season and higher stocking rates were required.

Ranchers with over 1000 animals were thought to be generally in the same situation as medium-sized farmers, having unimproved pasture and no storage of feed for the dry season. Other basic problems with the industry identified by the report included poor communication, lack of credit facilities, difficulties in obtaining agricultural supplies, inadequate technical advice, pests and diseases. Fluctuating prices for beef and high capital costs kept profits very low. The export market, which was mainly the USA, was variable, depending on conditions in the USA economy. Caribbean markets, which were available to Belize as a member of CARICOM, were difficult to develop due to local disease problems, variability in supply and demand, financing difficulties and management problems in the meat packing plant.

The report stated that the beef industry could become the country's main foreign exchange earner and presented detailed calculations as to how this could be achieved. It recommended that the better soils used mainly for cattle be used for both cattle and crops in a ley system, while the poorer soils, unutilized, could be used for cattle production. In 1970 the industry was based in the Belize Valley on 6,800 ha of improved and 42,700 ha of native pasture; it recommended that these be increased to a total of 85,500 ha by 1999, which would allow the herd to increase from 38,400 to 195,100. These animals could support two packing plants, which would be slaughtering 5000 animals a year. In order to achieve this goal, the first research priority in the development of the Belize Valley would be in pasture fodder and browse crops, including grasses and legumes.

Certain Government regulations, while instituted for valid reasons, had a depressing effect on the development of the cattle industry. In order to prevent Belizean land from being held undeveloped by investors, Government regulation of the sale of larger areas of land was instituted, along with an overseer role in its development. This has had a particularly unsettling effect on investors to whom the impartiality of Government officials is always suspect and their opportunities for chicanery too great. In addition, Government-regulated prices for meat sold internally were low, discouraging investment in the industry, and the blanket prohibition of export of breeding animals has essentially eliminated the markets for graziers with breeding herds.

Because of a lack of meat grading, there was no incentive to improve the quality of the meat sold. One grazier (F. Norris, Cayo District, 1974, personal communication) managed to get higher prices for his meat by setting up his own abattoir and selling hotels good quality packaged meat. This operation did not survive a change of ownership of the property. Poor management, high overheads and over-investment in infrastructure and development are major causes of the failure of cattle ventures; investors were rarely able to hire and retain competent managers. Temperate experience, while valuable, is usually not sufficient to overcome the problems encountered in a tropical, developing country. Returns for investment are small, and foreign investors, while initially over-enthusiastic and over-confident, become rapidly discouraged. Few foreign investment ventures had been successful in the cattle industry in Belize.

The small size of the country and its low population density is reflected in the small numbers of research and extension staff in the Ministry of Agriculture. The attractions of private farming and opportunities in regional research organizations have resulted in resignations of experienced personnel and as a result in the frequent movement between jobs of the remaining staff, with young, relatively inexperienced staff doing the actual field work. During the mid-1970s there was one pasture officer for the country. He was abroad for a year getting his MSc. His responsibilities for extension and management of pastures on Government stations left him little time for research. He then was promoted to other jobs and the country was left without a graduate pasture officer.

Research was also particularly hampered by the small number of Government staff and their frequent shifts in position. Donors had a record of undertaking projects for too short a period and for neglecting to provide for a competent researcher to run ongoing activities after the project terminated Long-term projects, which are essential for agricultural research, have been very few in number.

12.3 Conclusion

Though cattle have long been raised in Belize, only from the 1950s has attention been paid to the development of a beef industry. By the mid-1970s, much had been done to identify the problems of the industry and some considerable success had been achieved in alleviating them. However, in order for the industry to achieve its considerable potential, Government regulations required modification, more research was needed on the identification and management of improved pasture species, and solutions were required for marketing problems.

13 Botanical Composition and Nutritive Value of Selected Native Pastures in Belize

J.R. Lazier[*1]
**Formerly International Livestock Centre for Africa*

Abstract
An initial assessment was undertaken of the productivity, nutrient value and botanical composition of eight representative native pastures on a range of clay soils, including three vertisols, in central Belize and the Belize River Valley under wet tropical conditions. Regular harvests indicated that the productivity at all sites was very low, that palatable native legumes were present at low levels and that the native species, particularly *Mesosetum angustifolium*, the dominant grass at the more infertile sites, did not respond to fertilization either in yield or nutrient content. At all sites some of the nutrients essential for animal productivity were below minimal levels, particularly Cu, but also P at all but one site. N, Ca, Mg and Zn also were limiting at some sites and harvests. The introduced legumes being tested in the trials at the sites also had lower than minimum levels of Cu but mainly higher than minimum levels of the other nutrients tested.

13.1 Introduction

With the development of logging and commercial farming in Belize over the last century, cattle raising has been a major industry. However, only limited research had been done on native and improved pastures in central Belize. This has been reviewed below for each of the three major environments in Chapter 15 (this volume). Lacking access to knowledge and appropriate productive fodder species, the industry has depended almost exclusively on native pastures.

In 1969–70 a native grass productivity trial was established on the Lowland Pine Ridge (see Fig. 12.1, Chapter 12, this volume) on Colonel English series, Puletan soils at mile 25.5 on the Western Highway (Rose-Innes, 1973) under wet tropical conditions. A study of the native vegetation determined that the basal cover was 33.5% and sedges and *Mesosetum angustifolium* made up 74% of the species present. After an initial cutover, the increasing mature vegetation was clipped, botanically separated and chemically analysed. Clipping increased the nutrients present and decreased the percentage dry matter. Of the clipping rates applied, monthly rates appeared optimal, with more frequent clips producing a decline in herbage yield.

Three cutting frequencies (4, 8, 12 weeks) were applied to the same quadrat locations plus single monthly cuts of increasingly mature vegetation for 100 sites per frequency. The more frequent clips produced higher nutrient levels but declining dry matter yields. The 8-week clips gave

[1] E-mail: john.lazier@sympatico.ca

the best nutrient levels and dry matter production of the three frequencies. However, all clipping rates produced very low yields, which declined with time. Generally the productivity and nutrient levels of the Coastal Pine Ridge vegetation were lower to much lower than that produced at a similar trial at Central Farm on alluvial terraces, and slightly higher than yet another trial on Mountain Pine Ridge. K, Mn and Mg were all considered to be at adequate levels for maintenance and production, but crude protein, P, Ca, Cu, Zn and Co were all below maintenance levels.

Bazan (1969) undertook pot trials with Coastal (Lowland) Pine Ridge soils from two sites near Hattieville, which determined that unfertilized soil gave very poor yields of tomato, hairy indigo and Pangola grass. P was found to be the critical element lacking; there was a 50-fold increase in yield with its application (between 150 and 240 kg/ha), and that element was needed to get yield responses from the application of N and K. Lime at 4500 kg/ha provided additional yield responses. The effect of the applied P in most cases lasted only 6–10 months. The soils were assessed as suitable for pastures, crops and forestry. However, the provision of improved drainage and mulch were recommended.

At the time this study commenced, in February 1976, there was only one hardy African grass, Jaragua grass (*Hyparrhenia rufa*), which had recently naturalized and was becoming widely spread in local pastures with and without human assistance. In addition, as a result of research undertaken by Government researchers and farmer initiatives, other productive Africa grasses were being established, replacing native pastures, but limits in planting material had been a constraint to their dissemination.

The University of the West Indies–Belize Forage Legume Programme was established in 1972 to identify productive and persistent forage legumes that could be introduced into native and improved pastures to improve grass stands and the quality of the fodder on offer. One of the activities undertaken by the programme was a series of trials in which unreplicated strips of promising legumes were introduced into pastures of exotic fodder grasses, and into native pastures (Chapters 15 and 17, this volume). Native pastures on infertile clay soils in and adjacent to the Belize River Valley were of particular interest in this regard and in particular those on the large areas of land referred to locally as Lowland Pine Ridge and Mountain Pine Ridge, which supported sparse vegetation of low nutrient value, and thus were essentially unutilized. To complement the testing of adaptation of potential forage legumes to these environments (Chapter 15, this volume), simple preliminary studies were undertaken to estimate their botanic composition and productivity.

13.2 Methods

13.2.1 The trial sites

Eight sites were selected, which were representative of the native pastures of central Belize and the Belize River Valley. The sites were mainly on clay soils and comprised one infertile native pasture derived from limestone (a shallow vertisol < 40 cm at site 4), four pastures on waterlogged Low Pine Ridge soils (planosols at sites 5 and 6; vertisols at sites 7 and 8), and three in contrasting environments on shallow elevated Mountain Pine Ridge soils (podzols at sites 11, 12 and 13) (Tables 13.1 and 13.2). Photographs of these sites can be found in Chapter 15 (this volume). At each site a 27 × 65 m trial area was laid out and soil samples were collected at randomly selected sites within their perimeters for analysis, 30 sites at 0–15 cm and 15–30 cm depths, and 15 sites at 30–45 cm and 45–60 cm depths. The soil samples were analysed by the Central Analytical Laboratory at the University of the West Indies in St Augustine Trinidad. Strip trials to test the adaptation and productivity of a range of promising forage legumes were established at these sites. The performances of the legumes at the sites are presented in Chapter 15 (this volume), as well as the analyses of the soils (Appendix 15.1).

The fertilizer treatments were: before planting the 15-cm wide strips at site 4 had 63 kg/ha TSP and KMag respectively applied before planting, while all Low and Mountain Pine Ridge soils (Puletan and Pinol soils, respectively) had a one-time application of 63 kg/ha Fritted Trace Element (FTE) micronutrient mixture, and 200 kg/ha $MgSO_4$ and 63 kg/ha each of KCl and TSP.

Thereafter, one-half of each plot and the corresponding part of the trial area of all trial

Table 13.1. Trial locations, altitudes and rainfall of native pastures studied.

Trial no.[a]	Map reference[b]	Location	Alt. (m)	Rainfall (mm)	Pasture
4	16 QCQ 107 107	Orange Walk Estate, Western Highway near Roaring Creek	40	1500	Infertile, low productive pasture, soil < 40 cm deep
5	16 QCQ 329 153	Western Highway, km	40	2200	Level
6	16 QCQ 329 155	33, Low Pine Ridge	40	2200	Sloping
7	16 QCQ 302 131	Western Hwy, km 35,	40	2200	Shallow
8		Low Pine Ridge	40	2200	Shallow, harrowed
11	16 QBP 968 857	Bull Run Ranch, Mountain Pine Ridge	560	1600	Under mature pine
12	16 QBP 930 788	Rio On Line, Mountain Pine Ridge	560	1600	Under short pine
13	16 QCP 087 805	Baldy Beacon, Mountain Pine Ridge	970	2900	Sparse shrubby vegetation

[a]Trial number as given in Chapter 15 (this volume). [b]Map series E755 DOS 4499 Edition 3.

Table 13.2. Trial soils.

Trial no.[a]	Soil type	LRD[b] classification			USDA[c] Classification system
		Suite	Sub-suite	Series	
4	Vertisol	Yaxa	Yalbec	Spanish Lookout	Lithic Rendolls, clayey, montmorillonitic
5	Planosol	Puletan	Yobo	Western Highway mile 33	Abruptic Tropaqualfs, fine, kaolinitic
6	Planosol				
7	Vertisol	Yaxa	Yalbec	Beaver Dam	Aquentic pelluderts, very fine, montmorillonitic
8	Vertisol				
11	Podzol	Stopper	Stopper	Pinol cse s cl	Plinthic Tropudults, clayey, kaolinitic
12	Podzol	Ossary	Ossary	Pinol cl	Typic Tropudults, fine-loamy, mixed
13	Podzol	Ossary	Ossary	Pinol ls	Lithic Dystropts, loamy, mixed

[a]Trial number as given in Chapter 15 (this volume). [b]Land Resources Division (Jenkin *et al.*, 1976). [c]All are isohyperthermic (W. McKinzie, 1977, personal communication).

sites received the basic fertilizer rate of 63 kg/ha of TSP and KMag each per annum in ¼ portions every 12 weeks, after alternate observations. The other half received fertilizer on the same schedule but at four times the rate, creating low and high fertilized strips across the width of the trial area. The same part of all plots received the same fertilizer rate. Because the three sites on Mountain Pine Ridge had very slow growth on establishment and signs of severe nutrient deficiencies had appeared (as indicated by leaf coloration), a single application of marl (97.1–100% $CaCO_3$) was applied at 1790 kg/ha in April 1976.

13.2.2 Sampling

Botanical sampling was undertaken once using three 1 × 0.5 m^2 random quadrats harvested at ground level from the unfertilized perimeters of each trial at the end of the growing season in January and early February 1976. The samples were sorted botanically and dry matter yields were determined. Thereafter, periodic surveys were made of the trials and their environs to collect herbaceous plants that were flowering in order to assist in the identification of those sampled, and to obtain a more complete catalogue of those present at the sites. The Baldy Beacon site (site 13) was not sampled for dry matter but only was sampled for botanical composition. Herbarium specimens were made of the collected and sampled plants and these were identified at the Royal Botanic Gardens at Kew[2] and returned to Central Farm Belize for use as reference specimens.

Sampling for grass yield was undertaken within pegged 5 × 3 m plots established across two legume strips in which the legume treatments had not survived. Three 0.5 × 0.5 m quadrats were harvested at random from each pegged site at 15 cm height on four occasions at 6-week intervals commencing in February 1977.

In March and May of 1977, sampling for nutrient contents were undertaken from the plots, the grasses from high, low and no fertilizer treatments and leaves of two or three of the most vigorous legumes treatments plucked by hand from both fertilizer treatments. The analyses were undertaken by the Central Farm Analytic Laboratory. Photographs of the environments and the trial sites are provided in Chapter 15 (this volume).

[2] Kindly undertaken by Drs Bernard Verdcourt, Roger Polhill and Steve Renvoize at Royal Botanic Gardens, Kew, Richmond, Surrey TW9 3AB, UK.

13.3 Results and discussion

Rainfall for the period of the trials is given in Fig. 13.1. The dry season was relatively severe in 1976, with negligible rainfall in March and April, but was much less stressful in 1977 when there were frequent showers. The Low Pine Ridge sites, due to their waterlogged condition, were little affected by lack of rainfall.

13.3.1 Dry matter yields

The yields of the vegetation sampled for botanic analysis were the accumulated growth at the time of sampling at all sites (Table 13.3; Appendix 13.1, Tables 13.8–13.14). Only two of the sites had had livestock on them, Orange Walk and Bull Run; Orange Walk had been heavily grazed, while most of the Mountain Pine Ridge sites had signs of light grazing by wildlife, mainly deer. The highest dry matter yields were obtained at Rio On and Orange Walk. The Rio On site had unusually dense, tall (60 cm) clumps of dry *Mesosetum angustifolium* grass, indicating that the site may have been burned and vigorous growth had resulted from the released nutrients. The relatively

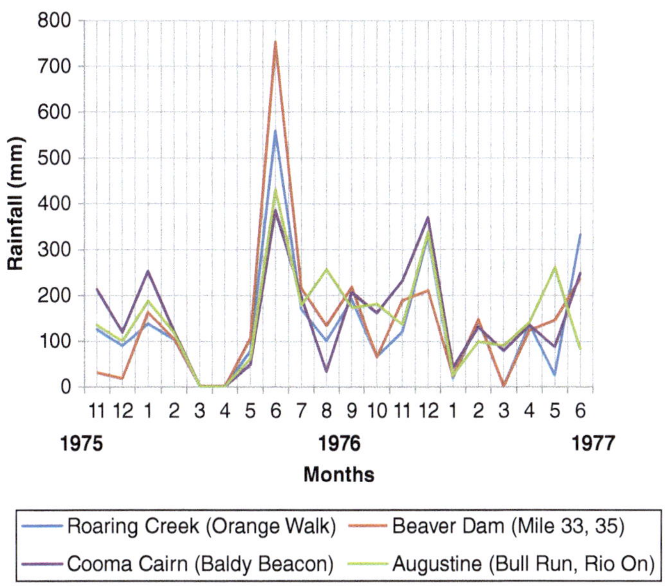

Fig. 13.1. Rainfall at native pasture sites.

high dry matter of the grass at Orange Walk is probably due to the higher nutrient and pH levels of the soils (see analyses presented in Table 15.7, Chapter 15, this volume) and the more productive grass species present.

The results for the four grass harvests for dry matter yield are presented in Table 13.4. At all sites the initial yields were high because the first harvest was of the accumulated growth of plants that had not been cut previously. The subsequent harvests clearly show that the productivity of the native pastures was very low until the final harvest in June. While the Orange Walk and Mountain Pine Ridge sites would have been affected by the low rainfall received during that period, for the Low Pine Ridge sites moisture availability would

Table 13.3. Unfertilized native grassland dry matter yields harvested at ground level in February 1976.

		Harvest yields (kg/ha × 10^{-1})					
Site	Location	Legume	Grass	Weed	Total	*Pinus caribea* leaves	Previous grazing
4	Orange Walk	299	1176	353	1828	0	very heavy
5	Mile 33, Low Pine Ridge, level	0	775	474	1247	0	none
6	Mile 33, Low Pine Ridge, sloping	2	688	213	901	0	none
7	Mile 35, Low Pine Ridge	0	387	104	491	0	none?
8	Mile 35, Low Pine Ridge, harrowed	0	165	89	254	0	none
11	Bull Run, Mountain Pine Ridge	210	764	37	801	3214	heavy
12	Rio On Line, Mountain Pine Ridge	0	4123	0	4153	58	none

Table 13.4. Native pasture dry matter yields under three fertilizer regimes harvested at 15 cm in 1977.

		Time of harvest, fertilizer treatment[a] and dry matter yields (kg/ha × 10^{-1})											
		February			March			April–May			June		
Site	Location	High	Low	No	High	Low	No	High	Low	No	High	Low	No
4	Orange Walk	175	n.d.	n.d.	21	1	19	0	0	0	0	0	0
5	Mile 33, Low Pine Ridge, level	514	564	493	456	761	254	61	53	34	128	71	45
6	Mile 33, Low Pine Ridge, sloping	610	493	364	166	188	72	54	39	19	72	52	25
7	Mile 35, Low Pine Ridge	592	508	574	241	183	133	46	26	14	61	35	19
8	Mile 35, Low Pine Ridge, harrowed	531	522	488	251	287	196	49	57	41	65	76	55
12	Bull Run, Mountain Pine Ridge	n.d.	n.d.	n.d.	265	137	141	25	22	30	33	29	52
13	Rio On Line, Mountain Pine Ridge	n.d.	n.d.	n.d.	522	454	164	60	50	44	80	67	59
14	Baldy Beacon, Mountain Pine Ridge	n.d.	n.d.	n.d.	118	75	116	40	32	45	59	43	60

[a]High, low and no fertilizer applied; n.d., no data.

not have been a limiting factor. The zero yields at the Orange Walk site are also due to the cutting height being too high for the prostrate grass species present. At the Low Pine Ridge sites there appeared to be small yield responses to added fertility, but this was not evident at the Mountain Pine Ridge sites. *Mesosetum angustifolium* thus does not appear to be very responsive to the nutrients applied. This is not unexpected as the grass is known to be hardy but low in productivity. In addition, soil N levels at all sites were very low and none was applied with the fertilizers.

13.3.2 Botanical composition

The bulk of the dry matter harvested in the botanic sampling at the Orange Walk site was composed of *Axonopus compressus*, *Mimosa prudens* and an *Acanthaceae* species (Table 13.8). The main grass, *Axonopus compressus*, is considered a pasture species that is resistant to heavy grazing and forms a good ground cover. However, it is a short species with low productivity and nutrient content (Cook *et al.*, 2005). *Setaria gracilis* undoubtedly forms part of the available fodder as well. Most of the native legume present was *Mimosa prudens*, which is relatively unpalatable to cattle. The *Acanthaceae* present were not identified, but it is likely that they are grazed because many in Mexico are reported as being palatable (Daniel, 2004). Considered to be an infertile site and unproductive, the site had been used as a pasture and had been heavily grazed for many years previous to the establishment of the trial.

Mesosetum angustifolium, a fibrous grass adapted to low nutrient soils and to frequently burned environments, dominated the vegetation at all Low and Mountain Pine Ridge sites. Higher grass yields were obtained from the Mile 33 (Rough Mile) site than the Mile 35 (Beaver Dam) site, and at the less densely shaded, ungrazed Rio On site compared with Bull Run, which had been grazed (Table 13.4).

Of the two sites at Mile 33, the sloping, better drained site 6 (Table 13.10) had less vegetation than site 5 (Table 13.9), a level site with the water table near the surface. The figures can be viewed as representing the normal amount of accumulated dry matter on such savanna sites because the trials had not been cut over or grazed, although previous to the establishment of the trials they had been burnt during most dry seasons. One noteworthy plant present at site 5 was *Schwenkia americana*, which has received research attention due to its containing chemicals with anti-microbial, anti-inflammatory and analgesic properties (e.g. Jimoh *et al.*, 2011). The species present at the Mile 35 site (Table 13.11) were similar to those at the Mile 33 sites, but the grass and total dry matter yields were much lower, perhaps due to having been recently burnt, and to the harrowing of site 8 in late June 1975.

On the Mountain Pine Ridge, all sites had legumes of fodder potential present; *Stylosanthes guianensis* was found at two sites, Bull Run and Rio On, *Zornia reticulata* at Bull Run and *Desmodium barbatum* at Baldy Beacon (Tables 13.12, 13.13 and 13.14). The presence of numerous, mainly palatable legumes (15) identified at Bull Run may have been a result of fertility from the dung deposited by cattle because the site was adjacent to the ranch buildings; in addition, the animals may have been importing the seeds from other environments. Other reasons for their presence are that the site could have been burnt relatively recently, though there was no obvious sign of this, or due to better drained soil conditions because the top 15 cm of the soil was 11% coarse sand. Although a dung pat was found at the Rio On site, its relatively remote location would probably have resulted in little previous grazing pressure; its high biomass may be due to the soil having higher C and N levels than the other two Mountain Pine Ridge sites.

13.3.3 Plant nutrient analyses

While there is no consensus on the precise requirements of nutrient levels in fodder for livestock production, estimates have been made of levels, such as those of the ARC (1980), and they provide a convenient if very rough yardstick to assess the value of the nutrient levels obtained in the trials for livestock production.

The nutrient analyses of the native grasses when compared with the estimated nutrient requirements for beef production (Tables 13.5 and 13.6) indicate that at all sites Cu was below the minimum levels required for maintenance and growth. The Orange Walk site had more than the minimum requirements of all of the nutrients analysed at all fertility rates, except for

Table 13.5. Summary of nutrient value analyses at or above estimated minimum and maximum herbage values required for beef production.

Treatment	Recommended nutrition level for beef production	Analyses at or above recommended nutrition levels							
		N	P	K	Ca	Mg	Mn	Cu	Zn
Grass	minimum	Orange Walk site only	few	all	most	most	almost all	none	some (1/3)
	maximum		none	most	some	few	almost all	none	none
Legume	minimum	all	most	all	all	all	most	some	all but one
	maximum	all	very few	some	all	most	most	none	very few

Cu and P. The Low and Mountain Pine Ridge sites, however, were deficient in N and, with the exception of the highest fertilizer level at the Low Pine Ridge sites, most of the sites were deficient in P at all fertilizer levels. In addition, Ca, Mg and Zn were below minimum requirements at some sites' harvests and fertilizer rates. With reference to the maximum levels required, these being mainly for larger animals, N and P were below at all sites, while Mn and Mg were above the maximum at all sites in March, and most sites in May. Regarding the other elements, the analyses were at or above the maximum requirements for K, and Mn (except Baldy Beacon in the May harvest), and many sites for Ca, Mg and Zn (in about two-thirds of the analyses). There was no clear nutrient response by the native grasses to the application of the fertilizers. An earlier foliar analysis study (Rose-Innes, 1973) on native pasture on Low Pine Ridge soil at mile 25 had reported that vegetation had K, Mn and Mg all at adequate levels for maintenance and production, but crude protein, P, Ca, Cu, Zn and Co were all below maintenance levels.

The results of the analyses of the leaves of vigorous legumes under the two fertilizer treatments (Tables 13.5 and 13.7) showed no consistent difference in the results between the two levels of fertilizers, and the levels of the elements studied were generally higher than in the native grasses. The maximum requirement levels of N and Ca were available at all sites and harvests. K, Mg and Zn analyses (with three exceptions) analyses were all above the minimum, P and Mn were mostly above the minimum, while most Cu analyses were below. In addition, most analyses for Mg and Mn, some for K, and a few of P and Zn were at or above the maximum levels.

The vigorous growth of legumes on the Low and Mountain Pine Ridge soils under the higher fertility treatment in particular (Chapters 16, 17 and 18, this volume) suggest that fodder banks of legumes are feasible in these environments. The higher N and P content of the legumes could assist in alleviating some of the nutrient shortages that livestock would experience when restricted to grazing only the natural vegetation.

13.4 Conclusions

The botanic sampling indicated that a more numerous and different range of plant species of higher nutrient value were present at the shallow heavy clay soil on limestone at Orange Walk compared with the two other environments. The botanic sampling was not exhaustive and more detailed rigorous sampling is needed to obtain a more accurate assessment of the species present and their contribution to pasture dry matter production. More species of native legumes were present at some sites than others, and most of these were palatable. Under regular cutting the productivity on all soils was similar and very low, and there was little discernible response to the fertilizer applied in the absence of N. Analysis of the fodder indicates that if native pastures on these soils are to be successfully utilized for beef production, supplementary higher quality fodder and minerals would be required. The establishment in these environments of fodder banks of fertilized legumes is a possible option. The success of the introduced legumes in the strip trials at the sites are reported later (Chapter 15, this volume). The dry matter and nutrient analysis results are similar to those obtained from an adjacent trial at Mile 33 on the Western Highway (Chapters 17 and 18, this volume).

Table 13.6. Nutrient analyses of native grasslands taken in March (A) and May (B) 1977. Shaded cells indicate less than minimum levels of nutrients for beef production.

Site	Location	Fert.[a]	N (%)		P (%)		K (%)		Ca (%)		Mg (%)		Mn (ppm)		Cu (ppm)		Zn (ppm)	
			A	B	A	B	A	B	A	B	A	B	A	B	A	B	A	B
4	Orange Walk	H	1.26	–	0.16	–	2.80	–	0.49	–	0.31	–	29	–	2.7	–	–	–
		L	1.28	–	0.13	–	2.20	–	0.45	–	0.34	–	47	–	4.5	–	–	–
		N	1.31	–	0.08	–	2.35	–	0.41	–	0.36	–	27	–	4.5	–	–	–
5	Mile 33 level, Low Pine Ridge	H	1.01	–	0.13	0.12	1.70	1.40	0.43	0.32	0.30	0.31	145	109	2.6	–	15	–
		L	0.65	0.70	0.05	0.04	1.00	1.80	0.17	0.22	0.27	0.32	141	94	1.6	2.2	12	14
		N	0.69	–	–	0.10	0.75	1.20	0.20	0.23	0.18	0.31	107	94	–	–	12	11
6	Mile 33 sloping, Low Pine Ridge	H	1.08	0.99	0.21	0.15	1.40	1.60	0.27	0.30	0.26	0.30	127	81	2.7	2.8	15	18
		L	0.73	–	0.07	0.07	1.20	1.50	0.27	0.17	0.24	0.22	136	67	1.3	3.0	17	20
		N	1.01	0.97	0.04	0.02	0.80	1.00	0.17	0.15	0.21	0.17	56	90	1.2	1.5	15	17
7	Mile 35, Low Pine Ridge	H	0.62	0.71	0.18	0.17	1.05	1.40	0.18	0.20	0.25	0.09	211	138	1.6	1.3	18	11
		L	0.69	0.77	0.06	–	0.80	–	0.26	–	0.15	–	280	–	1.3	–	18	–
		N	1.11	0.70	0.04	–	0.95	–	0.21	–	0.19	–	205	–	1.3	–	32	–
8	Mile 35 harrowed, Low Pine Ridge	H	0.79	0.85	0.23	0.11	1.20	1.20	0.14	0.19	0.18	0.09	150	188	2.1	2.3	16	22
		L	0.61	0.80	0.06	0.08	0.95	1.10	0.21	0.19	0.16	0.07	114	144	4.7	1.8	21	21
		N	0.76	0.73	0.05	0.03	0.97	0.70	0.11	0.13	0.12	0.07	127	–	2.3	1.8	12	12
11	Bull Run, Mountain Pine Ridge	H	0.67	0.97	0.14	0.12	0.90	1.45	0.30	0.33	0.14	0.16	102	46	3.8	1.5	11	14
		L	0.65	1.17	0.10	0.11	1.10	1.25	0.32	0.28	0.16	0.15	74	60	2.7	2.5	7	–
		N	0.68	0.71	0.04	0.08	0.90	1.80	–	0.15	0.19	0.16	92	109	1.8	4.0	9	16
12	Rio On, Mountain Pine Ridge	H	0.89	0.78	0.09	0.13	1.35	1.30	0.26	0.32	0.23	0.11	106	24	2.3	1.4	15	10
		L	0.82	0.88	0.05	0.07	0.95	1.30	0.19	0.32	0.18	0.12	60	56	5.0	1.8	16	12
		N	0.74	0.75	0.07	0.02	1.15	0.90	0.31	–	0.27	0.12	100	69	2.3	1.5	16	14
13	Baldy Beacon, Mountain Pine Ridge	H	0.73	0.71	0.20	0.09	0.80	0.75	0.25	0.17	0.16	0.13	28	11	2.7	–	22	–
		L	0.75	0.77	0.08	0.03	0.90	0.80	–	0.20	0.18	0.15	69	16	2.3	1.5	19	11
		N	0.58	0.87	0.07	0.12	0.60	1.05	–	–	0.14	0.26	75	–	1.8	2.0	27	17
Nutrients required for beef maintenance and growth[b]			**N** 1.15		**P** 0.11–0.25		**K** 0.47–0.84		**Ca** 0.17–0.28		**Mg** 0.12–0.21		**Mn** 20–25		**Cu** 8–15		**Zn** 13–28	

[a]Fert., fertilizer added: H, high; L, low; N, none. [b]Extracted from ARC (1980).

Table 13.7. Nutrient analyses of legumes from strips planted in native grasslands taken in March (A) and May (B) 1977. Shaded cells indicate less than minimum levels of nutrients for beef production. (See Table 13.5 for key.)

				N (%)		P (%)		K (%)		Ca (%)		Mg (%)		Mn (ppm)		Cu (ppm)		Zn (ppm)	
Site	Location	Legume[a]	Fert.[b]	A	B	A	B	A	B	A	B	A	B	A	B	A	B	A	B
4	Orange Walk	CF 29	H	3.01	2.06	0.29	0.13	1.50	1.10	1.00	1.10	0.24	0.41	52	19	7.4	3.1	20	12
			L	–	2.12	0.12	0.24	0.60	1.20	0.68	1.13	0.33	0.45	15	12	8.1	7.1	23	18
		CF 3-1	H	2.96	1.97	0.17	0.13	0.60	0.90	0.92	1.33	0.22	0.53	41	13	6.3	6.6	30	24
			L	2.86	2.20	0.16	0.08	0.60	0.70	0.69	1.22	0.23	0.46	16	13	5.9	6.6	26	24
		CF 4-5	H	2.93	2.16	0.21	0.09	0.75	0.90	–	1.48	–	0.46	16	19	–	7.6	–	25
			L	2.00	2.19	0.18	0.15	0.75	0.94	1.17	1.73	0.30	0.51	44	–	5.4	8.0	24	24
5	Mile 33 level Low Pine Ridge	CF 29	H	3.28	2.27	0.19	0.15	1.03	1.25	1.80	1.13	0.15	0.37	49	81	10.3	7.8	31	26
			L	2.61	2.24	0.23	0.12	1.20	1.55	1.01	0.85	0.38	0.30	96	69	6.5	8.8	26	28
		CF 4-5	H	2.70	–	0.27	0.11	1.05	1.50	1.40	1.73	0.39	0.58	–	79	6.1	8.3	26	35
			L	3.35	2.25	0.29	0.19	2.00	1.05	1.30	1.73	–	0.68	73	84	8.6	8.3	30	24
6	Mile 33 sloping Low Pine Ridge	CF 29	H	–	2.54	–	0.17	–	1.60	–	0.53	–	0.31	–	68	–	9.8	–	20
			L	–	2.09	–	0.13	–	1.70	–	0.35	–	0.21	–	39	–	12.3	–	21
		CF 4-5	H	–	2.59	–	0.27	–	1.70	–	1.35	–	0.71	–	49	–	6.8	–	28
			L	–	2.65	–	0.19	–	1.60	–	1.18	–	0.63	–	79	–	10.8	–	33
7	Mile 35 Low Pine Ridge	CF 29	H	2.48	–	0.17	–	1.25	–	1.33	–	0.38	–	215	–	7.4	–	16	–
			L	2.22	2.05	0.13	0.13	1.15	1.60	1.18	1.14	0.34	0.19	218	88	6.1	8.8	16	18
		CF 4-5	H	2.53	1.87	0.27	0.13	1.35	1.00	1.29	1.43	0.65	0.52	118	105	8.6	5.5	24	24
			L	–	2.21	–	0.12	–	1.60	–	1.46	–	0.43	–	89	–	7.3	–	31
8	Mile 35 harrowed Low Pine Ridge	CF 29	H	2.77	2.13	0.17	0.16	1.35	1.55	1.20	1.62	0.34	0.19	213	78	7.2	8.8	20	14
			L	2.39	2.35	0.11	0.12	1.16	–	0.97	1.13	0.29	0.18	–	75	6.1	9.0	17	16
		CF 4-5	H	2.88	2.21	0.25	0.12	1.40	1.40	1.07	1.31	0.79	0.51	136	86	7.4	9.3	28	23
			L	2.92	2.02	0.20	0.10	1.60	1.30	1.42	1.60	0.54	0.44	138	104	9.0	6.3	39	26
11	Bull Run Mountain Pine Ridge	CF 29	H	2.54	2.75	0.21	0.21	1.10	1.30	1.83	1.55	0.30	0.24	33	23	7.2	4.0	14	11
			L	2.64	2.32	0.20	0.12	0.30	0.69	1.96	1.88	0.28	0.26	15	20	5.4	6.0	12	9
		CF 4-5	H	2.97	2.76	0.19	0.24	0.88	1.10	2.36	2.10	0.57	0.41	28	26	5.1	4.0	16	17
			L	3.05	2.00	0.21	0.24	1.20	1.60	2.06	1.78	0.55	0.43	25	46	11.0	10.0	18	25
		CF 50	H	2.96	2.70	0.22	0.21	1.65	1.90	0.80	1.33	0.26	0.23	54	45	5.4	6.0	18	16
			L	2.70	2.32	0.20	0.15	0.90	1.90	0.57	1.16	0.36	0.22	40	38	3.6	5.5	19	18

12	Rio On Mountain Pine Ridge	CF 29	H	2.43	2.54	0.18	0.17	1.15	1.10	1.13	1.65	0.38	0.20	75	29	5.4	3.0	32	14
			L	2.43	2.29	0.15	0.10	0.95	0.81	1.42	1.15	0.22	0.14	60	43	4.5	4.0	23	–
		CF 4-5	H	2.44	2.68	0.12	0.16	–	1.60	1.79	2.21	–	0.44	104	44	7.2	5.5	21	16
			L	2.15	2.34	0.10	0.10	0.70	0.70	1.69	–	0.52	0.35	119	53	8.1	8.0	26	15
		CF 50	H	2.46	2.50	0.22	0.17	1.45	1.30	1.29	1.43	0.18	0.23	63	40	4.1	6.0	12	16
			L	2.61	2.38	0.18	0.14	1.10	0.84	1.61	1.82	0.22	0.16	32	55	1.2	3.0	23	13
13	Baldy Beacon Mountain Pine Ridge	CF 29	H	2.71	2.58	0.21	0.14	1.10	0.70	1.10	2.13	0.52	0.23	28	25	4.5	2.5	17	15
			L	2.15	–	0.19	–	0.70	–	2.01	–	0.35	–	36	–	2.7	–	18	–
		CF 4-5	H	3.17	0.68	0.23	0.03	0.95	0.81	1.46	0.30	0.60	0.12	25	28	4.1	–	23	17
			L	3.00	2.80	0.23	0.19	0.75	0.95	1.61	2.13	0.42	0.44	23	18	6.3	3.5	27	15

Nutrients required for beef maintenance and growth[c]	N	P	K	Ca	Mg	Mn	Cu	Zn
	1.15	0.11–0.25	0.47–0.84	0.17–0.28	0.12–0.21	20–25	8–15	13–28

[a]CF 3-1 *Macroptilium atropurpureum* cv. Siratro, CF 4-5 *Stylosanthes guianensis*, CF 29 *Codariocalyx gyroides*, CF 50 *Desmodium intortum* "Greenleaf". [b]Fert., fertilizer added: H, high; L, low. [c]Extracted from ARC (1980).

Appendix 13.1

Botanical analyses and yields

Table 13.8. Site 4, Orange Walk, native grassland botanic composition.

Code[a]	Species	Dry matter (kg/ha × 10^{-1})
Legumes		
27	*Acacia spadicigera* Cham. & Schul.	6
22	*Cassia mimosoides* L.	17
33	*Cassia tora* L.	trace
10	*Desmodium canum* (Sw) DC.	6
13	*Desmodium triflorum* DC.	4
16	*Mimosa pudica* L.	266
	Total	299
Grasses		
21	*Axonopus compressus* (S.W.) P. Beauv.	971
4	*Paspalum convexum* Humb. & Bonpl.	94
36	*Paspalum langei* (Fourn.) Nash	trace
31	*Paspalum* sp.	trace
19	*Setaria gracilis* H.B.K.	111
	Total	1176
Other vegetation		
6	*Acanthaceae*	172
12	*Asclepias curassaarica* L.	Trace
17	*Chaptalia nutans* L. Polak.	Trace
28	*Cochorus siliquosus* L.	56
24	*Cuphea* cf. *carthagenensis* (Jacq.) Macbr.	Trace
13	*Elvira biflora* L.	Trace
15	*Euporbia hirta* L.	11
26	*Phyla nodiflora* (L.) Greene	44
29	*Stachytarpheta cayennensis* (L.C.Rich.) Vahl.	Trace
1	Unidentified grass	Trace
14	Unidentified herb	6
23	Unidentified herb	56
14?	Unidentified herb	Trace
18	Unidentified sedge	8
	Total	353

[a]Site-specific sample and herbarium specimen number.

Table 13.9. Site 5 Mile 33 level, native grassland botanic composition.

Code[a]	Species	Dry matter (kg/ha × 10^{-1})
Legumes		
6	*Cassia diphylla* L.	
7	*Cassia tagera* L.	
5	Unidentified legume (sterile)	
Grasses		
8	*Andropogon virgatum* Desf.	
21 (36)	*Axonopus ciliatifolius* Swallen	29
4	*Echinolaena gracilis* Swallen	
40	*Mesosetum angustifolium* (Swallen) Swallen	746
35	*Panicum boliviense* Hack.	
26	*Panicum fusiforme* Hitchc.	
10 (34)	*Panicum olivaceum* Hitchc. & Chase	
41	*Paspalum pulchellum* Kunth.	
3 (38)	*Paspalum maculosum* Trin.	
	Two sterile grasses	
	Total	775
Other vegetation		
1 (19)	*Borreria* sp.	4
24	*Calea* sp.	
17	cf. *Nemastylis sylvestris* Loes.	
23	*Evolvulus* sp.	
20	*Hyptus conferta* Pohl. Ex Benth.	119
27	*Piriqueta cistoides* (L.) Meyer	
18	*Polygala longicaulis* Kunth.	
9	*Polygala panuculata* L.	
13	*Pterolepis trichotoma* Cogn.	
14	*Rubiaceae* sp.	4
29	*Sauvagesia* sp.	
15	*Sauvagesia erecta* L.	
12	*Schultesia* sp.	
22	*Schwenkia americana* L.	287
11	Sedge	60
6	*Widelia hispida* H.B.K.	
	Eight unidentifiable	
	Total	474

[a]Site-specific sample and herbarium specimen number.

Table 13.10. Site 6 Mile 33 sloping, native grassland botanic composition.

Code[a]	Species	Dry matter (kg/ha × 10^{-1})
Legumes		
14	*Cassia diphylla*	
8	*Cassia tagera* L.	2
	Total	2
Grasses		
4	*Andropogon virgatum* Desf.	
42	*Axonopus ciliatifolius* Swallen	
42	*Digitaria* sp.	
35, 45	*Eragrostis* sp.	
36	*Leptochloa virgate* (L.) P. Beauv.	
39, 41	*Leptochoryphium lanthanum* (H.B.K.) Nees	
33, 33A	*Mesosetum angustifolium* (Swallen) Swallen	623
12, 29	*Panicum olivaceum* Hitchc. & Chase	
38, 16, 40	*Paspallum pulchellum* Kunth.	10
31, 31A	*Paspalum maculosum* Trin.	
44	*Schizachyrium condensatum* (H.B.K.) Nees	
1	Sterile grass	14
2	Sterile grass	41
28	Sterile grass	
	Total	688
Other vegetation		
26	*Ageratum ellipticim* Robinson	
7, 10	*Borreria* spp.	
5	*Calea trichotoma* J.D. Sm.	
17	cf. *Nemastylis sylvestris* Loes.	
15	*Evolvulus sericeus* (Sw).	
24	*Hyptus conferta* Pohl. Ex Benth.	
16	*Oxalis* sp.	
13	*Oxalis yucatanensis* (Rose) Sprague & Riley	
21	*Polygala longicaulis* Kunth.	
20	*Pterolepis trichotoma* Cogn.	
11, 25	*Sauvagesia* sp.	
18	*Schultesia* sp.	
23	*Schwenkia americana* L.	
6	Unidentifiable	23
9	Unidentifiable	190
17, 22, 27, 32, 34, 37, 43	Unidentifiable	
	Total	213

[a]Site-specific sample and herbarium specimen number.

Table 13.11. Sites 7 and 8 Low Pine Ridge soils, native grassland botanic composition.

Code[a]	Species	Dry matter (kg/ha × 10^{-1}) Site 7	Site 8
Legumes			
25	Sterile		
Grasses			
7	*Aristida* sp.	none	
8-19	*Digitaria ciliaris* (Retz.) Koel.		
22	*Mesosetum angustifolium* (Swallen) Swallen	372	133
8-20	*Paspallum pulchellum* Kunth.	15	29
4, 19	*Schizachyrium sangiuneum* (Retz.) Alst.	none	3
20	Sterile grass	?	
22	Sterile grass	?	
27	*Trachypogon spicatus* (L.f.) Kuntze	1	none
	Totals	388	165
Other vegetation			
8	*Borreria* spp.	8	6
14	*Calea* sp.		
7	*Hyptus conferta* Pohl. Ex Benth.		
13	*Oxalis* sp.	4	2
18	*Piriqueta cistoides* (l.) Meyer		1
10	*Polygala longicaulis* Kunth.		1
4, 15	*Rubiaceae* (= 10-5 no 14)		
1	*Sauvagesia* sp.		
11, 17	*Schwenkia americana* L.		2
6	Sedge	4	25
12	Sedge	86	11
24	Sedge	2	
2	Sedge		3
21	Sedge		38
9, 26	*Wedelia hispida* H.B.K.		
	Fifteen unidentifiable		
	Totals	104	89

[a]Site-specific sample and herbarium specimen number.

Table 13.12. Site 11 Bull Run Ranch Mountain Pine Ridge: native grassland botanic composition.

Code[a]	Species	Dry matter (kg/ha × 10^{-1})
Legumes		
17	*Aeschynomene elegans* Cham. & Schlecht.	
11, 47	*Cassia aeschynomene* DC.	
14, 20	*Cassia diphylla L.*	
28	*Clitoria guianensis* Benth.	
15	*Crotalaria sagittalis* L.	
55	*Crotalaria vitellina*	
26	*Desmodium adscendens* DC.	
19	*Eriosema crinitum* Benth. var. *crinitum*	
7, 57	*Galactia anomala* Lundell	
13	*Pachyrrhizus vernalis* Clausen	
2	*Phaseolus linearis* H.B.K.	5
8	*Stylosanthes guianensis* (Aulb.) Sw.	74
16	*Tephrosia nitens* Benth.	
5	*Zornia reticulata* J.E. Sw.	131
18, 53	*Zornia* sp.	
	Total	210
Grasses		
6	*Axonopus purpusii* (Mez.) Chase	126
40	*Digitaria ciliaris* (Retz.) Koel.	
21	*Digitaria fuscescens* (Presl.) Henr.	
36	*Eragrostis maypurensis* (H.B.K.) Stead.	
30	*Eragrostis* sp.	
39	*Mesosetum angustifolium* (Swallen) Swallen	
1	*Panicum fusiforme* Hitchc.	47
32, 35	*Paspalum pectinatum* Nees.	47
42	*Paspalum pulchellum* Kunth.	
43	*Paspalum* sp.	
27	Sterile grass probably *Mesosetum angustifolium*	436
41	*Trachypogon spicatus* (L.f.) Kuntze	
3	Sterile grass	52
4	Sterile grass	32
9	Sterile grass	8
29	Sterile grass	16
	Total	764
Other vegetation		
11, 22	*Borreria* sp.	
51	*Borreria* sp.?	
10	*Calea trichotoma* J.D. Sm.	37
49, 61, 62	*Chrysopsis graminifolia* (Michx.) Nutt	

Continued

Table 13.12. Continued.

Code[a]	Species	Dry matter (kg/ha × 10^{-1})
52, 54	*Cuphea* cf. *carthagensis* (Jacq.) Macbr.	
47	*Diclieuxia fruiticosa*	
?	*Diodia sarmentosa* Sw.	
24	*Dividia sarmentosa* Sw.	
23	*Evolvulus* sp.	
48	*Orchidaceae*	
60	*Polygala aparinoides* H.&A.	
34	*Richardia scabra* L.	
50	*Scoparia dulcis* L.	
25	*Sida linifolia* Cav.	
33, 58, 59	*Turnera ulmifolia* L.	
	Total	37
	Pinus caribea (leaves)	3214

[a]Site-specific sample and herbarium specimen number.

Table 13.13. Site 12 Rio On, Mountain Pine Ridge native grassland botanic composition.

Code[a]	Species	Dry matter (kg/ha × 10^{-1})
Legumes		
27	*Stylosanthes guianensis* (Aulb.) Sw.	
16	Sterile	
	Total	0
Grasses		
4	*Aristida adscensionis* L.	
12, 18	*Axonopus purpusii* (Mez.) Chase	
26	*Digitaria ciliaris* (Retz.) Koel.	
10	*Elusine indica* (L.) Gaertn.	
12, 22?	*Leptocoryphium lanatnum* (H.B.K.) Nees	?
9	*Mesosetum angustifolium* (Swallen) Swallen	3773
15, 24	*Paspalum pectinatum* Nees.	
19	*Paspalum pulchellum* Kunth.	
20	*Paspalum* sp.	
23	*Sporobolus cubensis* A.C. Hitchc.	
5, 6, 21	*Trachypogon spicatus* (L.f.) Kuntze	380
13	Sterile grass	
	Total	4153
Other vegetation		
8	*Byrsonima crassifolia* H.B.K.	
1	*Calea longipedicellata* Robinson & Greenman	

Continued

Table 13.13. Continued.

Code[a]	Species	Dry matter (kg/ha × 10^{-1})
14	*Clidemia rubra* (Aubl.) Mart.	
2	Fern	
3	*Miconia albicans* (Sw) Triana	
11, 25	Two unidentifiable	
	Total	0
	Pinus caribea (leaves)	58

[a]Site-specific sample and herbarium specimen number.

Table 13.14. Site 13 Baldy Beacon, Mountain Pine Ridge native grassland botanic composition.

Code[a]	Species
Legumes	
7	*Cassia* cf. *tetraphylla* Desv.
20	*Desmodium barbatum* Benth.
Grasses	
42	*Aristida* sp.
36	*Axonopus purpusii* (Mez.) Chase
1	*Eleusine indica* (L.) Gaertn.
25	*Eragrostis maypurensis* (HBK) Stead.
24	*Mesosetum angustifolium* (Swallen) Swallen
2, 4, 28	*Panicum olivaceum* Hitchc. & Chase (in dung)
3	*Panicum strigosum* Muhl.
17	*Panicum viscidellum* Scribn.
14	*Panicum*?
8, 35	*Paspalum pulchellum* Kunth.
41	*Paspalum* sp.
27	*Thrasya campylostrachya* Chase
42	*Trachypogon spicatus* (L.f.) Kuntze
15	Sterile grass
Other vegetation	
40	*Asclepias curassarica* L.
13	*Borreria* sp.
	Borreria sp.?
19	*Coccocypselum hirsutum* (Bartl.)
19?	*Coccocypselum glabrum* DC.
11	*Emilia fosbergii* Nicholson
31	*Erechtites hieraciifolia* (L.) DC.
29	*Gerardia harperi* (Penn.) Penn.

Continued

Table 13.14. Continued.

Code[a]	Species
18 (13)?	*Hemidiodia ocymifolia* (Willd.) K. Schum.
12	*Hypericum fasciculatum* Lam.
5	*Hypericum styphelioides* A. Rich.
16, 30	*Hyptis lantanifolia* Poit.
9	*Marsypianthes chamaedrys* (Vahl.) Kuntze
33, 45	cf. *Nemastylis sylvestris* Loes.
32	*Orchidaceae*
22, 38	*Polygala* sp.
34	*Rubiaceae*?
10, 39	*Sauvagesia*
	Six unidentifiable

[a]Site-specific sample and herbarium specimen number.

14 Single Plant Trials of Potential Forage Legumes for Belizean Pastures on Clay Soils of the Upper Belize River and the Lowland Pine Ridge

J.R. Lazier*[1]

**Formerly International Livestock Centre for Africa*

Abstract

Fertilized replicated single plant trials were established on two major soil types in wet tropical conditions in central Belize, a heavy cracking clay and a waterlogged kaolinitic clay. Thirty-five accessions (32 species and 20 genera) were harvested by clipping at 6-week intervals. Survival was better on the wetter site, but yields were lower. *Macroptilium atropurpureum* cv. Siratro, *Indigofera hirsute*, *Pueraria phaseoloides* and *Desmodium cinerium* were the most productive on the cracking clay, while *P. phaseoloides*, *Vigna caracalla*, *Desmodium ovalifolium*, *Stylosanthes guianensis* and Siratro performed best on the kaolinitic clay. *D. cinerium* and *D. ovalifolium* in particular appeared to be worthy of further study for these soils.

14.1 Introduction

There are two major land forms with potential for livestock production in central Belize, the upper Belize River terraces and the Puletan soils of the lower Belize River Valley (Fig. 12.1, Chapter 12, this volume). The upper Belize River terraces and adjacent areas are characterized by soils with vertic properties, heavy, dark and cracking clays, on which medium-sized modern farms are situated and where livestock have been a major source of income. Despite the wet tropical conditions productivity was not high due to the low yields of native grasses and low nitrogen levels of the fodder available. The kaolinitic Puletan soils (planosols) of the lowland savanna, on the other hand, were little utilized for agriculture, although at the time of these trials in the 1970s some ranchers were testing the practicability of grazing them. Their sparse vegetation, waterlogged character and low nutrient status were all barriers to their use. The native vegetation of some representative pastures on these soils was reported earlier (Chapter 13, this volume). The identification of adapted legumes could raise the nutrient level of the natural vegetation, improve livestock productivity on these soils, and make livestock raising a more profitable and sustainable enterprise. Before this project, there had been very limited research on the establishment of pasture grasses and legumes on the clay soils of the upper Belize River Valley and on the Low Pine Ridge, for example Rose-Innes (1973), Blue (1969) and Government reports. However, there had been no immediate follow up to the results, which are summarized

[1] E-mail: john.lazier@sympatico.ca

in Chapter 15 (this volume) under the various environments.

Research projects are usually funded for brief periods, about 5 years. In that time there is usually insufficient opportunity to effectively screen collected native or acquired germplasm for adaptability and productivity. Seed of such material is usually available only in very small quantities, and there is not enough time to bulk the seed to the level needed for replicated plots. Additionally, resources to multiply the seed and undertake pure or mixed sward replicated trials may be limited. Such trials are expensive and methods of initially testing persistence and productivity of numerous pasture plants of unknown value, most of which will be of little utility, are laborious. An alternative, inexpensive approach is to screen such limited seed stock as single plants in a replicated trial, which allows large numbers of plants to be observed and clipped to provide an initial assessment of relative pasture character.

In mid-1977 the University of the West Indies–IDRC Forage Legume Project in Belize had some 670 accessions of leguminous native and exotic plants of fodder potential. The native legumes had been collected mainly in the vicinity of Central Farm from the terraces of the Belize River, and from the two adjacent nutrient-poor environments, the Low Pine Ridge and the Mountain Pine Ridge, and from Guatemala and the Yucatan. Additional experimental and commercial accessions were obtained from the Caribbean islands, the University of Florida, CIAT and CSIRO. Unreplicated introduction plots involving 99 species and 36 genera were planted as spaced seedlings, six per plot as the material became available on Central Farm, the government research station in Cayo District. A minimal-level basic fertilizer regime was followed. The site received abundant rainfall, was on deep soils with vertic properties and good moisture holding capacity, and had a relatively short dry season, so moisture stress was not severe. Detailed observations were taken on 13 parameters on a 6-weekly basis to assess their adaptation, productivity, persistence and other relevant agronomic characteristics.

The data from 501 of the plots, which were established over periods ranging from 9 to 31 months, were grouped into eight planting times and subjected to pattern analysis using four of the parameters observed (plot cover, vigour, height, spread and leafiness). As expected, many of the introductions performed well under such favourable environmental conditions (Lazier, unpublished data).

The project then faced the problem of having insufficient funding and time to multiply seed and screen all lines using traditional sown replicated plots. Some of the promising materials had sufficient seed to be screened using replicated plots, and trials were undertaken that have been reported previously (see Lazier 1978, 1980a, 1981a). However, for selected successful lines for which little seed was available, it was deemed appropriate to acquire better data on persistence, yield and adaptability by further testing them as single plants in a replicated trial on two important contrasting soils of central Belize: a high terrace of the Belize River on basic black cracking soils characterized by montmorillonitic clays which receives an average of 1460 mm rainfall per annum and on a Lowland Pine Ridge soil, a typical waterlogged site characterized by acid kaolinitic clays and 2200 mm annual rainfall.

14.2 Methods

Two sites were chosen for the trial; the first was on the Government Research Farm 'Central Farm' in Cayo District on a nearly level, well to moderately well drained native grass pasture situated on the old alluvium of a high terrace of the Belize River. This site represented the basic, black clay soils that dominate the higher terraces of the Belize River and adjacent areas in central Belize. They have been characterized by Wright *et al.* (1959) as Melinda Suite, Redbank Subsuite, Meditation series pale phase, and by Jenkin *et al.* (1976) as a silty clay of the Morning Star Series. A study of the trial site by McKinzie (1977) determined the soils to be Mollic Tropudalfs, fine, mixed isohyperthermic that are similar to Jenkin *et al.*'s Esperanza series.

A series of pot trials with Morning Star Series soils have given responses to phosphorous, copper and magnesium (Jenkin *et al.*, 1976). As Esperanza Series soils are generally higher terraces and thus generally more weathered with poor drainage and mottling, they are probably no more fertile than the Meditation Series of Wright *et al.* (1959).

The second site was located just west of the mile 33 post on the Western Highway, about 30 m off the road on the north side and was chosen to represent the sterile waterlogged old alluvium soils of the lower Belize River. It was adjacent to the strip trial sites 5 and 6 and a single plant trial (Chapters 14 and 15, this volume) that were established later. The environment was commonly referred to as the Low or Coastal Pine Ridge. The site chosen sloped gently to the north and was a clear area of savanna with a sparse ground cover of sedges and wire grass (*Mesosetum angustifolium*), a few clumps of palmetto (*Acoelorraphe wrightii* (Griseb.) Wendl.) and widely scattered pine (*Pinus caribea* Morelet), crabboe (*Brysonma crassifolia* (L.) DC) and gourd trees (*Cresedentia cujete* L.). The area was ungrazed but had been burnt over in most years. The profile was wet throughout and the lower layers 34–60 cm had Fe and Mg in clusters of soft dusty red accumulations and a few small concretions. The area's soils had been classified by Wright *et al.* (1959) as Puletan Suite, Yobo subsuite, Rough Mile Series and by Jenkin *et al.* (1976) as Yaxa Suite, Yalbec Subsuite, Beaver Dam Series, a planosol with infertile topsoil, low cation exchange capacity, potassium, sodium and phosphorous. An examination of the trial site by McKinzie (1977) described the soils as Aeric Tropaqualfs, fine kaolinitic, isohyperthermic and strongly acid in reaction.

The soils at the two sites were sampled for analysis at 30 sites at 0–15 cm and 15–30 cm, and 15 sites at 30–45 cm and 45–60 cm. The lower two layers were not sampled at the Low Pine Ridge site because the soil was so waterlogged at those depths that it flowed, and could not be sampled by auger or tube. The samples were bulked mixed for each depth, and subsampled for analysis. The analyses were undertaken at the Regional Analytic Laboratory in St Augustine, Trinidad. The results are presented in Table 14.1.

The Central Farm site chosen was a native pasture. It was prepared for planting by ploughing and harrowing then using a rotary tiller to produce a fine tilth. At the Lowland Pine Ridge site bushes and trees were cleared and the grass was mown; the plants were planted without further site preparation due to the very low vigour of the native grasses and sedges.

Fertilizer was applied initially at both sites at 63 kg/ha each of triple superphosphate and muriate of potash with the savanna site receiving the same amount of fritted trace element mixture. Further applications of the P and K fertilizers were made in split applications after alternate harvests at the same rate per annum. The fertilizers were broadcast by hand over the entire trial areas.

The 35 legume accessions chosen for the trials comprised 32 species belonging to 20 genera (Table 14.2) and were a mixture of common commercial pasture species, imported legume accessions that had been shown to hold promise in trials in other countries, and local accessions from the upper Belize River Valley terraces, the Low Pine Ridge and from the adjacent Mountain Pine Ridge. Siratro, which had performed well in earlier trials on Central Farm (Lazier, 1978, 1980a), was selected as the control treatment.

Not all of the legumes planted at the alluvial site on Central Farm were planted in the Low

Table 14.1. Chemical and physical analyses of soils at the two trial sites, Central Farm and Low Pine Ridge.

			Oven dry soil (%)					Oven dry soil (me/100 g)				Oven dry soil (%)			
Depth (cm)	pH	Elec cond mhosx 10^{-6}	Coarse sand	Fine sand	Silt	Clay	CEC meq/ 100 g	Ca	Mg	K	Na	C	N	C/N ratio	P ppm
Central Farm															
0–15	6.2	69	10	35	44	9	5.00	5.00	1.09	0.27	0.23	2.08	0.18	11.6	7
15–30	6.3	35	7	16	68	9	8.48	5.00	1.59	0.24	0.28	1.04	0.13	8.0	2
30–45	7.1	35	3	15	28	57	9.43	5.83	2.28	0.15	0.25	0.52	0.10	5.2	2
45–60	6.9	35	2	16	28	53	8.57	5.00	1.88	0.16	0.40	0.32	0.09	3.6	3
Low Pine Ridge															
0–15	5.8	29	24	45	4	29	6.67	3.33	1.39	0.08	0.44	1.92	0.13	14.8	2
15–30	5.4	24	19	25	30	25	9.95	4.17	2.78	0.09	0.26	0.80	0.08	10.0	2

Table 14.2. Species and varieties planted in the trials, and harvest details.

Accession number CF[a]	Species	Variety/ other accession no.	Trial	Origin		Harvest	
				Country	Soil series[b]/ character	Height (cm)	Diameter (cm)
199	*Calopogonium caeruleum*		AB	Commercial		15	30
42	*Calopogonium mucunoides*		AB	Belize	Acid	15	30
523	*Cassia diphylla*		B	Belize	Puletan	10	20
524	*Cassia tagera*		B	Belize	Puletan	10	20
16-1	*Centrosema plumieri*		B	Belize	Quamina	15	30
177	*Centrosema pubescens*		B	Belize	Puletan	15	30
204	*Clitoria rubignosa*		B	Belize	Puletan	15	30
51	*Clitoria ternatea*		AB	Belize	Regosol	15	30
122	*Crotalaria pumila*		AB	Belize	Pinol	15	30
175	*Crotalaria sagittalis*		A	Belize	Road margin	10	10
157	*Desmodium cinereum*		AB	Belize	Redbank	40	30
188	*Desmodium heterocarpon*	IRFL588	AB	cv. Hetero?		15	20
200	*Desmodium ovalifolium*		AB	Commercial		15	20
7	*Desmodium scorpiurus*		A	Belize	Quamina	5	20
198	*Flemingia congesta*		AB	Commercial		20	30
40	*Galactia striata*		AB	Belize	Acid	15	30
184	*Glycine wightii*	Tinaroo	AB	Commercial		15	30
156	*Indigofera hirsute*		AB	Commercial		15	30
181	*Lablab purpureus*	Highworth	AB	Commercial		15	30
187	*Lotononis bainesii*		AB	Commercial		2	30
3-1	*Macroptilium atropurpureum*	Siratro	AB	Commercial		15	30
178	*Macroptilium axillare*	IRFL1804	A	Costa Rica?		15	30
185	*Medicago sativa*	Hunter River	AB	Commercial		5	30
284	*Pachyrrhizus vernalis*		B	Belize	Puletan	15	30
180	*Pueraria phaseoloides*	Puero	AB	Commercial		15	30
93	*Rhynchosia longeracemosis*		A	Guatemala	Sotz	15	30
4-6	*Stylosanthes guianensis*		AB	Belize	Pinol acid	15	30
131	*Stylosanthes guianensis*		AB	Belize	Pinol	10	30
201	*Stylosanthes guianensis*	Oxley	AB	Brazil		15	20
12	*Stylosanthes hamata*		AB	Antigua		15	20
4-2	*Stylosanthes ingrata*		AB	Belize	Skeletal, acid	15	20
161	*Stylosanthes ingrata*		AB	Belize	Pinol	15	20
173	*Stylosanthes viscosa*	CPI 34094	A			15	20
196	*Tephrosia vogelii*		AB	Commercial		15	30
189	*Vigna caracalla*	IRFL1806	AB	Florida		15	30

[a]Central Farm accession number; [b]following Wright *et al.* (1959).

Pine Ridge soil; those not expected to succeed under the wet conditions were eliminated. These included *D. scorpiurus* (CF 7), *C. plumieri* CF 16-1, *R. longeracemosis* CF 93, *S. viscosa* CF 173, *C. sagittalis* CF 175 and *T. vogelii* CF 196 (insufficient planting material). These were replaced by *C. rubignosa* CF 204, which was collected in the Lowland Pine savanna, and two legumes that volunteered in the trial once it was fertilized, *Cassia diphylla* CF 523 and *Cassia tagera* CF 524. These were abundant enough to have one of each appropriately situated in each block without transplanting.

The plants were grown in pots in the propagation shed from cuttings or scarified seeds inoculated with appropriate strains of *Rhizobium* where known and otherwise with the cowpea strain.

The trial design was the same at both sites, with three blocks or replicates, and a single plant of each of 29 species and varieties planted randomly in each block spaced at 2-m intervals (Fig. 14.1a). Both sites were planted in mid-November 1975. Weeds and grasses were permitted to grow in the spaces between the plants. At the Central Farm site, weeding was done where needed immediately around the individual plants until they were established, and the intervening spaces between plants were regularly mown by a motorized rotary lawnmower in order to keep competition between the test plants and the weeds (mainly Bermuda grass *Cynodon dactylon*) to a minimum. At the Central Farm site this was normally done immediately after a harvest and again 3 weeks later. A small perimeter about each plant was hand weeded to prevent accidental mowing of the plants (Fig. 14.1b). In contrast, competition from native vegetation was low at the Low Pine Ridge site and mowing was only done after each harvest.

Harvesting of the plants for dry matter production was done at a height and radius from the main stem deemed appropriate to the growth form of each accession in an attempt to obtain the maximum amount of herbage without reducing the plant's vigour or survival (Table 14.2). Harvesting was commenced at Central Farm from mid-February 1976 and continued at 6-week intervals until July 1977. At the savanna site the plants were slow to establish because the initial fertilizer was not applied until January 1976 due to a technical error. The first harvest at that site was done in April of that year and poor regeneration and growth delayed the second harvest 15 weeks until July. Thereafter it was harvested at regular 6-weekly intervals until June 1977.

14.3 Results and discussion

The seasonal variations of temperature, sunshine and rainfall have marked effects on pasture productivity (Lazier, 1978, 1980a). The first year's dry season was of normal duration, but severe in April 1976 when there was negligible rain in March and none in April (Fig. 14.2). May rains were scattered and light. At the Central Farm site the soil profile was quite dry because only one rain near the end of the month was of any consequence and the soil profile was probably not wet until into June. The 1977 dry season was a very mild one, with frequent showers with the vegetation remaining green through most of it.

14.3.1 The Central Farm site

The dry matter yields at the Central Farm site (Tables 14.3, 14.4) indicate that *Macroptilium atropurpureum* cv. Siratro CF 3-1 was the most productive legume, followed by *Stylosanthes guianensis* CF 4-6 and *Desmodium cinereum* CF 157. *Glycine javanica* cv. Tinaroo CF 184 and *Pueraria phaseoloides* (Puero or Kudzu) CF 180 also did well. *Flemingia congesta* CF 198 produced lower but quite consistent yields.

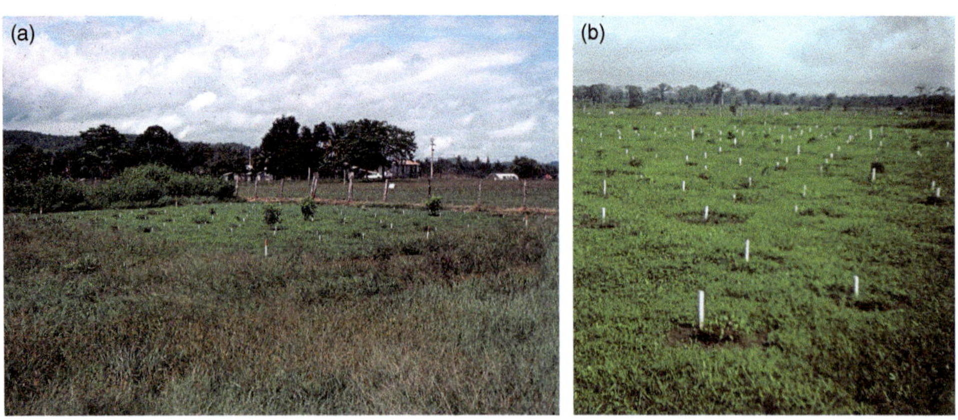

Fig. 14.1. Thirty-five accession single plant trial site at Central Farm. (a) Trial site before harvesting 1977. (b) Trial showing weeded plant perimeters.

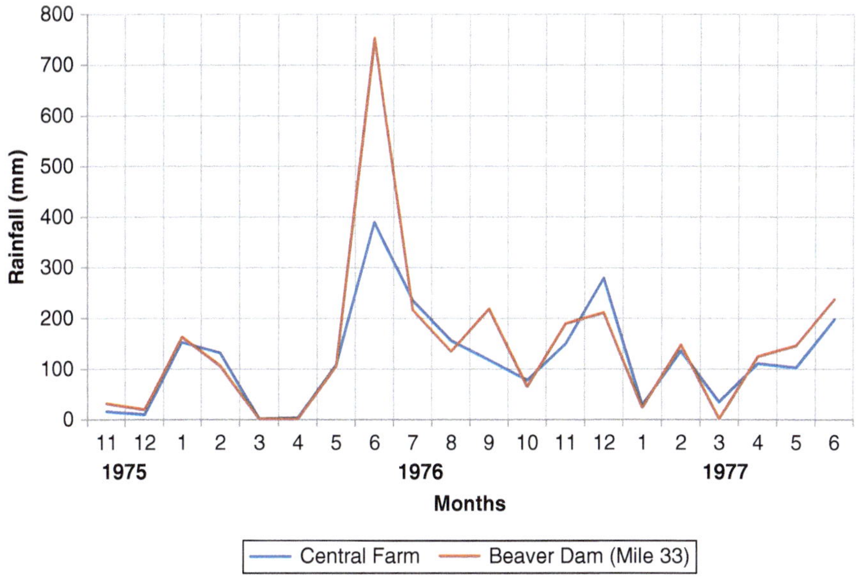

Fig. 14.2. Rainfall at trial sites.

Over the period of the trial, Siratro produced significantly more dry matter than all legumes except *S. guianensis* CF 4-6. Through the two dry periods (harvests 2 through 4, and 10 through 12), its production remained satisfactory (Table 14.3). Over a 1-year period (harvest 4 to harvest 13), it produced 826 g/plant or 8260 kg/ha at 1 plant/m^2. The yield is almost double that of the legume in pure sward (Lazier, 1980a), which is not surprising because the plant was not competing with other Siratro plants but only with *Cynodon dactylon* and short weeds that grew quite densely between mowings in the wetter periods.

S. guianensis CF 4-6, an accession from acid soils at 600 m elevation on the Mountain Pine Ridge in Belize, produced almost as much dry matter as Siratro. The plant had been noted to be a difficult one to establish, and in the introduction plots where, when it was not cut in the vegetative state, it acted as an annual. In the trial the plants in all three blocks died before the first harvest and, because there were no plants to replace them, seed was planted instead in February. Only one plant germinated and survived. Its slow growth through the dry season resulted in the treatment not being harvested until September, by which time the plant had suddenly became very large, demonstrating the vigour previously shown in the introduction plots. Despite having to recover from the severe September clipping, considerable dry matter was produced in the cool season, though little was produced in the dry season that followed. A definite response to higher rainfall was noted at the end of both dry periods.

Desmodium cinereum D.C. (synonyms *D. chiapense* T.S., *Hedysarum cinereum* HBK., *Meibomia cinerea* HBK. (standard 1)) has been mentioned as a legume that should be tested as a forage plant in Belize (Stehle, 1956). Stehle described it as a West Indian and Central American exotic, though in fact it is native to Central America (Molina, 1975). In the French West Indies, though it was thought to be too sparsely leafed to be used as a green manure, it showed promise as a forage shrub. Referred to as 'Cousin' in French and 'Tick Clover' in English, Stehle recommended it be planted through the Caribbean in wetter areas in light lateritic soils, and areas with semi-arborescent savannas and unused thickets. Dry matter production over the whole trial was significantly less (0.05) than that of Siratro, but over a year, May through May, the difference was no longer significant. The plant was not cut in the first harvest, because it was suspected to not be a legume, however it flowered soon thereafter and despite having grown well over 1 m, it survived cutting back to a stump at 40 cm at the second harvest. Regular cutting produced lateral branches at the top of the stump that eventually formed a flat table of twigs at cutting height, with

Table 14.3. Central Farm site dry matter yields and plant survival.

		Harvest date, number, dry matter (g/plant)																
		1976							1977								Plant numbers	
		16/2	30/3	10/5	22/6	3/8	14/9	25/10	6/12	18/1	1/3	13/4	26/5	4/7		Year DM total[b] H3–12		
Accession no. CF[a]	Species	1	2	3	4	5	6	7	8	9	10	11	12	13	Trial DM total (g)	(g)	Initial	Final
175	*Crotalaria sagittalis*	2	4	0	4	7	11	5	7	2	2	0	0	0	43	37	3	0
4-2	*Stylosanthes ingrata*	2	16	0	4	8	7	8	5	3	1	0	1	0	55	37	3	0
173	*Stylosanthes viscosa*	3	10	19	7	18	11	6	5	2	2	1	0	0	85	71	3	0
187	*Lotononis bainesii*	0	43	20	2	6	0	0	0	0	0	0	0	17	87	28	3	0
178	*Macroptilium axillare*	27	18	15	38	40	26	11	2	0	0	0	0	0	117	132	2	0
196	*Tephrosia vogelii*	56	15	6	18	14	4	0	0	0	0	0	0	21	123	31	3	0
161	*Stylosanthes ingrata*	5	17	2	19	20	19	23	10	4	5	3	1	0	129	107	3	0
201	*Stylosanthes guianensis*	3	8	8	25	39	23	8	12	3	2	0	0	0	130	119	3	0
7	*Desmodium scorpiurus*	–	1	0	6	12	21	55	10	5	2	6	8	9	135	126	2	1
42	*Calopogonium mucunoides*	22	11	0	7	30	22	25	6	5	0	0	0	9	136	94	3	1
12	*Stylosanthes hamata*	31	4	0	15	33	15	15	9	3	1	1	3	6	138	96	3	1
185	*Medicago sativa*	60	38	22	22	11	2	0	0	0	0	0	0	0	155	57	3	0
188	*Desmodium heterocarpon*	11	4	0	26	39	25	7	10	8	5	4	18	1	158	142	3	2
122	*Crotalaria pumila*	148	8	3	16	5	14	5	1	0	0	0	0	0	201	44	3	0
93	*Rhynchosia longeracemosis*	8	10	8	18	14	27	27	19	19	4	18	28	18	216	180	2	1
189	*Vigna caracalla*	37	14	12	32	66	15	27	9	7	0	14	3	0	235	183	3	1
131	*Stylosanthes guianensis*	4	10	14	35	26	38	36	16	13	8	7	18	13	238	211	3	2
181	*Lablab purpureus*	189	3	3	25	3	8	4	5	0	0	0	0	0	240	47	3	0
200	*Desmodium ovalifolium*	3	8	0	22	37	42	34	21	13	8	5	29	0	244	213	3	3
198	*Flemingia congesta*	2	13	0	6	16	34	33	32	32	21	25	22	18	245	222	3	2
51	*Clitoria ternatea*	5	7	0	17	26	62	91	20	5	1	2	5	3	246	230	3	1

139	*Calopogonium caeruleum*	13	20	12	50	38	46	35	26	13	13	5	25	14	309	263	3	2
40	*Galactia striata*	0	5	2	27	48	40	98	36	27	10	9	36	45	382	332	1	1
156	*Indigofera hirsute*	56	12	3	48	152	85	40	6	5	0	0	0	0	406	338	3	0
180	*Pueraria phaseoloides*	0	0	0	6	128	105	138	46	55	14	61	24	18	594	576	1	1
84	*Glycine wightii*	193	52	18	104	100	60	23	19	10	2	8	2	17	607	345	3	3
57	*Desmodium cinereum*	–	107	13	71	133	64	107	73	61	31	30	54	65	809	638	3	3
4-6	*Stylosanthes guianensis*	0	0	0	0	0	780	3	36	3	5	9	53	20	909	889	1	1
3-1	*Macroptilium atropurpureum*[c]	97	40	47	176	181	113	58	47	29	20	35	71	96	1010	777	3	3
LSD																		
P = 0.05		79	36	17	48	51	49	38	15	10	7	12	20	20	182	163		
P = 0.005		105	49	23	63	69	66	51	20	14	9	16	26	27	243	218		
P = 0.001		138	64	30	83	89	86	67	26	18	12	21	34	35	318	285		

[a]Central Farm accession number; [b]harvests 3 to 12; [c]control treatment.

regrowth forming vertical shoots 40–50 cm tall after periods of better rainfall.

Siratro produced more dry matter than *D. cinereum* in September, but by the October harvest the latter plant was producing more dry matter and it produced significantly more through the cool period in December and January (0.001) and to March (0.01). In the dry season, while *D. cinereum* produced well, Siratro produced more, but not significantly more until July after the rains began.

D. cinereum is a small perennial shrub that grows 2–3 m high if left uncut; the plants are little affected by disease and pests. Flowering is indeterminate and masses of seeds are produced. Compared with *Codariocalyx gyroides*, another vigorous and adapted bushy forage plant (e.g. Lazier, 1981c), it has a more flexible stem and suffers much less mortality with age and disease. *D. cinereum* is thus a forage plant worthy of further study.

14.3.2 The Low Pine Ridge site

Low Pine Ridge site (Fig. 14.3) dry matter yield and plant survival data are presented in Table 14.4. The two native legumes occurring spontaneously at the trial site that were used in the trial, *Cassia diphylla* CF 523 and *Crotalaria tagera* CF 534, and two native accessions from Mountain Pine Ridge, *Crotalaria pumila* CF 122 and *Pachyrrhizus vernalis* CF 284, were all particularly low dry matter producers; *Crotalaria pumila* CF 122, despite being found naturally at the site, died after the first harvest. On the other hand, *Centrosema pubescens* CF 177, collected previously nearby the trial site, was moderately productive. *Stylosanthes guianensis* CF 4-6, *Lotononis bainsii* CF 187 and *Medicago sativa* CF 185 did not survive until the first harvest, probably because of the waterlogged condition of the soil, and *S. hamata* CF 12, which is also from a drier environment, produced little growth and died out as well. The two *Stylosanthes ingrata* lines (CF 4-2 and CF 161) are from a higher elevation and better drained soils, and by the last harvest, both had died out: *S. hamata* CF 12 died much earlier than *S. ingrata* CF 161. All plants of the two legumes used as cover crops in Asia, *Flemingia congesta* CF 198 and *Tephrosia vogelii* CF 196, survived, but *F. congesta* produced negligible dry matter, while only modest amounts were produced by *T. vogelii*. *Desmodium heterocarpon* CF 188, a legume that had been found to be

Fig. 14.3. General view of the Low Pine Ridge trial site. Single plant trial in foreground (see also Fig. 17.3, Chapter 17, this volume).

Table 14.4. Low Pine Ridge site dry matter yields and plant survival.

Accession number CF[a]	Species	Harvest date, number, dry matter (g/plant)										Trial DM total (g)	Year DM[b] total H 2–10 (g)	Plant numbers	
		1976					1977								
		1/4	15/7	26/8	5/10	16/11	30/12	10/2	23/3	3/5	15/6				
		1	2	3	4	5	6	7	8	9	10			Initial	Final
4-6	*Stylosanthes guianensis*[b]	0	0	0	0	0	0	0	0	0	0	0	0	3	0
185	*Medicago sativa*	0	0	0	0	0	0	0	0	0	0	0	0	3	0
187	*Lotononis bainesii*	0	0	0	0	0	0	0	0	0	0	0	0	3	0
12	*Stylosanthes hamata*	1	1	0	1	2	1	0	0	0	0	6	5	3	0
122	*Crotalaria pumila*	14	0	0	0	0	0	0	0	0	0	14	1	1	0
181	*Lablab purpureus*	6	1	0	0	2	0	1	0	1	11	22	16	3	3
51	*Clitoria ternatea*	3	1	0	2	3	2	3	5	3	3	27	24	3	2
16-1	*Centrosema plumieri*	4	9	1	2	5	1	1	2	3	8	36	32	3	1
523	*Cassia diphylla*	–	28	5	5	1	0	0	0	0	0	39	39	3	0
204	*Clitoria rubignosa*	0	4	7	5	12	6	4	2	4	5	48	48	3	2
524	*Cassia tagera*	–	31	3	8	4	3	1	0	2	1	53	52	3	1
196	*Tephrosia vogelii*	9	5	3	2	4	2	4	6	7	14	57	48	3	2
284	*Pachyrrhizus vernalis*	4	26	6	6	6	2	1	0	2	5	62	58	2	2
157	*Desmodium cinereum*	0	0	0	4	2	2	11	15	16	25	75	75	3	2
4-2	*Stylosanthes ingrata*	17	17	4	12	8	6	6	11	5	13	98	81	3	2
161	*Stylosanthes ingrata*	5	23	19	18	19	7	10	11	3	0	117	112	3	0
199	*Calopogonium caeruleum*	3	10	2	8	4	5	15	39	18	38	141	138	3	3
201	*Stylosanthes guianensis*	2	27	15	22	14	10	8	18	14	16	147	145	2	2

Continued

Table 14.4. Continued.

Accession number		Harvest date, number, dry matter (g/plant)													
		1976					1977							Plant numbers	
		1/4	15/7	26/8	5/10	16/11	30/12	10/2	23/3	3/5	15/6		Year DM[b]		
CF[a]	Species	1	2	3	4	5	6	7	8	9	10	Trial DM total (g)	total H 2–10 (g)	Initial	Final
198	*Flemingia congesta*	1	8	9	13	12	8	18	27	23	29	147	146	3	3
184	*Glycine wightii*	13	12	6	9	6	6	22	41	15	39	169	156	3	3
177	*Centrosema pubescens*	1	15	3	14	16	7	11	23	26	55	171	170	3	3
188	*Desmodium heterocarpon*	5	42	31	11	6	7	14	18	17	39	189	184	3	3
40	*Galactia striata*	0	15	16	15	29	10	21	26	20	45	198	198	3	3
42	*Calopogonium mucunoides*	24	47	6	20	17	10	39	18	5	29	214	190	3	3
156	*Indigofera hirsute*	2	5	15	35	31	16	20	45	15	34	218	216	3	2
180	*Pueraria phaseoloides*	1	16	34	44	20	46	7	30	32	35	265	264	3	2
189	*Vigna caracalla*	14	31	9	15	18	19	32	51	35	47	271	257	3	3
200	*Desmodium ovalifolium*	12	76	28	21	22	11	15	31	30	43	291	279	3	3
131	*Stylosanthes guianensis*	9	117	24	28	21	10	28	23	23	32	316	307	3	3
3-1	*Macroptilium atropurpureum*[c]	30	116	63	37	38	21	47	68	57	88	567	537	3	3
LSD															
P = 0.05		15	66	17	21	16	10	23	22	19	26	136	127		
P = 0.005		20	88	23	28	22	13	31	29	25	35	181	170		
P = 0.001		26	115	30	37	28	17	40	37	32	45	236	221		

[a]Central Farm accession number; [b]*S. guianensis* died before the first harvest; [c]Control treatment.

successful on fine sands in Florida (A.E. Kretschmer, Florida, 1970s, personal communication), had moderate production in the trial, as did *Galactia striata* CF 40, *Indigofera hirsute* CF 156 and *Calopogonium mucunoides* CF 42. The latter three are normally associated with more fertile sites. *Vigna caracalla* CF 189 is an invasive species in South Florida and it produced good yields in the trial, as did *Pueraria phaseoloides* CF 180. *Desmodium ovalifolium* CF 200 and the local accession of *Stylosanthes guianensis* CF131 also had relatively good dry matter figures but their yields were significantly lower (0.001) than that of Siratro. All but the two legumes *Pueraria phaseoloides* CF 180 and *Vigna caracalla* CF 189 had reduced production in the cool season in December; the yield of *P. phaseoloides* continued to increase.

By the end of the trial in July 1977 it was noted that native legumes and those self-sown from the plant treatments were abundant both in the trial and in a block of an adjacent trial, probably due to the fertilizer added. It is noteworthy that the two native *Cassia* species CF 523 and CF 524 had low yields and poor survival under lenient cutting, indicating that they may contribute little dry matter to grazed pastures on these soils. The other self-sown legumes included *Macrotyloma axillare*, *Crotalaria retusa*, *S. guianensis* cvs Cook and Endeavour, *S. guianensis* CF 131, *Leucaena leucocephala*, *S. hamata* cv. Verano and *Clitoria rubignosa* (native to the area). Quail were sheltering in the large stylo bushes of the adjacent trial and were spreading seed of legumes in both trials.

14.3.3 Relative productivity at the two sites

Tables 14.5 and 14.6 data indicate that while plants at the Low Pine Ridge site had better survival than the river terrace site, perhaps due to there being no moisture stress in the dry season, the yields of most of the legumes planted at both sites were lower at the Low Pine Ridge site, undoubtedly due in part to the 3-month delay in the application of fertilizer and the additional harvests at the Central Farm site (three for the trial total, and one for the year). However, six legumes had better results on the Low Pine Ridge soils: *Clitoria ternatea* CF 51, *Stylosanthes guianensis* CF 131, *Desmodium heterocarpon* CF 188, *Vigna caracalla* CF 189, *Desmodium ovalifolium* CF 200 and *Stylosanthes guianensis* cv. Oxley cv. 201. *Stylosanthes ingrata* CF 161 had similar yields at both locations.

14.4 Conclusions

The trials provided only an initial screening of the accessions tested, and longer screening in pure and mixed grass-legume swards over a variety of seasons is needed to determine survival, spread, persistence, productivity under varying environmental conditions and management. For example, in central Belize, while Siratro is initially productive and abundant in a pure sward or mixed with grasses, a fungal disease commonly attacks it in the cool season, resulting in a significant deterioration of the stand.

While *M. atropurpureum* had the best productivity at the Central Farm site, relatively good productivity was also achieved by *Indigofera hirsute*, *Pueraria phaseoloides* and *Desmodium cinerium*. *Desmodium cinerium* in particular appears to warrant further investigation as a promising shrubby pasture legume.

At the Low Pine Ridge site, under the conditions of the trial plant, survival was generally better than at Central Farm, indicating that given adequate additional fertility, growth of a range of forage legume species is possible on these soils. *Pueraria phaseoloides*, *Vigna caracalla*, *Desmodium ovalifolium*, *Stylosanthes guianensis* and Siratro had the best productivity. However, while Siratro was the most productive legume at the Low Pine Ridge site, it and other low trailing and shrubby legumes may not be well adapted for use as pasture plants in this environment because grazing animals would trample them into the soft soil. A bushy legume such as *Desmodium ovalifolium* would be better suited for these soil conditions.

Table 14.5. Plant survival of 29 lines and 87 possible individual plants at the trial sites.

Plants surviving per treatment	Central Farm		Low Pine Ridge	
	Initial	Final	Initial	Final
3	23	4	27	12
2	3	4	1	9
1	3	9	1	2
0	0	12	0	6
Total plants	78	29	84	56

Table 14.6. A comparison of yields[a] at the Central Farm and the Lowland Pine Ridge sites.

Accession number CF[b]	Species	Variety/ other accession no.	Trial	Central Farm: Trial DM total (g)	Central Farm: Year DM total H 3–12 (g)	Low Pine Ridge: Trial DM total (g)	Low Pine Ridge: Year DM total[c] H 2–10 (g)	Low Pine Ridge vs Central Farm yields (those not lower)
3-1	*Macroptilium atropurpureum*[d]	Siratro	AB	**1010**	**777**	**567**	**537**	
4-2	*Stylosanthes ingrata*		AB	55	37	98	81	
4-6	*Stylosanthes guianensis*		AB	**909**	**889**	–	–	–
7	*Desmodium scorpiurus*		A	135	126	–	–	–
12	*Stylosanthes hamata*		AB	138	96	36	5	
16-1	*Centrosema plumieri*		B	–	–	36	32	–
40	*Galactia striata*		AB	382	332	198	198	
42	*Calopogonium mucunoides*		AB	136	94	214	190	higher
51	*Clitoria ternatea*		AB	246	230	27	24	
93	*Rhynchosia longeracemosis*		A	216	180	–	–	–
122	*Crotalaria pumila*		AB	201	44	0	0	
131	*Stylosanthes guianensis*		AB	238	211	316	**307**	higher
156	*Indigofera hirsute*		AB	406	338	218	216	
157	*Desmodium cinereum*		AB	**809**	**638**	75	75	
161	*Stylosanthes ingrata*		AB	129	107	117	112	similar
173	*Stylosanthes viscosa*	CPI 34094	A	85	71	–	170	–
175	*Crotalaria sagittalis*		A	43	37	–	–	–
177	*Centrosema pubescens*		B	–	–	171	–	–
178	*Macroptilium axillare*	IRFL1804	A	117	132	–	–	–
180	*Pueraria phaseoloides*	Puero	AB	594	**576**	265	**264**	
181	*Lablab purpureus*	Highworth	AB	240	47	22	16	
184	*Glycine wightii*	Tinaroo	AB	607	345	169	156	
185	*Medicago sativa*	Hunter River	AB	155	57	0	0	
187	*Lotononis bainesii*		AB	87	28	0	0	
188	*Desmodium heterocarpon*	IRFL588	AB	158	142	189	184	higher
189	*Vigna caracalla*	IRFL1806	AB	235	183	271	257	higher
196	*Tephrosia vogelii*		AB	123	31	57	48	
198	*Flemingia congesta*		AB	245	222	147	146	
199	*Calopogonium caeruleum*		AB	309	263	141	138	
200	*Desmodium ovalifolium*		AB	244	213	291	**279**	higher
201	*Stylosanthes guianensis*	Oxley	AB	130	119	147	145	higher
204	*Clitoria rubignosa*		B	–	–	48	48	–
284	*Pachyrrhizus vernalis*		B	–	–	62	58	–
523	*Cassia diphylla*		B	–	–	39	39	–
524	*Cassia tagera*		B	–	–	53	52	–
LSD								
P = 0.05				182	163	136	127	
P = 0.005				243	218	181	170	
P = 0.001				318	285	236	221	

[a]Best yields in bold and shaded, good yields shaded; [b]Central Farm accession number; [c]harvests 3 to 12; [d]control treatment.

15 Forage Legume Adaptation Strip Trials in Belize

J.R. Lazier*[1]
**Formerly International Livestock Centre for Africa*

Abstract

Twenty-one plantings of 24 accessions of 18 promising forage species belonging to 13 genera were established as strips under two fertilizer levels in 15 native and improved pastures and plots in contrasting environments (vertisols, planosols and podzols) under wet tropical conditions in central Belize to obtain an initial assessment of their potential under regular cutting and grazing. Observational methods were used to obtain data. At the seven upper Belize River Valley sites, *Leucaena leucocephala* and *Codariocalyx gyroides* had the best performance, followed by *Calopogonium caeruleum*, *Centrosema plumieri* and *C. pubescens*. At the four Low Pine Ridge sites the plants most consistently successful across the sites and fertilizer levels were the *S. guianensis* accessions and *C. gyroides*. However, *C. caeruleum*, *C. pubescens* and *D. intortum*, under the high fertilizer rate generally persisted and performed very well. On the Mountain Pine Ridge soils, performance at the three sites was poor without the application of marl. The most productive legumes were *C. gyroides*, *D. intortum*, *P. phaseoloides* and *S. guianensis* cv. Endeavour. The other *S. guianensis* lines also performed well at most sites and fertilizer levels, as did *C. pubescens*. *M. atropurpureum* performed best under the higher fertilizer regimes.

15.1 Introduction

Prior to the establishment of the University of the West Indies Forage Legume Programme in 1972 there had been relatively little research done on the productivity and persistence of native and exotic legumes in Belizean pastures. While some forage research was reported on the fertile terraces of the upper Belize River, much of the work that had been undertaken had emphasized the Low Pine Ridge planosols and the more elevated Mountain Pine Ridge podzols, areas that were unutilized because of their low fertility status and other edaphic constraints.

The IDRC/UWI/Belize research programme was based at the main national agricultural experimental station, Central Farm, which is located on terraces of the upper Belize River. There, by 1975, through its initial evaluation plots and ongoing formal replicated trials on heavy clay soils, the programme had identified indigenous and exotic legume species and genotypes of potential value for use in local pastures to improve the nutritional value of the fodder (Lazier, 1978, 1980a,b). A number of these accessions were available commercially because they were utilized as pasture species elsewhere, but had originated in the Americas and are native to Belize.

It thus was deemed appropriate to assess their relative productivity and persistence in native pastures and established swards of exotic, productive forage grasses of the three major

[1] E-mail: john.lazier@sympatico.ca

environments, the upper Belize River Valley, the Low Pine Ridge and the Mountain Pine Ridge. Preliminary studies were also undertaken on the botanic composition and nutritive value of the native pastures involved, the results of which are reported in Chapter 13 (this volume). For ease of reference, the details and results of each of the three environments are reported separately below. Twenty-four species and genotypes were selected as promising for testing based on their performance in introduction plots and trials on terraces of the upper Belize River and the Low Pine Ridge. Thirteen were accessions collected by the programme in the Belize River Valley and from the adjacent Mountain Pine Ridge, six were promising lines from Australia, three were commercially available genotypes, and one originated from South-east Asia.

15.2 Methods

Unreplicated strip trials were chosen as the method of assessing the adaptation and fodder potential because the number of environments and plants to be tested was large, and more detailed trials were not warranted at this initial stage. Care was taken that the sites chosen were on soils representing significant areas in the Valley and its vicinity. When possible, commercial lines were compared with accessions of the same species collected in Belize.

Fifteen strip trials consisting of unreplicated single strips of legumes were undertaken in three important Belizean environments: basic alluvial and heavy clay soils on the terraces of the upper Belize River in central Belize, low elevation acid waterlogged infertile clay deposits and vertisols toward the coast termed Low Pine Ridge soils, and higher elevation acid podzols adjacent to the Belize River Valley on Mountain Pine Ridge. They were established in a variety of native and improved pastures, with average annual rainfall ranging from 1500 to 2900 mm.

The methods of planting and management described here apply with minor variations to the plantings at sites 1 to 14 in all three environments. The National Show Grounds site (site 15 in the upper Belize Valley) varies in that it had the strips imposed on established grass plots; the size of these restricted the number of accessions that could be planted. The data on the locations and their environments are presented in Tables 15.1 and 15.2.

For each site a 27 × 65 m trial area was laid out and soil samples were collected at randomly selected locations within their perimeters for analysis: 30 sites at 0–15 cm and 15–30 cm depths, and 15 sites at 30–45 and 45–60 cm depths. The samples were analysed by the Central Analytical Laboratory at the University of the West Indies in St Augustine, Trinidad and the results are given in Appendix 15.1.

Sixteen plots were established in a 4 × 4 pattern in the following manner. Four 0.5-m wide strips spaced at 5 m were created along the entire length of the trial area by ploughing and harrowing, or hoeing. A 5-m margin was left around the perimeter of the trial. Along each of the four cultivated strips, four plots of 15 cm × 10 m were marked out with a 5-m gap between plots in the strip, and 5 m at each end, which was the trial margin. The resulting plots were thus 5 m × 10 m.

The upper Belize Valley sites had initial fertilizer applied along the 15-cm wide strips at 63 kg/ha TSP and KMag respectively before planting, while all Low Pine Ridge and Mountain Pine Ridge soils (Puletan and Pinol soils respectively; Wright *et al.*, 1959) had a one-time application of 63 kg/ha Fritted Trace Element (FTE) micronutrient mixture, 200 kg/ha $MgSO_4$ and 63 kg/ha each of KCl and TSP. The forage treatments were sown at the time of initial fertilizer application.

High and low fertilizer applications were broadcast by hand to half of each 10 × 5 m plot. Pegs were placed at the mid-point of each 10-m plot to mark the separation of the fertilizer application rates; the low rate was always on the same side of the dividing point. Pegs were placed at a distance of 1 m on either side of this central peg to define a border zone between the fertilizer rates. The 4 m of strip remaining within each fertilizer treatment within the plot were then the areas from which data for each forage treatment were recorded; no data were recorded from the border area.

Commencing 6 weeks after the initial fertilizer application and sowing of the strips, one-half of each plot and the corresponding part of the trial area of all trial sites received the low fertilizer rate of 63 kg/ha of TSP and KMag each per annum in quarter portions each 12 weeks after alternate observations and any imposed grazing. The other half received fertilizer on the same schedule but at four times the rate, creating low and high fertilized strips across the width of the trial area. The same side of all plots received

Table 15.1. Trial locations, altitudes, rainfall, pasture types and analysis groups.

Trial no.[a]	Map reference[b]	Location	Altitude (m)	Rainfall (mm)	Pasture/treatment	Site	Analysis group
1	16 QBQ 858 006	Norland Farm, Western Highway, km 70	60	1500	*Hyparrhenia rufa*	High alluvial terrace	2
2	16 QBQ 845 010		50	1500	*Pennisetum purpureum* var. Merker	Low alluvial terrace	2
3	16 QBP 738 915	Santiago Juan Farm, San Lorenzo, Western Highway	100	1500	*Hyparrhenia rufa*	Low alluvial terrace	1
4	16 QCQ 107 107	Orange Walk Estate, Western Highway near Roaring Creek	40	1500	Native	Shallow	2
5	16 QCQ 329 153	Western Highway, mile 33, Low Pine Ridge	40	2200	Native	Level, waterlogged	4
6	16 QCQ 329 155		40	2200	Native	Sloping, waterlogged	4
7	16 QCQ 302 131	Western Highway, mile 35, Low Pine Ridge	40	2200	Native	Shallow, waterlogged	4
8			40	2200	Native	Shallow, water-logged, harrowed	4
9			40	2200	Native	Shallow on a cay, waterlogged	n/a
10	16 QBQ 068 872	Klaas Fresian Farm, Spanish Lookout	80	1500	*Digitaria eriantha*	High alluvial terrace, shallow	1
11	16 QBP 968 857	Bull Run Ranch, Mountain Pine Ridge	560	1600	Native	Under mature pine	4
12	16 QBP 930 788	Rio On Line, Mountain Pine Ridge	560	1600	Native	Under short pine	4
13	16 QCP 087 805	Baldy Beacon, Mountain Pine Ridge	970	2900	Native	Exposed high ridge top	4
14	16 QBQ 864 015	Central Farm, Western Highway, mile 68	50	1500	*Cynodon nlemfluensis*	Alluvial terrace	3
15	16 QCQ 099 086	Show Grounds, Roaring Creek	40	1800	A range of exotic grasses	High alluvial terrace	5

[a]Original numbers for trials 11 to 15 were one greater; [b]Map series E755 DOS 4499 Edition 3; n/a, trial abandoned.

Table 15.2. Soils at trial sites.

Trial no.	Soil type	LRD[a] classification			USDA classification[b]
		Suite	Sub-suite	Series	
1	Alluvial	Melinda	Redbank	Meditation	Mollic Tropudalfs, fine, mixed
2	Alluvial				Mollic Tropufluvents, coarse loamy, non-acid, mixed
3	Rendzina	Yaxa	Yalbec	Piedregal	Vertic Eutropepts, fine, mixed
4	Vertisol	Yaxa	Yalbec	Spanish Lookout	Lithic Rendolls, clayey, montmorillonitic
5	Planosol	Puletan	Yobo	Rough Mile	Abruptic Tropaqualfs, fine, kaolinitic
6	Planosol				
7	Vertisol				
8	Vertisol	Yaxa	Yalbec	Beaver Dam	Aquentic Pelluderts, very fine, montmorillonitic
9	Vertisol				
10	Vertisol	Yaxa	Yalbec	Spanish Lookout	Vertic Rendolls, very fine, montmorillonitic
11	Podzol	Stopper	Stopper	Pinol cse s cl	Plinthic Tropudults, clayey, kaolinitic
12	Podzol	Ossary	Ossary	Pinol cl	Typic Tropudults, fine-loamy, mixed
13	Podzol	Ossary	Ossary	Pinol ls	Lithic Dystropets, loamy, mixed
14	Alluvial	Melinda	Redbank	Esperanza	Abruptic Tropudalfs, very fine mixed
15	Rendzina	Yaxa	Cuxu	Camalote	Entic Chromuderts, fine, montmorillonitic

[a]Land Resources Division (Jenkin *et al.*, 1976); [b]all are isohyperthermic (McKinzie, 1977).

the same fertilizer rate. Because the three sites on Mountain Pine Ridge had very slow growth on establishment and signs of severe nutrient deficiencies (as indicated by leaf coloration), a single application of marl (97.1–100% $CaCO_3$) was applied at 1790 kg/ha in April 1976. The performance of the legumes was much improved thereafter and nutrient-deficiency symptoms disappeared.

Half of the seed of each of the species to be planted was scarified and all was inoculated with the appropriate rhizobium, where known, and otherwise with the cowpea strain. *Leucaena* was pelleted with $CaCO_3$ before planting in the Low Pine Ridge and Mountain Pine Ridge soils. Sowing (at a rate of 10 kg/ha) and fertilization were done at the same time for all sites, mainly in mid-October 1975, but a few sites were planted later, in November and December.

Twenty-four legumes (Table 15.3) were sown in strips within the grass swards (see Fig. 15.4). Fifteen of the legumes were sown in individual strips and one strip was planted with a mixture of the remaining legumes; the same accessions were planted at all sites. The mixture was used to test accessions that had little seed available. After germination, seedling counts were done, followed by gap filling in December 1975 using the same accessions, with the exception of *C. pubescens* CF 177, which was replaced by CF 195 *Centrosema pubescens* cv. Belalto at eight of the sites due to lack of seed. Thirteen detailed observations on the pastures, focusing on legume and grass performance and involving 43 parameters, were taken at 6-week intervals over one and a half years (Table 15.5). When grazing was involved, 19 post-grazing parameters were recorded. The data on most parameters are not reported here but are available from the author.

The upper Belize Valley sites also had cutting or grazing imposed where appropriate and feasible in order to ascertain their utility under farming conditions (Table 15.4). The other two environments had no management treatments imposed and were used to test the adaptation of the legume treatments to the two stressful environments, the infertile waterlogged clay soils of the Low Pine Ridge and the shallow infertile acid soils of the Mountain Pine Ridge.

Table 15.3. Species/accessions planted in the trials. Shaded cells denote species/accessions planted. Site 15 had smaller plots and fewer treatments. All sites had one strip planted with a mixture of accessions.

Species	Accession no.[a]	Sites 1–14		Site 15		Location collected/source
		Strip	Mixture	Strip	Mixture	
Calopogonium caeruleum	CF 18					Central Farm[b]
Centrosema plumieri	CF 16-1					Central Farm[b]
Centrosema pubescens cv. Belalto[c]	CF 195					Australia[d]
Centrosema pubescens	CF 6-1					Central Farm[b]
Centrosema pubescens 'Centro'	CF 6-2					Commercial[d]
Centrosema pubescens[c]	CF 177					Central Farm[b]
Codariocalyx gyroides	CF 29					South-east Asia
Crotalaria retusa	CF 20					Central Farm[b]
Crotalaria vitellina	CF 15					Central Farm[b]
Desmodium canum	CF 1					Central Farm[b]
Desmodium intortum 'Greenleaf'	CF 50					Commercial[d]
Indigofera mucronata	CF 9					Central Farm[b]
Leucaena leucocephala cv. Peru	CF 179					Australia[d]
Macroptilium atropurpureum cv. Siratro	CF 3-1					Australia[d]
Macroptilium lathyroides	CF 14					Central Farm[b]
Pueraria phaseoloides 'Puero'	CF 180					Commercial
Rhynchosia minima	CF 13					Central Farm[b]
Stylosanthes guianensis cv. Oxley	CF 201					Commercial
Stylosanthes guianensis	CF 4-1					Mountain Pine Ridge[b]
Stylosanthes guianensis cv. Cook	CF 4-4					Australia[d]
Stylosanthes guianensis cv. Endeavour	CF 4-5					Australia[d]
Stylosanthes hamata cv. Verano	CF 186					Australia[d]
Teramnus uncinatus	CF 2					Central Farm[b]
Vigna vexillata	CF 43-2					Central Farm[b]

[a]CF: Central Farm accession number; [b]collected in Belize; [c]CF 195 replaced CF 177 at replanting at sites 1, 2, 8, 10, 11, 12, 13 and 14; [d]species native to Belize, seed imported.

15.2.1 Analysis

The trials were organized into four groups for analysis, each of which had the same sown legume treatments. Site 15, which had fewer legumes, was analysed separately (Table 15.5).

Seven observation periods over the trial period were selected as representing the critical times of the growth cycle (Tables 15.6 and 15.7). Selected observational data (Table 15.8) from each of the seven periods from each site were subjected to MULCLAS analysis followed by GROUPER analysis. Separate analyses were done of each fertilizer treatment at each period, thus producing 14 sets of data for each of the groups of sites (Table 15.5). For each of the analyses, six groups were found to give adequate division of the data. The presence of seedlings was not very common, and was either overridden by other factors where they occurred, or the plots containing plants with seedlings were placed in a separate grouping.

The groupings tended to reflect the vigour and survival of the plants in the strips. Detailed analysis of the groups and plants were done and

Table 15.4. Treatments imposed after observations taken in the following cycles.

Site		Pasture type	Treatment and observation cycle			Site type
			Grass cut	Grass and legume cut	Grazed[a]	
1	Norland Farm, Western Highway, km 70	Jaragua	1, 2	12, 13	5, 6, 7	River terrace
2		Merker	1, 2	8, 10, 12, 13	3	
3	San Lorenzo, Western Highway	Jaragua	1, 2	5, 6, 7, 8, 9, 10, 11, 12, 13		
4	Orange Walk Estate, Western Highway	Native	1, 2	5, 6, 7, 8, 9, 10, 11, 12, 13		Shallow on limestone (40 cm)
5	Western Highway, mile 33	Native	2			Low Pine Ridge
6		Native	2		7[b]	
7	Western Highway, mile 35	Native	2			
8		Native	2			
9		Native	–	–	–	
10	Klaas Fresian Farm, Spanish Lookout	Pangola	2, 4	5, 6, 13	8, 9, 10, 11	Shallow on limestone (150 cm)
11	Bull Run Ranch, Mountain Pine Ridge	Native	2		10, 11, 12, 13[d]	Shallow on ironstone (140 cm)
12	Rio On Line, Mountain Pine ridge	Native	2[c]		3[d]	Shallow on granite (96 cm)
13	Baldy Beacon, Mountain Pine Ridge	Native				Shallow on slate (37 cm)
14	Central Farm, Western Highway, mile 68	African Star	2		3, 4, 6, 7, 8,9, 10, 11, 12	High river terrace
15	Show Grounds, Roaring Creek	Pangola, Hamil, Para, Nunbank, Guinea, Jaragua	2, 3, 4, 5	6, 7, 8, 9, 10, 11, 12		Shallow on limestone (140 cm)

[a]Post-grazing harvests were also done; [b]grazed very lightly by sheep; [c]one-quarter only cut; [d]light unauthorized grazing (by deer?) before observations.

the means obtained for the seven analysed characters. The six groups produced by the analyses were reduced by truncation of the hierarchical structure to four, a level convenient for the needs of producing levels of success. The analysis results apply to all three environments and are reported under the upper Belize River Valley.

15.3 Upper Belize River Valley sites

15.3.1 Introduction

The upper Belize River Valley is characterized by alluvial terraces that are bordered by shallow soils over limestone on hills to the south and rolling land to the north (Fig. 15.1). The soils in this environment are generally basic to strongly basic, thus analyses of the pH in the upper 15 cm in the upper river terrace trial sites was between pH 6.2 and pH 6.7 (Appendix 15.1). Downstream, toward the east around Belmopan, the land beyond the river terraces becomes more level, shallow soils over limestone still predominate and they remain basic in reaction (sites 4 and 15, pH 6.5 to 7.1). The development of agriculture in central Belize has focused on the soils adjacent to the river in the upper Belize River Valley due to their depth and relative fertility.

Table 15.5. Groupings of trial sites for analysis (on the basis of species planted). See also Table 15.1.

Trial analysis sets	Sites
1	3, 10
2	1, 2, 4
3	6, 14
4	5, 6, 7, 8, 11, 12, 13
5	15

Research on the introduction, testing and relative productivity of fodder grasses in this environment has been undertaken since 1939. Studies on the potential of fodder legumes began in the mid-1950s (DOA, 1954). The first mixed grass-legume sward research was reported in 1966, and thereafter a considerable amount of testing was done on the pasture productivity of a fairly restricted range of tropical legumes. By the end of the 1960s the advent of new species of tropical fodder legumes being tested in Australia, and the rich native leguminous flora of the region prompted the establishment of the University of the West Indies Forage Legume Programme in Belize in 1973. This project, the results of which are partly reported in this volume, utilized introduction plots and a series of formal trials, mainly under cutting, but some under grazing, and established that, at least on the research station there were forage legumes available that could improve the quality of pasture fodder on the terraces of the upper Belize River (Lazier, 1978; 1980a,b; 1981a,b,c).

15.3.2 Trial sites in the upper Belize Valley

Site 1 Norland Farm

The cooperation of the farmer who promised to put his cattle in to graze the site as requested and

Table 15.6. Commencement dates of observation cycle data analysed at 14 sites.

	Observation cycles/seasons analysed and date commenced[a]												
	1976								1977				
Sites	1	2	3	4	5	6	7	8	9	10	11	12	13
1, 2, 3, 4, 10, 14, 15		20/2		13/5		5/8		28/10		19/1		14/4	25/5
5, 6, 7, 8, 11, 12, 13	27/1		22/4		13/7		5/10			5/2		28/4	14/6
	Establishment and pre-dry season		Late dry season		Wet season		Late wet season			Cold season		Late dry season	

[a]All sites had observation data recorded in all 13 cycles. Alternate cycle observation data (cells with day and month recorded) were analysed. Most cycles of observation were completed within 12 days.

Table 15.7. Seasons of observation data used in analysis.

Observation no.	Season
1–2	Establishment and pre-dry season
3–4	Dry season
5–6	Wet season
7–8	Late wet season
10	Cold season
11–12	Late dry season
3	Early wet season

Table 15.8. Observational data used in the analyses.

Factor	Description	Range/units
1. Plot cover:[a] legume	Legume cover of planted 15-cm wide strip	0–10
2. Plot cover:[a] grass	Grass cover of planted 15-cm wide strip	0–10
3. Height:[a] legume	Legume: Estimated average (only in 15-cm wide strip)	cm
4. Height: grass	Grass: Estimated average	cm
5. Spread: legume	Legume: Estimated average	cm
6. Tagged legumes alive	Number surviving plants of the 5 tagged at the first observation	0–5
7. Legumes alive	Number of plants in strip (maximum 20)	0–20
8. Plot cover: seedlings	Estimate of plot cover of seedlings (entire 4 × 5 m plot)	0–10

[a]The plot here refers to the 15 cm × 4 m cultivated strip.

Table 15.9. Groups formed and the mean observation ranges selected for each.

	Legume		
Vigour group	Strip cover	Tagged plants	Plants alive
A	7–10	4–5	15–20
B	4–7	2–4	10–15
C	2–4	1–2	5–10
D	< 2	1	0–5

the presence of a good stand of *Hyparrhenia rufa* (Jaragua grass) on a soil of local importance led to the site being chosen. Situated to the west of, and adjacent to, Central Farm, the soil was described by McKinzie (1977) as Mollic Tropudulf fine, mixed isohyperthermic consisting of old alluvium on a high terrace of the Belize River. It belongs to what has been termed Meditation series (Jenkin *et al.*, 1976), which is also represented in site 2 in this series as well as the two replicated grazed strip trials in Para and Pangola pastures on Central Farm (Chapter 16, this volume). This repetition of adjacent trials provides an opportunity to examine variations in legume response on the same soil type but with different grasses. The paddock used had not been fertilized and had been under Jaragua grass for some years. Its management had been restricted to being slashed four or five times a year to reduce the growth of woody species and being rotationally grazed by cattle that were circulated around three other pastures.

Site 2 Norland Farm

Chosen for the same reasons as site 1, the location was on the same farm, and again adjacent to the fields of Central Farm. For the previous 5 years the field had been planted to *Pennisetum purpureum* var. Merker, then to 1 year of maize that had been fertilized with a mixed fertilizer. Also on Meditation series soil (Jenkin *et al.*, 1976), it was classified as Mollic Tropufluvents, coarse loamy, non-acid, mixed isothermic by McKinzie (1977), who assessed it as appearing more similar to the Garbutt series than the Meditation series.

Site 3 San Lorenzo (Don Santiago Juan Farm)

The site was located near to and to the east of the community of Benque Viejo, about 19 km west of Central Farm along the Western Highway. The site was on a 5–7% slope below a steep limestone hill and was chosen to represent this very common landform. The site is a native pasture and had been under grazing. Jenkin *et al.* (1976) mapped the area as Piedregal series, but McKinzie (1977) found it did not match any of the series described by Jenkin *et al.* (1976). He classified it as a Vertic Eutropepts, fine mixed isohyperthermic derived from limestone colluvium but acid (pH 5.2). The acidity was confirmed by testing at the Central Farm Soils Laboratory (pH 5.7) (Appendix 15.1).

Fig. 15.1. A general view of the upper Belize Valley and a native pasture at San Lorenzo (site 3).

Site 14 Central Farm

The trial site was situated on a middle terrace of the Belize River at Central Farm on old alluvium and had been under an established stand of *Cynodon nlemfluensis*, African star grass. It was chosen because local farmers were very enthusiastic about this grass, rating its productivity and palatability as very high. As the grass has a relatively open canopy, it was judged to have good potential for growth of associated forage legumes. The Land Resources Division (LRD) soil survey classified it as belonging to the Camalote series (Jenkin *et al.*, 1976). McKinzie (1977) classified the soil as Abruptic Tropudalfs, very fine, mixed isohyperthermic and similar to the Meditation series of the LRD study but differing in the lack of well-expressed slickensides in the lower B and C horizons.

Site 4 Orange Walk Estate

Located just east of the community of Roaring Creek, the site has shallow black cracking soils over limestone and it was mapped as Spanish Lookout series by Jenkin *et al.* (1976), but McKinzie (1977) found it to be very similar to Pedrigal series and he described it as Lithic Rendalls, clayey, montmorillonitic, isohyperthermic. It has a slight slope to the north, improving drainage in what otherwise would be a waterlogged site. The native pasture is an old one and was considered to have very low productivity. Cohune palms (*Orbignya cohune*) are common.

Site 10 Spanish Lookout (Klaas Freisan Farm)

Located just north of Central Farm across the Belize River in Spanish Lookout, the site selected was on the farm of Klaas Freisan. The soil was marl derived and the marl in turn was derived from the limestone cap of the Mountain Pine Ridge peneplain within the area mapped by Jenkin *et al.* (1976) as Yaxa suite, Yalbec sub-suite, Spanish Lookout series, representing 20,600 ha in the Belize River Valley, 11% of the total area surveyed. An earlier survey (Wright *et al.*, 1959) indicated it represents probably 46,500 ha. McKinzie (1977) described the soil at the trial site as a Vertic Rendolls, very fine, montmorillonitic, isohyperthermic and having the top 15 cm as black heavy clay, grading into white sticky marly clay at 41 cm, above a white friable marl at 107 cm.

Jenkin *et al.* (1976) recommended that these soils be the next in the Belize Valley to be developed, because the Mennonite farmers' experience had indicated that they could be made productive. However, nutritional and productivity problems remain for pastures. It was hoped that locating a trial here could give some insight into which legumes might survive and persist under grazing.

The selected site was gently sloping at one end of a paddock that, since it was cleared of forest, had been planted to maize and beans with alternate rest years. Moderate amounts of mixed fertilizers had been applied. Immediately before the trial was planted, the field had been planted to maize. The farmer was in the process of planting the field to Pangola grass (*Digitaria eriantha*) but time and transport problems had made it a very protracted business and the field was dominated by weeds. The farmer had commented that relatively mature trees bordering the field were dying back ('growing backward'), indicating a possible nutrient deficiency.

Site 15 National Show Grounds

The site was situated in the National Show Grounds, located at the beginning of the Hummingbird Highway on the west side, close to the Belmopan turnoff and to the community of Roaring Creek. A grass trial had been terminated about a year previously and it presented an opportunity to test the introduction of several legumes into a number of exotic grasses at one time. The trial situation also offered the opportunity to demonstrate the feasibility of mixed swards to those visiting the show grounds. The plots held six grasses that had been irregularly fertilized with an NPK fertilizer: Pangola (*Digitaria eriantha*), Guinea (*Panicum maximum*, *Panicum maximum* cv. Hamil), Para (*Brachiaria mutica*), Jaragua (*Hyparrhenia rufa*) and Buffel (*Chloris nlemfluensis*).

As the plots were fairly small, only a few strips of legumes could be accommodated. Five strips were tilled in each grass plot; four of these were planted to individual legumes known to be adapted to the soil, and the remaining strip was planted to a mixture of 9 species and cultivars thought to have some potential in the environment (Table 15.3).

The soil at the site had been mapped by Wright *et al.* (1959) as Yaxa suite, Cuxa subsuite, Camalote series and was described as a Rendzina; McKinzie (1977) termed it an Eutric Chromuderts, fine montmorillonitic isohyperthermic and suggested that it bore closer association to the Society Hall series vertisols.

Weeding was required only once and harvester ants were a constant problem, as they were at the trial site at Orange Walk (site 4) on a similar soil. The grasses alone were cut fairly severely at 15 cm after observations every 6 weeks in March, April, May and July, and the cut material removed until the legumes were established. After the August observation both grass and legume were cut over regularly, at 30 cm for all grasses except Pangola, which was cut at 15 cm. All legumes were cut at 30 cm, except for *Leucaena leucocephala* and *Codariocalyx gyroides*, which were cut at 50 cm.

15.3.3 Results and discussion upper Belize Valley

The available rainfall data (Fig. 15.1) are from Benque Viejo, near site 3, Central Farm, representing sites 1, 2, 10 and 14, and Roaring Creek, near sites 4 and 15. Rainfall over the 18-month trial period, judging by the data from Central Farm, was higher than the 29-year average (2393 mm vs 2055 mm, 16% higher) both during 1976 (1789 mm vs 1542 mm, 16% higher) and during the first half of 1977 (604 mm vs 508 mm, 19% higher). The bulk of the excess rainfall was received in June and December 1976 and to a lesser extent in February 1977. Although rainfall was adequate for growth over much of the year, there was maximum water stress in March and April, when there was essentially no rainfall, and in addition the rainfall in August and September was less than normal. The rainfall patterns for Roaring Creek and Benque Viejo generally followed a similar pattern, with the peak rainfall in June being higher in Roaring Creek and lower in Benque Viejo. The rainfall stations are not far apart (Roaring Creek and Benque Viejo are 27 km east and 20.5 km west of Central Farm respectively), so similar rainfall patterns would be expected. As the rain bearing winds are from the east, the June rainfall is lower for sites further inland. Otherwise rainfall patterns and amounts are similar.

15.3.4 Analysis of results

The analysis groupings that are applicable to trials in all three edaphic environments showed very close relationships between the three main factors, height, spread and survival, thus it was unusual for a high plot cover to correlate with few plants in the strip or a low number of

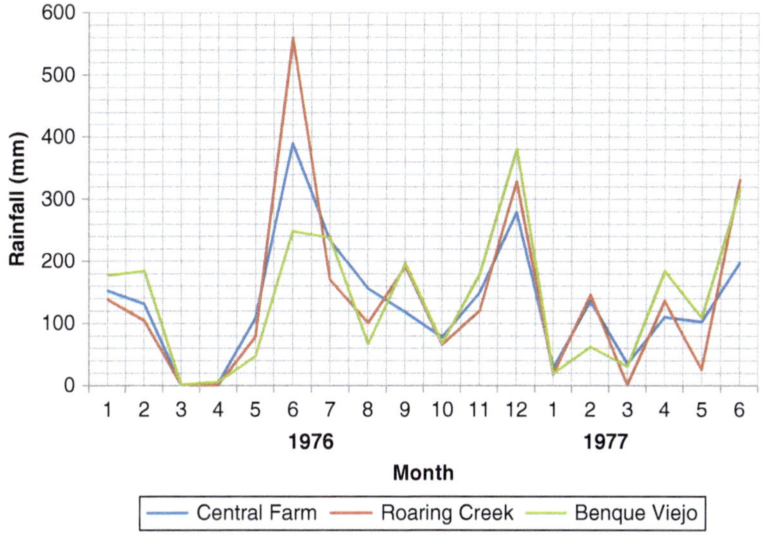

Fig. 15.2. Rainfall, upper Belize Valley 1976–1977.

survival of tagged plants. Height and spread reflected the morphology of the plants and varied with the cover. These relationships are presented in the detailed analyses but have little bearing on the generalized levels presented here. The variation through the seasons can be noted in the tables of results for the individual sites. Where establishment was good at the beginning of the trial, performance rankings frequently dipped in the first dry season, rose into the first wet season, then declined with declining rainfall and the cool weather and dry periods that followed. Some surviving legumes showed increased performance in the final observation in the early wet season of 1977. The general success or failure of the legumes at each site and the most successful will be discussed for each environment separately.

Generally little grass was present in the strips at most sites, thus that factor had little bearing on the final groupings. However, this factor did influence the groupings more at the sites with denser grass swards, in which cases there was greater emphasis placed on the factor, particularly in later observations when more grass had spread back into the tilled strips.

Site 1 Norland Farm, Hyparrhenia rufa (Fig. 15.3a)

Management during the period that the trial was observed included cutting of the grass (15 cm) after observations 1 and 2 to reduce grass competition, grazing following observations 5, 6 and 7, and cutting over of the grass (15 cm) and legumes (at appropriate heights) after observations 12 and 13. Establishment was not strong for this site, with only a few species showing much vigour, even initially (Fig. 15.3, Tables 15.10 and 15.11). *M. atropurpureum*, *Leucaena leucocephala* and the *S. guianensis* lines Cook and Endeavour were the most vigorous, but not consistently so. The ratings of these rose into the wet season and then declined, possibly as the result of reduced rainfall and of the two grazing periods before observation cycle 8. No plants remained vigorous in the late dry season at the end of the trial; however, *Leucaena* and *C. caeruleum* had highest ratings. There appeared to be little difference in establishment or productivity between the two fertilizer rates, perhaps because Jaragua grass does not respond strongly to additional fertilizer or the site may be relatively fertile. A second season would have been necessary in order to assess how the legumes responded to the next rainy season.

Site 2 Norland Farm, Pennisetum purpureum cv. Merker grass (Fig. 15.3b)

The grass was cut over after the first two observations and grazed very heavily, with much trampling of the plants after the third observation in

Fig. 15.3. (a) Site 1 Norland Farm, harvesting legume in *H. rufa* pasture. (b) View of site 2 Norland Farm, *P. purpureum* trial site illustrating the dense grass growth.

April 1976 in the late dry season, when cattle broke into the trial. The ratings of most legumes remained relatively high at the next observation in May (not shown in Tables 15.12 and 15.13). However, with the advent of the rainy season, very vigorous and tall grass growth severely inhibited legume growth (Fig. 15.3). The cutting of the tall grass at 30 cm after the observations were taken would also have affected the climbing legumes. Only *S. hamata*, Endeavour Stylo and *Leucaena* had recovered somewhat by the end of the trial. Establishment of the legumes was weaker under higher fertilizer, perhaps due to greater competition from the grass.

Site 3 San Lorenzo (Don Santiago Juan's farm) H. rufa pasture (Fig. 15.1)

Jaragua grass at site 3 was cut over after the initial two observations and both grass and legumes were cut after most observations were made. Both fertilizer levels had good establishment and persistence of the legumes (Tables 15.14 and 15.15), with some starting to recover at the end of the trial with higher rainfall. The most productive legumes, as judged by the parameters utilized in the analysis, were *C. caeruleum*, *C. gyroides*, *Leucaena* and the *S. guianensis* line cv. Endeavour. Others that performed relatively well but had not recovered by the early 1977 wet season included *C. plumieri*, local accessions of *C. pubescens* (CF 177 and 6-1), *D. intortum*, *P. phaseoloides*, *S. guianensis* cv. Cook and S. *hamata* cv. Verano. The higher fertilizer rate appeared to reduce the vigour of the legumes to some extent. These sites contrast with those on sites 1 and 2 at Norland, where the legumes performed much less vigorously. Grazing at site 1 and the taller grass at site 2 may have inhibited legume establishment and persistence. Cutting of the legumes at favourable heights helped ensure the survival of the plants.

Site 14 Central Farm, Cynodon nlemfluensis (African Star grass) (Fig. 15.4)

African Star grass is a fairly open-canopy grass and at the time of the trial it was a popular pasture species among local ranchers. Establishment was better under the low fertility rate, probably due to less competition from the grass (Tables 15.16 and 15.17). Once the legumes were established, the site was grazed by the same group of animals after most observations and many legumes survived and produced well under grazing until after the 8th observation, when the rain had tapered off and the plots were particularly heavily grazed; the ratings of most legumes dropped as a result. There was a vigorous invasion of the planted strips by the grass in the wet season after the fifth observation. After the 10th observation in mid-January, a different group of animals was used and was not supervised well, resulting in the trial being grazed very heavily for a second time and most of the legumes were severely reduced in the plots. Continued grazing through the cool and dry seasons resulted in high levels of trampling of the legumes. Many legumes disappeared from the strips. The woody legumes *Leucaena* and *C. gyroides* were the most persistent and productive at the end of the trial, despite having most limbs torn off during heavy grazing.

Table 15.10. Site 1, Norland Farm, *Hyparrhenia rufa*. Accessions' membership of vigour groups (Table 15.9) determined from observational data: A best, to D poor. Darker shading has been used to identify the better results.

		Observation cycle													
		Low fertilizer							High fertilizer						
		2	4	6	8	10	12	13	2	4	6	8[a]	10	12	13
Legumes	Accession no.	cool	dry	wet	late wet	cool	dry	wet	cool	dry	wet	late wet	cool	dry	wet
Calopogonium caeruleum	CF 18	D	D	C	D	D	C	C	D	D	**A**	D	D	C	C
Centrosema plumieri	CF 16-1	D	D	C	D	D	D	D	D	D	D	D	D	D	D
Centrosema pubescens cv. Belalto	CF 195	C	D	**A**	D	D	D	D	D	D	**A**	D	D	D	D
Centrosema pubescens	CF 6-1	D	D	–	D	–	–	–	D	C	–	–	–	–	–
Centrosema pubescens 'Centro'	CF 6-2	D	D	D	D	–	–	–	D	C	D	D	D	D	D
Codariocalyx gyroides	CF 29	D	D	B	C	D	C	D	D	D	D	–	–	–	–
Crotalaria retusa	CF 20[a]	C	D	D	–	–	–	–	D	–	–	–	–	–	–
Crotalaria vitellina	CF 15[a]	–	D	D	D	D	–	–	–	–	–	–	–	–	–
Desmodium canum	CF 1[a]	–	–	D	D	D	D	D	–	–	D	D	D	D	D
Desmodium intortum 'Greenleaf'	CF 50	D	D	B	D	–	–	–	–	–	–	–	–	–	–
Indigofera mucronata	CF 9[a]	–	–	–	–	–	–	–	–	–	–	–	–	–	–
Leucaena leucocephala cv. Peru	CF 179	**A**	D	B	C	D	B	B	D	C	D	D	D	C	D
Macroptilium atropurpureum cv. Siratro	CF 3-1	**A**	**A**	**A**	C	D	D	D	B	**A**	**A**	D	D	D	D
Macroptilium lathyroides	CF 14	–	–	–	–	–	–	–	–	–	–	–	–	–	–
	CF 14[a]	–	–	–	–	–	–	–	–	–	–	–	–	–	–
Pueraria phaseoloides 'Puero'	CF 180	**A**	D	**A**	D	D	D	D	B	C	**A**	D	D	D	D
	CF 180[a]	–	–	–	–	–	–	–	–	–	–	–	–	–	–
Rhynchosia minima	CF 13[a]	D	D	D	–	–	–	–	D	D	D	D	D	D	D
Stylosanthes guianensis cv. Oxley	CF 201[a]	–	–	–	–	–	–	–	–	–	–	–	–	–	–
Stylosanthes guianensis	CF 4-1	C	D	D	D	D	D	D	D	D	D	D	D	D	D
Stylosanthes guianensis cv. Cook	CF 4-4	D	C	**A**	D	–	–	–	B	D	**A**	D	–	–	–
Stylosanthes guianensis cv. Endeavour	CF 4-5	D	D	B	D	D	D	D	B	C	**A**	D	D	D	D
Stylosanthes hamata cv. Verano	CF 186	C	D	B	C	D	D	D	D	C	**A**	D	D	D	D
Teramnus uncinatus	CF 2[a]	C	D	D	–	–	–	–	D	D	D	D	D	D	D
Vigna vexillata	CF 43-2[a]	C	D	–	–	–	–	–	B	C	D	D	D	–	–

[a]In legume mixture plot.

Table 15.11. Observation number after which management was imposed on Site 1.

Vegetation	Grass cut	Grass and legume cut	Grazed
Jaragua	1, 2	12, 13	5, 6, 7

Site 10 Spanish Lookout Klaas Fresian Farm, Digitaria eriantha (Pangola) (Fig. 15.5)

Establishment and growth was very good at this site (Tables 15.18 and 15.19), with better legume establishment at the higher fertilizer rate. The grass and weeds were cut at 15 cm after observations 2 and 4, and both grass/weeds and legumes were cut after observations 5, 6 and 13. Pangola density was patchy, with only a few plots with high grass plot covers, and few planted strips had significant invasions of grass. Grazing was imposed after observations 8–11. The legumes generally stood up very well to the imposed management, with the *Calopogonium* and *Centrosema* accessions increasing their ratings, particularly in the high-fertility treatment. *P. phaseoloides*, *M. atropurpureum* (at the high fertilizer level) and *C. gyroides* performed well, while *Leucaena* did not. This may be due in part to the relatively high palatability of *Leucaena*. The legumes also survived heavy trampling and fairly severe grazing; *C. gyroides* was less damaged by grazing where it was in thicker stands. The good survival of the legumes under cutting would be expected because the cutting heights were chosen to ensure legume survival. Competition from weeds and grass was apparently not an important factor, even where there was significant invasion of the sown strips by the grass. The lack of competition was likely the reason for the legumes' persistence and high ratings.

Site 4 Orange Walk, native pasture (Fig. 15.6)

At Orange Walk, where the trial was planted in natural pasture, establishment of most legumes was weak (Tables 15.20 and 15.21). However, *M. atropurpureum* cv. Siratro (Fig. 15.6), *S. guianensis* cvs Cook and Endeavour, and *S. hamata* generally performed well, though establishment of the *Stylosanthes* accessions was slow. *Centrosema pubescens* persisted and was still present in 1981. The native pasture was cut over three times in the early period of the trial in order to encourage establishment of the legumes. Productivity of the native grasses (90% *Axonopus compresus*, 10% *Paspalum convexum*) was very weak inside the fence over the period of trial, both with and without fertilizer (Fig. 15.6). As the grass provided relatively little competition to the legumes, growth under the low fertilizer rate was similar to or only slightly less than that of the higher rate. *M. lathyroides* and *C. plumieri* also showed some promise.

Site 15 Agricultural Show Grounds, various exotic grasses (Fig. 15.7)

At site 15 on the Agricultural Show Grounds, the six grass plots were cut over after each observation cycle to 15 cm and the legumes after each observation from cycle 6. There was generally little invasion of the planted strips by the grasses, except for Pangola and Hamil grasses, which achieved considerable plot cover in most of the strips; however, the legumes were apparently not severely affected. Generally none of the legumes had consistently high Vigour Grouping ratings in any of the grasses (Tables 15.22–15.28). While invasion of the planted strips was generally not a factor, the long-established dense swards may have had well-developed root systems that competed for water, and their rapid growth may have shaded the strips. *C. gyroides* was the best legume at the site (Fig. 15.7), followed by *M. atropurpureum* cv. Siratro. *Leucaena* did well in two of the grasses and *S. hamata* cv. Verano and *D. intortum* had some success in two plots and one plot respectively. Legume growth appeared to be best in the more open Para grass and weakest in the Pangola plot. The other grasses had intermediate but generally weak legume ratings.

15.3.5 Summary of the upper Belize River Valley results

The final observation results from the sites (Table 15.29) shows that while many of the legumes survived until the end of the trial, legume performance was poor at most sites. The two exceptions were site 3, San Lorenzo and site 10, Spanish Lookout. The results contrast with the

Table 15.12. Site 2, Norland Farm, *Pennisetum purpureum* cv. Merker. Accessions' membership of vigour groups (Table 15.9) determined from observational data: A best, to D poor. Darker shading has been used to identify the better results.

		Observation cycle													
		Low fertilizer							High fertilizer						
		2	4	6	8	10	12	13	2	4	6	8	10	12	13
Legumes	Accession no.	cool	dry	wet	late wet	cool	dry	wet	cool	dry	wet	late wet	cool	dry	wet
Calopogonium caeruleum	CF 18	C	C	D	D	D	D	D	B	C	D	D	D	D	D
Centrosema plumieri	CF 16-1	–	–	–	–	–	–	–	D	–	–	–	–	–	–
Centrosema pubescens cv. Belalto	CF 195	C	C	–	–	–	–	–	D	D	–	–	–	–	–
Centrosema pubescens	CF 6-1	C	C	D	D	D	D	D	B	C	D	D	D	D	D
Centrosema pubescens 'Centro'	ChF 6-2	C	C	D	–	–	–	D	D	C	D	D	D	D	–
Codariocalyx gyroides	CF 29	**A**	B	D	–	–	–	D	B	**A**	–	D	–	D	–
Crotalaria retusa	CF 20[a]														
Crotalaria vitellina	CF 15[a]	D	–	–	–	–	–	–	–	–	–	–	–	–	–
Desmodium canum	CF 1[a]	C	D	D	D	D	D	D	D	D	D	D	D	D	D
Desmodium intortum 'Greenleaf'	CF 50	**A**	**A**	B	C	D	D	D	B	D	**A**	D	D	D	D
Indigofera mucronata	CF 9[a]	D	D	–	D	–	–	–	–	–	D	–	–	–	–
Leucaena leucocephala cv. Peru	CF 179	C	D	D	D	D	D	D	D	C	D	D	D	C	D
Macroptilium atropurpureum cv. Siratro	CF 3-1	**A**	**A**	C	D	–	–	D	B	**A**	D	D	D	D	–
Macroptilium lathyroides	CF 14	**A**	B	D	D	–	–	–	**A**	B	D	–	D	D	–
	CF 14[a]	C	D	–	D	D	–	D	D	D	D	D	D	D	–
Pueraria phaseoloides 'Puero'	CF 180	**A**	**A**	D	D	–	–	D	B	C	–	D	D	D	–
	CF 180[a]	D	D	–	–	D	D	D	D	D	D	D	–	D	D
Rhynchosia minima	CF 13[a]	D	D	–	D	D	–	D	D	D	D	–	–	–	D
Stylosanthes guianensis cv. Oxley	CF 201[a]	–	–	–	–	–	–	–	–	–	–	–	–	–	–
Stylosanthes guianensis	CF 4-1	C	D	–	D	–	–	–	B	C	–	–	–	–	D
Stylosanthes guianensis cv. Cook	CF 4-4	**A**	C	–	–	–	–	D	B	**A**	–	D	D	D	–
Stylosanthes guianensis cv. Endeavour	CF 4-5	**A**	**A**	D	D	D	C	D	B	**A**	D	D	D	D	C
Stylosanthes hamata cv. Verano	CF 186	**A**	**A**	D	C	D	D	B	B	**A**	D	D	D	B	D
Teramnus uncinatus	CF 2[a]	C	D	D	D	D	D	D	B	D	D	D	D	D	D
Vigna vexillata	CF 43-2	–	–	–	–	–	–	–	–	–	–	–	–	–	–

[a]In legume mixture plot.

Table 15.13. Observation number after which management was imposed on Site 2

Vegetation	Grass cut	Grass and legume cut	Grazed
Merker grass	1, 2	8, 10, 12, 13	3

multiple trials undertaken on Central Farm river terrace soils (e.g. Lazier, 1978), which have shown that these legumes are adapted and productive under cutting, and once established have done well under grazing with careful management. The soil type does not appear to be a factor here because site 10 is relatively shallow soil over limestone, while site 3 is a deep river terrace soil. Similarly, the Norland sites are on deep river terrace soils and are very close to the successful Central Farm trial sites. The better legume performance in the formal replicated trials on Central Farm may be attributable to the better establishment of the legumes. Good establishment occurred because (i) there was less competition from the grasses since they were planted at the same time as the legumes; (ii) the trials were irrigated through the initial dry season that occurred shortly after planting; (iii) the plots were cut over in a manner that ensured that the legumes were not stressed, but there was optimal off take of the grasses; and (iv) the plants had two complete wet seasons to become established before being grazed. There, Siratro had been the best performer, but *C. gyroides* and *Leucaena* also performed well in the last year of the trial under grazing. However, *C. gyroides* was prone to physical damage by cattle and did poorly in the dry season (Lazier, 1978). In the strip trials, *Leucaena* and *C. gyroides* had the best performance, followed by *C. caeruleum* and the *Centrosema* species.

15.4 Low Pine Ridge sites

15.4.1 Introduction

The infertile coastal sand deposits are planosols, are locally termed Lowland Pine Ridge soils, and were classified as Puletan by Wright *et al.* (1959). They are important nationally since they cover about 170,000 ha or about 9% of the country and include relatively large areas along the Belize River west of Belize City. Sparsely vegetated, the area is mainly covered with coarse grass (*Mesosetum angustifolium*), with scattered islands or 'cays' of palmettos (*Acoelorraphe wrightii* and *Schippia concolor*), various *Quercus* species, crabboo (*Brysonima crassifolia*), and associated understory plants (Fig. 15.8). Unutilized due to their low fertility and high water table, they are attractive areas for development because there is little vegetation to be cleared, and due to their proximity to markets. As a result, considerable effort had been made since the 1950s in testing the growing of crops such as rice and the adaptation of productive pasture grasses. In the early 1950s, Guatemala (*Trypsacum laxum*) and Guinea (*Panicum maximum*) grasses were planted at Hector Creek (now Hattieville) with the following treatments: no lime, lime (rate not given), phosphate at 126 kg/ha alone and with 112 kg/ha of sulphate of ammonia, sub-soiled and not sub-soiled. One harvest was taken of the Guatemala grass and two of the Guinea grass. There was no response to lime or sub-soiling, but there was a small uneconomical increase with the application of nitrogen (DOA, 1953).

As the early tests on growing food crops were not successful because of the waterlogged condition of the soil, the Department of Agriculture, in hopes of finding some use for the land, refocused on developing the land as pasture.

In 1956 Pangola-phosphate trials were established at six sites in the country, one of which was at Hattieville. TSP was applied at 0, 110, 220 and 450 kg/ha in 1956 and 1957, and at 900 kg/ha in 1956 only (DOA, 1956, 1957). A sulphate of ammonia topdressing was given after each 3-weekly clipping. The harvest intervals were found to be too short because the grass stands became increasingly thin. Six-weekly cuttings were then imposed and by the end of the first year higher DM yields were being obtained and these increased with increasing levels of phosphate fertilizer applied. It was also noted that DM yields were lower November to January, probably due to short days, foggy conditions and cool temperatures.

The same year, two Pangola trials were planted on a deep Puletan soil (DOA, 1957). The first trial had four replicates of the following treatments: control, P, PK, NP, NPK and lime (no rates were given). One clip was taken after 14 weeks of growth. The treatment NP and NPK had similar fresh weights, which were significantly more than

Table 15.14. Site 3, San Lorenzo (Don Santiago Juan Farm), *Hyparrhenia rufa*. Accessions' membership of vigour groups (Table 15.9) determined from observational data: A best, to D poor. Darker shading has been used to identify the better results.

		Observation period													
		Low fertilizer							High fertilizer						
		2	4	6	8	10	12[a]	13	2	4	6	8	10	12[a]	13
Legumes	Accession no.	cool	dry	wet	late wet	cool	dry	wet	cool	dry	wet	late wet	cool	dry	wet
Calopogonium caeruleum	CF 18	B	B	**A**	**A**	B		**A**	B	C	**A**	**A**	B		**A**
Centrosema plumieri	CF 16-1	B	D	**A**	**A**	B		C	**A**	D	**A**	D	C		D
Centrosema pubescens	CF 177	**A**	C	**A**	**A**	B		–	B	B	**A**	B	C		–
Centrosema pubescens	CF 6-1	**A**	C	**A**	**A**	D		D	B	B	**A**	B	C		D
Centrosema pubescens 'Centro'	CF 6-2	**A**	C	**A**	D	D		D	B	B	**A**	**A**	D		–
Codariocalyx gyroides	CF 29	**A**	C	**A**	**A**	B		**A**	B	B	**A**	**A**	**A**		**A**
Crotalaria retusa	CF 20[b]	C	D	B	D	D		D	B	D	B	–	–		–
Crotalaria vitellina	CF 15[b]	D	D	C	D	D		D	D	D	D	D	D		–
Desmodium canum	CF 1[b]	C	D	C	D	D		D	D	D	D	C	C		D
Desmodium intortum 'Greenleaf'	CF 50	**A**	C	**A**	**A**	D		D	B	B	**A**	**A**	C		D
Indigofera mucronata	CF 9[b]	–	–	–	–	–		–	–	–	D	D	–		–
Leucaena leucocephala cv. Peru	CF 179	**A**	C	**A**	**A**	B		**A**	B	B	**A**	**A**	**A**		**A**
Macroptilium atropurpureum cv. Siratro	CF 3-1	**A**	C	C	D	D		D	**A**	B	**A**	C	D		D
Macroptilium lathyroides	CF 14	B	D	–	D	–		–	B	B	D	–	–		D
	CF 14[b]	C	D	–	–	–		–	D	D	–	–	–		D
Pueraria phaseoloides 'Puero'	CF 180	**A**	C	**A**	**A**	B		C	**A**	B	**A**	**A**	C		D
	CF 180[b]	C	D	C	D	C		D	D	D	D	D	D		D
Rhynchosia minima	CF 13[b]	C	D	C		D	–	D	**A**	B	D	D	D		D
Stylosanthes guianensis cv. Oxley	CF 201[b]	–	–	–	–	–		–	–	–	–	–	–		–
Stylosanthes guianensis	CF 4-1	C	D	C	D	D		C	D	D	D	D	C		D
Stylosanthes guianensis cv. Cook	CF 4-4	**A**	C	**A**	**A**	B		**A**	B	B	**A**	**A**	B		D
Stylosanthes guianensis cv. Endeavour	CF 4-5	**A**	C	**A**	**A**	B		D	B	B	**A**	**A**	B		D
Stylosanthes hamata cv. Verano	CF 186	**A**	B	**A**	**A**	D		–	B	B	D	**A**	–		–
Teramnus uncinatus	CF 2[b]	C	B	D	D	D		D	D	D	D	D	–		–
Vigna vexillata	CF 43-2	–	–	–	–	–	–	–	–	–	–	–	–		–

[a]Observation not taken; [b]in legume mixture plot.

the other treatments. There was no response to the application of lime.

The second trial had six replicates of N and P treatments at the following rates (in kg/ha): 0:40, 12:40, 32:40, 64:80. After a 6-month establishment period, three clippings were taken over the next half year. Generally, the yield increased with the fertilizer rate and the highest rates gave about double the fresh weight yield of the next lower rate.

A further Pangola trial was established on a shallow Puletan soil (DOA, 1956); all plots were given N at 90 kg/ha/annum in 11.25 kg/ha increments after each 6-weekly harvest. The trial had four replicates of plots receiving the P treatments at the following levels: 0, 45, 90, 180, 360 kg/ha/annum. The fresh weight yields of the P treatments did not differ significantly from each other, though they were greater than the control. Apparently some other unidentified factor had limited productivity.

Observation trials on six grasses (Pangola, Jaragua, Guinea, Buffel, Pensacola Bahia and Bermuda) were established on four sites on the lower Pine Ridge at Hector Creek and Boom (DOA, 1961, 1962). Half of each unreplicated grass plot was fertilized with a single application of 225 kg/ha of 18-46-0 fertilizer. No germination occurred in the unfertilized halves of the plots. Jaragua and Buffel did best initially, while in the second season all plots except Bermuda and Buffel had good, but declining, plot coverage despite little invasion by native species. Sulphate of ammonia was applied in the second year at 674 kg/ha but little response was noted.

Table 15.15. Observation number after which management was imposed on Site 3.

Grass	Grass cut	Grass and legume cut	Grazed
Jaragua	1, 2	5, 6, 7, 8, 9, 10, 11, 12, 13	

A pot trial was established using Pangola grass planted in Puletan soil in an attempt to determine the nutrients and levels required for vigorous growth (Blue, 1969). An initial unreplicated trial applied varying levels of N, P, K, Ca, Mg, lime and micronutrients (as frittered trace elements) as treatments. Two harvests were taken and the results indicated that the addition of fertilizer elements increased their concentration in the leaves. The second harvest had markedly less K in the foliage, indicating that there was limited K present in the soil. Foliar recovery of N and P was frequently in excess of the nutrients applied as the addition of lime and fertilizer apparently increased the rate of decomposition of soil organic matter. Dry matter yield generally increased with P application with the lowest rate of Ca applied (200 ppm), and as well with N, when P was added.

Fig. 15.4. Site 14 Central Farm, *C. nlemfluensis*. (a) Early establishment phase of strip layout with high fertility lower in photo, *Centrosema pubescens* January 1976. (b) *Codariocalyx gyroides* post grazing January 1977.

Table 15.16. Site 14, Central Farm, *Cynodon nlemfluensis* (African Star grass). Accessions' membership of vigour groups (Table 15.9) determined from observational data: A best, to D poor. Darker shading has been used to identify the better results.

		Observations													
		Low fertilizer							High fertilizer						
		2	4	6	8	10	12	13	2	4	6	8	10	12	13
Legumes	Accession no.	cool	dry	wet	late wet	cool	dry	wet	cool	dry	wet	late wet	cool	dry	wet
Calopogonium caeruleum	CF 18	C	C	D	B	D	D	D	B	B	B	D	D	D	D
Centrosema plumieri	CF 16-1	C	C	D	B	C	D	–	B	C	C	D	D	–	–
Centrosema pubescens cv. Belalto	CF 195	C	D	D	D	D	D	D	D	D	C	D	D	–	D
Centrosema pubescens	CF 6-1	**A**	D	**A**	B	C	D	D	B	B	B	D	D	D	D
Centrosema pubescens 'Centro'	CF 6-2	**A**	**A**	**A**	B	C	D	D	B	B	B	D	D	D	D
Codariocalyx gyroides	CF 29	**A**	**A**	B	B	B	B	B	B	B	B	B	B	D	D
Crotalaria retusa	CF 20[a]	–	–	–	–	–	–	–	D	–	–	–	–	–	–
Crotalaria vitellina	CF 15[a]	–	–	–	–	–	–	–	–	–	–	–	–	D	–
Desmodium canum	CF 1[a]	C	D	C	D	D	D	D	B	C	D	D	D	D	D
Desmodium intortum 'Greenleaf'	CF 50	**A**	**A**	C	D	D	D	D	B	B	D	–	–	–	–
Indigofera mucronata	CF 9[a]	C	D	D	D	D	–	–	–	D	D	D	D	D	–
Leucaena leucocephala cv. Peru	CF 179	**A**	C	B	B	B	B	B	B	B	B	B	B	B	**A**
M. atropurpureum cv. Siratro	CF 3-1	**A**	**A**	**A**	–		–	–	B	**A**	B	B	D	–	–
Macroptilium lathyroides	CF 14	–	–	–	–	–	–	–	–	–	–	–	–	–	–
	CF 14[a]	**A**	D	–	–	–	–	–	B	C	D	–	–	D	D
Pueraria phaseoloides 'Puero'	CF 180	C	D	D	B	D	D	D	B	B	B	D	D	–	–
	CF 180[a]	C	–	D	D	–	–	–	D	D	D	D	D	D	–
Rhynchosia minima	CF 13[a]	C	D	D	D	D	–	–	D	D	D	D	–	–	–
Stylosanthes guianensis cv. Oxley	CF 201[a]	–	–	–	–	–	–	–	–	–	–	–	–	–	–
Stylosanthes guianensis	CF 4-1	C	C	**A**	**A**	D	–	–	B	D	D	D	D	–	–
Stylosanthes guianensis cv. Cook	CF 4-4	**A**	**A**	**A**	B	D	–	–	B	B	B	B	–	–	–
Stylosanthes guianensis cv. Endeavour	CF 4-5	**A**	**A**	**A**	**A**	C	–	–	B	C	B	B	–	–	–
Stylosanthes hamata cv. Verano	CF 186	**A**	**A**	**A**	B	D		–	–	B	B	B	B	D	–
Teramnus uncinatus	CF 2[a]	**A**	D	D	D	D	–	–	B	C	B	–	D	D	D
Vigna vexillata	CF 43-2	**A**	**A**	**A**	D	D	D	D	B	B	B	D	D	D	D
	CF 43-2[a]	–	–	–	–	–	–	–	–	–	–	–	–	–	–

[a]In legume mixture plot.

Table 15.17. Observation number after which management was imposed on Site 14.

Grass	Grass cut	Grass and legume cut	Grazed
African Star	2		3, 4, 6, 7, 8, 9, 10, 11, 12

A second pot trial involved 16 treatments with four replicates. N was added to all treatments at 50 ppm, and P, K, Mg, S and lime added at varying rates and three clips were taken. The dry matter response to lime was linear and was possibly dependent on the presence of P. P and lime together significantly increased the forage yields, up to 13-fold when added with substantial quantities of other nutrient elements. In the first harvest, 3 months after planting, K, Mg and S did not alter production and the foliage had fairly low nutrient concentrations as it was mature. P concentrations were probably deficient for cattle. By the third harvest, 8 months after planting, N concentrations had dropped. It was noted in the first trial that only at the higher applications of P was its level adequate for cattle. Blue noted that forage legumes would probably have to be maintained with the grass before cattle production would become economic.

Twelve grasses in four replicates were planted on Puletan soil at Hattieville and given a fairly heavy regular application of NPK. The trial required replanting due to waterlogging. Waterlogging and nutrient stress were noted between fertilization periods and were assumed to be the cause of poor production. Kazangula Setaria, Para, Pangola and, big blue stem produced the best fresh weights (DOA, 1972, 1973, 1974).

Several legume trials were done of the Puletan soils by Dr L. S. Snook during his stay in Belize. Observation plots of legumes and grasses were planted at mile 33 on the Western Highway and a cowpea (*Vigna sinensis*) macro- and micronutrient trial was established in 1967. On the basis of the results a more refined cowpea trial was planted in 1968, having three replicates and 39 treatments (NPK at 200 kg/ha alone and TSP at the same rate with either 1.1, 2.2 or 3.4 kg/ha of Zn, Cu, Mo, Fe and Mn, either alone or in various combinations). The control plots grew poorest, perhaps due to waterlogging. Plots that had micronutrients applied showed no significant differences in their dry matter yields because of wide variations in growth between blocks, again perhaps due to waterlogging. Yields did increase significantly though with increasing application of Mn and Mo with TSP and Fe, Mn and Mo with TSP, though there was no significant difference between these treatments and TSP alone.

15.4.2 Trial sites on the Low Pine Ridge

Five sites were chosen to represent the low-altitude Pine Ridge savanna area (Tables 15.1 and 15.2). The sites were undeveloped but had been burned over in drier years with the fires helping to reduce bush encroachment from the scattered 'cays' (islands of trees, cut grass and palmetto) in the otherwise rather thinly vegetated soils. All of these sites had been burned about 7 months previous to the start of the research. The reactions of the sterile waterlogged Low Pine Ridge Puletan soils at the sites chosen were strongly acid (pH 5.4–5.6).

Fig. 15.5. Site 10 *D. eriantha* trial at Spanish Lookout. (a) View across site and the Belize River valley. (b) Site 10 *Leucaena leucocephala* September 1978.

Table 15.18. Site 10, Spanish Lookout (Klaas Fresian Farm), *Digitaria eriantha* (Pangola). Accessions' membership of vigour groups (Table 15.9) determined from observational data: A best, to D poor. Darker shading has been used to identify the better results.

		Observation period													
		Low fertilizer							High fertilizer						
		2	4	6	8	10	12	13	2	4	6	8	10	12	13
Legumes	Accession no.	cool	dry	wet	late wet	cool	dry	wet	cool	dry	wet	late wet	cool	dry	wet
Calopogonium caeruleum	CF 18	C	D	C	B	B		C	D	**A**	**A**	B	**A**		**A**
Centrosema plumieri	CF 16-1	D	D	**A**	B	B		C	D	C	**A**	**A**	B		**A**
Centrosema pubescens	CF 195	C	B	**A**	B	B		**A**	B		**A**	**A**	**A**		**A**
Centrosema pubescens	CF 6-1	C	D	**A**	B	B		**A**	B	D	**A**	**A**	B		**A**
Centrosema pubescens 'Centro'	CF 6-2	**A**	C	**A**	**A**	B		B	B	B	C	B	B		**A**
Desmodium gyroides	CF 29	C	C	**A**	**A**	B		**A**	B	B	**A**	**A**	**A**		**A**
Crotalaria retusa	CF 20[a]	–	–	–	–	–		–	D	–	–	–	–		–
Crotalaria vitellina	CF 15[a]	D	D	C	D	D		–	D	D	D	–	–		–
Desmodium canum	CF 1[a]	D	D	C	D	D		D	D	D	D	D	D		D
Desmodium intortum 'Greenleaf'	CF 50	**A**	B	**A**	**A**	B		C	B	B	**A**	**A**	B		D
Indigofera mucronata	CF 9[a]	C	D	**A**	**A**	D		–	D	D	**A**	C	C		D
Leucaena leucocephala cv. Peru	CF 179	C	C	**A**	D	B		B	D	D	C	D	D		D
M. atropurpureum cv. Siratro	CF 3-1	**A**	B	**A**	B	B		C	**A**	B	**A**	**A**	B		**A**
Macroptilium lathyroides	CF 14	B	B	**A**	D	–		D	B	B	**A**	D	–		D
	CF 14[a]	B	B	**A**	–	–		–	B	B	**A**	–	–		–
Pueraria phaseoloides 'Puero'	CF 180	C	–	**A**	**A**	B		B	–	B	**A**	**A**	B		**A**
	CF 180[a]	–	B	C	–	–		–	**A**	–	–	–	D		D
Rhynchosia minima	CF 13[a]	C	D	**A**	D	–		–	D	D	D	D	–		D
Stylosanthes guianensis cv. Oxley	CF 201[a]	–	–	–	–	–		–	D	D	–	–	–		–
Stylosanthes guianensis	CF 4-1	C	D	C	**A**	B		C	D	D	D	**A**	C		D
Stylosanthes guianensis cv. Cook	CF 4-4	C	C	**A**	**A**	B		C	B	B	**A**	**A**	B		D
Stylosanthes guianensis cv. Endeavour	CF 4-5	C	C	**A**	**A**	B		C	D	D	D	C	C		D
Stylosanthes hamata cv. Verano	CF 186	C	B	**A**	**A**	B		D	B	B	**A**	**A**	B		D
Teramnus uncinatus	CF 2[a]	D	D	**A**	**A**	D		D	D	D	**A**	**A**	B		D
Vigna vexillata	CF 43-2[a]	–	–	–	–	–		–		–	–	–	–		–

[a]In legume mixture plot.

Table 15.19. Observation number after which management was imposed on Site10.

Grass	Grass cut	Grass and legume cut	Grazed
Pangola	2, 4	5, 6, 13	8, 9, 10, 11

Sites 5 and 6, mile 33 Western Highway

This location was chosen because it had been the site of plantings by L. Snook in 1967. The details of the more recent trials are lost but a field worker remembered that *Calopogonium caeruleum* (probably a native line), *Codariocalyx gyroides*, *Stylosanthes guianensis* (probably cv. Schofield) were planted with three grasses 'Blue Gama' (*Trypsacum laxum* rather than *Bouteloua gracilis?*) and big blue stem (*Andropogon gerardii*). There were 45 plots, so there may have been three replications or three fertilizer treatments. Fertilizer had been applied. Single plots and strips of plants were established and establishment was tested in wet and dry conditions. Drains had to be dug at one end of the trial to help the plants survive. The trial was abandoned in 1968 and although it had presumably been burnt over several times since, and the site had been reportedly used for grazing by an occasional horse, many of the legumes survived until 1976. These included a hairy-leafed *Centrosema pubescens*, which may have been locally collected, and *C. caeruleum*; both were climbing over invading trees and shrubs. There were a couple of patches of *S. guianensis*, and quite a lot of *C. gyroides*, the latter mainly in the trial but also extending in a parabolic shape downwind, outside the trial. This external area may have been a fertilized site where the strips and single plants were grown. There had also been quite a significant invasion by savanna trees and shrubs, indicating the effects of soil fertilization. The particularly good performance of *C. gyroides* indicates that the shrub may have potential as a fodder species on these soils. *Andropogon gerardii* and *Setaria sphacelata* cv. Kazungula were also still to be found on the site.

The soils in the area are old coastal alluvium planosols with a thin A horizon (4–50 cm thick) of black to grey clay, below which is an abrupt change to a very heavy, light-grey C horizon that becomes white with depth. The site, in common with much of the Low Pine Ridge soils, is wet to within a few centimetres of the surface, and remains so even in exceptionally long dry seasons when the soils still retain a moist plastic character. McKinzie (1977) classified them as Abruptic Tropaqualfs, fine, kaolinitic. The pH of sites 5 and 6, according the UWI analysis, were 5.6 and 5.5 respectively (Appendix 15.1), but McKinzie found all horizons at a soil pit at the site to be more acid, 4.4 at the A horizon. Site 6 was slightly less acid at 5.0.

The persistence of the legumes under such adverse environmental conditions led to the selection of the location for two trials. One site was selected on more level, wetter land (site 5) and the other on a slightly sloping site (site 6); the two sites were selected for their contrasting drainage and to ensure that, if the lower situation was too wet and survival was

Fig. 15.6. Site 4 Orange Walk, natural pasture. (a) *M. atropurpureum* cv. Siratro; (b) a year's grass growth in the trial site compared with the continually grazed pasture outside.

Table 15.20. Site 4, Orange Walk, natural pasture. Accessions' membership of vigour groups (Table 15.9) determined from observational data: A best, to D poor. Darker shading has been used to identify the better results.

		Observation period													
		Low fertilizer							High fertilizer						
		2	4	6	8	10	12	13	2	4	6	8	10	12	13
Legumes	Accession no.	cool	dry	wet	late wet	cool	dry	wet	cool	dry	wet	late wet	cool	dry	wet
Calopogonium caeruleum	CF 18	D	–	–	–	–	–	D	D	D	D	D	D	D	D
Centrosema plumieri	CF 16-1	D	D	C	C	B	D	D	D	D	D	C	D	D	D
Centrosema pubescens	CF 177	D	D	D	D	–	–	D	D	D	D	D	D	D	D
Centrosema pubescens	CF 6-1	D	D	C	C	D	D	D	B	D	D	C	D	D	D
Centrosema pubescens 'Centro'	CF 6-2	D	D	C	C	D	D	D	D	D	D	C	D	D	C
Codariocalyx gyroides	CF 29	D	D	D	D	D	D	D	D	D	D	D	D	C	C
Crotalaria retusa	CF 20[a]	D	–	–	–	–	–	–	D	–	–	–	–	–	–
Crotalaria vitellina	CF 15[a]	D	D	D	D	–	–	–	D	D	D	D	D	–	–
Desmodium canum	CF 1[a]	D	D	D	D	D	D	D	D	D	–	D	–	D	D
Desmodium intortum 'Greenleaf'	CF 50	D	D	D	D	D	D	D	D	D	D	D	D	C	D
Indigofera mucronata	CF 9[a]	–	–	–	–	–	–	–	D	D	D	D	D	D	–
Leucaena leucocephala cv. Peru	CF 179	D	D	D	D	D	D	D	D	D	D	D	D	C	C
Macroptilium atropurpureum cv. Siratro	CF 3-1	**A**	**A**	**A**	**A**	B	B	C	D	**A**	D	**A**	**A**	B	B
Macroptilium lathyroides	CF 14	**A**	B	**A**	**A**[a]	C	D	D	**A**	B	C	**A**	B	B	D
	CF 14[a]	**A**	B	**A**	C	C	D	D	**A**	B	C	**A**	B	C	–
Pueraria phaseoloides 'Puero'	CF 180	C	C	**A**	**A**	B	B	C	D	–	D	C	D	D	D
	CF 180[a]	–	–	–	–	–	–	–	D	D	D	D	D	D	D
Rhynchosia minima	CF 13[a]	D	D	D	D	D	D	D	D	D	D	D	D	C	D
Stylosanthes guianensis cv. Oxley	CF 201[a]	D	D	D	D	–	–	D	–	–	–	–	–	–	–
Stylosanthes guianensis	CF 4-1	D	D	D	D	D	D	D	D	D	D	D	D	C	D
Stylosanthes guianensis cv. Cook	CF 4-4	D	D	D	**A**	B	B	B	D	D	D	**A**	**A**	C	C
Stylosanthes guianensis cv. Endeavour	CF 4-5	D	D	B	**A**	B	B	B	B	D	D	**A**	**A**	B	B
Stylosanthes hamata cv. Verano	CF 186	D	D	B	**A**	B	D	D	D	D	D	**A**	**A**	–	D
Teramnus uncinatus	CF 2[a]	D	D	D	D	D	D	D	D	D	D	D	D	C	D
Vigna vexillata	CF 43-2	–	–	–	–	–	–	–	–	–	–	–	–	–	–

[a]In legume mixture plot.

Table 15.21. Observation number after which management was imposed on Site 4.

Grass	Grass cut	Grass and legume cut	Grazed
Native	1, 2, 5		

poor, then survival might be better on the sloping site where drainage was assumed to be better.

Sites 7, 8 and 9, mile 35 Western Highway, Gulf American Property

This area was chosen because of the importance of the soil type present, the ease of access and the interest of the management of the firm that owned the land, Gulf American, which wished to undertake grazing on these extensive soils. They were already grazing one paddock on the savanna and were convinced that periodic harrowing of the soils resulted in improved grass growth. Thus, at the time the trials were established, they had started harrowing a large area of the savanna.

The area chosen was flat, sloping gently to the north. The three trials were established in close proximity. To determine if any differences did exist in vegetation growth between harrowed and undisturbed areas, native undisturbed savanna was selected as one site (site 7), and land next to it was harrowed as the second site (site 8). The third site was located on a 'cay', an area of denser vegetation that included woody species (site 9).

Jenkin *et al.* (1976) classified the site as Yaxa suite, Yalbec sub-suite, Beaver Dam series, making up some 31,000 ha in the Belize River Valley – about 17% of the area of the Low Pine Ridge soils in the LRD Belize Valley Survey. Wright *et al.* (1959) mapped the general area as Puletan sandy loam, representing 37,000 ha or about 22% of the Low Pine Ridge soils. However, McKinzie (1977) found the actual site was a clay. Wright *et al.* (1959) mapped mainly by vegetation relationships and by land form, so it can be assumed that the area is similar in these characteristics to the 37,000 ha he matched it with. McKinzie characterized the soil as Aquentic Pelluderts, very fine montmorillonitic, isohyperthermic. The profile is normally wet through in the wet season, and only the A horizon dries in the dry season. The reaction was neutral to slightly acidic with depth.

Examination of the soil profile where grass was present indicated that, due to the high moisture content of the lower soil layers, roots were confined to the A horizon, about 10 cm at the top of the soil profile. Vegetation largely consisted of sedges and the coarse grass *Mesosetum angustifolium*. Where the A horizon was thinner, about 5 cm, only *M. angustifolium* and sedges were present, but as the A horizon became thicker, the grass height and density increased, more species were present and root profiles were

Fig. 15.7. Site 15 Agricultural Show Grounds, six oversown grass plots. (a) Uncut plots with *C. gyroides* strips clearly seen (1978). (b) Plots during harvesting period, again *C. gyroides* strips are evident.

Table 15.22. Site 15, Agricultural Show Grounds, *Panicum maximum* cv. Hamil. Accessions' membership of vigour groups (Table 15.9) determined from observational data: A best, to D poor. Darker shading has been used to identify the better results.

		Observations													
		Low fertilizer							High fertilizer						
		2	4	6	8	10	12	13	2	4	6	8	10	12	13
Legumes	Accession no.	cool	dry	wet	late wet	cool	dry	wet	cool	dry	wet	late wet	cool	dry	wet
Calopogonium caeruleum[a]	CF 18	D	D	D	D	C	D	D	D	D	D	D	D	D	D
Centrosema pubescens[a]	CF 16-1	D	D	D	D	C	D	D	D	D	D	D	D	D	D
Centrosema pubescens 'Centro'[a]	CF 6-2	D	D	D	D	D	D	D	D	D	D	D	–	–	D
Codariocalyx gyroides	CF 29	D	D	D	C	C	B	B	D	D	D	D	**A**	**A**	B
Desmodium intortum 'Greenleaf'[a]	CF 50	D	D	D	D	D	–	D	D	C	D	D	D	–	D
Leucaena leucocephala cv. Peru	CF 179	D	D	C	C	D	B	B	D	D	D	D	C	D	C
M. atropurpureum cv. Siratro	CF 3-1	**A**	C	**A**	B	C	D	C	**A**	C	**A**	B	D	D	D
Pueraria phaseoloides 'Puero'[a]	CF 180	D	D	D	D	–	–	–	D	D	D	B	D	–	D
Stylosanthes guianensis cv. Cook[a]	CF 4-4	D	D	D	D	D	D	D	D	–	D	D	D	D	D
Stylosanthes guianensis cv. Endeavour[a]	CF 4-5	[b]	–	D	D	D	D	D	[b]	D	D	D	D	D	D
Stylosanthes hamata cv. Verano	CF 186	D	D	D	D	D	D	D	D	D	D	D	–	–	–

[a]In legume mixture plot; [b]not observed.

Table 15.23. Site 15, Agricultural Show Grounds, Buffel grass *Cenchrus ciliaris* var. Nunbank (A best, to D poor).

		Observations													
		Low fertilizer							High fertilizer						
		2	4	6	8	10	12	13	2	4	6	8	10	12	13
Legumes	Accession no.	cool	dry	wet	late wet	cool	dry	wet	cool	dry	wet	late wet	cool	dry	wet
Calopogonium caeruleum[a]	CF 18	D	D	D	D	D	D	D	D	D	D	D	D	D	D
Centrosema pubescens[a]	CF 16-1	D	D	D	D	C	D	D	D	D	D	D	D	C	D
Centrosema pubescens 'Centro'[a]	CF 6-2	D	D	D	D	D	D	D	D	D	D	D	D	D	–
Codariocalyx gyroides	CF 29	D	D	D	D	C	B	B	D	D	D	D	C	B	B
Desmodium intortum 'Greenleaf'[a]	CF 50	D	D	D	D	D	–	–	D	D	D	–	–	–	–
Leucaena leucocephala cv. Peru	CF 179	D	D	C	C	D	B	D	D	D	D	D	D	C	C
M. atropurpureum cv. Siratro	CF 3-1	**A**	**A**	**A**	B	B	B	C	**A**	C	**A**	B	C	C	C
Pueraria phaseoloides 'Puero'[a]	CF 180	D	D	D	D	D	D	D	D	D	D	D	D	D	–
Stylosanthes guianensis cv. Cook[a]	CF 4-4	D	D	D	D	D	D	D	D	D	D	D	D	D	C
Stylosanthes guianensis cv. Endeavour[a]	CF 4-5	[b]	D	D	D	D	–	–	[b]	D	D	D	D	D	D
Stylosanthes hamata cv. Verano	CF 186	D	D	C	C	D	–	–	D	D	**A**	B	D	–	–

[a]In legume mixture plot; [b]not observed.

Table 15.24. Site 15, Agricultural Show Grounds, Guinea grass *Panicum maximum*. Accessions' membership of vigour groups (Table 15.9) determined from observational data: A best, to D poor. Darker shading has been used to identify the better results.

		Observations													
		Low fertilizer							High fertilizer						
		2	4	6	8	10	12	13	2	4	6	8	10	12	13
Legumes	Accession no.	cool	dry	wet	late wet	cool	dry	wet	cool	dry	wet	late wet	cool	dry	wet
Calopogonium caeruleum[a]	CF 18	–	–	–	–	–	–	–	D	C	D	D	D	C	D
Centrosema plumieri[a]	CF 16-1	D	D	D	D	C	D	D	D	D	D	D	D	D	D
Centrosema pubescens 'Centro'[a]	CF 6-2	–	D	–	–	–	–	–	D	D	D	D	D	D	–
Codariocalyx gyroides	CF 29	D	D	D	C	D	B	B	D	C	D	B	**A**	**A**	B
Desmodium intortum 'Greenleaf'[a]	CF 50	D	D	C	D	D	D	D	D	C	–	–	D	D	D
Leucaena leucocephala cv. Peru	CF 179	D	D	D	D	–	D	D	D	D	–	–	–	D	D
M. atropurpureum cv. Siratro	CF 3-1	**A**	C	D	D	D	D	D	**A**	C	D	D	D	D	D
Pueraria phaseoloides 'Puero'[a]	CF 180	D	–	D	–	–	–	–	–	–	–	–	–	–	–
Stylosanthes guianensis cv. Cook[a]	CF 4-4	D	D	–	–	–	–	–	D	D	D	D	–	–	
Stylosanthes guianensis cv. Endeavour[a]	CF 4-5	[b]	–	–	–	–	–	–	[b]	–	–	–	–	–	–
Stylosanthes hamata cv. Verano	CF 186	–	D	–	–	–	–	–	D	–	–	–	–	–	–

[a]In legume mixture plot; [b]not observed.

Table 15.25. Site 15, Agricultural Show Grounds, Pangola *Digitaria eriantha*. Accessions' membership of vigour groups (Table 15.9) determined from observational data: A best, to D poor. Darker shading has been used to identify the better results.

		Observations													
		Low fertilizer							High fertilizer						
		2	4	6	8	10	12	13	2	4	6	8	10	12	13
Legumes	Accession no.	cool	dry	wet	late wet	cool	dry	wet	cool	dry	wet	late wet	cool	dry	wet
Calopogonium caeruleum[a]	CF 18	D	D	D	D	D	D	D	D	D	D	–	–	–	–
Centrosema plumieri[a]	CF 16-1	D	–	–	–	–	–	–	D	D	D	D	D	D	D
Centrosema pubescens[a]	CF 6-1														
Centrosema pubescens 'Centro'[a]	CF 6-2	D	D	–	–	–	D	–	D	D	D	D	D	D	D
Codariocalyx gyroides	CF 29	D	D	D	D	C	B	B	D	C	C	B	**A**	**A**	B
Desmodium intortum 'Greenleaf'[a]	CF 50	–	–	–	–	–	–	–	D	D	D	–	–	–	–
Leucaena leucocephala cv. Peru	CF 179	D	D	D	D	C	B	B	D	D	C	D	C	B	B
M. atropurpureum cv. Siratro	CF 3-1	D	C	D	D	C	D	–	D	C	D	B	D	C	D
Pueraria phaseoloides 'Puero'[a]	CF 180	D	D	D	D	–	–	–	D	–	D	D	–	–	–
Stylosanthes guianensis cv. Cook[a]	CF 4-4	D	D	D	D	D	D	D	[b]	D	D	D	D	D	D
Stylosanthes guianensis cv. Endeavour[a]	CF 4-5														
Stylosanthes hamata cv. Verano	CF 186	D	D	C	B	C	–	–	D	C	C	B	C	D	D

[a]In legume mixture plot; [b]not observed.

Table 15.26. Site 15, Agricultural Show Grounds, Jaragua grass *Hyparrhenia rufa*. Accessions' membership of vigour groups (Table 15.9) determined from observational data: A best, to D poor. Darker shading has been used to identify the better results.

		Observations													
		Low fertilizer							High fertilizer						
		2	4	6	8	10	12	13	2	4	6	8	10	12	13
Legumes	Accession no.	cool	dry	wet	late wet	cool	dry	wet	cool	dry	wet	late wet	cool	dry	wet
Calopogonium caeruleum[a]	CF 18	D	–	D	–	D	D	D	–	–	–	–	D	–	–
Centrosema plumieri[a]	CF 16-1														
Centrosema pubescens[a]	CF 6-1	D	D	D	D	C	D	D	–	–	–	–	–	–	–
Centrosema pubescens 'Centro'[a]	CF 6-2	–	–	–	–	–	D	D	D	–	D	D	D	D	D
Codariocalyx gyroides	CF 29	D	D	D	C	C	B	B	D	C	D	D	D	B	D
Desmodium intortum 'Greenleaf'[a]	CF 50	D	D	**A**	C	D	D	D	–	–	–	–	–	–	–
Leucaena leucocephala cv. Peru	CF 179	D	D	D	D	D	D	D	D	D	D	D	D	D	D
M. atropurpureum cv. Siratro	CF 3-1	**A**	D	**A**	B	C	B	C	D	D	**A**	B	D	C	D
Pueraria phaseoloides 'Puero'[a]	CF 180	–	–	–	–	–	–	–	D	–	D	D	D	D	D
Stylosanthes guianensis cv. Cook[a]	CF 4-4	–	–	–	–	–	–	–	D	D	D	D	–	D	D
Stylosanthes guianensis cv. Endeavour[a]	CF 4-5	–	–	–	–	–	–	–	[b]	D	D	D	D	D	D
Stylosanthes hamata cv. Verano	CF 186	D	D	C	C	D	D	–	D	D	**A**	B	C	–	–

[a] In legume mixture plot; [b] not observed.

Table 15.27. Site 15, Agricultural Show Grounds, Para (Carib) *Brachiaria mutica*. Accessions' membership of vigour groups (Table 15.9) determined from observational data: A best, to D poor. Darker shading has been used to identify the better results.

		Observations													
		Low fertilizer							High fertilizer						
		2	4	6	8	10	12	13	2	4	6	8	10	12	13
Legumes	Accession no.	cool	dry	wet	late wet	cool	dry	wet	cool	dry	wet	late wet	cool	dry	wet
Calopogonium caeruleum[a]	CF 18	D	D	D	D	D	D	D	D	C	D	D	D	C	D
Centrosema plumieri[a]	CF 16-1														
Centrosema pubescens[a]	CF 6-1	D	D	D	D	C	D	D	D	D	D	D	D	C	D
Centrosema pubescens 'Centro'[a]	CF 6-2	D	D	C	D	D	D	D	D	D	D	D	D	D	D
Codariocalyx gyroides	CF 29	D	D	**A**	B	C	B	B	D	D	D	B	C	**A**	B
Desmodium intortum 'Greenleaf'[a]	CF 50	**A**	**A**	**A**	B	B	B	D	**A**	C	**A**	B	C	D	D
Leucaena leucocephala cv. Peru	CF 179	D	D	D	D	D	D	D	D	D	D	D	D	D	C
M. atropurpureum cv. Siratro	CF 3-1	**A**	**A**	**A**	B	B	B	C	**A**	**A**	**A**	B	C	C	C
Pueraria phaseoloides 'Puero'[a]	CF 180	D	D	D	D	D	D	D	D	C	D	D	D	D	D
Stylosanthes guianensis cv. Cook[a]	CF 4-4	D	D	–	D	–	–	–	–	–	D	D	–	–	D
Stylosanthes guianensis cv. Endeavour[a]	CF 4-5	[b]	–	D	–	–	–	–	[b]	D	–	–	D	D	–
Stylosanthes hamata cv. Verano	CF 186	D	D	C	D	D	D	D	D	D	**A**	**A**	D	–	–

[a]In legume mixture plot; [b]not observed.

Table 15.28. Site 15, Summary of Agricultural Show Grounds results: most persistent and productive legumes of those tested assessed at end of trial.

Legumes	Accession no.	Guinea, *P. maximum*	Pangola, *D. eriantha*	cv. Hamil, *P. maximum*	cv. Nunbank, *C. ciliaris*	Jaragua, *H. rufa*	Para, *B. mutica*
Calopogonium caeruleum[a]	CF 18						
Centrosema plumieri[a]	CF 16-1						
Centrosema pubescens[a]	CF 6-1						
Centrosema pubescens 'Centro'[a]	CF 6-2						
Codariocalyx gyroides	CF 29	best	best	best	best	best	best
Desmodium intortum 'Greenleaf'[a]	CF 50					moderate	best
Leucaena leucocephala cv. Peru	CF 179		best	best	best		
M. atropurpureum cv. Siratro	CF 3-1		moderate	best	best	best	best
Pueraria phaseoloides 'Puero'[a]	CF 180						
Stylosanthes guianensis cv. Cook[a]	CF 4-4						
S. guianensis cv. Endeavour[a]	CF 4-5						
Stylosanthes hamata cv. Verano	CF 186		moderate		moderate	moderate	moderate

best moderate

deeper. This was particularly noticeable in a section near a cay where the A horizon was 10–20 cm thick and was associated with fairly dense grass stands; within the cay there was 20–60 cm of A horizon above the white clay, and a dense thicket of palmetto, cut grass, crabboe (*Brysonima crassifolia*) and other plants was supported.

The trial site situated on the cay itself took up roughly half of its 0.25 ha area. The trial portion was cleared of woody vegetation and the strips were laid out and planted with the legume treatments. Once the rainy season began, it became clear that there was a slight gilgai effect of 5–10 cm at the surface of the cay that was deep enough to trap water and to cause flow in definite channels. After the legumes were planted, regrowth of the natural bush vegetation was very rapid and dense. The project had insufficient time and labour to deal with this additional constraint, so the trial at the site was abandoned, though general observations were occasionally made.

15.4.3 Results and discussion Low Pine Ridge sites

The rainfall data (Fig. 15.9) indicate that the lower Belize Valley weather stations at the airport and Beaver Dam (adjacent to the mile 33 site)

Table 15.29. Summary of the final observations at upper Belize Valley sites. Accessions' membership of vigour groups (Table 15.9) determined from observational data: A best, to D poor. Darker shading has been used to identify the better results.

Legumes	Accession no.	Sites and fertilizer levels: final observations												
		1 Norland		2 Norland		3 San Lorenzo		4 Orange Walk		10 Spanish Lookout		14 Central Farm		15 Show Grounds
		low	high	low	high	low	high	low	high	low	high	low	high	
Calopogonium caeruleum	CF 18	C	C	D	D	**A**	**A**	D	D	C	**A**	D	D	
Centrosema plumieri	CF 16-1	D	D	–	–	C	D	D	D	C	**A**	–	–	
Centrosema pubescens	CF 177[a]					–	–	D	D					
Centrosema pubescens cv. Belalto	CF 195[a]	D	D	–	–					**A**	**A**	D	D	
Centrosema pubescens	CF 6-1	–	–	D	D	D	D	D	D	**A**	**A**	D	D	
Centrosema pubescens 'Centro'	CF 6-2	–	D	D	–	D	–	D	C	B	**A**	D	D	
Codariocalyx gyroides	CF 29	D	–	D	–	**A**	**A**	D	C	**A**	**A**	B	D	B
Crotalaria retusa	CF 20[b]	–	–			D	–	–	–	–	–	–	–	
Crotalaria vitellina	CF 15[b]	–	–	–	–	D	–	–	–	–	–	–	–	
Desmodium canum	CF 1[b]	D	D	D	D	D	D	D	D	D	D	D	D	
Desmodium intortum 'Greenleaf'	CF 50	–	–	D	D	D	D	D	D	C	D	D	–	B
Indigofera mucronata	CF 9[b]	–	–	–	–	–	–	–	–	–	D	–	–	
Leucaena leucocephala cv. Peru	CF 179	B	D	D	D	**A**	**A**	D	C	B	D	B	**A**	
M. atropurpureum cv. Siratro	CF 3-1	D	D	D	–	D	D	C	B	C	**A**	–	–	B
Macroptilium lathyroides	CF 14	–	–	–	–	–	D	D	D	D	D	–	–	
	CF 14[b]	–	–	D	–	–	D	D	–	–	–	–	D	
Pueraria phaseoloides 'Puero'	CF 180	D	D	D	–	C	D	C	D	B	**A**	D	–	
	CF 180[b]	–	–	D	D	D	D	–	D	–	D	–	–	
Rhynchosia minima	CF 13[b]	–	D	D	D	D	D	D	D	–	D	–	–	
Stylosanthes guianensis cv. Oxley	CF 201[b]	–	–	–	–	–	–	D	–	–	–	–	–	
Stylosanthes guianensis	CF 4-1	D	D	–	D	C	D	D	D	C	D	–	–	
Stylosanthes guianensis cv. Cook	CF 4-4	–	–	D	–	**A**	D	B	C	C	D	–	–	
S. guianensis cv. Endeavour	CF 4-5	D	D	D	C	D	D	B	B	C	D	–	–	
Stylosanthes hamata cv. Verano	CF 186	D	D	B	D	–	–	D	D	D	D	–	–	C
Teramnus uncinatus	CF 2[b]	–	D	D	D	D	–	D	D	D	D	–	D	
Vigna vexillata	CF 43-2	–	–	–	–	–	–	–	–	–	–	D	D	–

[a]Either CF 177 or CF 195 were planted at a given site. Neither was planted at the Show Grounds; [b]in legume mixture plot.

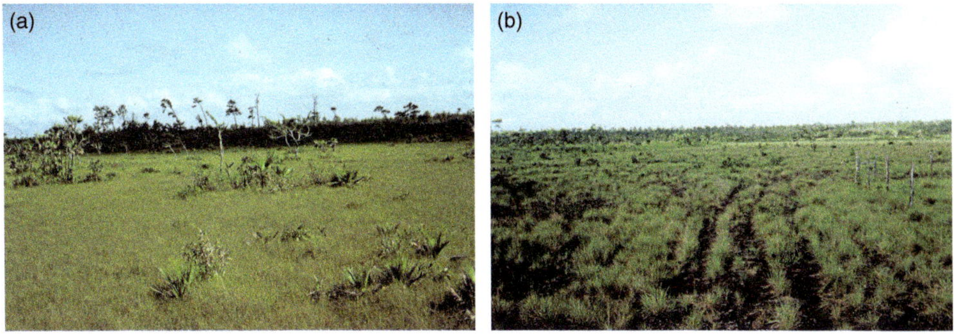

Fig. 15.8. Two views of the Lowland Pine Ridge *Mesosetum angustifolium* savanna.

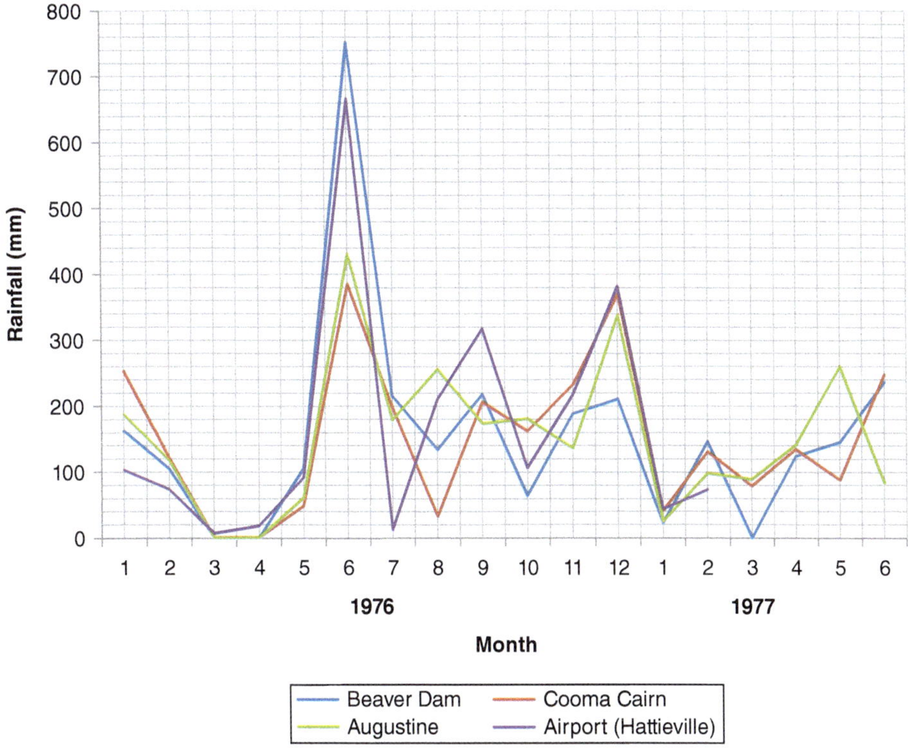

Fig. 15.9. Rainfall for the Lower Belize Valley and Mountain Pine Ridge.

had higher rainfall at the beginning of the rainy season in June than any of the sites to the west. The rainfall patterns for the remainder of the year were similar to the other sites. Rainfall data from Mountain Pine Ridge were available from Cooma Cairn relatively near site 14 Baldy Beacon, and Augustine which is near sites 12 Bull Run and 13 Rio On; they had rainfall amounts and patterns similar to that of Central Farm (Fig. 15.1).

The analysis results are presented for each site separately (Tables 15.30–15.37) and in

Table 15.30. Site 5, mile 33, Hattieville, level. Accessions' membership of vigour groups (Table 15.9) determined from observational data: A best, to D poor. Darker shading has been used to identify the better results.

		Observations													
		Low fertilizer							High fertilizer						
		1	3	5	7	10	12	13	1	3	5	7	10	12	13
Legumes	Accession no.	cool	dry	wet	late wet	cool	dry	wet	cool	dry	wet	late wet	cool	dry	wet
Calopogonium caeruleum	CF 18	B	B	C	B	C	C	**A**	D	D	**A**	**A**	**A**	**A**	**A**
Centrosema plumieri	CF 16-1	D	B	C	B	C	C	C	**A**	D	**A**	**A**	**A**	C	**A**
Centrosema pubescens	CF 177	D	D	D	D	C	C	D	D	D	C	**A**	**A**	**A**	**A**
Centrosema pubescens	CF 6-1	B	B	**A**	B	C	**A**	**A**	**A**	**A**	**A**	**A**	**A**	**A**	**A**
Centrosema pubescens 'Centro'	CF 6-2	B	B	**A**	B	C	**A**	**A**	**A**	**A**	**A**	**A**	**A**	**A**	**A**
Codariocalyx gyroides	CF 29	B	B	**A**	**A**	**A**	**A**	**A**	**A**	**A**	**A**	**A**	**A**	**A**	**A**
Crotalaria retusa	CF 20[a]	C	D	D	C	D	C	D	D	D	D	D	D	D	D
Crotalaria vitellina	CF 15[a]	D	D	D	D	D	–	–	D	D	D	D	D	D	D
Desmodium canum	CF 1[a]	D	D	D	D	D	C	D	D	D	D	C	D	D	D
Desmodium intortum 'Greenleaf'	CF 50	B	B	**A**	C	**A**	**A**	**A**	B	**A**	**A**	**A**	**A**	**A**	**A**
Indigofera mucronata	CF 9[a]	–	–	–	–	–	–	–	D	D	D	D	–	–	–
Leucaena leucocephala cv. Peru	CF 179	B	B	D	D	**A**	**A**	**A**	D	D	**A**	**A**	**A**	**A**	**A**
M. atropurpureum cv. Siratro	CF 3-1	B	B	**A**	B	C	D	C	**A**	**A**	**A**	**A**	**A**	**A**	C
Macroptilium lathyroides	CF 14	C	C	**A**	C	D	D	D	B	**A**	B	B	B	B	C
	CF 14[a]	C	D	**A**	C	–	–	–	**A**	D	B	B	D	D	D
Pueraria phaseoloides 'Puero'	CF 180	B	B	B	B	**A**	**A**	**A**	**A**	**A**	**A**	**A**	**A**	**A**	**A**
	CF 180[a]	D	D	C	D	D	C	D	D	D	**A**	C	D	D	–
Rhynchosia minima	CF 13[a]	D	–	D	C	–	C	D	–	D	D	–	D	D	D
Stylosanthes guianensis cv. Oxley	CF 201[a]	D	D	D	D	D	C	D	D	D	C	C	–	–	–
Stylosanthes guianensis	CF 4-1	D	D	D	B	**A**	**A**	**A**	**A**	D	**A**	**A**	**A**	**A**	**A**
Stylosanthes guianensis cv. Cook	CF 4-4	B	B	**A**	**A**	**A**	**A**	**A**	**A**	**A**	**A**	**A**	**A**	B	**A**
Stylosanthes guianensis cv. Endeavour	CF 4-5	B	B	**A**	**A**	**A**	**A**	**A**	**A**	**A**	**A**	**A**	**A**	**A**	**A**
Stylosanthes hamata cv. Verano	CF 186	B	B	**A**	**A**	**A**	D	D	**A**	**A**	**A**	B	D	–	B
Teramnus uncinatus	CF 2[a]	D	D	D	D	D	–	D	D	D	**A**	**A**	D	D	D
Vigna vexillata	CF 43-2[a]	–	D	D	D	C	C	D	D	D	D	D	D	C	–

[a]In legume mixture plot.

Table 15.31. Observation numbers after which management was imposed on Site 5.

Grass	Grass cut	Grass and legume cut	Grazed
Native	2	–	–

summary in Table 15.38. The accessions that were sown together in the seed mixture strips generally were given low Vigour Group ratings. The seed density may not have been high enough to raise the plot cover sufficiently to create significant plot cover. The higher fertilizer rates produced very vigorous growth in the adapted legumes, with the low fertilizer treatment plants having only a fraction of the bulk of the high rate. Higher fertilizer rates appeared to have most effect on the *Calopogonium* and *Centrosema* lines. The plants most consistently successful across the sites and fertilizer levels were the *S. guianensis* accessions and *C. gyroides* (Fig. 15.10). However, *C. caeruleum*, *C. pubescens* and *D. intortum*, under the high fertilizer rate, generally persisted and performed very well. By the end of the trial the woody legumes *C. gyroides* and *Leucaena* achieved heights of 2.5 m and at times more than 3 m in both fertilizer treatments, *C. gyroides* had luxuriant growth under the high fertilizer treatment, but otherwise it and *Leucaena* were tall but not very leafy. Harrowing of the soils did not appear to give any advantage to establishment and persistence of the legumes. At the Hattieville, mile 33 sites, the sloping site 6 appeared from the ratings to provide slightly better results than the more level site 5. The legume growth on the abandoned cay site 9 at mile 35 was good when last viewed in November 1976, but luxuriant natural shrubby vegetation was in the process of crowding it out.

However, the system of analysis used showed no differences between the fertilizer treatments because the more vigorous legumes were all in the highest Vigour Grouping. The detailed observations indicated considerably better growth with the higher fertilizer levels. For example, at the final observation *S. guianensis* cv. Endeavour had average spreads of 83–202 cm and heights of 52–78 cm under low fertilizer and spreads of more than 2 m and heights of 51–74 cm under high fertilizer treatment.

The results indicate that a low fertilizer application will stimulate acceptable levels of growth of those legumes that are well adapted to the edaphic conditions. The plants that were adapted grew very well because they were not limited regarding soil moisture, and there was no competition from the native savanna grass and weed species. It should be noted, however, that in contrast with the management of the upper Belize River Valley sites, grazing and cutting management were not imposed. Informal observations in later years showed invasion by other grasses and weeds, and vigorous growth of savanna legumes.

15.5 Mountain Pine Ridge sites

15.5.1 Introduction

Three trial sites (11, 12 and 13) were located on the Mountain Pine Ridge, an old peneplain (Fig. 15.11) rising to the south of and adjacent to the Belize River Valley with an area of about 800 km^2 and elevations of 400 m asl in the north to over 1000 m asl in the south.

History

Mountain Pine Ridge lies next to the relatively well-developed Belize River Valley and is accessible by a road built to aid forestry. Because of its cooler temperatures and savanna woodland vegetation dominated by *Pinus caribea* it has always attracted interest as an area with potential for development for cattle raising.

First opened up by a road in the 1920s in order to provide access for logging to the south of the Ridge, it served as a grazing area for draft animals used in the logging industry, and for stock from the Belize River Valley during the dry season and during droughts. It was declared a forest reserve by the Forest Department in 1944, and in 1949 after fire swept the whole ridge a network of roads was constructed with the aim of improving management of fires. In the same year the Commonwealth Development Corporation began a cattle and sheep operation with an aim to stock at one-quarter beast/ha, but it soon failed in part due to overcapitalization and low productivity. During the 1960s a private operator ran cattle in the area, but poor calving, and mustering and rustling problems, caused it to close down.

Table 15.32. Site 6, mile 33, Hattieville, sloping. Accessions' membership of vigour groups (Table 15.9) determined from observational data: A best, to D poor. Darker shading has been used to identify the better results.

		Observations													
		Low fertilizer							High fertilizer						
		1	3	5	7	10	12	13	1	3	5	7	10	12	13
Legumes	Accession no.	cool	dry	wet	late wet	cool	dry	wet	cool	dry	wet	late wet	cool	dry	wet
Calopogonium caeruleum	CF 18	D	B	C	B	C	A	A	D	D	A	A	A	A	A
Centrosema plumieri	CF 16-1	D	D	C	D	C	C	B	D	D	A	D	A	B	A
Centrosema pubescens	CF 177	B	B	C	B	A	A	C	D	D	A	A	A	A	A
Centrosema pubescens	CF 6-1	B	B	C	B	C	A	C	D	D	A	A	A	A	A
Centrosema pubescens 'Centro'	CF 6-2	B	B	A	B	A	A	A	A	A	A	A	A	A	A
Codariocalyx gyroides	CF 29	B	B	A	A	A	A	A	A	A	A	A	A	A	A
Crotalaria retusa	CF 20[a]	D	D	D	C	–	–	–	D	D	C	B	–	–	–
Crotalaria vitellina	CF 15[a]	D	D	D	D	–	–	–	–	–	D	D	–	–	–
Desmodium canum	CF 1[a]	D	D	D	C	C	D	D	D	D	C	C	D	D	D
Desmodium intortum 'Greenleaf'	CF 50	B	C	A	A	A	A	A	D	A	A	A	A	A	A
Indigofera mucronata	CF 9[a]	D	D	D	D	–	–	–	–	–	–	–	–	–	–
Leucaena leucocephala cv. Peru	CF 179	D	D	D	C	C	D	D	D	D	A	A	A	A	A
M. atropurpureum cv. Siratro	CF 3-1	B	B	A	B	A	A	A	A	A	A	A	A	A	A
Macroptilium lathyroides	CF 14	C	C	A	C	A	A	C	B	A	B	B	A	A	A
	CF 14[a]	C	D	D	C	D	D	D	D	D	C	B	D	D	D
Pueraria phaseoloides 'Puero'	CF 180	B	B	A	B	A	A	A	A	A	A	A	A	A	A
	CF 180[a]	B	D	C	D	C	C	C	D	A	D	D	D	C	D
Rhynchosia minima	CF 13[a]	B	C	–	–	–	–	–	–	–	–	–	–	–	–
Stylosanthes guianensis cv. Oxley	CF 201[a]	D	D	D	D	D	D	D	D	D	C	C	–	D	–
Stylosanthes guianensis	CF 4-1	B	D	A	B	A	A	A	D	D	A	A	A	A	A
Stylosanthes guianensis cv. Cook	CF 4-4	B	B	A	A	A	A	A	A	A	A	A	A	A	A
Stylosanthes guianensis cv. Endeavour	CF 4-5	B	B	A	A	A	A	A	A	A	A	A	A	A	A
Stylosanthes hamata cv. Verano	CF 186	B	B	A	A	A	A	C	A	A	A	B	A	D	D
Teramnus uncinatus	CF 2[a]	D	D	D	C	C	D	D	D	D	A	A	–	D	D
Vigna vexillata	CF 43-2[a]	D	D	C	B	C	D	C	D	D	A	C	A	A	A

[a]In legume mixture plot.

Table 15.33. Observation numbers after which management was imposed on Site 6.

Grass	Grass cut	Grass and legume cut	Grazed
Native	2	–	7 very lightly

In 1977, few people lived in the area. A small Forestry Department station maintained the reserve and one lumber mill survived, sourcing timber from the lowlands to the south. A hotel ran for a few years, catering to overseas tourists. While no logging of timber was being done due to the low density of trees of useful size (Johnson and Chaffey, 1973), most of the more mature trees were being tapped for resin. Occasional military exercises were held on it by the British and Belgian armed forces.

Geology

The peneplain is higher in the east than the west, with all drainage systems running east to west. The area originally possessed a limestone cap over basement granite, quartzite and shales, but much of it has been eroded and in the process has contributed to the Low Pine Ridge soils of the Belize River Valley and coast. Only a few short hills at the western edge remain. Soils of the feature therefore have been derived from the limestone, itself secondarily deposited, and from the shales, quartzites and granites. As a result they are quite senescent, low in nutrients, high in iron and show features of podsolization (Hunt, 1970).

Climate

Only a few years of climatic data were available and these from only two sites, Cooma Cairn on the eastern, higher side of the area and Augustine on the lower western edge (Walker, 1973). Their annual rainfall patterns (Fig. 15.2) are similar to those in the lowlands in the Belize River Valley. The eastern site, which is more elevated, gets higher rainfall (2100 mm) than the western side (1600 mm), perhaps due to an orthographic effect as the rainy season winds are mainly easterly from the coast. Due to its altitude, the climate of the Mountain Pine Ridge is cooler than that of much of the country, which is generally low lying. Its climate fits Papadakis' (1966) climatic grouping of 2.1 Semi-tropical Terra Fria, in contrast to the 11.2 Humid Tropical climate of the adjacent lowlands. Johnson and Chaffey (1973) note that the climate is subtropical according to Holdridge's system (Holdridge, 1947; Holdridge and Hooper, 1966) because it falls just below the 24°C lower limit for tropical climax vegetation.

Soils

The soils of the area were first mapped by Charter (1941) as a separate suite, 'Croja', with good drainage and lying on low lime material. Daniel also did a survey in 1952, but his work has been unobtainable. Wright *et al.* (1959) classified the soils as red-yellow podzols, Pinol series and divided them into eight sets based on their parent material. Birchall (Johnson and Chaffey, 1973) took a fresh look at the soils and studied three traverses that were selected to include as wide a variety of vegetation as possible. The classification by Wright *et al.* (1959) was found to be essentially in agreement with the findings, except that slate-derived soils in the east of the area were not clays and clay loams but rather silty loams and clay loams.

Darcel (1952) was reported by Birchall to have noted the granite and slate-derived soils were acid (4.0 to 5.3) and the acidity increased with depth. Marked deficiencies included very low P and K, low soluble Mn and, as would be expected, high to very high soluble Al. Wright noted that the upper well-drained layers of Pinol coarse sandy clay hardened in the dry season to a depth of 15–30 cm. The lower layers were heavy and slow to drain. Field and pot trials indicated that successful grass and legume growth could be maintained if fertilizer with 2 cwt TSP, 1.5 cwt sulphate of ammonia, 1 cwt muriate of potash in split dressings was applied. Guinea, Para, Molasses and Bahia grasses were suggested along with Mediterranean clover, *Indigofera fruticosa* and *Desmodium* species.

Vegetation

The vegetation of the Mountain Pine Ridge contrasts strongly with that found at lower elevations and this has attracted a number of researchers who have described the species present (Bartlett, 1935; Lundell, 1940; Beard, 1953; Hunt, 1962; Furley, 1968). Generally pine is the

Table 15.34. Site 7, mile 35, Gulf American, undisturbed. Accessions' membership of vigour groups (Table 15.9) determined from observational data: A best, to D poor. Darker shading has been used to identify the better results.

		Observations													
		Low fertilizer							High fertilizer						
		1	3	5	7	10	12	13	1	3	5	7	10	12	13
Legumes	Accession no.	cool	dry	wet	late wet	cool	dry	wet	cool	dry	wet	late wet	cool	dry	wet
Calopogonium caeruleum	CF 18	D	D	C	C	C	C	C	D	D	C	C	A	C	A
Centrosema plumieri	CF 16-1	D	D	B	C	C	C	D	D	D	C	C	D	C	A
Centrosema pubescens	CF 177	D	D	D	D	D	D	C	D	D	A	C	A	A	A
Centrosema pubescens	CF 6-1	D	C	C	B	C	C	C	D	D	A	A	A	A	A
Centrosema pubescens 'Centro'	CF 6-2	D	D	D	C	D	C	C	D	D	C	C	B	A	A
Desmodium gyroides	CF 29	B	B	A	A	C	A	A	D	A	A	A	A	A	A
Crotalaria retusa	CF 20[a]	–	–	D	D	D	D	D	D	D	D	D	D	–	–
Crotalaria vitellina	CF 15[a]	B	D	D	D	D	D	D	B	D	C	D	D	D	D
Desmodium canum	CF 1[a]	D	D	D	C	D	D	C	D	D	C	C	D	A	B
Desmodium intortum 'Greenleaf'	CF 50	D	D	D	D	D	D	D	D	D	D	C	D	B	A
Indigofera mucronata	CF 9[a]	–	–	D	D	D	D	D	D	D	D	D	D	D	D
Leucaena leucocephala cv. Peru	CF 179	B	C	A	A	C	A	A	D	D	A	A	A	A	A
M. atropurpureum cv. Siratro	CF 3-1	B	C	A	B	A	A	A	A	A	A	A	A	A	C
Macroptilium lathyroides	CF 14	B	C	A	C	C	D	A	A	D	A	B	A	A	B
	CF 14[a]	C	D	A	C	D	D	D	B	D	B	B	A	A	D
Pueraria phaseoloides 'Puero'	CF 180	D	C	B	A	A	D	D	D	D	A	A	A	A	A
	CF 180[a]	B	D	C	D	C	D	D	D	D	A	A	A	C	D
Rhynchosia minima	CF 13[a]	D	D	D	C	D	D	D	D	D	D	C	D	D	D
Stylosanthes guianensis cv. Oxley	CF 201[a]	D	–	D	D	D	D	D	D	D	D	D	–	–	–
Stylosanthes guianensis	CF 4-1	B	D	A	B	C	A	A	D	D	A	A	A	A	A
Stylosanthes guianensis cv. Cook	CF 4-4	B	C	A	A	C	A	A	A	A	A	A	A	A	A
Stylosanthes guianensis cv. Endeavour	CF 4-5	B	B	A	A	C	A	A	A	A	A	A	A	A	A
Stylosanthes hamata cv. Verano	CF 186	B	D	A	B	A	A	A	D	D	A	B	A	A	B
Teramnus uncinatus	CF 2[a]	C	D	D	C	D	D	D	D	D	C	C	A	D	D
Vigna vexillata	CF 43-2[a]	D	D	D	–	D	D	D	D	D	–	–	–	–	–

[a]In legume mixture plot.

Table 15.37. Observation numbers after which management was imposed on Site 8.

Grass	Grass cut	Grass and legume cut	Grazed
Native	2	–	–

seven times a year, with N applied after each cut and P and K twice a year. No detailed results are available. It was observed that growth was very poor in the dry season and the leaves showed nutrient-deficiency symptoms. Plots receiving the complete fertilizer treatment gave best DM yields; the yields increased with the rate of fertilizer applied and the effect of P was greater than that of K. Based on the results, the possible stocking rates with the application of 790 kg/ha of urea and 110 kg/ha of TSP were estimated as 1.5 head/ha. The results were not analysed for statistical significance.

A further randomized split-plot trial was undertaken on Mountain Pine Ridge to study the effects of cutting frequency on eight grasses (4, 6 and 8 weeks). Fertilizer was added at 168 kg/ha urea and 229 kg/ha TSP. Rhodes, Dallis grass and Molopo Buffel did not establish well. *Pennisetum purpureum* cv. Merker gave the highest total fresh weight under 4-weekly cutting, while *Panicum maximum* var. Colonaio gave the highest at 6 weeks. *Cynodon dactylon* and *Cynodon nlemfluensis* did not do as well, but their yields were best in the 6-weekly cycle.

Snook (1968) attempted to determine the nutrient status of Mountain Pine Ridge soils by using inoculated cowpea (*Vigna sinensis*) as a test crop and applying TSP (46% P_2O_5) at 0, 56, 112 and 225 kg/ha, a micronutrient mixture at 0 and 36 kg/ha which supplied S and recommended rates of Zn, Cu, Mn, Mo and Bo. One treatment received only an NPK mixture (TSP and muriate of potash to supply nutrients at 56:112:56 kg/ha) to see if establishment would be aided by the addition of urea. It was thought that crushed lime would not be practical for pastoralists to apply, so it was not used in the trial.

A dry spell after germination inhibited growth. The NPK treatment was superior to all other treatments, which were similar. Three weeks after planting, the NPK plants had no nodules; the trace element mixture had small white nodules; and those receiving trace elements plus TSP had numerous active pink nodules. The TSP treatments were soon superior to the others, with the highest TSP treatment appearing to produce the most vigorous plants. Seeds and plants were harvested and dried on maturity. It was concluded that TSP is required for good dry matter and seed yields; though there was a marked effect at 56 kg/ha, the higher the rate (up to 200 kg/ha was applied), the better the yields. The response to added P was thought to be due to the fixation of P by the acid soil. The NPK treatment gave very good results for the first 6 weeks, but it was not clear which elements were responsible. It was noted that the mixture was superior to P alone at the same P rate.

Test plots of 13 tropical legumes were planted at the same time and place as the cowpea trial, presumably with the same fertilizer rates but no record has been found of the results. A more detailed study on the response of cowpea to fertilizers was planted the following year, but the results of this are also unavailable.

Potential for livestock raising

The areas of open savanna and of thin tree cover present an attractive situation for cattle grazing, but the low fertility status of the soils and forage suggest that weight gains would be very low, and that cobalt might be deficient enough to cause the reported calving problems. The area's steep slopes, with sparsely covered coarsely textured soils, result in erosion being common after burning in the Bald Hills, and where fire roads are constructed. A denser, more vigorous soil cover is needed to hold the soil and if a vigorous, adapted legume is identified it could do much to stabilize the soil, provide nutrients for associated vegetation including grass and trees, and provide improved quality grazing for livestock. Any livestock nutrient deficiencies could be offset by salt licks.

The vegetation of the area includes many legumes, many of them with pasture potential. Some of these have been listed in Appendix 13.1 (Chapter 13, this volume), but mention should be made of the various *S. guianensis*, *S. viscosa* and *S. ingrata* genotypes that can be locally abundant, particularly after burning. Snook mentions the profusion of native legumes appearing after fertilizers, particularly if P was applied, and he suggested that it was not necessary to look further: the indigenous legumes would be sufficient.

Table 15.38. Summary of final observation of four Low Pine Ridge trial sites: low and high fertilizer. Accessions' membership of vigour groups (Table 15.9) determined from observational data: A best, to D poor. Darker shading has been used to identify the better results.

Legumes	Accession no.	Site 5 Mile 33, level		Site 6 Mile 33, sloping		Site 7 Mile 35, undisturbed		Site 8 Mile 35, harrowed	
		low	high	low	high	low	high	low	high
Calopogonium caeruleum	CF 18	A	A	A	A	C	A	A	A
Centrosema plumieri	CF 16-1	C	A	B	A	D	A	C	D
Centrosema pubescens	CF 177[a]	D	A	C	A	C	A		
Centrosema pubescens cv. Belalto	CF 195[a]							C	A
Centrosema pubescens	CF 6-1	A	A	C	A	C	A	A	C
Centrosema pubescens 'Centro'	CF 6-2	A	A	A	A	C	A	A	A
Codariocalyx gyroides	CF 29	A	A	A	A	A	A	A	A
Crotalaria retusa	CF 20[b]	D	D	–	–	D	–	D	D
Crotalaria vitellina	CF 15[b]	–	D	–	–	D	D	–	–
Desmodium canum	CF 1[b]	D	D	D	D	C	B	C	D
Desmodium intortum 'Greenleaf'	CF 50	A	A	A	A	D	A	A	A
Indigofera mucronata	CF 9[b]	–	–	–	–	D	D	D	D
Leucaena leucocephala cv. Peru	CF 179	A	A	D	A	A	A	D	A
M. atropurpureum cv. Siratro	CF 3-1	C	C	A	A	A	C	C	A
Macroptilium lathyroides	CF 14	D	C	C	A	A	B	A	A
	CF 14[b]	–	D	D	D	D	D	D	D
Pueraria phaseoloides 'Puero'	CF 180	A	A	A	A	D	A	A	A
	CF 180[b]	D	–	C	D	D	D	C	D
Rhynchosia minima	CF 13[b]	D	D	–	–	D	D	D	D
Stylosanthes guianensis cv. Oxley	CF 201[b]	D	–	D	–	D	–	D	D
Stylosanthes guianensis	CF 4-1	A	A	A	A	A	A	A	A
Stylosanthes guianensis cv. Cook	CF 4-4	A	A	A	A	A	A	A	A
Stylosanthes guianensis cv. Endeavour	CF 4-5	A	A	A	A	A	A	A	A
Stylosanthes hamata cv. Verano	CF 186	D	B	C	D	A	B	A	A
Teramnus uncinatus	CF 2[b]	D	D	D	D	D	D	D	D
Vigna vexillata	CF 43-2[b]	D	–	C	A	D	–	C	–

[a]Only one of CF177 or CF 195 was sown at a trial site; [b]in legume mixture plot.

Fig. 15.10. There was similar growth at all four Low Pine Ridge sites. The high productivity was due to abundant soil moisture year round and the fertilizer added. (a) Site 6, mile 33, low fertilizer on left, from front *P. phaseoloides*, *S. guianensis* cv. Endeavour, *S. hamata* cv. Verano, *C. gyroides* October 1976. (b) Site 7, Low Pine Ridge, mile 35. *Leucaena* in front of *C. gyroides*. Low fertilizer rate on left half of strips, 1979.

Fig. 15.11. Mountain Pine Ridge topography and vegetation. (a) The Bald Hills from Baldy Beacon (1000 m asl), the easterly portion with few trees. (b) Further west, scattered trees east of Cooma Cairn. (c) Seen from the D'Silva (Augustine) Forestry Station there are denser stands of trees at the western part of the plateau.

To test recent accessions identified as having potential for pastures in Belizean environments for persistence and productivity on Mountain Pine Ridge, three strip trials were established at Bull Run Ranch (site 12), Rio On Line (site 13) and Baldy Beacon (site 14). The reactions of the soils at the trial sites tested were strongly acid (pH 5.0–5.3).

15.5.2 Trial sites on the Mountain Pine Ridge

Site 11 Bull Run Ranch (Fig. 15.13)

The site was on the property of the only ranch on the Mountain Pine Ridge so that any successful growth would appear applicable to a pasture

Fig. 15.12. Mountain Pine Ridge, fires and burned vegetation between Bradley line 1971 and Cooma Cairn.

Fig. 15.13. Site 11 Mountain Pine Ridge, Bull Run Ranch. (a) Typical vegetation prior to planting trial. (b) *C. gyroides* and *P. phaseoloides* 1981.

situation. It was situated at 560 m asl on a soil that was mapped by Wright *et al.* (1959) as Pinol Clay, but that had a fine loamy A horizon. Notable amounts of iron and manganese concretions were present. This soil is found on about 1.7% of the ridge. McKinzie (1977) classified it as a Plinthic Tropudult, clayey, kaolinitic isohyperthermic.

The site was chosen more for the ranch situation than for the type of soil. The ranch was the original site of the Commonwealth Development Bank's property on the Mountain Pine Ridge. Established in the 1950s, it was, among other projects, attempting to raise cattle and sheep on the ridge. The project failed and a private operator took over the property on the north edge of the ridge and proceeded to run cattle there and through the forestry reserve. The operation was closing down at the time the trial was established. It was felt that having the trial on the property within a few hundred metres of the ranch buildings would encourage an interest in the use of legumes on the ranch and perhaps revive interest in grazing in the area once again.

The site had been grazed closely for some years, particularly because it was close to the ranch buildings where salt and other supplements were fed to the cattle. Some angling of the trial strips was required to avoid trees because the site held an evenly distributed stand of mature *Pinus caribea* trees, which had a fairly complete canopy cover. Undergrowth in the form of oak or young pine was removed when they were within the planted strips, but otherwise was left.

Site 12 Rio On Ridge (Fig. 15.14)

A slightly sloping situation near the top of a ridge at 560 m asl was chosen for this site. It had scattered young *Pinus caribea* trees and a dense stand of tall *Mesosetum angustifolium*. Wright *et al.* (1959) identified the soil as Pinol Clay with 10 cm of finely granular loam on coarse sandy clay loam. The situation suggests that the top 10 cm was transported from further up the slope and deposited. McKinzie (1977) described the soil as an Orthic Tropudult, fine loamy, mixed, hyperthermic.

The site had been grazed by animals from the extensive cattle operation less than a year before the trial was established, but the grazing must have been light because there was an unusually tall stand (60 cm) of *Mesosetum angustifolium* grass and relatively few dung pats were present; the pats remained for some period during the trial.

Site 13 Baldy Beacon (Fig. 15.15)

Located on the extensive hilly area of shallow treeless, grass-covered soils characterized by steep slopes at the north-east edge of the Mountain Pine Ridge, the site has shallow soils over metamorphosed rocks. An atypically level situation for the trial was chosen on a short spur ridge running roughly east–west immediately below and to the south of Baldy Beacon. Level ground was chosen for ease of access, to avoid causing soil erosion and to reduce transport of the planted seed by water. Wright *et al.* (1959) termed the soil Pinol loamy sand, while McKinzie (1977) described it as a Lithic Dystropet, loamy, mixed isohyperthermic. At 920 m, on the most elevated easterly portion of the peneplain, the site is subjected to strong cool winds and frequent showers from the moisture-bearing easterly winds. Management of the site involved removal of small trees from within the planted strips, otherwise the vegetation was left intact.

Fig. 15.14. Site 12 Rio On. (a) *Desmodium intortum* 'greenleaf' (left in photo) and *C. gyroides*, 1977. (b) *C. gyroides* 1978, (c) *P. phaseoloides* 1978. Both browsed by wildlife. Significantly better growth of trees within trial compared with outside (see also Figure in title pages, this volume).

Fig. 15.15. Site 13 Baldy Beacon, Mountain Pine Ridge. (a) *Macroptilium atropurpureum* cv. Siratro 1977. (b) Substantial biomass produced by *C. gyroides*. Browsed by wildlife 1978. (c) Significantly better pine growth within trial than outside, 1981.

15.5.3 Results and discussion Mountain Pine Ridge sites

The three Mountain Pine Ridge sites were not subjected to cutting or grazing, so the results of the analyses reflect the adaptation and productivity of the legumes to the environment under the fertility conditions provided (Tables 15.39–15.44 and summarized in Table 15.45). Initially the plants were stunted and yellow, indicating that the standard trial fertility conditions were not adequate. With the application of marl in late April 1976, after the third observation the deficiency symptoms at the Bull Run and Rio On sites generally disappeared and the plants grew vigorously (Figs 15.13 and 15.14). However, deficiency symptoms lingered on some legumes to the end of the trial at the Baldy Beacon site.

The Rio On site 12 ratings were slightly higher than those at site 11 Bull Run Ranch, which had the lowest ratings; this may be due, at least in part, to grazing of the plots by unknown herbivores, perhaps deer, from late 1976. The best adapted legumes, those that performed well under both low and high fertilizer regimes at all sites, were *C. gyroides*, *D. intortum*, *P. phaseoloides* and *S. guianensis* cv. Endeavour. However, the other *S. guianensis* lines in full plots performed well at most sites and fertilizer levels, as did *C. pubescens*. *M. atropurpureum* performed best under the higher fertilizer regimes. Productivity under the high fertilizer rate was consistently much higher than that under the low fertilizer rate, but over time with continued small applications of fertilizer the stands of plants under the low rate greatly improved. At Rio On and Baldy Beacon, the sites with younger trees, growth of the pines was noted to have been very significantly improved as well (Fig. 15.15). Fertility thus appears to be the main constraint to adequate fodder growth on the Mountain Pine Ridge soils tested. The application of marl appears to be necessary on the Puletan soils tested, however it can be readily sourced along the access road to the plateau.

The performance data figures for the productive legumes were somewhat lower than

Table 15.39. Site 11, Bull Run Ranch. Accessions' membership of vigour groups (Table 15.9) determined from observational data: A best, to D poor. Darker shading has been used to identify the better results.

		Observations													
		Low fertilizer							High fertilizer						
		1	3	5	7	10	12	13	1	3	5	7	10	12	13
Legumes	Accession no.	cool	dry	wet	late wet	cool	dry	wet	cool	dry	wet	late wet	cool	dry	wet
Calopogonium caeruleum	CF 18	D	D	D	C	**A**	C	C	D	D	C	D	**A**	**A**	C
Centrosema plumieri	CF 16-1	D	D	D	D	C	C	C	–	–	D	D	D	D	D
Centrosema pubescens cv. Belalto	CF 195	D	C	D	C	C	C	C	D	D	C	C	**A**	**A**	**A**
Centrosema pubescens	CF 6-1	D	C	D	D	D	D	D	D	D	C	**A**	**A**	D	C
Centrosema pubescens 'Centro'	CF 6-2	B	C	D	**A**	**A**	**A**	C	D	D	C	C	**A**	D	C
Codariocalyx gyroides	CF 29	D	D	D	**A**	**A**	**A**	**A**	B	D	**A**	**A**	**A**	**A**	**A**
Crotalaria retusa	CF 20[a]	–	–	–	D	D	D	D	–	–	D	D	D	D	D
Crotalaria vitellina	CF 15[a]	–	–	D	D	–	D	–	–	D	D	D	–	–	–
Desmodium canum	CF 1[a]	D	D	D	D	D	D	D	D	–	–	–	–	–	–
Desmodium intortum 'Greenleaf'	CF 50	B	D	**A**	**A**	**A**	**A**	**A**	D	D	C	**A**	**A**	**A**	**A**
Indigofera mucronata	CF 9[a]	–	–	–	D	D	D	D	–	–	–	–	–	–	–
Leucaena leucocephala cv. Peru	CF 179	D	D	D	C	D	D	D	D	D	C	C	D	D	B
M. atropurpureum cv. Siratro	CF 3-1	B	D	D	C	C	C	**A**	D	D	C	C	**A**	**A**	**A**
Macroptilium lathyroides	CF 14	B	D	**A**	C	**A**	**A**	D	D	D	C	B	B	**A**	B
	CF 14[a]	D	D	D	C	D	D	D	D	D	C	B	D	D	D
Pueraria phaseoloides 'Puero'	CF 180	B	D	**A**	B	**A**	**A**	**A**	D	D	D	C	**A**	C	**A**
	CF 180[a]	–	D	D	D	D	C	C	D	D	D	D	D	D	D
Rhynchosia minima	CF 13[a]	–	–	D	–	–	–	–	–	–	–	–	–	–	–
Stylosanthes guianensis cv. Oxley	CF 201[a]	D	D	D	D	D	D	D	–	–	D	–	–	–	–
Stylosanthes guianensis	CF 4-1	D	D	D	C	**A**	**A**	C	D	D	C	C	**A**	D	**A**
Stylosanthes guianensis cv. Cook	CF 4-4	B	C	**A**	**A**	**A**	**A**	D	D	D	C	**A**	**A**	**A**	**A**
Stylosanthes guianensis cv. Endeavour	CF 4-5	B	C	**A**	**A**	**A**	**A**	**A**	D	D	**A**	**A**	**A**	**A**	**A**
Stylosanthes hamata cv. Verano	CF 186	B	D	**A**	**A**	**A**	**A**	D	D	D	C	**A**	**A**	**A**	B
Teramnus uncinatus	CF 2[a]	D	D	D	D	D	D	D	D	D	D	–	–	–	–
Vigna vexillata	CF 43-2[a]	–	–	–	–	–	–	–	–	–	–	–	–	–	–

[a]In legume mixture plot.

Table 15.40. Observation numbers after which management was imposed on Site 11.

Grass	Grass cut	Grass and legume cut	Grazed
Native	2	–	–

those recorded for the Low Pine Ridge trial sites and data for *S. guianensis* cv. Endeavour are provided as an example (Table 15.46). The accumulation of grass dry matter at the Rio On site provides evidence of the accumulation of dry matter that occurs below the *P. caribea* upper storey vegetation under natural conditions and provides fuel for fires caused by lightning or humans. The introduction of forage legumes in fodder banks could do much to improve the productivity of livestock and the economic viability of ranching on these soils. Grazing by reducing accumulation of dry matter would reduce the incidence and severity of fires and damage to Mountain Pine Ridge tree stock. Evidence of browsing appeared in 1978 after the trials were abandoned, suggesting that the legumes were attracting wildlife (Figs 15.14 and 15.15).

15.5.4 Persistence of the plantings: observations in 1981

Informal observations were made on a number of the IDRC-UWI/Belize strip trial sites in 1981 to assess the persistence of the legume planting 4 years after most of the trials had been abandoned. The results, along with those on three other trials not reported in this volume, are presented in Table 15.47 and Figs 15.13 and 15.15. It is notable that several legumes persisted well at almost all sites, and at those sites where they failed (sites 10 and 12) it was apparently due, at least in part, to unmanaged grazing. At protected sites, where moisture was not limiting and grazing pressure was low or absent, such as on the Low Pine Ridge, many species had self-sown and spread within the fertilized trial area. However, in all cases in the Low Pine Ridge and Mountain Pine Ridge, nutrient deficiencies apparently prevented any significant spread beyond the fertilized areas. A summary of all the most promising species at all sites and trials in the project is presented in Chapter 19 (this volume).

15.6 General discussion

In the upper Belize River Valley under cutting and grazing, legume performance was generally weak; grazing pressures and competition from established grasses were the likely major causes. *Leucaena* and *C. gyroides* had the best performance, followed by *C. caeruleum*, and the *Centrosema* species. In the Low Pine Ridge soils the plants most consistent across the sites and fertilizer levels were the *S. guianensis* accessions and *C. gyroides*. However, *C. caeruleum*, *C pubescens*, *D. intortum* and *Leucaena* under the high fertilizer rate generally persisted and performed very well because water availability was not a limiting factor. On the Mountain Pine Ridge soils, performance was poor and nutrient-deficiency symptoms were present without the application of marl. With adequate fertility, legume growth was good on all sites and the most productive legumes were *C. gyroides*, *D. intortum*, *P. phaseoloides* and *S. guianensis* cv. Endeavour. The other *S. guianensis* lines in full plots performed well at most sites and fertilizer levels, as did *C. pubescens*. *M. atropurpureum* performed best under the higher fertilizer regimes.

This analysis reports the data for one and a half calendar years of growth; it involved only one complete growing season, two dry seasons and the beginning of a second growing season. The legumes were not given the opportunity to fully express their potential at the end of the second dry season. As perennial species cannot be considered properly tested after only one growing season, the results reported here are only very preliminary, particularly in regard to the upper Belize Valley sites. Informal observations were taken of the abandoned trials in 1981; these are summarized in Appendix 15.2 below and in Chapter 19 (this volume).

The Low Pine Ridge and Mountain Pine Ridge trials were only able to test legume adaptation to the environments and to gain an assessment of their relative potential. Given adequate fertility, legumes were shown to be productive in these environments, thus consideration should be given to establishing livestock operations utilizing the native grasses with productive legumes grazed in carefully managed fodder banks. Care would have to be taken that on both of these infertile soil types that micronutrient needs of the livestock are met. The lowland Puletan soils

Table 15.41. Site 12, Rio On, native pasture. Accessions' membership of vigour groups (Table 15.9) determined from observational data: A best, to D poor. Darker shading has been used to identify the better results.

		Observations													
		Low fertilizer							High fertilizer						
		1	3	5	7	10	12	13	1	3	5	7	10	12	13
Legumes	Accession no.	cool	dry	wet	late wet	cool	dry	wet	cool	dry	wet	late wet	cool	dry	wet
Calopogonium caeruleum	CF 18	D	D	D	C	C	C	C	A	D	C	C	A	A	A
Centrosema plumieri	CF 16-1	D	D	D	D	D	D	C	D	D	C	C	A	C	A
Centrosema pubescens cv. Belalto	CF 195	D	D	A	B	A	A	A	D	D	A	A	A	A	A
Centrosema pubescens	CF 6-1	B	D	A	B	C	A	A	D	D	C	A	A	A	A
Centrosema pubescens 'Centro'	CF 6-2	D	D	D	C	D	A	D	D	D	C	A	D	D	D
Codariocalyx gyroides	CF 29	B	D	D	A	A	D	A	D	D	C	A	A	A	A
Crotalaria retusa	CF 20[a]	–	–	–	–	–	–	–	–	–	–	–	–	–	–
Crotalaria vitellina	CF 15[a]	D	D	D	D	C	A	D	D	D	D	D	D	D	B
Desmodium canum	CF 1[a]	D	D	D	C	C	A	D	D	D	C	C	D	D	D
Desmodium intortum 'Greenleaf'	CF 50	B	D	A	A	A	A	A	D	D	C	A	A	A	A
Indigofera mucronata	CF 9[a]	–	–	–	D	D	D	D	–	–	–	A	D	D	D
Leucaena leucocephala cv. Peru	CF 179	D	D	D	C	C	D	D	D	D	C	C	A	A	B
M. atropurpureum cv. Siratro	CF 3-1	B	D	A	B	A	A	C	D	D	A	A	A	A	A
Macroptilium lathyroides	CF 14	B	D		C	A	A	A	D	D	A	A	A	A	A
	CF 14[a]	D	D	D	C	A	D	D	D	D	C	C	A	D	A
Pueraria phaseoloides 'Puero'	CF 180	B	D	A	B	A	A	A	A	D	C	A	A	A	A
	CF 180[a]	D	D	D	C	C	D	C	D	–	D	D	D	C	D
Rhynchosia minima	CF 13[a]	D	–	–	D	D	D	D	–	–	D	D	D	–	–
Stylosanthes guianensis cv. Oxley	CF 201[a]	D	D	D	D	D	D	D	–	–	D	D	D	–	–
Stylosanthes guianensis	CF 4-1	B	D	A	A	A	A	A	D	D	C	A	A	A	A
Stylosanthes guianensis cv. Cook	CF 4-4	B	D	A	A	A	A	A	A	D	A	A	A	A	A
Stylosanthes guianensis cv. Endeavour	CF 4-5	B	D	A	A	A	A	A	A	D	A	A	A	A	A
Stylosanthes hamata cv. Verano	CF 186	B	D	A	A	A	A	A	A	D	A	B	A	A	B
Teramnus uncinatus	CF 2[a]	D	D	D	C	C	A	A	D	D	C	C	A	A	A
Vigna vexillata	CF 43-2[a]	–	–	–	–	–	–	–	–	–	–	–	–	–	–

[a]In legume mixture plot.

Table 15.42. Observation numbers after which management was imposed on Site 12.

Grass	Grass cut	Grass and legume cut	Grazed
Native	2	–	–

have a fragile nature and cattle hooves can sink deeply into them, so management must ensure that grazing does not eliminate the native pasture and any introduced legumes through trampling. It was noted on a visit to the Rio On trial site in 1985 that the growth of young *P. caribea* trees within the plot was much improved, presumably due to the fertilizer applied, and the presence of substantial stands of legumes.

15.7 Conclusions

The trials in the upper Belize Valley demonstrated good establishment of legumes both in native pastures and those planted to productive grasses, but the legumes were generally not productive, probably because of competition from the grass and the effects of heavy grazing (trampling and consumption). Carefully managed grazing and the establishment of fodder banks of leguminous species for cut-and-carry feeding or for brief periods of well-managed grazing may provide better results. Woody legume species, once well established, may be better able to compete and survive than low herbaceous or viney ones.

Provided with adequate fertility, including marl in the case of the Pinol soils on the Mountain Pine Ridge, leguminous fodder species were shown to be productive on the Low Pine Ridge Puletan soils and on the shallow Pinol soils of the Mountain Pine Ridge. The introduction of leguminous species to these neglected environments would provide the opportunity for improved productivity through utilization by ruminant animals. This in turn may reduce the severity of fires.

Table 15.43. Site 13, Baldy Beacon, native pasture. Accessions' membership of vigour groups (Table 15.9) determined from observational data: A best, to D poor. Darker shading has been used to identify the better results.

		Observations													
		Low fertilizer							High fertilizer						
		2	4	6	8	10	12	13	2	4	6	8	10	12	13
Legumes	Accession no.	cool	dry	wet	late wet	cool	dry	wet	cool	dry	wet	late wet	cool	dry	wet
Calopogonium caeruleum	CF 18	D	D	C	C	**A**	C	C	D	D	C	C	**A**	C	**A**
Centrosema plumieri	CF 16-1	–	–	D	B	**A**	C	C	D	D	D	D	D	C	C
Centrosema pubescens cv. Belalto	CF 177	D	D	D	B	C	C	**A**	D	D	C	**A**	**A**	**A**	**A**
Centrosema pubescens	CF 6-1	D	D	D	B	**A**	**A**	**A**	D	D	C	**A**	**A**	C	**A**
Centrosema pubescens 'Centro'	CF 6-2	B	D	C	B	C	C	C	D	D	C	**A**	**A**	D	**A**
Codariocalyx gyroides	CF 29	D	D	**A**	**A**	**A**	**A**	**A**	D	D	C	**A**	**A**	**A**	**A**
Crotalaria retusa	CF 20[a]	–	–	D	D	D	D	–	–	–	–	D	D	D	–
Crotalaria vitellina	CF 15[a]	–	–	D	D	–	–	–	–	–	D	D	–	–	–
Desmodium canum	CF 1[a]	D	D	D	C	D	D	D	D	D	C	C	D	D	D
Desmodium intortum 'Greenleaf'	CF 50	B	D	**A**	B	**A**	**A**	**A**	D	C	**A**	**A**	**A**	**A**	**A**
Indigofera mucronata	CF 9[a]	–	–	–	D	D	D	D	–	–	D	C	D	D	D
Leucaena leucocephala cv. Peru	CF 179	D	D	D	C	C	**A**	C	D	D	C	C	D	B	C
M. atropurpureum cv. Siratro	CF 3-1	D	D	**A**	B	**A**	**A**	C	D	D	**A**	**A**	**A**	**A**	**A**
Macroptilium lathyroides	CF 14	–	–	–	–	–	–	–	–	–	–	–	–	–	–
	CF 14[a]	D	D	**A**	C	**A**	D	D	D	D	C	D	D	D	–
Pueraria phaseoloides 'Puero'	CF 180	D	D	**A**	B	**A**	**A**	**A**	D	D	**A**	**A**	**A**	**A**	**A**
	CF 180[a]	D	D	D	D	C	C	C	D	D	C	**A**	D	C	D
Rhynchosia minima	CF 13[a]	–	–	–	D	–	–	–	–	–	D	D	–	–	–
Stylosanthes guianensis	CF 4-1	D	D	D	B	**A**	**A**	**A**	D	D	C	**A**	**A**	**A**	**A**
Stylosanthes guianensis cv. Cook	CF 4-4	D	D	**A**	**A**	**A**	**A**	**A**	D	D	**A**	**A**	**A**	**A**	**A**
Stylosanthes guianensis cv. Endeavour	CF 4-5	D	D	**A**	**A**	**A**	**A**	**A**	D	D	**A**	**A**	**A**	**A**	**A**
Stylosanthes hamata cv. Verano	CF 186	D	D	**A**	**A**	C	**A**	**A**	D	D	C	**A**	**A**	**A**	B
Teramnus uncinatus	CF 2[a]	D	D	**A**	C	**A**	**A**	**A**	D	D	**A**	**A**	**A**	**A**	**A**
Vigna vexillata	CF 43-2	D	D	**A**	B	**A**	**A**	**A**	D	D	C	C	**A**	B	C

[a]In legume mixture plot.

Table 15.44. Observation numbers after which management was imposed on Site 13.

Grass	Grass cut	Grass and legume cut	Grazed
Native	–	–	–

Table 15.45. Summary Mountain Pine Ridge sites. Accessions' membership of vigour groups (Table 15.9) determined from observational data: A best, to D poor. Darker shading has been used to identify the better results.

		Final observation					
		Site 11		Site 12		Site 13	
Legumes	Accession no.	Low	High	Low	High	Low	High
Calopogonium caeruleum	CF 18	C	C	C	A	C	A
Centrosema plumieri	CF 16-1	C	D	C	A	C	C
Centrosema pubescens cv. Belalto	CF 195	C	A	A	A		
Centrosema pubescens	CF 177					A	A
Centrosema pubescens	CF 6-1	D	C	A	A	A	A
Centrosema pubescens 'Centro'	CF 6-2	C	C	D	D	C	A
Codariocalyx gyroides	CF 29	A	A	A	A	A	A
Crotalaria retusa	CF 20[a]	D	D	–	–	–	–
Crotalaria vitellina	CF 15[a]	–	–	D	B	–	–
Desmodium canum	CF 1[a]	D	–	D	D	D	D
Desmodium intortum 'Greenleaf'	CF 50	A	A	A	A	A	A
Indigofera mucronata	CF 9[a]	D	–	D	D	D	D
Leucaena leucocephala cv. Peru	CF 179	D	B	D	B	C	C
M. atropurpureum cv. Siratro	CF 3-1	A	A	C	A	C	A
Macroptilium lathyroides	CF 14	D	B	A	A	–	–
	CF 14[a]	D	D	D	A	D	–
Pueraria phaseoloides 'Puero'	CF 180	A	A	A	A	A	A
	CF 180[a]	C	D	C	D	C	D
Rhynchosia minima	CF 13[a]	–	–	D	–	–	–
Stylosanthes guianensis cv. Oxley	CF 201[a]	D	–	D	–		
Stylosanthes guianensis	CF 4-1	C	A	A	A	A	A
Stylosanthes guianensis cv. Cook	CF 4-4	D	A	A	A	A	A
Stylosanthes guianensis cv. Endeavour	CF 4-5	A	A	A	A	A	A
Stylosanthes hamata cv. Verano	CF 186	D	B	A	B	A	B
Teramnus uncinatus	CF 2[a]	D	–	A	A	A	A
Vigna vexillata	CF 43-2					A	C
	CF 43-2[a]	–	–	–	–		

[a]In legume mixture plot.

Table 15.46. Range of observation data for *S. guianensis* cv. Endeavour at the end of the trial (cm).

	Low fertilizer		High fertilizer	
Environment	Spread	Height	Spread	Height
Low Pine Ridge	83–202	52–78	above 200	51–74
Mountain Pine Ridge	116–133	27–41	139–165	36–57

Table 15.47. Persistence of legumes in strip and other concurrent trials. 1981 observations.

Trial	Management	Observations in 1981
Site 3, Santiago Juan	Grazed for 4 years	High fertilizer rate legumes were vigorous: *C. plumieri*, *D incanum*, *Rhynchosia* sp. In the low fertilizer area much less. Some *Stylosanthes guianensis* cv. Endeavour left and a row of *Leucaena* 4 m in high fertility, but only 1.2 m in low fertility. Some grazed to 2-cm sticks. A few legumes had self-sowed out of the trial area.
Site 4, Orange Walk	Trial cut over and abandoned, not grazed	Dense *Desmodium incanum* in high fertilizer area, two short *Leucaena* trees (30–40 cm), also *C. caeruleum*, *C. plumieri* especially and some *M. lathyroides*, *C. pubescens*, *M. atropurpureum*, *P. phaseoloides*.
Site 5, Western Highway mile 33	Trial continued, with grazing	Plants remain vigorous.
Site 6, Western Highway mile 33	Trial continued	Legumes generally have spread through seeding and become mixed: *C. gyroides*, *C. plumieri*, *C. pubescens*, *C. caeruleum*, *D. intortum* especially under high fertility.
Sites 7, 8, Western Highway mile 35	Continued	*C. gyroides* spread widely, *P. phaseoloides*, *C. caeruleum*, and *C. plumieri* did well. Four large *Leucaena* trees.
Site 10, Klaas Freisan, Spanish Lookout	Abandoned	Legumes were reported to have spread throughout farm but had been grazed out.
Site 11, Bull Run	Moderately grazed by native fauna, presumably deer	Vigorous *Pueraria phaseoloides* covers trial area, big patches of *Centrosema pubescens*, tall *C. gyroides* leafless as high as deer could graze, seeding Outside trial site a few *Pueraria* seedlings with nutrient deficiency on leaves.
Site 12, Rio On	Heavily grazed by native fauna, presumably deer	Only seedlings of *Centrosema plumieri*, *C. pubescens*, *Pueraria phaseoloides*, *Codariocalyx gyroides*, two 2.5-m *Leucaena leucocephala* trees, a few *Macroptilium atropurpureum* plants.
Site 15, Show Grounds		*Leucaena*, *C. gyroides*, *C. caeruleum*, *M. atropurpureum* all did well. Only *D. intortum* did poorly.
Other concurrent trials		
A[a]	Heavy mob grazing	No legumes, no Para grass.
D[a]		*Leucaena* persisted, became tall bushes, some Siratro present. Pangola most persistent, Guinea only 10% left, very little Jaragua.
E[a]		Legumes persisted well for 3 years before the trial was abandoned. Some plants of *C. pubescens* and *M. atropurpureum* seen.

[a]See Appendix 15.2, Chapter 15 (this volume). A. Lazier (1981 a, b); D. Lazier (1978); E. Lazier (1978).

Appendix 15.1: Soil analysis results

Table 15.48. Soil analysis results (no analyses available for site 13 at Baldy Beacon).

		EEC	Sand					Ca	Mg	K	Na	C	N	C/N	P
		Mohs	coarse	fine	Silt							% oven			
Depth (cm)	pH	(×10⁻⁶)	Oven dry soil (%)			Clay	CEC	Oven dry soil (me/100g)				dry soil		ratio	ppm
Site 1. Norland, Jaragua, upper Belize River terrace															
0–15	6.4	50	11	28	44	16	2.81	4.84	1.33	0.35	0.07	2.88	0.19	15.2	9
15–30	6.4	30	8	37	44	10	9.19	3.23	0.95	0.24	0.07	1.64	0.10	16.4	6
30–45	6.7	22	2	12	80	8	3.95	3.23	1.14	0.20	0.04	0.54	0.08	6.8	4
45–60	6.5	19	1	10	35	55	8.14	4.03	1.33	0.19	0.04	0.48	0.09	5.3	4
Site 2. Norland, Merker, upper Belize River terrace															
0–15	6.2	53	12	35	31	19	4.33	0.40	0.38	0.28	Trace	1.28	0.12	10.7	7
15–30	6.3	26	12	39	31	19	8.43	2.82	0.76	0.20	0.05	0.72	0.09	8.0	6
30–45	6.5	16	3	48	19	27	3.57	2.42	0.95	0.20	0.05	0.28	0.05	5.6	7
45–60	6.1	25	4	38	28	28	10.76	1.21	1.04	0.19	0.04	0.28	0.05	5.6	2
Site 3. Santiago Juan, native pasture, upper Belize River terrace															
0–15	6.3	71	0	29	27	43	9.48	8.06	1.89	0.49	0.09	1.28	0.20	6.4	6
15–30	6.2	36	3	34	19	45	7.48	4.84	1.14	0.22	0.04	1.28	0.12	10.7	3
30–45	6.3	21	0	26	11	59	10.62	10.48	1.70	0.37	0.11	0.52	0.10	5.2	2
45–60	5.8	20	1	14	28	54	15.67	9.68	1.52	0.17	0.10	0.44	0.08	5.5	3
Site 4. Orange Walk, native pasture, central Belize River terrace															
0–15	6.5	136	4	20	12	61	55.00	28.23	17.52	0.29	0.26	2.48	0.30	8.6	6
15–30	6.5	92	1	21	12	65	54.05	28.63	14.68	0.20	0.35	1.60	0.18	8.9	3
30–45	7.2	149	1	20	12	65	53.57	66.53	18.94	0.34	0.38	0.88	0.10	8.8	1
45–60							43.81	56.45	18.94	0.28	0.26	2.88	0.19	15.2	2
Site 5. Mile 33, native pasture, level, Low Pine Ridge															
0–15	5.6	35	1	20	12	65	15.19	5.65	2.18	0.08	0.16	2.72	0.20	13.6	3
15–30	5.4	27	49	5	12	33	9.81	4.69	1.98	0.08	0.15	1.40	0.12	11.7	4

Site 6. Mile 33, native pasture, sloping, Low Pine Ridge															
0–15	5.5	24	45	8	12	33	6.67	2.34	1.29	0.04	0.09	1.60	0.12	13.3	4
15–30	5.4	22	53	16	12	21	8.48	2.34	2.58	0.04	0.17	0.52	0.06	8.7	2
30–45	5.6	30	8	28	6	54	18.29	4.69	6.15	0.08	0.35	0.16	0.03	5.3	2
Sites 7 and 8. Mile 35, native pasture, level, Low Pine Ridge															
0–15	5.4	62	2	7.0	26	62	36.29	11.33	8.33	0.17	0.34	2.98	0.25	11.9	4
15–30	6.0	104	12	2.5	10	50	34.52	18.75	8.18	0.26	0.36	1.24	0.11	11.3	5
Site 11. Bull Run, native pasture, Mountain Pine Ridge															
0–15	5.2	31	11	27	48	9	2.71	0.42	0.12	0.05	Trace	1.76	0.08	22.0	3
15–30	5.3	19	2	47	40	8	2.67	0.42	0.10	0.03	0.24	0.76	0.04	19.0	2
Site 12. Rio On, native pasture, Mountain Pine Ridge															
0–15	5.0	32	2	43	44	8	2.86	0.83	0.13	0.09	0.15	2.84	0.17	16.7	3
15–30	5.1	27	11	39	40	9	1.71	0.42	0.08	0.06	trace	1.32	0.09	14.7	3
Site 14. Central Farm, African Star, upper Belize River terrace															
0–15	6.7	69	9	34	6	46	8.86	19.53	1.29	0.24	0.08	1.88	0.17	11.1	12
15–30	7.0	37	10	43	2	42	6.57	7.03	1.09	0.12	0.08	0.96	0.11	8.7	8
30–45	6.7	25	3	24	15	54	10.14	5.08	1.98	0.12	0.10	0.56	0.10	5.6	4
45–60	6.0	19	4	28	3	62	13.14	5.47	2.28	0.15	0.09	0.44	0.09	4.9	5
Site 15. Agricultural Show Grounds, central Belize River terrace															
0–15	7.1	223	3	21	8	66	27.24	21.09	4.56	0.26	0.18	2.36	0.23	10.3	10
15–30	–	–	11	25	12	54	26.38	32.03	4.96	0.17	0.23	1.08	0.10	10.8	4
30–45	7.4	122	5	10	28	58	24.76	32.81	2.38	0.19	0.24	0.68	0.07	9.7	4
45–60	7.3	124	8	17	8	66	20.14	33.20	5.16	0.14	0.24	0.40	0.05	8.0	5

Appendix 15.2

Observations were also made on other trials undertaken by the project that had also been abandoned in 1977. Trial D, involving five commercially available legumes (*L. leucocephala*, *S. guianensis* cv. Schofield, *Desmodium intortum*, *Centrosema pubescens* and *Neonotonia wightii* cv. Tinaroo) and three sown grasses (Pangola, *Digitaria eriantha*; Jaragua, *Hyparrhenia rufa*; Guinea, *Panicum maximum*) (Lazier, 1978) had been under heavy mob grazing since the trial was terminated. Leucaena had persisted and had grown to 3.5 m trees, but of the herbaceous legumes only occasional *M. atropurpureum* plants were found. Even the grasses were greatly reduced. Pangola persisted best, but only about 10% of the Guinea grass remained, and very little Jaragua grass. In an adjacent trial with 19 native legumes and *C. gyroides* in Para grass (*Brachiaria mutica*) (Lazier, 1981a,b), both the legumes and planted grass had been eliminated. A grazing management trial of legumes in Pangola grass, in which legumes had survived well (Lazier, 1978), only occasional plants of *C. pubescens* and *M. atropurpureum* remained after 3 years without management. In summary, it can be concluded that forage legumes can grow, persist and spread with the application of relatively low levels of additional fertility in the three environments in which they were tested by the project. However, it appears that grazing management is essential for legume and improved grass survival.

16 Initial Screening for Persistence and Productivity of 20 Promising Native and Exotic Pasture Legume Species in Two Productive Contrasting Forage Grasses in Belize

J.R. Lazier*[1]

*Formerly International Livestock Centre for Africa

Abstract

In order to assess persistence and productivity of promising fodder legumes under the wet tropical conditions of central Belize on a black heavy clay soil, strips of 20 mainly locally collected accessions of fodder legumes (13 genera, 15 species) of potential were planted in three replicates on an upper terrace of the Belize River in established pastures of two grasses (Pangola grass, *Digitaria eriantha* and Para grass, *Brachiaria mutica*) under low and high levels of applied fertilizer. Observational data were acquired at 6-week intervals. Once the legumes were well established the plots were mob grazed at 6-week intervals and observations were taken immediately before and after. Legume establishment and persistence was best in Para grass, and *Leucaena leucocephala* cv. Peru and *Stylosanthes guianensis* were the most promising materials tested. Common Centro and three locally collected accessions (*C. pubescens* CF 6-1, *Calopogonium caeruleum* CF 18 and *Centrosema plumieri* CF 16-1) also showed some promise.

16.1 Introduction

In the 1970s cattle production was an important industry in the Belize River Valley; however, productivity was low because of lack of land for expansion, unproductive stock, poor management and mediocre pasture. Few pastures were planted to productive grasses, and these were generally badly managed and suffered from lack of nutrients. The introduction of leguminous fodder species into local pastures could do much to improve grass growth and livestock productivity.

While research to identify adapted grass and legume species had been underway in Belize for some years, there had not been a systematic testing of adapted legume species under farm conditions. Early testing in the selection process often consists of assessing unreplicated strips in selected environments but due to environmental variability within a test plot such methods can produce false results. Replication of the strips is likely to produce more reliable results. In addition, such selection is often done in plots without grass or weed competition and under such conditions plants can be successful that would not persist in pastures.

It was thus decided to assess the productivity and persistence of a number of fodder legumes that had performed particularly well in introduction plots by assessing their ability to persist and

[1] E-mail: john.lazier@sympatico.ca

spread in replicated strip trials planted in established pastures under rotational grazing. Two contrasting grasses well adapted to Belizean conditions were chosen for the study, *Digitaria eriantha* (Pangola grass), which is densely leafy, and *Brachiaria mutica* (Para grass), which has an open plant architecture.

At the time of the trial (1975–1977), Pangola grass (*Digitaria eriantha*) pastures were found on Central Farm, the Department of Agriculture's main research farm, and occasionally elsewhere in the country. A highly productive species, it requires good fertility to persist. When it is productive, its dense leafy canopy inhibits the successful introduction of associated fodder legumes, although a naturally occurring *Desmanthus* plant had been found in one pasture. Para grass (*Brachiaria mutica*, known locally as Carib grass) is readily established and was a popular grass for improved pastures at the time. In contrast with Pangola grass, it has a thick stems and the canopy is much less dense than Pangola, thus offering a good opportunity for the introduction of associated leguminous fodder species. A trial was established in order to identify species of productive leguminous fodder plants that would persist in swards of these species and that would improve the productivity of the associated grass and the nutritional value of the fodder present. In order to assess legume persistence under normal farming conditions, the plots were grazed on a rotational system every 6 weeks.

16.1.1 The environment

The sites selected were in pastures 3 and 5 on Central Farm, Cayo District in Belize (topographic map reference 16 QBQ 865 007 and 866 007 respectively[2]). They were located on upper terraces of the Belize River at an elevation of 60 m asl and received an average annual rainfall of about 1500 mm. The adjacent trial sites had established 85 m × 85 m pastures of Pangola grass (*Digitaria eriantha*) and Para grass (*Brachiaria mutica*).

The sites chosen were on black heavy clay soils representative of the old alluvium of the Belize River Valley. Classified as Meditation series (Jenkin *et al.*, 1976), they were described as deep, silty clay loams or clay loams with slightly acid top soils, medium to high CEC and base saturation, low total and available P. McKinzie (1977) characterized the upper Pangola site as Vertic Tropudalfs very fine mixed isohyperthermic, a mixture of colluvium and alluvial sediments sloping at 1% to 2%. He found the upper horizon of the lower, Pangola grass trial site to be altered, perhaps as far back as Mayan times. The surface layers were highly calcareous silty clay loam over an original soil similar to that of the *B. mutica* site. The soils were sampled before the trial was commenced, each site having 30 random samples taken at 0–15 and 15–30 cm, and 15 samples at 30–45 and 45–60 cm. The samples were analysed at the University of the West Indies Soils Department laboratory and the results are presented in Appendix 16.1 (Tables 16.7 and 16.8). They indicate the sites have neutral pH and vary in clay content, with the lower Pangola site having higher clay content (47%) than that of the Para grass site (36%). The analytic data indicated that the sites were similar otherwise.

Rainfall at Central Farm is normally about 1500 mm (Fig. 16.1). Growth is negatively affected by a dry season from March to May, which may be completely dry. Growth is at its peak as the main rainy period begins in June and July, perhaps due to the accumulated soil N from the dry season and longer days; thereafter the rain tapers off. A short period of relatively dry weather occurs again for 2–5 weeks in August–September and in years when the reduction is more severe, growth can be affected. Another rainfall peak occurs before the cooler, drier weather of December and January, followed by a period of about 2 weeks of constant cloud cover that reduces growth in January–February just before the dry season returns. As temperatures rise in February and days lengthen, there can be a brief period of increased growth before the dry season begins.

Since the area was cleared, it was successively in pastures, pigeon peas, a cattle rotation trial, planted to maize, beans and tomatoes, and finally a pasture-crop rotation before being left as permanent pasture. Moderate amounts of fertilizer were applied during the later years when 1-ha paddock areas were established and were used for rotational grazing, initially by bullocks, and then by eight breeding bulls.

In order to test the production and persistence of promising forage legume accessions when

[2] Map series E755 DOS 4499 Edition 3.

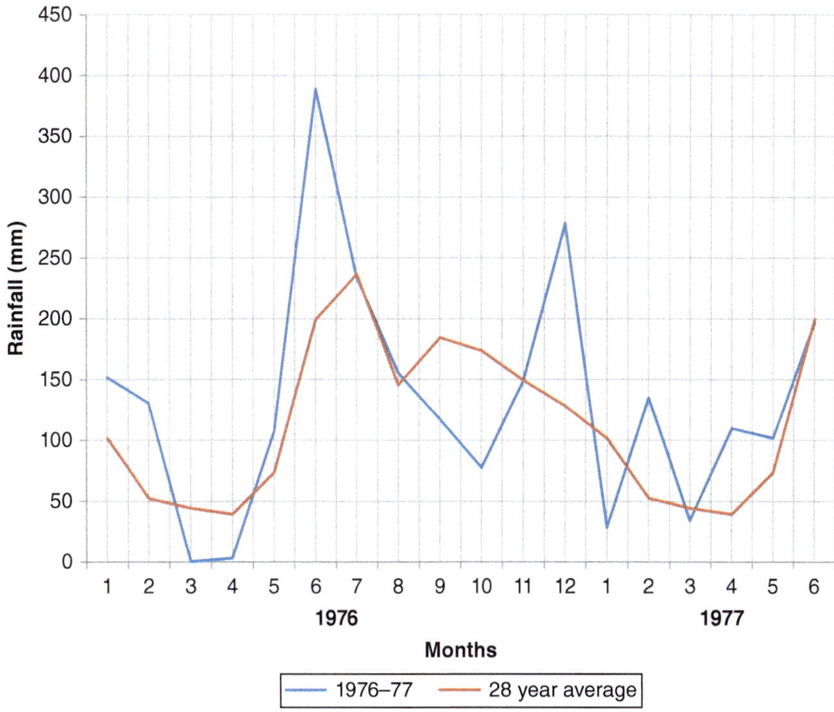

Fig. 16.1. Central Farm rainfall during the trial and over 28 years.

sown into established swards of improved pasture grasses, a trial was established with legumes sown into tilled strips under low and high fertilizer rates.

16.2 Methods

A total of 20 legume species and cultivars were selected comprised of 16 species (Table 16.1), all but two of which were native to Belize (*C. gyroides* and *P. phaseoloides*). Of those native to Belize, 10 accessions were genotypes collected within Belize, and eight were commercial or experimental genotypes acquired from outside of Belize. The indigenous genotypes were of species that had been found to occur naturally in local pastures and had shown promise in introduction plots and earlier trials (Lazier, 1978, 1980a,b, 1981a,b).

The trial design at each site consisted of three replicates of 16 plots organized in blocks. The plots were in a 4 × 4 pattern and consisted of strips 0.50-m wide and 15-m long cultivated in the pasture, spaced 5 m apart (between the centres of the parallel strips) with 5 m between the ends of the plots and with a 5 m border strip about the perimeter. The trial dimensions in each pasture were thus 55 m × 65 m. Each strip was ploughed and then cultivated using a mechanical cultivator (Fig. 16.2). The legumes were sown in the cultivated strips down the middle of each plot (Fig. 16.2). Fifteen of the 16 plots in each block were sown to only a single accession, the remaining plot was sown to a mixture of accessions of limited seed availability, or that were deemed of lower potential.

Pegs were placed at the mid-point of each strip to mark the separation between the high and low fertilizer application rates. The low rate was always on the same side of the dividing point. Two more pegs were placed 1 m apart from this central peg to act as a border area between the fertilizer rates, which was not observed. The 4 metres of strip remaining was then the area data was recorded from.

Half of the seed was scarified and all was inoculated with the appropriate rhizobia and the accessions were sown by hand in late October 1975 in prepared strips 10 m × 15 cm at 2.5 g/strip or 17 kg/ha. Each constituent of the mixture

Table 16.1. Species planted in the trials.

Species	Accession number[a]	Species in[b] Strip	Mixture	Source
Calopogonium caeruleum	CF 18	AB		Central Farm[c]
Centrosema plumieri	CF 16-1	AB		Central Farm[c]
Centrosema pubescens cv. Belalto	CF 195	AB		Australia[d]
Centrosema pubescens	CF 6-1	AB		Central Farm[c]
Centrosema pubescens 'Centro'	CF 6-2	AB		Commercial[d]
Codariocalyx gyroides	CF 29	AB		South-east Asia
Crotalaria retusa	CF 20		AB	Central Farm[c]
Desmodium incanum	CF 1		AB	Central Farm[c]
Desmodium intortum 'Greenleaf'	CF 50	AB		Commercial[d]
Indigofera mucronata	CF 9		B	Central Farm[c]
Leucaena leucocephala cv. Peru	CF 179	AB		Australia[d]
Macroptilium atropurpureum cv. Siratro	CF 3-1	AB		Australia[d]
Pueraria phaseoloides 'Puero'	CF 180	AB		Commercial
Rhynchosia minima	CF 13		AB	Central Farm[c]
Stylosanthes guianensis	CF 4-1	AB		Mountain Pine Ridge[c]
Stylosanthes guianensis cv. Cook	CF 4-4	AB	B	Australia[d]
Stylosanthes guianensis cv. Endeavour	CF 4-5	AB		Australia[d]
Stylosanthes hamata cv. Verano	CF 186	AB		Australia[d]
Teramnus uncinatus	CF 2		AB	Central Farm[c]
Vigna vexillata	CF 43-2	AB		Central Farm[c]

[a]CF: Central Farm accession number; [b]A: *Digitaria eriantha*; B: *Brachiaria mutica*; [c]collected in Belize; [d]species native to Belize, seed imported.

Fig. 16.2. Establishment of the trial in the *Digitaria eriantha* pasture.

strip was sown at 1.5 g/strip or 10 kg/ha. *Centrosema plumieri* was planted at 0.4 m spacing as nursery grown seedlings because of a lack of seed. Only the tilled strips were fertilized, with half of each set of strips receiving fertilizer at the rate of one quarter of the annual rate (63 kg/ha each of TSP and muriate of potash), and the other half of the strip receiving four times that amount. All the plot low fertilizer application rates were on the same side of the centre pegs.

A germination count in late November showed that very few seeds had germinated, probably because of a combination of effects: very vigorous competition from the grasses, particularly the Pangola grass, rough strip preparation, and very heavy rains that occurred soon after planting. The strips were replanted in December after better soil preparation. They were tilled repeatedly with a small rotor tiller to a width of 1 m to reduce grass competition, and the planted strips 0.15 × 5 m hoed to a depth of 10 cm. The trial was again fertilized, but this time the entire plot areas were fertilized at the same low and high rates as was done initially. Fertilization was continued for the remainder of the trial at the same rate quarterly, after alternate 6-weekly observation-grazing periods.

After replanting, germination and growth were satisfactory. The strips were weeded once after germination to aid establishment. During the period of the trial, woody weeds and invading weed grasses such as sour grass (*Paspalum virgatum*) were removed. Observations on the major characteristics of each plot were taken on both the high and low fertilizer rates of each plot (Table 16.2) at 6-week intervals; the paddocks were grazed after observation numbers 4–6 and 8–13 (Table 16.2) and further observations were then taken. The length of time the animals were left in the paddocks was determined by the degree the grass on offer had been reduced, not by the effects of the grazing on the legumes.

The 6-weekly observations began in early May 1976. Once the plants were sufficiently robust to withstand grazing (late June), grazing was commenced and this was followed by post grazing observation. The pre- and post-grazing observations were designed to present as accurate a picture of the state of the grass and individual legumes within each plot as possible with the least effort.

Cattle were initially put in for a week at a time, but because the herd was rather large for the amount of feed present, the period was shortened, and once the grass was moderately heavily to heavily grazed the cattle were removed. One grazing period (October 1976) was omitted because of overgrazing in the previous cycle, which had weakened the condition of the legumes. After post-grazing observation numbers 2–5, the grasses were cut over, and after post-grazing observations 6–13 both grasses and legumes were cut. Grazing and observations were continued until June of the following year.

16.2.1 Analysis

Seven observation periods over the trial period were selected as representing the critical times of the growth cycle (Table 16.3). Selected observational data (Table 16.4) from each of the seven periods from each grass site were subjected to MULCLAS analysis followed by GROUPER analysis. Six groups were found to give adequate division of the data. The presence of seedlings was not very common, and in the analysis was either overridden by other factors where they occurred, or the plots containing plants with seedlings were placed in a separate grouping. The groupings

Table 16.2. Observation dates of replicated legume strips planted in two pastures under grazing.

	Observation number and date												
	1976								1977				
Pasture grass	1[a]	2	3[a]	4[b]	5[a,b]	6[b]	7[a]	8[b]	9[b]	10[a,b]	11[b]	12[a,b]	13[a,b]
Pangola	11/2	23/3	3/5	15/6	22/7	2/9	14/10	24/11	7/1	17/2	31/3	11/5	24/6
Para	11/2	23/3	4/5	17/6	29/7	8/9	22/10	2/12	13/1	22/2	6/4	23/5	29/6

[a]Observations used in analysis; [b]grazed after initial observations.

Table 16.3. Seasons of observation data used in analysis.

Observation number	Season	Pangola	Para
1	Establishment and pre-dry season	2 Feb	2 Feb
2	Late dry season	3 May	4 May
3	Wet season	22 July	29 July
4	Late wet season	14 October	22 October
5	Cold season	17 February	22 February
6	Late dry season	11 May	23 May
7	Early wet season	24 June	29 June

Table 16.4. Data selected for cluster analysis.

Factor	Range/units
Legume plot cover of the planted strip	Estimated out of 10
Legume height	Average over the entire planted plot in cm
Legume spread	Average over the entire plot in cm
Legume plants surviving	Number of those tagged at the first observation
Legume plants in the strip	Number to a maximum of 20
Legume seedlings	Plot cover out of 10

tended to reflect the vigour and survival of the plants in the strips. Detailed analysis of the groups and plants were done and the means obtained for the seven analysed characters.

The six groups produced by the analysis were reduced by truncation of the hierarchical structure to four, a level convenient for the needs of producing levels of success.

The detailed analyses of the harvests are available on request, but for the purpose of simplified presentation, four levels or Vigour Groups, A to D, were chosen based on the groups the analyses produced. Level A represents the most successful plants and D the least successful. To be in a given category the strip cover, number of tagged plants and the number of plants alive all had to fall within the range chosen (Table 16.5). If one factor was less than the minimum the group was placed at the next lower level. Frequently all four levels were produced during one observation period, but in some instances all six groups were similar and only one vigour grouping resulted.

16.3 Results and discussion

Rainfall over the 18-month trial period was higher than normal (2393 mm vs 2055 mm) both during 1976 (1789 mm vs 1542 mm, 16% higher) and during the first half of 1977 (604 mm vs 508 mm, 19% higher). Higher rainfall was received in June and December 1976 and to a lesser extent in February 1977. Although rainfall was adequate for growth over much of the year, there was maximum water stress in March and April when there was essentially no rainfall, and in addition, in August and September the rainfall was less than normal.

The detailed results of the analyses are presented in Appendix 16.1, Table 16.9 (*Brachiaria mutica* pasture) and Table 16.10 (*Digitaria eriantha* pasture). Almost all of the accessions planted established and grew well initially, at least at the Vigour Group level C, with the exception of *Indigofera mucronata* and *Desmodium incanum* in the mixture plots. *D. incanum* is an abundant constituent of Belizean native pastures and its poor performance in the trial may indicate a lack of ability to establish rapidly and to compete with vigorous grasses. The same factors may have affected *I. mucronata* because it is commonly found in areas where there is little competition from other species.

The cluster analysis groupings showed very close relationships between height, spread and survival. Thus it was unusual for a high plot cover to correlate with few plants in the strip or a low survival of tagged plants. Height and spread reflected the morphology of the plants

Table 16.5. Categories of groups formed and their observation ratings.

	Legume		
Vigour group	Strip cover (out of 10)	Tagged plants (no. of 5)	Plants alive (max. 20)
A	7–10	4–5	15–20
B	4–7	2–4	10–15
C	2–4	1–2	5–10
D	< 2	1	0–5

and varied with the plot cover. These relationships are presented in the detailed analyses, but have little bearing on the generalized levels presented here. The tabulated results are presented in Appendix 16.1 and the variation through the seasons can be noted therein. The general success or failure of the legumes at each site and the most successful will be discussed in this section.

As the results reflect the plants' growth and the effects of grass competition and animal grazing, it is difficult to assign the success of the plants to their productivity. Factors such as relative palatability and the effects of the applied fertilizer levels on plant palatability also need to be considered. Numerous accessions did not persist and were eliminated by the end of the trial. The tables also indicate that higher fertility rates did not result in higher survival, or in higher scores, but rather the reverse. Scores of even the better plants under low fertility tended to be lower under high fertility. Increased grass competition and legume palatability are among the factors that may be involved.

A summary (Table 16.6) and an examination of the detailed observation cluster analysis scores in Tables 16.9 and 16.10 in the Appendix show clearly that for the individual accessions establishment was better and Vigour Group scores were higher initially, and higher for longer, in the Para Grass pasture compared with the Pangola pasture. This is probably due to lower competition from Para grass, but the clay content of the soil was also higher at the Pangola site (Table 16.8) and this could be an additional factor.

Leucaena leucocephala grew particularly well, but it was greatly favoured by the grazing animals, which when let into the paddock would go first to these plots, completely stripping the leaves and consuming branches up to 1 cm in diameter. However, the plants did survive and it was the most vigorous and persistent legume present in both pastures.

Codariocalyx gyroides, which has a similar morphology, is also very palatable but is more susceptible to damage from trampling and grazing, and so performed poorly, and did not persist well. The next best plants to *Leucaena* were the *Stylosanthes guianensis* lines, which established and persisted well in both pastures.

The viney and more prostrate plants, many of which established well, were not too productive or persistent under the conditions of the trial. Two of the *Centrosema pubescens* accessions, the commercial line and a line collected on Central Farm, maintained a moderate presence in the Para grass paddock at both fertilizer levels, but were weak in the Pangola grass pasture. *Macroptilium atropurpureum* cv. Siratro also established well, but was not as persistent as the two *C. pubescens* lines. *Centrosema plumieri* and *Calopogonium caeruleum* also maintained some presence in the Para grass pastures; both accessions were collected on Central Farm and are commonly found in the native pastures. The heavy black clays become soft when wet, so grazing severely affects the growth and survival of viney and prostate legumes by coating them with soil and trampling them into the mud. Restricting grazing to drier periods should enhance legume survival and productivity. Taller bushy legumes such as *L. leucocephala* and *C. gyroides* are less prone to these two problems, but on the other hand can suffer physical damage by animals breaking them while grazing or moving through the pasture.

The trial illustrates the conundrum of the grazier in managing mixed grass–legume swards. A productive grass tends to crowd out associated legumes, and the more legumes present, the greater the vigour of the grass due to the higher soil N levels. Compounding this problem is grazing

Table 16.6. Summary of cluster analysis groupings of the accessions showing the greatest promise as pasture plants. Darker shading indicates better plant performance.

		Pasture							
		Para				Pangola			
		Low fertilizer		High fertilizer		Low fertilizer		High fertilizer	
Species	Accession number[a]	Initial	Final	Initial	Final	Initial	Final	Initial	Final
Centrosema pubescens[b]	CF 6-1	**Very Good**	Fair	**Very Good**	Fair	Good	Poor–fair	Good	Poor–fair
Centrosema pubescens 'Centro'	CF 6-2	**Very Good**	Fair	**Very Good**	Fair–poor	**Very Good**	Poor	**Very Good**	Poor
Leucaena leucocephala cv. Peru	CF 179	**Very Good**	Good	**Very Good**	Good	Good	Good–poor	Good	Good
Macroptilium atropurpureum cv. Siratro	CF 3-1	**Very Good**	Poor–fair	**Very Good**	Poor–fair	**Very Good**	Poor	**Very Good**	Poor
Stylosanthes guianensis[b]	CF 4-1	**Very Good**	Good					Good	Poor
Stylosanthes guianensis cv. Cook	CF 4-4	**Very Good**	Good	Good	Fair	Good	Poor	Good	Poor
Stylosanthes guianensis cv. Endeavour	CF 4-5	**Very Good**	Good	**Very Good**	Fair	**Very Good**	Fair	**Very Good**	Poor–fair
Stylosanthes hamata cv. Verano	CF 186	**Very Good**		Good	Fair				
Teramnus uncinatus[b,c]	CF 2					Good	Poor	**Very Good**	Poor
Vigna vexillata[b]	CF 43-2	**Very Good**	Poor	**Very Good**	Poor	**Very Good**	Poor	**Very Good**	Poor

[a]CF: Central Farm accession number; [b]locally collected germplasm; [c]only in mixed seed treatments.

and the relative palatability of the grass and the legume. A less palatable legume, such as perhaps the *Stylosanthes* species, can contribute to the diet to some extent and still provide fertility to the soil. But even a shrubby legume, which once established grows above competition from the grass, will soon be grazed out if it is highly palatable, like *L. leucocephala*, if management is not imposed. Fodder banks of shrubby and arboreal legume species that stock can be put into briefly and taken out before the plants are too severely grazed is one option. The fodder bank approach would also assist in maintaining viney legumes, which suffer from physical damage from trampling and by fouling with mud in rainy weather. A further alternative is to focus on providing N to the grass by planting a legume with low palatability that is strong enough to withstand physical damage by stock. Smallholder farmers should be able to maintain mixed swards, and particularly those with arboreal legumes such as *Leucaena*, by using cut-and-carry methods for productive and valuable stock such as dairy animals.

The methodology utilized here, replicated strips in established grass pastures combined with rotational grazing and regular observations, is less costly in effort and labour compared with replicated pure legume and mixed legume–grass treatment trials where success is assessed through determining dry matter yields. The replications in this trial have shown that the scores of the accessions were reasonably consistent between replicates, and within pastures. However, there was still sufficient variation to indicate that single strips may not be sufficient to judge the potential of an accession. Additionally, although the two sites were adjacent, the soils differed in particle content, with the Pangola pasture having higher clay content, and this factor cannot be overlooked.

16.4 Conclusions

The screening of promising legumes by planting them in replicated strips in established pastures and subjecting them to rotational grazing was successful in identifying *Leucaena leucocephala* cv. Peru and *Stylosanthes guianensis* as good prospects for further study as pasture legumes on clay soils under Belize River Valley conditions. *Centrosema pubescens*, common Centro, and three locally collected accessions (*C. pubescens* CF 6-1, *Calopogonium caeruleum* CF 18 and *Centrosema plumieri* CF 16-1) also showed some promise. Establishment and persistence was better in Para grass, which has an open canopy, than in Pangola grass, which has dense foliage and is more competitive. While ratings for growth and survival of individual accessions were relatively uniform across replicates, they were not entirely so and thus replication in such screening trials would appear to be advantageous. Observational ratings appeared to be adequate for identification of differences between the performances of accessions at this level of screening.

Appendix 16.1

Table 16.7. Results of chemical and physical analyses of soil samples from the *Digitaria eriantha* pasture.

Depth (cm)	pH	EEC (Mohs × 10^{-6})	Sand: Coarse	Sand: Fine	Silt (%)	Clay (%)	CEC (meq/100 g)	Ca	Mg	K	Na	C	N	C/N ratio	P (ppm)
								Oven dry soil (me/100g)				% oven dry soil			
6-1 Block A															
0–15	6.4	51	10	11	32	51	15.71	6.05	1.59	0.31	0.25	1.12	0.17	6.7	7
15–30	6.3	31	2	4	30	63	14.05	5.24	1.59	0.19	0.22	2.24	0.13	17.2	4
30–45	6.4	26	2	4	28	67	13.81	4.44	1.59	0.13	0.11	0.52	0.11	4.7	5
45–60	6.9	66	0	4	28	67	11.33	5.24	1.49	0.10	0.30	0.36	0.10	3.6	4
6-1 Block B															
0–15	6.5	37	10	1	44	47	13.24	6.05	1.59	0.37	0.06	2.04	0.18	11.3	8
15–30	6.3	22	10	16	40	39	10.29	4.03	1.39	0.20	0.20	1.12	0.13	8.6	5
30–45	5.9	5	10	4	24	67	12.57	4.03	1.59	0.16	0.30	0.52	0.10	5.2	4
45–60	5.4	7	7	4	28	63	20.29	7.26	2.08	0.19	0.18	0.48	0.08	6.0	5
6-1 Block C															
0–15	6.1	22	12	12	36	43	10.29	4.03	1.39	0.45	0.27	2.00	0.16	12.5	7
15–30	6.1	22	10	3	40	51	8.95	3.63	1.59	0.28	0.12	0.92	0.11	8.4	4
30–45	5.7	22	12	6	20	67	19.14	4.44	1.98	0.21	0.19	0.40	0.09	4.4	4
45–60	5.9	22	9	3	20	71	19.62	1.21	1.29	0.08	0.08	0.28	0.09	3.1	3

Table 16.8. Results of chemical and physical analyses from the *Brachiaria mutica* pasture.

Depth (cm)	pH	EEC (Mohs × 10^{-6})	Sand Coarse	Sand Fine	Silt (%)	Clay (%)	CEC (meq/100 g)	Ca	Mg	K	Na	C	N	C/N ratio	P (ppm)
								Oven dry soil (me/100g)				% oven dry soil			
6-2 Block A															
0–15	6.6	31	15	15	40	35	11.10	5.24	1.39	0.24	0.08	1.44	0.12	12.0	6
15–30	6.5	30	10	16	32	43	11.52	8.47	1.67	0.15	0.11	0.96	0.08	12.0	6
30–45	6.6	31	11	5	28	59	13.57	8.06	1.87	0.17	0.14	0.60	0.07	8.8	3
45–60	6.5	15	4	3	32	63	9.52	6.05	1.67	0.13	0.11	0.40	0.08	5.0	3
6-2 Block B															
0–15	6.2	29	12	16	40	35	10.57	7.26	1.46	0.29	0.13	1.24	0.11	11.3	8
15–30	6.3	24	8	11	36	47	11.81	8.06	1.56	0.16	0.13	0.88	0.08	11.0	5
30–45	6.4	24	4	7	30	59	15.52	10.08	1.98	0.19	0.14	0.48	0.07	6.7	5
45–60	6.0	20	8	8	22	63	15.52	8.47	1.77	0.19	0.14	0.52	0.08	6.5	5
6-2 Block C															
0–15	6.0	31	10	12	42	39	13.33	8.06	1.67	0.39	0.13	1.72	0.14	12.3	8
15–30	6.1	25	17	5	34	43	13.86	8.06	1.67	0.18	0.14	1.08	0.11	9.8	4
30–45	6.1	26	3	9	34	55	14.19	8.87	1.98	0.19	0.13	0.48	0.08	6.0	3
45–60	6.1	19	8	8	26	59	13.57	8.06	1.77	0.17	0.13	0.36	0.07	5.1	3

Table 16.9. Cluster analysis of observational ratings of legume performance in a *Brachiaria mutica* pasture taken before seven regular 6-weekly grazings (A, best performance to D, poor performance).

			Observation periods													
			Low fertilizer							High fertilizer						
Legumes	Accession no.	Block	1	2	3	4	5	6	7	1	2	3	4	5	6	7
Calopogonium caeruleum	CF 18	1	**A**	**A**	**A**	D	C	C	C	**A**	B	B	D	C	D	C
		2	C	C	C	D	D	D	D	C	D	C	D	C	D	D
		3	C	C	D	D	D	D	D	C	D	D	D	D	D	C
Centrosema plumieri	CF 16-1	1	C	C	D	D	C	D	D	C	D	D	D	D	D	C
		2	C	**A**	D	B	B	C	C	C	**A**	B	D	C	D	C
		3	C	C	D	D	C	C	C	C	D	B	D	C	D	C
Centrosema pubescens cv. Belalto	CF 195	1	C	C	D	D	D	D	D	C	D	C	D	–	D	D
		2	C	C	D	D	D	D	D	C	D	C	D	C	D	D
		3	C	C	D	D	D	–	D	C	D	D	D	–	–	–
Centrosema pubescens 'Centro'	CF 6-2	1	**A**	**A**	**A**	**A**	B	C	C	**A**	B	B	B	C	**A**	C
		2	**A**	**A**	**A**	B	B	C	C	**A**	**A**	B	B	C	**A**	C
		3	**A**	**A**	**A**	B	B	B	C	**A**	**A**	B	B	C	**A**	C
Centrosema pubescens	CF 6-1	1	**A**	**A**	**A**	**A**	B	B	C	**A**	**A**	B	D	C	D	D
		2	**A**	**A**	**A**	B	B	C	C	**A**	**A**	B	B	C	D	C
		3	**A**	**A**	**A**	B	C	C	C	**A**	**A**	B	B	C	**A**	C
Codariocalyx gyroides	CF 29	1	C	**A**	D	–	–	–	–	C	D	D	D	D	D	D
		2	C	**A**	D	D	D	D	D	C	B	C	D	C	D	D
		3	C	–	–	–	D	D	D	C	–	–	–	–	–	–
Crotalaria retusa	CF 20[a]	1	C	D	D	D	D	D	C	**A**	B	D	B	D	D	D
		2	D	–	D	D	D	D	D	C	B	D	D	D	D	D
		3	–	–	–	–	–	–	–	D	D	D	–	D	–	D
Desmodium canum	CF 1[a]	1	D	–	–	–	–	–	–	–	–	–	–	–	–	–
		2	D	–	–	–	–	–	–	–	–	–	–	–	–	–
		3	D	–	–	–	D	D	D	C	D	D	D	D	D	D
Desmodium intortum 'Greenleaf'	CF 50	1	C	C	D	D	D	D	D	C	B	C	D	D	–	–
		2	**A**	**A**	D	–	D	–	–	**A**	B	C	D	–	–	–
		3	C	C	D	D	–	–	–	C	D	C	–	–	–	–

Indigofera mucronata	CF 9[a]	1	–	–	–	–	–	–	–	–	–	–	–	–	–	
		2	–	–	–	–	–	–	–	–	–	D	–	–	–	–
		3														D
Leucaena leucocephala cv. Peru	CF 179	1	**A**	**A**	B	B	B	B	B	**A**	B	B	B	B	B	**A**
		2	**A**	**A**	B	B	B	B	B	**A**	B	B	B	B	B	C
		3	**A**	**A**	B	B	B	B	B	**A**	B	B	B	B	B	C
Macroptilium atropurpureum cv. Siratro	CF 3-1	1	**A**	**A**	**A**	B	C	C	C	**A**	**A**	B	D	C	D	D
		2	**A**	**A**	**A**	B	C	D	D	**A**	**A**	B	D	C	D	D
		3	**A**	**A**	**A**	B	D	D	D	**A**	**A**	B	D	D	D	C
Pueraria phaseoloides 'Puero'	CF 180	1	C	C	C	B	D	D	D	C	D	C	D	D	D	D
		2	C	C	D	D	D	D	D	C	D	–	D	C	C	D
		3	C	C	D	D	C	D	C	C	D	D	D	D	D	D
Rhynchosia minima	CF 13[a]	1	C	D	D	D	D	D	D	C	D	D	B	C	C	C
		2	D	–	–	–	D	D	D	C	D	D	D	D	D	D
		3	D	D	D	D	D	D	D	D	D	D	D	–	–	–
Stylosanthes guianensis	CF 4-1	1	C	**A**	**A**	**A**	B	B	B	C	B	C	B	**A**	**A**	C
		2	C	C	**A**	**A**	B	C	C	C	D	C	B	C	**A**	C
		3	C	**A**	**A**	B	B	B	B	C	D	C	B	**A**	**A**	C
Stylosanthes guianensis cv. Cook	CF 4-4	1	C	C	**A**	**A**	B	B	B	C	B	C	B	C	C	C
		2	**A**	**A**	**A**	**A**	B	B	B	**A**	B	B	B	C	**A**	C
		3	**A**	**A**	**A**	**A**	B	B	B	C	B	C	D	C	D	D
	CF 4-4[a]	1	**A**	**A**	**A**	**A**	C	C	D	**A**	B	C	B	C	C	C
		2	**A**	**A**	**A**	**A**	C	C	D	**A**	B	C	B	C	C	C
		3	**A**	**A**	D	B	C	C	D	C	B	C	B	C	C	C
Stylosanthes guianensis cv. Endeavour	CF 4-5	1	**A**	**A**	**A**	**A**	C	C	B	**A**	B	C	C	**A**	**A**	C
		2	**A**	**A**	**A**	**A**	B	C	B	**A**	B	B	B	C	**A**	C
		3	**A**	**A**	**A**	**A**	B	B	B	**A**	B	C	C	**A**	C	C
Stylosanthes hamata cv. Verano	CF 186	1	**A**	**A**	**A**	**A**	B	C	D	**A**	B	C	B	C	C	C
		2	**A**	**A**	**A**	**A**	C	C	D	**A**	B	C	B	C	C	C
		3	**A**	**A**	D	B	C	C	D	C	B	C	B	C	C	C
Teramnus uncinatus	CF 2[a]	1	**A**	**A**	**A**	D	D	D	D	**A**	B	C	D	C	C	D
		2	**A**	C	D	D	D	–	–	C	B	C	D	–	–	–
		3	C	C	**A**	D	D	D	D	C	D	C	D	D	D	–
Vigna vexillata	CF 43-2	1	**A**	**A**	**A**	D	D	D	D	**A**	**A**	B	D	–	–	–
		2	**A**	**A**	**A**	D	D	D	D	**A**	B	B	D	D	D	D
		3	**A**	**A**	**A**	D	D	D	D	**A**	**A**	B	D	–	D	D

[a]Accessions so labelled sown together as a mixture.

Table 16.10. Cluster analysis of observational ratings of legume performance in a *Digitaria eriantha* pasture taken before seven regular 6-weekly grazings (A, best performance to D, poor performance).

			Observation periods													
			Low fertilizer							High fertilizer						
Legumes	Accession no.	Block	1	2	3	4	5	6	7	1	2	3	4	5	6	7
Calopogonium caeruleum	CF 18	1	**A**	C	D	D	D	D	D	**A**	**A**	C	D	D	D	D
		2	C	C	D	D	D	D	D	**A**	**A**	D	D	D	D	D
		3	C	D	C	D	D	D	D	C	D	D	D	D	D	D
Centrosema plumieri	CF 16-1	1	C	**A**	C	D	D	D	–	C	**A**	C	D	D	D	D
		2	C	C	D	D	–	D	D	C	**A**	C	D	D	D	D
		3	C	C	D	–	–	–	–	C	**A**	C	D	D	–	–
Centrosema pubescens cv. Belalto	CF 195	1	C	C	D	D	D	D	D	C	D	C	D	D	D	D
		2	C	D	D	D	D	D	D	C	D	C	D	D	–	D
		3	C	C	C	D	C	D	C	C	D	C	D	D	D	D
Centrosema pubescens	CF 6-1	1	**A**	**A**	**A**	B	C	D	D	**A**	**A**	B	B	D	D	C
		2	**A**	**A**	**A**	B	B	D	C	**A**	**A**	B	B	D	D	D
		3	C	C	**A**	B	C	D	D	C	C	C	D	D	D	D
Centrosema pubescens 'Centro'	CF 6-2	1	**A**	C	**A**	D	D	D	D	**A**	**A**	B	D	C	D	D
		2	**A**	**A**	**A**	B	C	D	D	**A**	**A**	B	B	C	D	D
		3	**A**	**A**	D	B	C	D	D	**A**	**A**	B	D	D	D	D
Codariocalyx gyroides	CF 29	1	C	–	D	D	D	D	D	C	D	C	D	D	D	D
		2	C	C	C	D	C	D	D	C	B	C	D	C	D	C
		3	C	C	C	D	D	D	D	D	D	–	–	–	–	–
Crotalaria retusa	CF 20[a]	1	C	C	D	–	–	–	D	C	B	D	–	–	–	D
		2	C	C	B	–	–	–	–	**A**	B	D	D	–	–	D
		3	C	C	D	D	–	C	D	D	B	D	D	–	–	D
Desmodium canum	CF 1[a]	1	C	C	C	D	D	D	D	C	–	–	D	D	D	D
		2	D	–	–	D	D	D	–	D	–	–	D	D	D	D
		3	–	–	–	–	–	–	–	D	–	–	–	–	–	–
Desmodium intortum 'Greenleaf'	CF 50	1	C	C	C	D	–	–	–	C	D	D	D	D	–	–
		2	**A**	C	C	D	–	–	–	**A**	D	D	D	D	D	D
		3	C	C	D	–	–	–	–	C	B	D	D	–	–	–

Indigofera mucronata	CF 9[a]	1	–	–	–	–	–	–	–	–	–	–	–	–	–	–
		2	–	–	–	–	–	–	–	–	–	–	–	–	–	–
		3	–	–	–	–	–	–	–	–	–	–	–	–	–	–
Leucaena leucocephala cv. Peru	CF 179	1	**A**	**A**	B	B	B	B	B	**A**	B	B	B	B	B	**A**
		2	**A**	C	B	B	B	B	B	C	B	B	B	B	B	**A**
		3	C	C	B	D	B	D	D	**A**	B	B	B	B	B	C
M. atropurpureum cv. Siratro	CF 3-1	1	**A**	**A**	C	D	D	D	D	**A**	**A**	B	D	D	D	D
		2	**A**	**A**	**A**	D	D	–	–	**A**	**A**	C	D	D	D	D
		3	**A**	**A**	**A**	B	D	D	D	**A**	**A**	C	D	D	D	D
Pueraria phaseoloides 'Puero'	CF 180	1	**A**	C	C	D	D	D	D	**A**	B	C	D	D	D	D
		2	**A**	C	C	D	D	D	D	**A**	D	C	D	D	–	–
		3	**A**	C	D	D	D	D	D	**A**	D	C	–	D	–	–
Rhynchosia minima	CF 13[a]	1	D	C	C	D	D	D	D	C	D	D	D	D	D	D
		2	D	D	–	D	–	D	–	C	B	D	B	D	D	D
		3	D	D	D	D	–	D	D	C	B	D	B	D	D	D
Stylosanthes guianensis	CF 4-1	1	C	C	C	D	D	D	D	C	D	C	D	D	D	D
		2	**A**	C	D	D	D	D	D	**A**	B	C	D	C	D	D
		3	C	C	**A**	B	C	D	D	**A**	B	C	D	D	D	D
Stylosanthes guianensis cv. Cook	CF 4-4	1	**A**	**A**	C	D	D	D	D	**A**	B	C	D	D	D	D
		2	**A**	C	C	B	D	D	D	**A**	B	C	B	D	D	D
		3	C	D	D	D	–	D	D	C	D	D	D	D	–	D
	CF 4-4[a]	1	D	D	D	–	–	–	–	–	–	–	–	–	–	–
		2	–	–	–	–	–	–	–	–	–	–	–	–	–	–
		3	–	–	–	–	–	–	–	–	–	–	–	–	–	–
Stylosanthes guianensis cv. Endeavour	CF 4-5	1	**A**	**A**	C	B	B	D	B	**A**	B	C	B	D	D	D
		2	**A**	**A**	C	**A**	B	B	B	**A**	B	C	B	C	D	C
		3	**A**	**A**	**A**	**A**	B	D	D	**A**	B	C	B	D	D	D
Stylosanthes hamata cv. Verano	CF 186	1	**A**	C	C	B	C	D	C	**A**	B	C	B	C	D	D
		2	C	C	C	B	B	D	D	C	B	C	B	C	D	D
		3	C	C	**A**	**A**	C	D	D	C	B	C	B	**A**	D	D
Teramnus uncinatus	CF 2[a]	1	**A**	C	C	D	D	D	D	**A**	B	C	D	D	D	D
		2	**A**	C	C	D	D	D	D	**A**	B	C	D	D	D	D
		3	C	**A**	**A**	D	D	C	D	**A**	B	C	D	D	–	–
Vigna vexillata	CF 43-2[a]	1	C	**A**	C	D	D	D	D	**A**	B	C	D	–	D	D
		2	C	**A**	**A**	D	D	–	C	**A**	**A**	B	D	D	D	D
		3	C	**A**	**A**	D	D	D	D	**A**	B	B	D	–	–	–

[a]Accessions so labelled sown together as a mixture.

17 Productivity of 16 Forage Legumes Under Cutting in Belize on Contrasting Soils: a High Alluvial Terrace and A Lowland Pine Savanna I. Dry Matter Yields

J.R. Lazier*[1]

**Formerly International Livestock Centre for Africa*

Abstract

In order to determine the relative productivity of 16 promising forage legumes (12 species, 8 genera) in two major environments of central Belize (a neutral heavy cracking clay on an upper river terrace, and an acid waterlogged sandy loam on the Low Pine Ridge coastal deposits), trials with three replicates were established and 12 harvests for dry matter were made at 6-week intervals. The high terrace site was a long-established pasture of Coastal Bermuda grass (*Cynodon dactylon*). The grass and weeds proved to be aggressive and dominated the legumes tested, such that many low and viney legumes died out. *Codariocalyx gyroides*, *Crotalaria anagyroides* and *Leucaena leucocephala*, which grew above the grass, had the best total yields. The Bermuda grass production was very poor in the cool and dry seasons. At the waterlogged savanna site there was little competition from the native grass *Mesosetum angustifolium*, but very little legume growth until fertilizer was applied. Thereafter, *C. anagyroides* and *C. gyroides* were very vigorous, yielding well through the dry season and producing substantially more in the last 9 months of the trial than the high river terrace site. The wet soil conditions permitted best legume growth in the dry season. The legumes generally were more productive than at the heavy clay site. *M. angustifolium* yields were very low throughout the trial, and particularly in the cool season.

17.1 Introduction

Two areas of present and potential importance for cattle raising in Belize are the less fertile upper terraces of the Belize River and the extensive areas of infertile Low Pine Ridge savanna. These heavy clays soils are easy to access and currently under pressure for development, but have had little documented work on pasture productivity, particularly on productive legumes.

While high fertilizer costs have helped discourage investment in improved pastures, it is hoped that the identification of adapted, productive forage legumes might encourage greater pasture development by raising per hectare productivity and lowering the annual investment in fertilizer. The infertile lowland pine savanna soils particularly require a source of nitrogen in order that more productive pasture species might be established to replace the fibrous unproductive native grasses.

17.2 Methods

The high alluvial terrace site was at an altitude of 60 m on a heavy clay soil in pasture I.4 on

[1] E-mail: john.lazier@sympatico.ca

Central Farm, the Belize Department of Agriculture research station in Cayo District. The soil type was classified as Esperanza series, Melinda suite by Jenkin *et al.* (1976), and a fine, mixed isohyperthermic mollic Tropudalf by McKinzie (1977; USDA, 1975). The lowland pine savanna site was located on the north side of the Western Highway at mile 33.5 (56 km to the east of the high river terrace site) on a slightly sloping land at an altitude of 40 m and mapped as a vertic Ultisol termed Rough Mile series, Puletan suite (Jenkin *et al.*, 1976), but examination of the site revealed that it was a fine kaolinitic isohyperthermic aeric Tropaqualfs (McKinzie, 1977; USDA, 1975). Both trial sites had soil samples taken at 30 random locations at depths of 0–15 cm and 15–30 cm. At the terrace site, additional samples were taken at 15 locations in each block at 30–45 cm and 45–60 cm. At the savanna site the entire profile was wet and samples could not be taken below 30 cm because the soil was too wet to remain on the augers. Soil analysis of the high terrace site indicated a silty loam of pH 6.4, which was low in P (4 to 9 ppm) and K (0.1 to 0.4 ppm), and less fertile generally than the lower Belize River terraces. The savanna site was a sandy loam of pH 5.8, low in N (0.13%), K (0.9 ppm) and very low in P (2 ppm) (Table 17.1).

The trials were laid out in a randomized block design of three blocks, each with 16 plots (4 m × 4 m) containing nine plants spaced at 1 m, leaving plot margin of 0.5 m. One-metre paths were left between plots and there was a 2-m trial perimeter. Soil preparation at the alluvial site included ploughing, harrowing and rotavating, while at the savanna site the sparse grasses were simply mown before planting.

An initial fertilizer application of 12 kg/ha of P and 20 kg/ha of K was broadcast on the alluvial site in October 1975 and on the Low Pine Ridge site in January 1976. The same amount was applied as an annual rate in quarterly applications after alternate 6-weekly harvests. The savanna site received an additional one-time application of 53 kg/ha of a fritted trace element mixture in January 1976. The species planted (Table 17.2) included lines that had been persistent and productive on the lower, better drained, more fertile terraces of the Belize River (Lazier, 1980a,b, 1981a,b,c): *Macroptilium atropurpureum* cv Siratro, *Centrosema plumieri* CF 18, *Centrosema pubescens* CF 6-1 and Centro, *Calopogonium caeruleum* CF 18, *Codariocalyx gyroides* CF 29 and *Crotalaria retusa* CF 57. The remainder of the lines planted had performed well in introduction plots on a heavy clay soil.

Plants were raised in pots from scarified, appropriately inoculated seeds. A specific inoculum was used where known, otherwise cowpea inoculum was used. The trials at both sites were planted in early November 1975 and dead plants were replaced until February 1976, when all were well established. The alluvial site that had vigorous weed growth was weeded until the plants were well established, and from May 1976 grass and weeds were allowed to invade. The vegetation at the Low Pine Ridge savanna site provided little competition to the legumes and no weeding was done. Pathways and perimeters in both trials were kept mown to reduce legume invasion from adjacent plots. Taller weeds were removed periodically.

Harvesting was done at 6-weekly intervals from the end of March 1976 to early July 1977

Table 17.1. Particle size distribution and chemical properties of soils.

Depth (cm)	pH	EEC (Mohs × 10^{-6})	Sand		Silt (%)	Clay (%)	CEC (Meq/ 100 g)	Ca	Mg	K	Na	C	N	C/N ratio	P (ppm)
			Coarse	Fine				Oven dry soil (me/100g)				% oven dry soil			
Upper Belize River Valley terrace															
0–15	6.4	72	8	17	32	41	8.86	6.67	1.67	0.41	0.13	2.36	0.23	10.3	9
15–30	6.4	26	1	12	28	57	9.57	5.83	1.69	0.22	0.09	0.88	0.14	6.3	5
30–45	6.8	66	1	11	28	61	10.48	7.50	1.98	0.12	0.28	0.52	0.11	4.7	4
45–60	7.3	62	1	11	28	61	10.00	6.67	1.98	0.10	0.14	0.24	0.10	2.4	5
Low Pine Ridge															
0–15	5.8	29	24	45	4	29	6.67	3.33	1.39	0.88	0.44	1.92	0.13	14.8	2
15–30	5.4	24	19	25	30	25	9.95	4.17	2.78	0.09	0.26	0.80	0.08	10.0	2

Table 17.2. Accessions planted at the two trial sites.

Species	Accession number[a]	Seed source	Trial site[b]
Calopogonium caeruleum	CF 18	Belize	B
Centrosema plumieri	CF 16-1	Belize	AB
Centrosema pubescens	CF 6-1	Belize	AB
Centrosema pubescens	Centro	Commercial	AB
Centrosema pubescens	cv. Belalto	Commercial	A
Codariocalyx gyroides	CF 29 (CPI 13379?)		AB
Crotalaria anagyroides	CF 197	Commercial	AB
Crotalaria retusa	CF 57	Belize	AB
Leucaena leucocephala	cv. Peru	Commercial	AB
Macroptilium atropurpureum	CF 3-1	Commercial	AB
Neonotonia wightii	cv. Cooper	Commercial	AB
Stylosanthes guianensis	cv. Cook	Commercial	AB
Stylosanthes guianensis	cv. Endeavour	Commercial	AB
Stylosanthes hamata	cv. Verano	CSIRO[c]	AB
Stylosanthes scabra	CPI 34925	CSIRO[c]	AB
Stylosanthes scabra	CPI 40205	CSIRO[c]	AB
Stylosanthes scabra	cv. Seca	CSIRO[c]	AB

[a]CF Central Farm: Project accession number, CPI CSIRO accession number; [b]A High river terrace site; B Lowland pine savanna site; [c]CSIRO Commonwealth Scientific and Industrial Research Organization.

for the alluvial site and from early April 1976 until mid-June 1977 for the savanna site. In the first two harvests, the entire area within the 0.5 m plot margins was harvested by hand for dry matter yields, while in succeeding harvests, as the plots were covered with the legumes, three random 1 m^2 quadrats were sampled per plot. Sampling was done at heights judged appropriate for each legume (15 cm for all legumes except *C. retusa*, *C. gyroides* and *Crotalaria anagyroides*, which were harvested at 50 cm). After the first harvest, the cutting heights of all *Stylosanthes* lines were raised to 30 cm to enable the plants to compete with weeds, because the lower height had not been successful in an unreported earlier trial. After sampling, the plots were cut over at their harvesting heights and the cut material was removed.

The effect of the presence of a vigorous forage legume and fertilizer on the growth of the dominant savanna grass species *Mesosetum angustifolium* at the savanna site was determined by harvesting three randomly located permanent 1 m × 0.5 m quadrats at 15 cm height from each of the plots of *C. gyroides* and from three randomly selected unfertilized plots about the perimeter of the trial. An initial cut for uniformity was done in February 1977 and sampling for grass dry matter yields was done at the same time as the regular trial harvests in March, May and June 1977.

17.3 Results and discussion

The rainfall, about 1800 mm/year, was higher than the long-term average of 1500 mm at both sites (Fig. 17.1). The first March through May dry period was of normal length, but March and April were somewhat drier than the average year. The second dry season had frequent showers, resulting in much less water stress.

Once weeding was terminated after the second harvest, the high terrace plots were very rapidly invaded by weeds. The dominant grass was Coastal Bermuda grass (*Cynodon dactylon*) which spread from an adjacent pasture and densely covered over two-thirds of the plots; *Melampodium divaricatum* was the major non-graminaceous weed. *Ixophorus unisetus* and *Paspalum conjugatum* also made important contributions to the sward. It was noticeable that there were distinct boundaries between the *Cynodon* dominant plots and those with dominant native grasses and weeds. In botanical sorting of samples, only Coastal Bermuda grass was included in the grass fraction.

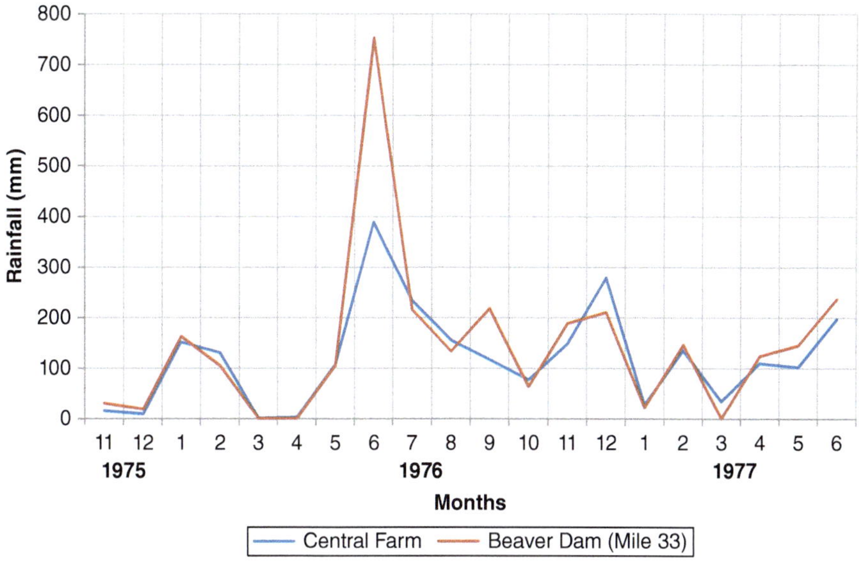

Fig. 17.1. Rainfall at trial sites.

17.3.1 The high terrace site

Legumes

The dense nature of the Coastal Bermuda and weed swards below the cutting heights provided strong competition for the legumes, particularly those cut at the lower levels. *C. retusa*, Verano and the three *S. scabra* lines did not persist (Fig 17.2, Table 17.3). All legume treatments but Verano did well initially, but with the advent of the rains and the cessation of weeding in May 1976 the yields of the remaining four legume treatments dropped to negligible levels and by January all yields had declined to or near zero. As these plants had done well in introduction plots that were partially weeded, competition by grasses and weeds was the likely cause of the yield declines. With the exception of *C. retusa*, these legume lines are recommended for lower rainfall areas or those with longer dry seasons in which they would have much less competition.

The vigour of the Coastal Bermuda grass is also indicated by the low yield of *C. retusa*, which had previously persisted well in introduction plots and with Para grass (*Brachiaria mutica*) (Lazier, 198la). The four *Centrosema* lines and Cooper *Neonotonia* produced similar low total dry matter yields, with *C. plumieri* highest and the native *C. pubescens* CF 6-l line and Cooper *Neonotonia* lowest. The *C. pubescens* lines and Cooper *Neonotonia* did poorly in the dry seasons and best in the early wet season. *C. plumieri* had better yields throughout the 1977 dry season than all lines except *C. gyroides* and *Leucaena*. Its cool season production was relatively high as well.

Siratro, and Cook and Endeavour Stylos produced similar total yields. The yields of all increased markedly after the first rains in May 1976. The Siratro yields continued to increase until August, then declined sharply and remained low for the rest of the trial. The stylos' yields decreased in August, perhaps in response to the large amount of dry matter that was removed in June, then reached a peak in September before declining again and remaining low through the remainder of the trial. Low total yields and in particular low yields in the second dry season for all three indicated low competitive ability with the dense Coastal Bermuda grass. Siratro was particularly affected: in other trials in Belize it had gained vigour during the dry season (e.g. Lazier, 1980a). The wetter than usual dry season in 1977 probably maintained a more vigorous grass sward than would normally occur in that period of the year.

Fig. 17.2. High alluvial terrace site. The two taller legume treatments on left are *Codariocalyx gyroides*, the one on the right is *Crotalaria anagyroides* (see also Fig. 2.3, Chapter 2, this volume).

The most productive legumes in the trial were, in order of declining yield: *C. gyroides*, *C. anagyroides* and *L. leucocephala* (Figure 17.2). The first two were significantly higher (P = 0.05) in total yield than the fourth highest yielding line, Cook Stylo; the total production of *Leucaena* was significantly higher (P = 0.05) than all the remaining legumes with the exception of Cook and Endeavour Stylos. More important than the total dry matter production was the cool and dry season yields, which were highest for *C. gyroides* followed by *Leucaena*; *C. anagyroides*, the second highest yielding legume, had low yields in the dry season, and highest yields in the early wet season.

Grasses and weeds

The grass and weed yields (Tables 17.4 and 17.5) reflected the cutting heights of the plots, yields being higher on the lower cut plots. The total yields (Table 17.6) were mainly determined by weed yields. Since the Coastal Bermuda grass was irregularly established in the trial, no comparison can be made between grass, weed or total yields of plots.

Coastal Bermuda grass production for three legumes cut at 15 cm (Cooper *Neonotonia*, Siratro and *C. plumieri*) showed markedly increased production in August 1976 after the beginning of the rainy season. The delayed response to the rains was probably due to the grass not being fully established in the plots. Production declined toward the end of the rains and into the cool period, as would be expected (Miller, 1960; Murata and Iyama, 1963). The slight January yield increase when rainfall was low may be due to increased sunshine, while cloudy skies at the end of January due to the cool north winds ('northers'), and higher rainfall in February probably caused the reduced yields of the early March harvest. The 1977 dry season (March, April and May) had occasional showers and thus stressed the sward little; yields increased as a result.

The weed dry matter yields of these three legume treatments (Table 17.4) paralleled those of the Coastal Bermuda grass, with a much smaller early rainy season growth rate in 1976, but a similar one in 1977. The dry matter yields were almost as high as those of the grass in the dry season.

Table 17.3. High river terrace site: legume dry matter harvest yields from mixed swards under 6-weekly cutting.

	Harvest yields (year and month) (kg/ha × 10^{-1})												Totals	
	1976							1977						
Legume	3	5	6	8	9	10	12	1	3	4	5	7	All trial	August to July
C. plumieri	0	1	13	29	25	39	15	16	15	22	28	46	250	236
C. pubescens CF 6-1	2	1	10	38	28	13	3	7	1	5	12	57	176	163
C. pubescens cv. Belalto	1	1	18	84	27	14	3	4	4	10	37	53	256	236
C. pubescens (Centro)	3	1	6	70	33	26	21	1	1	3	6	49	201	191
C. gyroides	22	28	166	156	115	41	22	41	12	21	51	13	689	473
C. anagyroides	56	8	143	146	54	64	40	36	7	2	8	5	569	362
C. retusa	128	23	1	4	0	0	0	1	1	5	1	0	175	13
L. leucocephala cv. Peru	155	40	90	49	24	13	14	17	6	19	30	87	544	259
M. atropurpureum cv. Siratro	12	18	65	140	46	13	5	1	1	0	5	23	329	234
N. wightii cv. Cooper	14	2	28	62	20	7	2	3	0	2	2	16	158	114
S. guianensis cv. Cook	4	7	103	45	116	64	12	3	1	7	5	11	377	263
S. guianensis cv. Endeavour	11	4	61	43	130	62	42	8	1	0	3	8	373	298
S. hamata cv. Verano	0	0	2	3	10	5	0	0	0	0	0	0	21	19
S. scabra CPI 34925	21	5	20	2	2	0	0	2	0	0	0	0	52	6
S. scabra CPI 40205	96	5	17	1	0	20	0	0	0	0	0	0	139	21
S. scabra cv. Seca	17	1	3	3	0	1	4	0	0	0	0	0	29	8
LSD ($P \geq 0.05$)	25	15	54	65	45	40	10	14	4	10	ns	37	185	

LSD between legumes and harvest ($P \geq 0.05$), same legume 32, between legumes 34. ns, not significant ($P \geq 0.05$).

17.3.2 The savanna site

While the rainfall (Fig. 17.1) was generally similar for the two sites, at the savanna site (Figure 17.3) it was higher in November 1975 and June 1976, and lower in July 1976.

The plants established very slowly with no initial fertilizer application, fertilizer being first applied in January 1976. The plots were not harvested until April, when the plants showed vigour despite the dry weather. In order to allow the plants time to become better established, the next harvest was done 12 weeks later (July) when regular 6-weekly harvests were commenced.

Legumes

Legume dry matter yields (Table 17.7) indicated slow establishment but rapid growth of the more productive legumes thereafter. The highest yielding legumes were *C. anagyroides* and *C. gyroides* (Figure 17.3b). These two plants produced more total dry matter than any of the other legumes at the site ($P = 0.05$). They also produced more in the 1977 dry season (May) and more in the preceding and following harvests than all but Siratro and *C. retusa*. All of the *C. gyroides* plants survived to the end of the trial.

Generally all the legumes did better in the dry season than in the cooler, more overcast period of shorter days that preceded it. Both the

Table 17.4. High river terrace site: Coastal Bermuda (*Cynodon dactylon*) grass dry matter harvest yields from mixed swards under 6-weekly cutting.

	Harvest yields (year and month) (kg/ha × 10^{-1})												
	1976							1977					Trial
Legume treatment	3	5	6	8	9	10	12	1	3	4	5	7	totals
C. plumieri	1	0	106	336	189	132	103	143	77	128	185	344	1745
C. pubescens CF 6-1	2	1	125	290	184	137	97	136	53	99	174	310	1608
C. pubescens cv. Belalto	1	1	132	328	164	133	91	148	62	95	178	290	1622
C. pubescens (Centro)	3	1	82	330	145	106	77	116	50	58	124	266	1357
C. gyroides	22	28	176	243	165	85	37	65	28	30	54	121	1052
C. anagyroides	56	8	143	189	56	74	51	49	19	27	41	94	806
C. retusa	138	23	3	66	21	12	11	30	15	27	17	38	401
L. leucocephala cv. Peru	155	40	175	325	185	116	78	108	56	103	104	305	1751
M. atropurpureum cv. Siratro	12	18	149	402	202	143	77	140	54	111	126	284	1719
N. wightii cv. Cooper	14	2	163	400	187	152	80	120	60	114	149	269	1711
S. guianensis cv. Cook	4	7	234	145	192	99	25	39	25	42	66	160	1037
S. guianensis cv. Endeavour	11	4	158	141	188	87	64	33	22	28	55	136	925
S. hamata cv. Verano	0	0	179	164	149	50	16	36	11	53	62	158	878
S. scabra CPI 34925	21	5	130	89	92	54	25	36	23	30	41	183	729
S. scabra CPI 40205	96	5	79	146	69	64	29	35	18	44	58	136	780
S. scabra cv. Seca	17	1	181	189	78	60	31	37	24	35	83	182	858
LSD ($P \geq 0.05$)	25	15	85	119	77	75	40	44	36	38	73	88	317

LSD between legumes and harvest ($P \geq 0.0\ 5$), same legume 62, between legumes 64.

character of the soil, an Ultisol, and the dry season showers probably contributed to the increased growth. Siratro, Cook Stylo and *S. scabra* CPI 40205 persisted, produced well through the dry season and were still increasing their yields when the trial terminated. Their production, however, was usually considerably less than *C. anagyroides* and *C. gyroides*.

The three *Centrosema* lines and *S. scabra* CPI 40292 form a less productive group of persistent legumes. Plants that were ill-adapted or very low producers of dry matter included Verano (which died out, then regenerated from self-sown seed), *C. caeruleum*, *S. scabra* CPI 34925, *Leucaena* and Cooper *Neonotonia*. The *Leucaena* would have been expected to be more productive if the cutting height had been higher, but the wet nature of the soil was probably a major cause of its low productivity as well as of the *S. scabra* lines and Verano. *C. retusa* produced relatively well in the trial; however, its utility is only as a green manure, for which it shows some promise.

Analysis of the proportion of leaves in the sampled material of the three bushy legumes *C. gyroides*, *Leucaena* and *C. anagyroides* produced very similar results over the period of the trial (68%, 65% and 61%, respectively), though it varied widely between harvests for each legume (65–90%, 49–100% and 52–80%, respectively). *C. gyroides* and *Leucaena* had higher proportions of leaves when yields were lower; the pattern of leaf production for *C. anagyroides* was generally similar but erratic, possibly indicating some loss of leaves in March, August and November.

Table 17.5. High river terrace site: weed dry matter harvest yields from mixed swards under 6-weekly cutting.

	Harvest yields (year and month) (kg/ha × 10^{-1})												
	1976							1977					Trial
Legume treatment	3[a]	5[a]	6	8	9	10	12	1	3	4	5	7	totals
C. plumieri	–	–	74	224	104	65	54	76	34	79	89	184	983
C. pubescens CF 6-1	–	–	86	153	34	32	19	35	18	33	38	74	521
C. pubescens cv. Belalto	–	–	65	149	38	46	33	24	17	42	88	140	643
C. pubescens (Centro)	–	–	48	144	6	11	5	3	1	3	7	15	244
C. gyroides	–	–	7	36	13	5	0	1	0	0	0	16	78
C. anagyroides	–	–	0	7	0	0	4	0	0	0	0	0	11
C. retusa	–	–	2	32	13	1	0	0	4	0	0	8	60
L. leucocephala cv. Peru	–	–	32	117	21	12	7	10	3	16	20	54	293
M. atropurpureum cv. Siratro	–	–	55	149	62	40	14	40	12	64	51	110	597
N. wightii cv. Cooper	–	–	115	271	109	97	46	77	31	70	84	164	1065
S. guianensis cv. Cook	–	–	99	71	25	11	2	6	2	14	16	64	309
S. guianensis cv. Endeavour	–	–	54	61	3	4	0	4	2	2	6	20	156
S. hamata cv. Verano	–	–	156	145	98	29	5	20	4	35	40	92	625
S. scabra CPI 34925	–	–	74	62	20	5	1	2	1	3	8	48	223
S. scabra CPI 40205	–	–	44	113	45	23	2	9	5	25	38	87	391
S. scabra cv. Seca	–	–	164	144	52	25	7	11	2	11	22	95	478
LSD ($P \geq 0.05$)	–	–	82	123	54	45	22	34	15	38	50	76	375

[a]Plots weeded. LSD between legumes and harvest ($P \geq 0.05$), same legume 49, between legumes 55.

Table 17.6. High river terrace site: total dry matter harvest yields from mixed swards under 6-weekly cutting.

	Harvest yields (year and month) (kg/ha × 10^{-1})												
	1976							1977					Trial
Legume treatment	3[a]	5[a]	6	8	9	10	12	1	3	4	5	7	totals
C. plumieri	–	–	19	82	61	28	34	52	28	27	67	114	512
C. pubescens CF 6-1	–	–	29	99	123	92	75	93	34	61	125	180	911
C. pubescens cv. Belalto	–	–	48	94	98	73	54	120	41	44	53	97	723
C. pubescens (Centro)	–	–	27	116	106	68	70	113	48	52	111	202	913
C. gyroides	–	–	2	51	36	38	14	23	16	9	3	92	286
C. anagyroides	–	–	0	36	2	10	7	14	11	25	32	89	227
C. retusa	–	–	0	30	8	11	11	29	10	22	16	31	166
L. leucocephala cv. Peru	–	–	53	158	140	91	57	81	47	68	54	164	913
M. atropurpureum cv. Siratro	–	–	28	113	94	90	58	99	42	47	70	151	793
N. wightii cv. Cooper	–	–	20	66	57	48	32	40	28	43	63	90	488
S. guianensis cv. Cook	–	–	33	29	51	24	11	30	22	21	46	85	351
S. guianensis cv. Endeavour	–	–	42	37	55	21	22	21	18	25	47	108	397
S. hamata cv. Verano	–	–	21	16	40	17	10	16	7	17	22	65	232
S. scabra CPI 34925	–	–	36	26	70	48	24	32	22	27	32	135	454
S. scabra CPI 40205	–	–	18	33	24	21	26	27	13	19	20	49	250
S. scabra cv. Seca	–	–	15	42	25	35	20	26	22	25	61	86	351
LSD ($P \geq 0.05$)			22	64	66	38	26	38	ns	27	46	66	217

[a]Plots were weeded. LSD between legumes and harvest ($P \geq 0.05$), same legume 36, between legumes 39; ns, not significant ($P \geq 0.05$)

Fig. 17.3. Low Pine Ridge trial site. (a) General view of trial site (behind single plant trial). See also Fig. 14.3, Chapter 14, this volume. (b) *Codariocalyx gyroides* after 10 harvests, July 1977.

Table 17.7. Lowland pine savanna site: legume dry matter harvest yields from mixed swards under 6-weekly cutting.

	Harvest yields (year and month) (kg/ha × 10^{-1})										Totals	
	1976						1977					
Legume treatment	4	7	8	10	11	12	2	3	5	6	Trial	July to June
C. caeruleum	0	1	3	3	4	3	6	35	25	20	101	101
C. plumieri	1	5	7	6	5	4	6	15	18	28	95	94
C. pubescens CF 6-1	1	2	5	6	10	6	15	34	31	28	136	135
C. pubescens (Centro)	1	1	2	3	9	2	8	29	25	25	105	104
C. gyroides	0	18	38	52	26	19	40	223	141	142	699	699
C. anagyroides	2	10	24	13	31	21	58	176	133	235	702	700
C. retusa	1	2	0	8	9	14	57	130	48	52	322	321
L. leucocephala cv. Peru	11	15	5	16	4	1	6	10	14	8	91	80
M. atropurpureum cv. Siratro	5	22	17	25	19	20	21	52	56	61	297	292
N. wightii cv. Cooper	1	10	6	8	7	6	7	40	21	8	114	113
S. guianensis cv. Cook	6	43	2	52	49	4	10	44	30	51	292	286
S. guianensis cv. Endeavour	0	20	1	9	23	4	2	1	1	4	65	65
S. hamata cv. Verano	0	9	0	4	9	0	0	0	0	4	28	28
S. scabra CPI 34925	2	6	0	7	1	2	4	6	10	7	47	45
S. scabra CPI 40205	11	70	8	24	9	12	19	28	40	50	270	259
S. scabra cv. Seca	5	37	8	16	18	7	12	12	11	36	162	157
LSD ($P \geq 0.05$)	5	26	14	17	10	11	20	53	31	87	35	148

LSD between harvests ($P \geq 0.05$) 35.

The figures for proportion of leaves in the harvested material over the trial period are similar to those obtained in a previous trial (Lazier, 1981c), in which for *C. gyroides* and *Leucaena* cut at 6-weekly intervals at 50 cm and 15 cm respectively, the proportions of leaves in the harvests were 71% and 68% respectively. The proportion of *C. gyroides* leaves was higher at shorter cutting intervals and taller cutting heights (81% for 2-week harvests cut at 50 cm).

The proportion of palatable browse of *C. gyroides* is somewhat higher than indicated because, unlike the other two browse species whose new stems rapidly lignify, the young stems of *C. gyroides* remain green and succulent and are normally grazed by cattle. Thus *C. gyroides* appears to have a somewhat higher proportion of total palatable material than either *Leucaena* or *C. anagyroides*.

Grass

Grass production, mainly *Mesosetum angustifolium*, was very low (Table 17.8). Generally the legumes that were harvested at higher cutting heights had no recorded grass, while lower cutting heights produced only low yields. The plots, which were all mown before planting in November 1975, had no grass recorded in the first harvest (April 1976) because it had not grown up to the lowest cutting height. The relatively high yields in the following harvest (July 1976) were the dry matter accumulated since the plots were mown 8 months before. Low grass yields in August 1976 may be due to a growth check after the July harvest; yields continued to be low in the following rainy and cool seasons and production rose only slightly through the 1977 dry season and into the following wet season. The native savanna grass yields then appeared to be low in the cool season and elevated in the dry and early wet seasons.

The harvest at 15 cm of unfertilized *M. angustifolium* plots and fertilized *M. angustifolium* from *C. gyroides* plots for grass productivity produced almost identical dry matter yields for March, May and June 1977 (70, 70; 140, 160; 160, 140 kg/ha, respectively). These yields are similar to those of the Siratro plots, which were also harvested at 15 cm. The lack of response of *M. angustifolium* to the applied fertilizer and the presence of a vigorous legume indicates the need for its replacement with a vigorous grass that will respond to fertilizers applied at the rates used in the trial, and that will compete with vigorous forage legumes such as *C. gyroides*.

Weeds

Weed production (Table 17.9) was dependent on cutting height, since the main weeds were two

Table 17.8. Lowland pine savanna site: *M. angustifolium* dry matter harvest yields from mixed swards under 6-weekly cutting.

	Harvest yields (year and month) (kg/ha × 10^{-1})										
	1976						1977				
Legume treatment	4	7	8	10	11	12	2	3	5	6	Trial totals
C. caeruleum	0	15	10	17	11	6	7	10	12	12	100
C. plumieri	0	46	18	23	29	7	10	13	7	28	181
C. pubescens CF 6-1	0	33	16	15	13	13	9	13	13	18	144
C. pubescens (Centro)	0	46	12	18	19	6	13	30	22	20	188
C. gyroides	0	0	0	0	0	0	0	0	0	0	0
C. anagyroides	0	1	0	11	0	0	0	0	0	0	12
C. retusa	0	0	0	0	0	0	0	0	0	0	0
L. leucocephala cv. Peru	0	55	25	20	27	11	11	15	23	21	208
M. atropurpureum cv. Siratro	0	30	9	24	16	6	11	12	12	17	137
N. wightii cv. Cooper	0	39	18	25	14	13	12	16	12	20	170
S. guianensis cv. Cook	0	29	0	1	1	1	1	2	1	4	41
S. guianensis cv. Endeavour	0	28	0	1	1	1	2	1	2	2	37
S. hamata cv. Verano	0	31	0	0	1	1	2	0	1	12	50
S. scabra CPI 34925	0	37	1	1	3	4	1	4	6	7	64
S. scabra CPI 40205	0	55	0	1	3	1	1	2	6	4	73
S. scabra cv. Seca	0	52	0	0	1	1	0	1	1	5	61
LSD ($P \geq 0.05$)		27	7	15	13	5	6	8	9	13	68

LSD between harvests ($P \geq 0.05$) same legume 37, between legumes 38.

Table 17.9. Lowland pine savanna site: weed dry matter harvest yields from mixed swards under 6-weekly cutting.

	Harvest yields (year and month) (kg/ha × 10^{-1})										
	1976						1977				
Legume treatment	4	7	8	10	11	12	2	3	5	6	Trial total
C. caeruleum	0	8	1	3	1	0	0	0	1	1	16
C. plumieri	0	8	4	2	4	0	1	0	3	4	28
C. pubescens CF 6-1	0	1	3	3	1	1	1	1	0	1	12
C. pubescens (Centro)	0	3	23	3	4	1	1	8	0	6	50
C. gyroides	0	0	0	0	0	0	0	0	0	0	0
C. anagyroides	0	0	0	3	0	0	0	0	0	0	3
C. retusa	0	0	0	0	0	0	0	0	0	1	2
L. leucocephala cv. Peru	0	2	6	1	2	0	1	1	0	1	14
M. atropurpureum cv. Siratro	0	1	9	1	4	5	0	0	0	1	18
N. wightii cv. Cooper	0	7	4	2	2	1	0	0	2	3	22
S. guianensis cv. Cook	0	2	0	0	2	0	0	0	0	0	4
S. guianensis cv. Endeavour	0	4	0	1	0	0	0	0	0	0	5
S. hamata cv. Verano	0	4	0	1	1	0	0	0	2	1	10
S. scabra CPI 34925	0	0	0	0	0	0	0	0	0	0	1
S. scabra CPI 40205	0	4	0	0	1	0	0	0	1	2	9
S. scabra cv. Seca	0	2	0	0	1	1	1	1	1	2	9
LSD ($P \geq 0.05$)		ns	ns	ns	ns	ns	ns	ns	ns	ns	ns

ns, not significant ($P \geq 0.05$).

low native legumes, *Cassia diphylla* and *C. tagera*. They visually showed markedly increased vigour in the fertilized plots but formed a very small proportion of the total dry matter harvested.

The yield totals of the treatments (Table 17.10) reflect the yields of the dominant fractions of the swards, the introduced legumes.

17.3.3 Relative productivity in the two pasture environments

Legumes

Comparing the legume yields of the last nine harvests of the two sites (Tables 17.3 and 17.7), it can be seen that without weed and grass competition the legumes that were adapted to the wetter environment of the Puletan Ultisol produced at least as much dry matter as those on the alluvial terrace, and *C. anagyroides*, *C. retusa* and *S. scabra* (CPI 40205 and CPI 40292) produced substantially more.

The more acid conditions of the savanna site may have been a factor in the better *S. scabra* performance there. Legumes that were apparently poorly adapted to the savanna site compared with the river terrace site were Endeavour Stylo, the *Centrosema* lines and *Leucaena*.

Growth patterns were similar at both sites in that yields decreased during the cool period and increased with the advent of the wet season. However, during the dry season yields decreased at the river terrace site while at the savanna site yields increased, particularly in the case of *C. gyroides* and *C. anagyroides*.

As trampling by grazing animals on the soft wet savanna soils can rapidly eliminate low forage plants, it is advantageous that the more successful lines in the trial were bushy and thus would be affected by trampling only in the early stages of growth. *C. gyroides* may have potential for the small Belizean sheep industry, because, due to its height and leaf retention in the dry season, fodder can be conserved above the grazing height of the animals. Cattle, on the other hand, can normally graze the entire bush.

Table 17.10. Lowland pine savanna site: total dry matter harvest yields from mixed swards under 6-weekly cutting.

	Harvest yields (year and month) (kg/ha × 10^{-1})										
	1976						1977				
Legume	4	7	8	10	11	12	2	3	5	6	Trial totals
C. caeruleum	0	25	15	22	15	10	13	46	38	33	216
C. plumieri	1	59	29	31	39	11	18	28	29	60	304
C. pubescens CF 6-1	1	36	25	24	24	19	25	48	44	46	292
C. pubescens (Centro)	1	50	38	25	31	9	23	66	48	51	342
C. gyroides	0	18	38	52	26	19	40	223	141	142	699
C. anagyroides	3	10	24	27	31	21	59	176	133	235	717
C. retusa	1	2	0	8	9	14	58	131	48	54	323
L. leucocephala cv. Peru	10	72	36	38	34	12	19	26	38	30	314
M. atropurpureum cv. Siratro	5	53	35	50	36	31`	32	64	68	79	452
N. wightii cv. Cooper	1	55	29	35	24	20	20	55	35	31	306
S. guianensis cv. Cook	6	74	2	54	51	5	11	46	32	55	336
S. guianensis cv. Endeavour	0	52	1	10	24	5	4	1	3	6	107
S. hamata cv. Verano	0	44	0	6	12	1	3	1	3	18	88
S. scabra CPI 34925	2	44	1	8	4	6	6	10	16	14	111
S. scabra CPI 40205	11	129	8	25	12	13	20	31	47	57	353
S. scabra cv. Seca	5	92	8	16	21	8	13	14	13	43	232
LSD ($P \geq 0.05$)	5	44	22	18	19	11	21	55	33	90	173

Grasses

The *M. angustifolium* at the savanna site produced more consistent but much lower yields than the Coastal Bermuda grass at the river terrace site. Due to the absence of productive grass, the total plot yields at the savanna site were much lower as well. With basal fertilizer application and N supplied by a vigorous associated legume such as *C. gyroides*, the native *M. angustifolium* sward could be replaced by more productive lines of water-tolerant grasses such as *Panicum maximum* cv. Makueni and *Brachiaria mutica* (Para), which have persisted in more recent studies by the Belize Department of Agriculture at sites nearby. On such soft, water-logged soils, careful grazing management would be required to prevent destruction of the grass and any herbaceous legume component by trampling.

Coastal Bermuda grass, while generally lower in yield than Para grass on a lower, more fertile river terrace (Lazier, 1981a), produced comparable yields in some plots (e.g. Cooper *Neonotonia*). Its persistence under heavy grazing and low fertility conditions on the research farm (Central Farm) and its ability to spread rapidly indicate that it may be suitable for pastures receiving low levels of management.

Weeds

Herbaceous weed yields were very low at the lowland pine savanna site compared with the river terrace site. The low yields are probably due to the presence of relatively few species of weeds adapted to the wet and infertile soils, and that those that were adapted were genetically unable to take full advantage of the higher soil fertility at the trial site. There was apparently a low incidence of weed seeds in the soil and a very low incidence of weeds adapted to more fertile wet soils.

This absence of herbaceous weed competition at the savanna site over the period of the trial indicated that pasture establishment should be relatively easy compared with on soils that are naturally more fertile, because weed competition would not be a problem until the pastures have had sufficient time to become established. The reduced management required should, at least in part, offset the higher fertilizer requirements of pasture establishment on such soils.

18 Productivity of 16 Forage Legumes in Mixed Swards Under Cutting in Belize on Contrasting Soils: a High Alluvial Terrace and A Lowland Pine Savanna II. Nutrient Levels and Feeding Value

J.R. Lazier*[1]

**Formerly International Livestock Centre for Africa*

Abstract

In Belize samples from the more productive treatments from two harvests of mixed swards of legumes and grasses in two trials were analysed for contents of N, P, K, Ca, Mg, Mn, Cu, Zn and S. The results were related to estimated minimal nutrient levels for maintenance and production of cattle. The trial sites were on a high river terrace on a heavy black clay soil where the grass component was *Cynodon dactylon* cv. Coastal and three legume treatments were analysed, and on a lowland pine savanna, a waterlogged Ultisol, where the grass was the native savanna grass *Mesosetum angustifolium* and six legume treatments were analysed. Cu was below minimal maintenance levels at both sites and in all fractions analysed. The mature savanna grass had below maintenance levels of N, P, K, Ca, Mg and Zn. Nutrient levels in frequently cut unfertilized savanna grass did not differ from those in frequently cut fertilized savanna grass in the legume plots, and had levels of N, P, Zn and perhaps Mg below estimated maintenance levels for all weights of cattle. Fertilized savanna weed nutrients were below maintenance levels of the largest animals for P in both harvests and N and Mg in the June harvest. P and Zn concentrations in the planted legumes of the savanna trial were below maintenance levels for larger animals. Only in the *Stylosanthes* lines was adequate Zn present for all levels of production. The high river terrace had fewer nutrition deficiencies. The grass had levels of Zn and Mg that were below maintenance and Ca and Mg too low for production. The Zn content of the legumes was below maintenance levels and P contents were below minimum levels required for production. The implications of these results on pasture management are briefly discussed.

18.1 Introduction

The nutritive value of feed from pastures is of considerable importance because low levels of nutritional elements can result in poor live-weight gains despite adequate dry matter intake. The higher alluvial terraces of the Belize River and the infertile waterlogged Low Pine Ridge soils are currently being developed for agriculture and little work has been done on the nutrient levels of forages grown on those sites with or without fertilizer.

An earlier comparative study of the nutrient composition of unfertilized Jaragua grass (*Hyparrhenia rufa*) swards on a low alluvial terrace of the Belize River at Baking Pot and on the

[1] E-mail: john.lazier@sympatico.ca

Low Pine Ridge soils at Retaliation 49 indicated at both sites the crude protein levels, Cu, Zn and Co were below maintenance levels, I, K, Mg and Mn were adequate at both sites, and P and Ca were adequate only at the alluvial site (Rose-Innis, 1973). A similar study on *Brachiaria mutica* on the two soil types gave similar results for most nutrient elements. Iron had also been analysed and was found to be adequate for maintenance and production (Wright *et al.*, 1959).

This study was undertaken as part of two separate trials comparing the dry matter productivities of fertilized grass–legume swards on two infertile soil types: a heavy black clay on a high river terrace and a waterlogged Ultisol on the lowland pine savanna. The details of the soils and the trials are presented in Chapter 17 (this volume). Samples were analysed of the legumes, grasses and weeds in order to determine their feed value.

18.2 Methods

The experiments and the sites were described previously (Chapter 17, this volume). The high alluvial terrace site was at an altitude of 60 m on a heavy clay soil on Central Farm, the Belize Department of Agriculture research station in Cayo District. The lowland pine savanna site was a waterlogged Ultisol located on the north side of the Western Highway at mile 33.5 (56 km to the east of the high river terrace site) on slightly sloping land at an altitude of 40 m. The fertilizer rates had been chosen so as to attempt to supply the minimum requirements for plant growth and to be an economically feasible investment for local farmers.

Both sites were planted in early November 1975. An initial fertilizer application of 12 kg/ha of P and 20 kg/ha of K as triple super phosphate and muriate of potash respectively was made at the time of planting at the alluvial site and in January 1976 at the lowland pine savanna site. The same amount was applied as an annual rate in quarterly applications after alternate harvests. The savanna site received an additional one-time application of 63 kg/ha of a fritted trace element mixture that was available locally. The constituent trace elements were applied at the following rates: Fe 9.4 kg/ha, Zn 2.5 kg/ha, B 0.8 kg/ha, Mo 0.42 kg/ha, Cu 1.0 kg/ha and Mn 9.5 kg/ha.

The samples analysed were harvested from the higher yielding treatments at both sites and were taken at the times of the regular 6-weekly harvests for yield in February and June 1977 at the savanna site, and January and March 1977 harvests from the alluvial terrace site. Young fully expanded leaves were plucked for analysis of legume and grass nutrient contents.

The samples were dried at 65°C in a forced draught oven and analysed at the Ministry of Agriculture Analytic Laboratory at Central Farm, Cayo District, Belize using the following techniques: N, microkjeldhal, followed by distillation in a Markham still; K, flame photometer using controlled dry ashing; P, spectrophotometry using the vanado-molybdate method (dry ashing); Ca and Mg, atomic absorption spectrophotometry; Mn, Fe, Cu and Zn, atomic absorption of controlled dry ashing; S, precipitation as barium sulphate with solid barium chloride and stabilizer, measured by visual spectrophotometry.

18.3 Results

The results of the analyses are presented in Tables 18.1 (grasses) and Tables 18.2 (legumes and weeds).

The recommended cattle nutrient levels presented are taken from the ARC review of nutrient requirements (ARC, 1980). Animal feed intake is related to the metabolizability of the energy of the diet (the q value). Because the q values were not determined in this study, the nutrient requirements were taken from the lowest q values presented in the ARC tables (0.6 unless otherwise stated).

18.3.1 Nitrogen

The recommended N level for maintenance and growth for all sizes of beef cattle is 1.15% of dry matter ($q = 0.5$), except for the lowest live weight presented, the 100 kg calf, which requires 1.48% N. For 10 kg/day milk production at the same q value, the N requirement varies between 1.17% N and 1.36% N for temperate dairy breeds, and for 20 kg/day milk production between 1.39 % and 1.52 % is required.

Table 18.1. Grass nutrient contents of fertilized mixed swards at low pine savanna and high river terrace sites (bold indicates below maintenance levels).

Legume	N %		P %		K %		Ca %		Mg %		Mn ppm		Cu ppm		Zn ppm		S %
Low Pine Savanna (*Mesosetum angustifolium*)																	
	Feb	Jun	Feb	Jun	Feb	Jun	Feb	Jun	Feb	Jun	Feb	Jun	Feb	Jun	Feb	Jun	Jun
C. gyroides	**0.94**	1.25	**0.09**	0.15	0.60	1.18	0.28	0.30	0.25	0.19	145	105	**3.7**	**3.3**	20	21	0.18
L. leucocephala cv. Peru	**1.03**	**0.78**	0.17	0.12	1.67	1.48	0.27	0.22	0.25	0.19	135	96	**2.3**	**2.4**	25	**12**	0.21
M. atropurpureum cv. Siratro	**1.06**	**1.06**	0.11	**0.08**	**0.09**	1.35	0.31	0.27	0.24	0.20	116	92	**2.3**	**4.0**	36	20	0.20
S. guianensis cv. Cook	**0.86**	**1.07**	**0.09**	**0.10**	1.13	1.93	0.33	0.34	0.23	0.19	135	133	**5.0**	**4.1**	22	19	0.25
S. guianensis cv. Endeavour	**1.06**	**0.89**	0.11	0.12	1.15	1.43	0.32	0.29	0.26	0.16	132	105	**2.4**	**3.6**	24	16	0.18
Control[a]	**0.61**	**0.94**	**0.02**	**0.08**	0.55	1.77	0.16	0.09	0.17	0.19	89	92	**1.8**	**2.0**	19	20	na
LSD ($P \geq 0.05$)	23	ns	0.05	ns	0.86	ns	ns	ns	0.05	ns	ns	ns	0.7	ns	ns	ns	ns
High River Terrace (*C. dactylon* cv. Coastal)																	
	Jan	Mar	Jan	Mar	Jan	Mar	Jan	Mar	Jan	Mar	Jan	Mar	Jan	Mar	Jan	Mar	Jan
Legume plots[b]	2.47	1.72	0.26	0.27	1.80	1.20	0.45	0.46	0.26	0.17	50	79	**2.5**	**5.2**	25	24	0.18

[a]Control is unfertilized *M. angustifolium*; [b]mean nutrient contents from *C. gyroides* and *L. leucocephala* plots; na, not analysed, ns, not significant ($P \geq 0.05$).

Table 18.2. Legume and weed nutrient contents of fertilized mixed swards at low pine savanna and high river terrace sites (bold indicates below maintenance levels).

Legume	N %		P %		K %		Ca %		Mg %		Mn ppm		Cu ppm		Zn ppm		S %
Low Pine Savanna																	
	Feb	Jun	Feb	Jun	Feb	Jun	Feb	Jun	Feb	Jun	Feb	Jun	Feb	Jun	Feb	Jun	Feb
C. gyroides	3.11	2.62	0.27	0.20	1.67	1.82	1.00	1.09	0.36	0.24	272	113	**6.1**	9.3	29	22	0.25
C. anagyroides	4.48	2.61	0.33	0.19	2.25	2.27	1.27	0.81	0.38	0.30	123	104	**4.2**	**6.4**	28	27	0.31
L. leucocephala cv. Peru	3.40	3.18	0.25	0.21	1.87	4.33	1.07	1.24	0.30	0.29	68	66	**4.4**	**4.1**	28	18	0.23
M. atropurpureum cv. Siratro	2.91	2.62	0.22	0.15	1.51	2.10	1.01	1.20	0.37	0.35	65	75	**5.8**	**5.3**	33	26	0.24
S. guianensis cv. Cook	2.59	1.68	0.21	0.14	1.15	2.22	1.39	1.58	0.36	0.31	105	81	**4.8**	**5.8**	35	30	0.20
S. guianensis cv. Endeavour	2.09	1.91	0.22	0.14	1.38	2.87	1.15	0.96	0.51	0.37	131	106	**5.1**	**5.7**	32	36	0.17
Weed[a]	1.84	1.26	0.24	0.22	1.86	2.32	0.46	0.60	0.34	0.20	166	126	**3.3**	**6.8**	32	33	0.25
LSD ($P \geq 0.05$)	1.50	0.61	0.04	0.06	ns	ns	0.18	0.57	ns	0.11	152	41	ns	2.9	ns	9	0.08
High River Terrace																	
	Jan	Mar	Jan	Mar	Jan	Mar	Jan	Mar	Jan	Mar	Jan	Mar	Jan	Mar	Jan	Mar	Jan
C. gyroides	2.98	3.02	0.27	0.24	1.60	1.31	1.18	1.68	**0.26**	**0.08**	83	62	11.1	**4.3**	26	23	0.19
C. anagyroides	4.92	–	0.29	–	2.42	–	1.09	–	**0.29**	–	38	30	**6.8**	**6.9**	28	24	0.21
L. leucocephala cv. Peru	4.79	4.40	0.28	0.29	2.35	1.80	1.47	1.94	**0.28**	**0.21**	33	35	**6.8**	**5.8**	20	24	0.39
Weed[a]	2.03	1.82	0.24	0.23	3.23	1.61	1.33	1.14	–	–	–	–	–	–	–	–	–

[a]Samples from *L. leucocephala* plots and includes native legumes; ns, not significant ($P \geq 0.05$); –, not analysed.

The savanna grass N contents were all below the minimal N levels for maintenance except for that of the June harvest of *C. gyroides* (1.25%). The unfertilized uncut savanna grass from the February harvest had less N than any of the other plots, which had all been cut previously. In June, after four harvests, the N content of the grass in the unfertilized plot did not differ from that of the grass in the fertilized plots. The river terrace grass had N contents that were adequate for all levels of production. The savanna legume N yields varied widely. *C. anagyroides* had the highest N contents in the February harvest and the two *Stylosanthes* lines had the lowest contents in both harvests. Legume N levels were adequate for all levels of production at both sites. Weed N levels at both sites were adequate for maintenance and only the June harvest at the savanna site was within the upper range of values for productivity.

18.3.2 Phosphorous

The recommended minimum intake P levels for maintenance varied from 0.11% of dry matter for 100 kg beef cattle to 0.25% of dry matter for 600 kg animals. For this range of live weights, production of 0.5 kg live weight gain per day (LWG/day) requires P levels from 0.27% to 0.29% of dry matter, respectively. Pregnant cattle require P levels of 0.24% to 0.26% of dry matter while lactating animals at those two live weights and 10 and 20 kg/day production require 0.23% to 0.30% and 0.31% to 0.35% P, respectively.

The mature unfertilized savanna grass analysed from the February harvest had a very low P content (0.02%), lower than any of the samples from the fertilized plots. The June harvest of the unfertilized plots that had been cut over and had regrown had similar P levels to those of the fertilized plots, indicating that the addition of P fertilizer at the levels used in the trial made no significant difference to the P content of the savanna grass. The P levels in the fertilized plots were never higher than the lowest levels required for maintenance, and were frequently below. The river terrace site had grass P contents approaching or at the levels considered adequate for pregnant animals or 0.5 kg LWG/day for larger animals.

The savanna legume and weed P levels were generally higher than those of the grasses and most legumes were within the range of levels required for maintenance in the February harvest; *C. anagyroides* (0.33% P) and *C. gyroides* (0.27% P) had P levels adequate for some levels of production. In the June harvest, however, the mean P levels of the legumes and weeds (0.18%) were lower ($P = 0.001$) than in the February harvest (0.24%) and were below the maintenance requirements of larger animals. The three lowest P levels in each harvest were the *Stylosanthes* lines and Siratro.

The river terrace legume P levels in January were similar to the levels in the same legumes in February at the savanna site, except for the *Leucaena* P levels which were higher, particularly in the March harvest. Only *Leucaena* at the terrace site had adequate P for 0.5 kg LWG/day for the full range of stock weights.

The weed P levels were similar at both sites and were just below the P levels required for maintenance of the largest animals.

18.3.3 Potassium

The K levels required for maintenance, growth and production range from 0.47% to 0.84% of dry matter. In February at the savanna site the grass in the unfertilized plots and the *C. gyroides* plots had a K content just above the lowest maintenance levels. The low K levels were probably due to the maturity of the grass, the unfertilized plot having not been cut previously, and the *C. gyroides* plot, harvested at 50 cm not having been cut since the establishment of the trial 15 months previously. With the exception of the two mature grass plots at the savanna site, all other analyses at both sites indicated sufficient K for all levels of production. The unfertilized savanna grass in the June harvest had as much K as the plots receiving K fertilizer. The analyses tended to be higher in the June harvest than in the February harvest at the savanna site and in the January harvest compared with the March harvest at the river site, but the differences were not significant. Weed K contents were similar to those of the legumes.

18.3.4 Calcium

Maintenance levels of Ca ranged from 0.17% of dry matter for a 100 kg beef animal to 0.28% for one of 600 kg. Growth at 0.5 kg LWG/day is 0.55% to 0.34% for the same animals. Pregnant cattle require 0.30% to 0.32% Ca in the dry matter, while cattle producing from 10 to 20 kg of milk per day require 0.32% to 0.33% and 0.34% to 0.35% respectively.

The Ca levels of the savanna grass were below maintenance levels in both harvests, while the savanna grass in the fertilized plots all had Ca levels adequate only for maintenance, with the exception of the *Leucaena* plots, which particularly in the June harvest did not have adequate Ca levels for maintenance of larger animals. The higher Ca levels in the fertilized plots are of interest because no Ca fertilizer was added. The river terrace site grasses had higher Ca levels than the savanna grass and these were generally adequate for slow growth (0.5 kg/day) of all but the smallest animals (100 kg). The legume Ca contents at both sites were adequate for all levels of production, as were the weeds.

18.3.5 Magnesium

Minimum levels of Mg for growth at 0.5 kg LWG/day, pregnancy and lactation up to 20 kg/day fall within the range of minimum maintenance levels for 100 kg to 600 kg animals, 0.12% to 0.21% of dry matter respectively.

With the exception of the anomalous March harvest of *C. gyroides* at the river terrace site, all of the legume analyses at both sites exceeded the Mg requirements for maintenance and production. All of the analyses of fertilized grasses in the first harvests exceeded minimum levels as well. In the second harvests the analyses were lower, significantly at the savanna site ($P = 0.001$), where the mean Mg contents of the legumes and weeds fell from 0.39% to 0.28%. Mg in mature savanna grasses in February could be limiting for maintenance of larger animals, as could the June contents of the fertilized and unfertilized savanna grass, savanna weed and the river terrace grass in March.

Differences between fertilized and unfertilized savanna grass Mg values in February are probably due to the difference in ages of the sampled material. In the June harvest, after several cuts, the unfertilized grass had Mg contents similar to that of the fertilized plots.

18.3.6 Manganese

The minimum recommended level of Mn in the diet for growth and skeletal development is 20–25 ppm. All of the analyses at both sites exceeded these values. *C. anagyroides* and *Leucaena* at the river terrace site had comparatively low, though still adequate, levels. The mean of the legume Mn contents in the February harvest at the savanna site (130 ppm) was higher ($P = 0.001$) than that of the June harvest (104 ppm). Poor drainage at the lowland pine savanna site may account for the somewhat higher Mn levels there (Viets and Lindsay, 1973).

18.3.7 Copper

Cu requirements vary with the q value of the feed, and for cattle they range between 8 and 15 ppm for slow-growing beef cattle and 12 to 19 ppm for maintenance of a 500 kg adult. Pregnant cattle require 13 to 20 ppm and lactating animals 10 to 14 ppm.

Only two of the samples exceeded the lowest minimum Cu requirements. *C. gyroides* at the river terrace in January and at the savanna site in June. The mean legume Cu levels (4.7 and 5.9 ppm) were apparently slightly higher than those of the grass (2.8 and 3.2 ppm) at the savanna site for the February and June harvests respectively.

Mo and S have an antagonistic effect on the absorption of Cu. S concentrations obtained in the analyses range from 0.17% to 0.31%, which in the presence of levels of Mo of 1 ppm or greater could seriously reduce the absorption of Cu. Mo levels were not determined in this trial; however, one value obtained for Mo in grass grown on the savanna soils, 0.19 ppm (Wright *et al.*, 1959), indicates a small S effect on Cu absorption.

18.3.8 Zinc

Herbage usually contains 15–60 ppm Zn and critical levels may be in the range of 21–49 ppm

(Whitehead, 1966). Though laboratory tests have indicated that 30 ppm may be a critical level for ruminants, the estimates of Zn requirements for animals in the field change with the q value of the feed; where the q value is 0.4–0.6, they lie within the range of 13–28 ppm for maintenance and slow growth, 13–21 ppm for pregnancy and 18–25 ppm for lactation at 10 kg milk/day.

The grass samples at both sites, with only one exception, were below 30 ppm but were within the range of minimal requirements. One sample from the savanna site had 12 ppm Zn, which is below the range of minimal requirements. The mean grass Zn content at the savanna site was lower in June (18 ppm) than in February (25 ppm) ($P = 0.01$), while the mean legume Zn levels did not differ but were apparently higher than those of the savanna grass (32 and 29 ppm for the February and June harvests). The Zn content of the weeds and the two *Stylosanthes* lines were above the minimal Zn levels in both harvests, but the remainder of the legume samples at both sites were within the range of minimal requirements.

18.3.9 Sulphur

The requirements of plants and animals for S are determined by the N content because both elements are utilized in amino acids. The N:S ratio in non-S deficient plants has been estimated to be approximately 17:1 (Dijkshoorn and Van Wijk, 1967), while the ratio required by animals has been estimated by utilizing the degradable portion of the two elements as 14.1:1 (from 1.1 to 1.6 g S/kg dry matter) (ARC, 1980). The S contents of the fractions analysed in this study all contained more than the minimum requirements for plant growth and for animal production when considered individually on a g/kg dry matter basis. Only the material from the fertilized plots was analysed, however: the unfertilized savanna grass remains to be examined.

18.4 Discussion

18.4.1 The savanna site

The data generally confirmed results obtained from the analyses of the strip trials reported in Chapter 13 (this volume). The elements Mn and S were apparently adequate for maintenance and all levels of production in all fractions and harvests analysed. The mature unfertilized savanna grass in February had N, P, Ca and Cu below maintenance requirements, the lowest minimal maintenance levels of K, adequate levels of Zn and Mg for maintenance and production of all but larger animals; only Mn was adequate for all levels of production for all sizes of cattle. Although legumes formed a major portion of the weed component of the unfertilized savanna, the weeds formed an insignificant fraction of the dry matter present. Cattle dependent solely on mature savanna vegetation would not be expected to thrive and would require other sources of feed if they were to maintain weight.

The savanna does have patches of dense vegetation where tree species dominate. These are on sites that are presumably better drained, for the soil is characterized by gilgai microrelief and a much broader range of species. However, the extent of these patches is limited and while undoubtedly providing better quality nutrition due to better productivity and the presence of native legumes, they would soon be impoverished by overgrazing. Removal of the native vegetation in such areas and replanting with vigorous pasture plants may be possible, but the vigour of the bush regrowth is considerable and clearing by hand to preserve the microrelief (rather than mechanically), necessary on such soft soils, and planting and control of the regrowth would require considerable labour.

Savanna grass, if used for grazing, should be considered as almost solely a source of energy. Most of the remainder of the ration would have to be provided in the form of supplements or through grazing on alternative, more fertile sites.

After three cuttings the nutrient levels in the unfertilized savanna grass in June were higher than those of February for N, P and K, and lower for Ca, but none significantly so. The nutrient levels did not differ from those of the fertilized plots of the legume treatments. The addition of fertilizers of the types and at the quantities used in the trial to either mature uncut or cut savanna grass thus had no effect on the nutrient levels, or the dry matter production (Chapter 17, this volume). The cut unfertilized and cut fertilized grasses had below maintenance levels of N and Cu, the lowest maintenance levels or below of P and maintenance levels of Ca. K was adequate for all levels of maintenance and production.

Cutting of the mature grass will stimulate new growth of higher nutritive value, so some similar form of management should be exercised on mature grass before grazing. On much of the savanna the wet nature of the soil necessitates care if machinery is to be used to cut the grass, for deep tracks may be left in the soil, or the machinery may simply bog down. Disc harrowing is used by some farmers to stimulate regrowth; fire is the simplest and the traditional method, but is destructive of fencing.

Native legumes were observed to form a major portion of the weed fraction of the fertilized swards and their nutrient levels were higher than those of the associated native grass. In newly established pastures on the lowland pine savanna, weeds appear to form a very small fraction of the total dry matter productivity initially (Chapter 17, this volume) and thus are unlikely to contribute significantly to animal nutrition; the main native legumes *Cassia diphylla* and *C. tagera* are of low palatability and being short plants are readily trampled into the wet soil, as are most commercial forage legumes. Leguminous browse species may provide a more successful alternative.

Forage legumes planted on these soils without fertilization would not be expected to be productive, as was indicated by the poor growth of the legumes at the savanna site in the period between planting in October and fertilizer application in January.

The legumes planted in the fertilized plots of the trial contained the highest levels of nutrients of the sampled fractions. They had adequate N, K, Ca and Mg for maintenance and production. P was adequate for maintenance for all but larger animals; Zn was within the range of minimum values for maintenance and production, except in the *Stylosanthes* lines where it was adequate for all levels of production. Thus only Cu was below maintenance levels.

Some of the legumes in the trial were very productive throughout the year, notably *C. gyroides* and *C. anagyroides*, which produced over three times the best *M. angustifolium* yields (Chapter 17, this volume). Such legumes planted in the lowland savanna with minimum fertilizer application could thus greatly increase the dry matter productivity and the nutrient value of the forage. For maintenance, Co (Rose-Innes, 1973), Zn and Cu would have to be given as salt licks, and for production some form of P supplement.

The addition of fertilizer, while stimulating legume growth, will also encourage the growth and propagation of vigorous fertilizer-responsive weeds, which with time will rapidly increase their populations and will tend to crowd out the savanna grasses, and eventually the introduced legumes. A vigorous water-tolerant grass should thus be sown to replace *M. angustifolium* in order to compete more effectively with the weeds and to provide more and higher quality forage dry matter.

18.4.2 The high river terrace

More fertile than the lowland pine savanna, the legumes and Coastal Bermuda grass from the fertilized swards all had adequate levels of N, P, K, Mn and S for maintenance and production. The Zn and Cu contents were similar to those of the savanna site, being within the range of maintenance levels and below maintenance levels, respectively. The Coastal Bermuda grass had levels of P just below the level required for 0.5 kg LWG/day and adequate for pregnant cattle. Ca was adequate for growth of all but the smallest animals. Levels of Mg were sufficient for all levels of maintenance and production in January, but in March they were sufficient for maintenance of larger, not smaller animals. Legume Ca and Mg were adequate for all levels of maintenance and production. P was generally above minimum maintenance levels, but only *Leucaena* achieved P contents approaching that required for 0.5 kg LWG/day.

Weed levels of N, Ca and K were adequate for all levels of maintenance and production. P was adequate for maintenance of all but the largest animals. No other elements were analysed. So, to maintain animals on fertilized mixed swards of Coastal Bermuda grass and legumes would require only the addition of Co (Rose-Innes, 1973), Cu and Zn to the diet. P supplementation would be necessary as well, if higher levels of production were desired.

Although the river terrace site reported here was not located within a savanna area, such more recent, and thus presumably more fertile, alluvial terrace soils are found along rivers throughout the lowland pine savannas (Wright *et al.*, 1959). The utilization of both areas in grazing systems would be advantageous. Cattle could be rotated between pastures on both soil types, thus reducing the amount of supplements required for animal grazing the savannas. Alternatively, the savannas could be utilized as areas of reserve feed, in which vigorous introduced legumes would be allowed to accumulate bulk in the wet season and then would be grazed in the dry season when growth in the river valleys would be minimal.

19 A Summary of the Results of the IDRC-UWI/Belize Forage Legume Adaptation and Productivity Trials, 1973–1977

J.R. Lazier*[1]
*Formerly International Livestock Centre for Africa

Abstract
The results of 4.5 years of research of the IDRC-UWI/Belize legume adaptation and productivity trials over the years 1973–1977 are summarized. Numerous replicated and unreplicated trials were conducted on the terraces of the upper, mid and lower Belize River Valley and on the Mountain Pine Ridge on a wide range of clay soils. Data were acquired through both observation and harvesting; grazing was imposed on many of the trials. Adapted and persistent herbaceous and woody legume species were identified for all soil types and numerous companion grasses. For vigorous growth, fertility was required for most sites; grazing management was critical for legume persistence.

19.1 Introduction

It is appropriate at this point to review the UWI-Belize Forage Legume Programme results reported herein, as well as its earlier results reported elsewhere. The environments involved were the heavy black clay low, mid- and high terraces of the upper Belize River Valley, the waterlogged heavy clays of the Low Pine Ridge of the lower Belize River Valley and the more acid heavy clays of the Mountain Pine Ridge. The area is under high rainfall conditions with relatively brief cool and dry seasons.

19.2 Methods

The environments, general characteristics and management details of the trials undertaken are presented in Tables 19.1 and 19.2.

19.3 Results and discussion

As the project time span was brief, only 4.5 years, most accessions utilized in formal trials were limited to lines for which seed was available: commercial materials or promising accessions acquired from other collections. While the project managed to accomplish considerable collection of genetic material in the region, there was inadequate time to multiply the seed of promising accessions to allow more advanced testing in more formal trials. However, the high rainfall, good moisture-holding capacity of the soil and the brevity of the dry season provided an environment in which many of the commercial species and advanced experimental lines developed in Australia for relatively benign conditions were able to grow well when adequate fertility was available. These thus constitute the bulk of the accessions that have proved to be

[1] E-mail: john.lazier@sympatico.ca

Table 19.1. Summary of the most promising species on the Low Pine Ridge and Mountain Pine Ridge to July 1977.

	Lowland Pine Ridge			Mountain Pine Ridge
Chapter, this volume	14	15	16	15
Legume accessions tested	35	24	16	24
Grass	–	Native pasture	Native pasture	Native pasture
Trial type	Single plant	Strips	Plots	Strips
Replicated	Yes	No	Yes	No
Management	Cut	Cut	Cut	Cut
Sampling	Clipping	Observation	Clipping	Observation
Species				
Calopogonium caeruleum		X		
Centrosema plumieri				
Centrosema pubescens		X		
Codariocalyx gyroides		X	X	X
Crotalaria anagyroides			X	
Desmodium intortum		X		X
Desmodium ovalifolium	X			
Desmodium incanum				
Indigofera hirsute				
Leucaena leucocephala		X		
Macroptilium atropurpureum	X			X[a]
Pueraria phaseoloides	X	X		X
Stylosanthes guianensis	X			X

[a]Under the higher fertility application.

promising in the trials reported here. This is not entirely unexpected because many of the species are native to Belize and the region. The species that have shown some initial promise are presented in Table 19.1 for the Low Pine Ridge and Mountain Pine Ridge, and Table 19.2 for the terraces of the upper Belize River. The tables present only a summary and because not all accessions of a given species were successful, it is important to refer to the accession numbers of the successful genotypes in the relevant chapter. In addition, not all of the species listed were included in all trials, thus the frequency of occurrence in the tables does not necessarily indicate greater frequency of success. Other species and accessions with somewhat lower productivity may prove in the longer term and under different management to be productive and persistent in pastures. The informal observations undertaken in 1981 of some of the trial sites 4 years after being abandoned give some indication of longer term survival under unmanaged conditions.

The brevity of the project has left the critical issue of productivity and persistence of these legumes over the longer term unanswered. The results herein mainly indicate the ability to establish rapidly and persist over a year or more, plus provide some information on response to grazing. As shown in Table 2.9 (Chapter 2, this volume), growth can be rapid initially, and that character can indicate possible success as a pasture species. However, other factors are involved such as palatability, seasonal growth pattern, seeding and seeding survival rates, ability to regrow after grazing and thus ability to compete with associated vegetation. Taller, woody species may be able to better stand poor pasture management, and even, as in the case of *Leucaena*, grow out of the reach of the animals, and provide additional nutrition through seedlings in the sward. On the other hand, shorter, herbaceous plants, particularly those with a low crown as is found in *Desmanthus* (Chapters 2, 4 and 5, this volume), or those that seed heavily may be more successful with replacement of plants removed by grazing or physical damage as a critical factor in survival. Thus many aspects of persistence and productivity have yet to be determined, particularly over the longer term.

Table 19.2. Summary of the most promising species on the upper Belize River Valley terraces to July 1977.

	Chapter, this volume									Other concurrent trials[a]				
	14	14	15				16	17		A[b]	B[c]	C[d]	D[e]	E[f]
Legume accessions tested	182	35	24			13	20	16		17 native, 1 exotic	1 exotic	6 commercial	5 commercial	4 commercial
Location	Middle terrace	Middle terrace	Middle terrace	LPR	MPR	Middle terrace	High terrace	Middle terrace	LPR	Low terrace	Low terrace	Low terrace	Low terrace	Middle terrace
Grass	–	–	Mainly native pastures	Mainly native pastures	Mainly native pastures	6 commercial plots	Para, Pangola pastures	Coastal Bermuda	Native	Para	–	–	Jaragua, Pangola, Guinea	Pangola pastures
Trial type	Single plant	Single plant	Strips	Strips	Strips	Strips	Strips	Plots	Plots	Plots	Plots	Plots	Plots	Over sown
Replicated	Yes	Yes	No	No	No	No	Yes	Yes	Yes	Yes		Yes	Yes	Two grazing pressures
Management	Cut	Cut	Cut	Cut	Cut	Cut over	Cut, grazed	Cut	Cut	Cut, grazed	Cut	Cut	Cut, grazed	Grazed
Sampling	Clipping	Clipping	Observation	Observation	Observation	Observation	Observation	Clipping	Clipping	Clipping	Clipping	Clipping	Clipping	Clipping
Species														
Calopogonium caeruleum			X	X	X					X				
Centrosema plumieri	X		X	X	X					X				
Centrosema pubescens			X	X	X									
Centrosema schottii	X													
Codariocalyx gyroides			X	X	X	X		X	X	X	X			
Crotalaria anagyroides								X	X					

Continued

Table 19.2. Continued.

	Chapter, this volume									Other concurrent trials[a]				
	14	14	15				16	17		A[b]	B[c]	C[d]	D[e]	E[f]
Legume accessions tested	182	35	24			13	20	16		17 native, 1 exotic	1 exotic	6 commercial	5 commercial	4 commercial
Desmanthus pubescens	X													
Desmanthus leptophyllus	X													
Desmodium cinerium		X												
Desmodium distortum	X													
Desmodium incanum										X				
Desmodium intortum				X	X									
Leucaena leucocephala	X		X	X		X	X	X					X	
Macroptilium atropurpureum	X	X	X	X	X	X		X				X	X	X
Macroptilium lathyroides					X									
Neonotonia wightii														
Pueraria phaseoloides		X	X	X	X									
Stylosanthes guianensis	X	X	X	X	X		X	X						X
Stylosanthes hamata					X									
Teramnus uncinatus					X									
Vigna speciosa	X													
Vigna vexillata					X									

[a]Trials undertaken by the UWI-Belize Forage Legume Project which are reported elsewhere; [b]Lazier (1981a,b); [c]Lazier (1981c); [d]Lazier (1980a,b); [e]Lazier (1978); [f]Lazier (1978). LPR, Low Pine Ridge; MPR, Mountain Pine Ridge.

Observations taken in 1981 on the strip trials and three other project trials not reported in this volume provide some information on persistence (see Appendix 15.2, Chapter 15, this volume). However, further research is required in order to provide more definitive answers to the identity of persistent and productive forage legumes for the environments of central Belize.

References

Abel, D.J. and Williams, W.T. (1981) NEBALL and FINGRP, new programs for multiple nearest neighbour analyses. *Australian Computer Journal* 13, 24–25.

Abel, D.J. and Williams, W.T. (1985) A re-examination of four classificatory fusion strategies. *Australian Computer Journal* 28, 439–443.

ACIAR (1984) *Proceedings of the International Workshop on Soils,* 12–16 September 1983. Australian Centre for International Agricultural Research, Townsville, Australia.

Adams, M.W. and Pipoly, J.J. (1980) Biological structure, classification and distribution of economic legumes. In: Summerfield, R.J. and Bunting, A.H. (eds) *Advances in Legume Science.* Volume 1 of the Proceedings of the International Legume Conference, Royal Botanic Gardens, Kew, UK, pp. 1–16.

Adjei, M.B. and Pitman, W.D. (1993) Response of *Desmanthus* to clipping on a phosphatic clay mine-spoil. *Tropical Grasslands* 27, 94–99.

Agrimix (n.d.) http://www.progardes.com.au

Ahmad, N. (1980) *Contributions of Forage Legumes to Pasture Production in the Caribbean.* FAO report (http://www.fao.org/wairdocs/ilri/x5488e/x5488e0h.htm) Accessed November 2015.

Ahn, J.H., Robertson, B.M., Elliott, R., Guttgeridge, R.C. and Ford, C.W. (1989) Quality assessment of tropical browse legumes: tannin content and protein degradation. *Animal Feed Science and Technology* 27, 147–156.

Allen, O.N. and Allen, E.K. (1981) *The Leguminosae. A Source Book of Characteristics, Uses and Nodulation.* Springer, Madison, Wisconsin.

Andrade-Lima, D. de (1982) Present day forest refuges in north-eastern Brazil. In: Prance, G.T. (ed.) *Biological Diversification in the Tropics.* Columbia Press, New York, pp. 245–51.

Anon (1979) *Sown Pasture and Seed Production in Queensland in 1977 and 1978.* Queensland Department of Primary Industries.

Anon (1992) B. Legumes, 23. *Clitoria ternatea* L. (butterfly pea) cv. Milgarra. *Australian Journal of Experimental Agriculture* 32, 547–548.

Anon (1996) Caatinga stylo *Stylosanthes* – 'Primar' and 'Unica'. *Plant Varieties Journal* 9, 19–20.

Anon (2006) Climate change. *The Economist* 25 November, 78.

Anon (2007) Improving the search for intelligence. *New Scientist* 2590, 22–23.

ARC (1980) *The Nutrient Requirements of Livestock.* CAB, Slough, UK.

Argel, P.J. and Paton, C.J. (1999) Overcoming legume hardseededness. In: Loch, D.S. and Ferguson, J.E. (eds) *Forage Seed Production, Volume 2: Tropical and Subtropical Species.* CAB International, Wallingford, pp. 247–265.

Armke, F.W. and Scott, C.B. (1999) Using cattle to disperse seeds for winter forage plants. *Texas Journal of Agriculture and Natural Resources* 12, 28–38.

Arnaldo, M. (1964) *Wat in het Grusit en Bloeit op Curacao, Aruba en Bonaire*, 2nd edn. Zakflora Utgauen van de Natuurwetensch – appelijka Werkgroup Nederlandse Antillen 16, 1–232.

Arnold, R.M. (1950) Report on certain aspects of the livestock industry in British Honduras. Mimeo, Agricultural Library, Central Farm, Belize.

Baker, H.G. (1974) The evolution of weeds. *Annual Review of Ecology and Systematics* 5, 1–24.

Bartlett, H.H. (1935) A method of procedure for field work in tropical American phytogeography based upon a botanical reconnaissance in parts of British Honduras and the Peten forest of Guatemala. *Publications of the Carnegie Institution* 461, 1–25.

Baskin, C.C. and Baskin, J.M. (1998) *Ecology, Biogeography and Evolution of Dormancy and Germination*. Academic Press, New York.

Batianoff, G.N. and Butler, D.W. (2002) Assessment of invasive naturalized plants in south-east Queensland. *Plant Protection Quarterly* 17, 27–34.

Bättig, M.B. (2007) Do results of climate change research facilitate the formation of successful international climate change agreements? A global analysis on a per country level. PhD thesis. ETH Zurich, DISS. ETHNo. 17151.

Bazan, R. (1969) The coastal Pine Ridge soils of British Honduras and their fertility status. PhD thesis. University of Florida, Gainesville, Florida.

Beard, J.S. (1953) The savanna vegetation of northern tropical America. *Ecological Monographs* 23, 149–215.

Beauchemin, K.A., Kreuzer, M., O'Mara, F. and McAllister, T.A. (2007) Nutritional management for enteric methane abatement: a review. *Australian Journal of Experimental Agriculture* 48, 21–27.

Bell, L. (2009) Building better feed systems. *Tropical Grasslands* 43, 199–206.

Bennett, S.J. and Virtue, J.G. (2004) Salinity mitigation versus weed risks – can conflicts of interest in introducing new plants be resolved? *Australian Journal of Experimental Agriculture and Animal Husbandry* 44, 1141–1156.

Bishop, H. (1988) Establishment and spread of the tropical legume Glen Joint Vetch (*Aeschynomene americana*) via the grazing animal. *Proceedings of the Australian Society of Animal Production* 17, 377.

Bishop, H. (n.d.) *Pastures MacKay Whitsunday region*. DPI, Queensland, Australia (https://futurebeefnew-daff.netdna-ssl.com/wp-content/uploads/Pastures_Mackay_Whitsunday_region.pdf) Accessed November 2015.

Bishop, H.G. (2003) *Managing Old (Discontinued) Plant Evaluation Sites (MOPES).* Final report of project MLA NAP3.225. Meat and Livestock Australia, North Sydney, Australia.

Bishop, H.G. and Hilder, T. (2005) *Backup Legumes for Stylos*. Final report of MRC project DAQ.083. Meat and Livestock Australia, North Sydney, Australia.

Blue, W.G. (1969) Fertilizer response with Pangola grass on Puletan loamy fine sand, British Honduras as indicated by a pot experiment. *Tropical Agriculture (Trinidad)* 88, 25–29.

Bogdan, A.V. (1977) *Tropical Pasture and Fodder Plants*. Tropical Agricultural Series. Longman, London.

Boschma, S.P., Crocker, G.J., Pengelly, B.C. and Harden, S. (2012) Potential of *Desmanthus* spp. in northern New South Wales. *Proceedings of the 16th ASA Conference,* UNE, October 2012. (http://www.regional.org.au/au/asa/2012/pastures/7975_boschmasp.htm#TopOfPage) Accessed November 2015.

Bowen, M., Buck, S. and Chudleigh, F. (2015) Feeding forages in the Fitzroy. A guide to profitable beef production in the Fitzroy river basin. Queensland Department of Agriculture and Fisheries (http://cdn.futurebeef.com.au/wp-content/uploads/Feeding-forages-in-the-Fitzroy.pdf) Accessed November 2015.

Brandon, N.J., Date, R.A., Jones, R.M., Clem, R.L., Bahnisch, G.A., Spies, P.R., Becerra, A., Robertson, B.A. and Graham, T.W.G. (1997) Rhizobial and Nutritional Responses of *Desmanthus virgatus* on Clay Soils of Queensland (http://www.internationalgrasslands.org/files/igc/publications/1997/1-10-019.pdf) Accessed November 2015.

Brolmann, J.B. and Sonoda, R.M. (1981) Occurrence and characteristics of two off-type plants in the progeny of a tetraploid *Stylosanthes hamata* native of Florida USA. *Tropical Grasslands* 15, 37–39.

Brown, A.H.D. (1978) Isozymes, plant population genetic structure and genetic conservation. *Theoretical and Applied Genetics* 52, 145–157.

Brown, A.H.D. and Spillane, C. (1999) Implementing core collections – principles, procedures, progress and promise. In: Johnson, R.C. and Hodgkin, T. (eds) *Core Collections for Today and Tomorrow*. International Plant Genetic Resources Institute (IPGR), Rome, pp. 1–9.

Brown, S., Cook, B. and Pengelly, B. (2008) Development of an Australian pasture selection tool using Lucid. *Proceedings of the 14th Australian Agronomy Conference*, 21–25 September 2008, Adelaide, Australia.

Bunting, A.H. and Lea, J.D. (1962) The soils and vegetation of the Fung, east-central Sudan. *Journal of Ecology* 50, 529–568.

Burr, E.J. (1970) Cluster sorting with mixed character types. II. Fusion strategies. *Australian Computer Journal* 2, 98–103.

Burrows, D.M. and Porter, F.J. (1993) Regeneration and survival of *Desmanthus virgatus* 78382 in grazed and ungrazed pastures. *Tropical Grasslands* 27, 100–107.

Burt, R.L. (1975a) The climatic background of *Stylosanthes* accessions. In: Williams, W.T. (ed.) *Pattern Analysis in Agricultural Science.* CSIRO, Melbourne, Australia, pp. 162–169.

Burt, R.L. (1975b) Description of introduced *Stylosanthes* material using morphological and performance attributes. In: Williams, W.T. (ed.) *Pattern Analysis in Agricultural Science*. CSIRO, Melbourne, Australia, pp. 139–150.

Burt, R.L. (1983) Observation, classification and description. In: McIvor, J.G. and Bray, R.A. (eds.) *The Genetic Resources of Forage Plants*. CSIRO, Melbourne, Australia, pp. 169–181.

Burt, R.L. (1984a) Natural variation in *Stylosanthes*. In: Stace, H.M. and Edye, L.A. (eds) *The Biology and Agronomy of Stylosanthes*. Academic Press, Sydney, Australia, pp. 103–123.

Burt, R.L. (1984b) *Zornia*. In: 'Plant Introduction and Evaluation Workshop' Queensland Department of Primary Industry Conference and Workshop Series QC 85005, pp. 161–162.

Burt, R.L. (1984c) *Desmanthus*. In: 'Plant Introduction and Evaluation Workshop' Queensland Department of Primary Industry Conference and Workshop Series QC 85005, pp. 129–130.

Burt, R.L. (1986) The response of *D. virgatus* to defoliation. In: *CSIRO Division of Tropical Crops and Pastures, Annual Report 1985–1986*. Commonwealth Scientific and Industrial Organization, Brisbane, Australia, pp. 21–22.

Burt, R.L. (1993a) *Desmanthus*: A tropical and subtropical forage legume, Part 1. General review. *Herbage Abstracts* 63, 401–413.

Burt, R.L. (1993b) *Desmanthus*: A tropical and subtropical forage legume. Part 2. Artificial key and species descriptions. *Herbage Abstracts* 63, 473–478.

Burt, R.L. and Ebersohn, J.P. (1969) Some comments on pasture plant introduction in Queensland. *Journal of the Australian Institute of Agricultural Science* 35, 124–127.

Burt, R.L. and Haydock, K.P. (1968) Yield stability as an aid to the preliminary assessment of introduced forage plants. *Journal of the Australian Institute of Agricultural Science* 34, 228–231.

Burt, R.L. and Miller, C.P. (1975) *Stylosanthes* – a source of pasture legumes. *Tropical Grasslands* 9, 117–123.

Burt, R.L. and Reid, R. (1976) Exploration for, and utilization of, collections of tropical pasture legumes: III. The distribution of various *Stylosanthes* species with respect to climate and phytogeographic regions. *Agro-Ecosystems* 2, 319–327.

Burt, R.L. and Williams, W.T. (1975) Plant introduction and the *Stylosanthes* story. *Australian Meat Research Council Review* 25, 1–26.

Burt, R.L. and Williams, W.T. (1979a) Strategy of evaluation of a collection of tropical herbaceous legumes from Brazil and Venezuela. II. Evaluation in the quarantine glasshouse. *Agro-Ecosystems* 5, 119–134.

Burt, R.L. and Williams, W.T. (1979b) Strategy of evaluation of a collection of tropical herbaceous legumes from Brazil and Venezuela III. The use of ordination techniques in evaluation. *Agro-Ecosystems* 5, 135–146.

Burt, R.L. and Williams, W.T. (1988) A history of plant introduction in Australia. In: Home, R.W. (ed.) *Australian Science in the Making. Bicentennial Volume*. Australian Academy of Science, Cambridge University Press, Cambridge, pp. 252–276.

Burt, R.L., Edye, L.A., Grof, B. and Williams, R.J. (1970) Assessing the agronomic potential of the genus *Stylosanthes* in Australia. *Proceedings of the 11th International Grassland Congress*, Surfers Paradise, Australia, pp. 219–223.

Burt, R.L., Edye, L.A., Williams, W.T., Grof, B. and Nicholson, C.H.L. (1971) Numerical analysis of variation patterns in the genus *Stylosanthes* as an aid to plant introduction and assessment. *Australian Journal of Agricultural Research* 22, 737–757.

Burt, R.L., Williams, W.T. and Compton, J.F. (1973) Variation within naturally occurring Townsville stylo (*Stylosanthes humilis*) populations; changes in population structure and some agronomic implications. *Australian Journal of Agricultural Research* 24, 705–713.

Burt, R.L., Edye, L.A., Williams, W.T., Gillard, P., Grof, B., Page, M., Shaw, N.H., Williams, R.J. and Wilson, G.P.M. (1974) Small sward testing of *Stylosanthes* in Northern Australia: preliminary considerations. *Australian Journal of Agricultural Research* 25, 559–575.

Burt, R.L., Reid, R. and Williams, W.T. (1976) Exploration for and utilization of, collections of tropical pasture legumes I. The relationship between agronomic performance and climate of origin of introduced *Stylosanthes* spp. *Agro-Ecosystems* 2, 293–307.

Burt, R.L., Isbell, R.F. and Williams, W.T. (1979) Strategy of evaluation of a collection of tropical herbaceous legumes from Brazil and Venezuela. I Ecological evaluation at the point of collection. *Agro-Ecosystems* 5, 99–117.

Burt, R.L., Pengelly, B.C. and Williams, W.T. (1980a) Network analysis of genetic resource data III. The elucidation of plant/soil/climate relations. *Agro-Ecosystems* 6, 119–127.

Burt, R.L., Williams, W.T. and Grof, B. (1980b) *Stylosanthes* – structure, adaptation and utilization. In: Summerfield, R.J. and Bunting, A.H. (eds) *Advances in Legume Science*. Royal Botanic Gardens, Kew, UK, pp. 537–551.

Burt, R.L., Williams, W.T. and Abel, D.J. (1983) A new graph theoretic technique for the analysis of genetic resource data. *Agro-Ecosystems* 8, 231–245.

Cameron, D.G. (1985) Tropical and subtropical legumes 8: Phasey bean (*Macroptilium lathyroides*). The predecessor of Siratro. *Queensland Agricultural Journal* 111, 211–214.

Carangal, V.C., Rebancos, E.T. Jr, Armada, E.C. and Tengco, P.L. (1994) Integration of food, forage and green manure production systems. In: Ladha, J.K. and Garrity, D. (eds) *Green Manure Production Systems for Asian Ricelands*. International Rice Research Institute, Los Baños, Philippines.

Carver, M., Gross, G.F. and Woodward, T.E. (1991) *Hemiptera*. In: *The Insects of Australia.* CSIRO Division of Entomology, Melbourne University Press, Carlton, Australia, pp. 429–444.

Castillo, A.C., Cuyugan, O.C., Fogarty, S. and Shelton, H.M. (1997) Growth, psyllid resistance and forage quality of *Leucaena leucocephala, L. pallida, L. diversifolia* and the F1 hybrid of *L. leucocephala x L. pallida*. *Tropical Grasslands* 21, 188–200.

Chakraborty, S. (ed.) (2004) *High-Yielding Anthracnose Resistant Stylosanthes for Agricultural Systems*. ACIAR, Canberra.

Chapman, R.E., Williams, O.B. and Moule, G.R. (1973) The wool industry. In: Alexander, G. and Williams, O.B. (eds) *Pastoral Industries of Australia*. Sydney University Press, Sydney, Australia, pp. 79–116.

Charmley, E., Stephens, M.L. and Kennedy, P.M. (2008) Predicting livestock productivity and methane emissions in northern Australia: development of a bio-economic modelling approach. *Australian Journal of Experimental Agriculture* 48, 109–113.

Charter, C.F. (1941) *A Reconnaissance Survey of the Soils of British Honduras North of the Central Metamorphic and Igneous Massif with Special Reference to their Utilization in Agriculture*. Government Printer, Port of Spain.

Chudleigh, P. and Bramwell, T. (1996) *Assessing the Impact of Introduced Tropical Pasture Plants in Northern Australia*. Report prepared for CSIRO taskforce interested in introduced pasture plants. Agtrans Research, Brisbane, Australia.

Clem, R.L. (2004) Animal production from legume-based ley pastures in south-eastern Queensland. In: Whitbread, A.M. and Pengelly, B.C. (eds) *Tropical Legumes for Sustainable Farming Systems in Southern Africa and Australia*. ACIAR Proceedings No. 115, Australian Centre of International Agricultural Research, Canberra, pp. 135–144.

Clem, R.L. and Cook, B.G. (2004) Identification and development of forage species for long-term pasture leys for the southern speargrass region of Queensland. In: Whitbread, A.M. and Pengelly, B.C. (eds) *Tropical Legumes for Sustainable Farming Systems in Southern Africa and Australia.* ACIAR Proceedings No. 115, Australian Centre of International Agricultural Research, Canberra, pp. 64–72.

Clem, R.L. and Hall, T.J. (1994) Persistence and productivity of tropical pasture legumes on three cracking clay soils (Vertisols) in north-eastern Queensland. *Journal of Australian Experimental Agriculture* 34, 161–171.

Clem, R.L. and Jones, R.M. (1996) *Legumes for Clay Soils*. Final report of MRC project DAQ.086. Meat and Livestock Australia, North Sydney, Australia.

Clem, R.L. and Jones, R.M. (1998) *Legumes for Clay Soils.* Report to MRC, June 1998. Queensland Government Internal Report.

Clem, R.L., Cook, S., French, A.V. and Turner, J.P. (1994) Establishment. In: Partridge, I.J., Burrows, W.H. and Weston, E.J. (eds) *Sown Pastures for the Brigalow Lands*. DPI Queensland, Brisbane, Australia.

Clements, R. and Williams, R.J. (1980) Genetic diversity on *Centrosema*. In: Summerfield, R.J. and Bunting, A.H. (eds) *Advances in Legume Science*. Royal Botanic Gardens, Kew, UK, pp. 559–567.

Coates, D.B. (2000) Faecal NIRS – what does it offer today's graziers? *Tropical Grasslands* 34, 230–239.

Coates, D., Miller, C., Hendricksen, R. and Jones, R. (1997) Animal production from *Stylosanthes* pastures. *Tropical Grasslands* 31, 494–502.

Cobley, L.S. (1956) *An Introduction to the Botany of Tropical Crops*. Longmans, Green & Co., London.

Cook, S.J., Clem, R.L., MacLeod, N.D. and Walsh, P.A. (1993a) Tropical pasture establishment. 7. Sowing methods for pasture establishment in northern Australia. *Tropical Grasslands* 27, 335–343.

Cook, B.G., Graham, T.W.G., Clem, R.L., Hall, T.J. and Quirk, M.F. (1993b) Evaluation and development of *Desmanthus virgatus* on medium- to heavy-textured soils in Queensland, Australia. Proceedings of the XVII International Grassland Congress, Rockhampton, 1993b, pp. 2148–2149.

Cook, B.G., Pengelly, B.C., Brown, S.D., Donnelly, J.L., Eagles, D.A., Franco, M.A., Hanson, J., Mullen, B.F., Partridge, I.J., Peters, M. and Schultze-Kraft, R. (2005) *Tropical Forages: An Interactive Selection Tool.* [CD-ROM] CSIRO, DPI&F (Qld) CIAT and ILRI, Brisbane, Australia.

Coradin, L. and Schultze-Kraft, A. (1990) Germplasm collection of tropical pasture legumes in Brazil. *Tropical Agriculture (Trinidad)* 67, 98–100.

Cowlishaw, S. (1969) Report on a visit to British Honduras from March 23 to March 29, 1969. Mimeo, Agricultural Library, Belize.

Cox, K.G. (1998) A study of seed production in *Desmanthus (Desmanthus virgatus L.)*. PhD thesis, Massey University Palmerston North, New Zealand.

Cox, K.G. (2005) Selective herbicide strategies for use in Australian *Desmanthus* seed crops. *Tropical Grasslands* 39, 171–181.

Cox, K.G. (2013) Recent development of pasture plants in Queensland. In: Revitalising grasslands to sustain our communities. *Proceedings of the 22nd International Grassland Congress*, Sydney, Australia, pp. 507–508.

Cox, K.G. (2014) *Managing Old Plant Evaluation Sites: Containment and Progressive Eradication*. Final report of Project MLA B.NBP.0706. Meat and Livestock Australia, North Sydney, Australia.

Cox, K., Clem, B., Hall, T., Hilder, T., Keating, M., Silcock, R., Gardiner, C. and McDonald, C. (2007) Pre-emptive control of four persistent forage legumes in grassland environments in Queensland – a case study of an evolving eradication program. *Proceedings of the 9th Queensland Weed Symposium*, Gold Coast 3–6 June 2007.

Cox, K.G., Scrivener, N. and Kilpatrick, F. (2009) Reversing the declining quality of the tropical forages collection. *Tropical Grasslands* 43, 259–262.

Cox, K., Keating, M., Dayes, S. and Gardiner, C. (2012) Low-input, high-quality legume hays for north Queensland. *Proceedings of the 16th ASA Conference*, 14–18 October 2012, Armidale, Australia (http://www.regional.org.au/au/asa/2012/pastures/8033_coxkg.htm) Accessed June 2015.

Creswell, K.J. (2007) Anthelmintic effects of tropical legumes in ruminant animals. PhD thesis, James Cook University, Townsville, Australia.

Curtis, M.D., Cameron, D.F. and Manners, J.M. (1995) Molecular evidence that diploid *Stylosanthes humilis* and diploid *Stylosanthes hamata* are progenitors of allotetraploid *Stylosanthes hamata*. *Genome* 38, 344–348.

Daehler, C. and Carino, D.A. (2000) Predicting invasive plants: prospects for a general screening system based on current regional models. *Biological Invasions* 2, 92–103.

Daehler, C. and Denslow, J. (2007) *Weed Risk Assessments for Hawaii and Pacific Islands.* (http://www.botany.hawaii.edu/faculty/daehler/wra).

Dalzell, S.A., Brandon, N.J. and Jones, R.M. (1997) Response of *Lablab purpureus* cv. Highworth, *Macroptilium bracteatum* and *Macrotyloma daltonii* to different intensities and frequencies of cutting. *Tropical Grasslands* 31, 107–113.

Daniel, T.F. (2004) *Acanthaceae* of Sonora: taxonomy and phytogeography. *Proceedings of the California Academy of Sciences* 55, 690–805.

Darcel, F.C. (1952) A reconnaissance soil survey of the northern section of the Mountain Pine Ridge. Unpublished report, Commonwealth Development Corporation, London.

Darwin, C.R. (1859) *On the Origin of Species by Means of Natural Selection* (http://www.literature.org/authors/darwin-charles/the-origin-of-species/) Accessed November 2015.

Date, R.A. (1991) Nitrogen fixation in *Desmanthus:* strain specificity of *Rhizobium* and response to inoculation in acidic and alkaline soil. *Tropical Grasslands* 25, 47–55.

Date, R.A., Burt, R.L. and Williams, W.T. (1979) Affinities between various *Stylosanthes* species as shown by rhizobial, soil pH and geographic relationships. *Agro-Ecosystems* 5, 57–67.

Date, R.A., Edye, L.A. and Liu, C.J. (1996) *Stylosanthes* sp. aff. *scabra* – a potential new forage plant for northern Australia. *Tropical Grasslands* 30, 133.

DeMeo, J. (2006) A climate of change. *American Scientist* 94, 196–197.

Dijkshoorn, W. and van Wijk, A.L. (1967) The sulphur requirements of plants as evidenced by the sulphur-nitrogen ratio in the organic matter. A review of published data. *Plant Soil* 26, 129–157.

DOA (1933–1974) *British Honduras/Belize Department of Agriculture Annual Reports*. Government Printer, Belize City, Belize.
Dobson, N. (1973) *A History of Belize*. Longman Caribbean Ltd, Jamaica.
Dudal, R. and Bramao, D.L. (eds) (1965) Dark clay soils of tropical and subtropical regions. *FAO Agricultural Development Paper No. 84*. FAO, Rome.
Eckard, R.J., Grainger, C. and de Klein, C.A.M. (2010) Options for the abatement of methane and nitrous oxide from ruminant production: a review. *Livestock Science* 130, 47–56.
Edye, L.A. (1988) *Stylosanthes Collecting in Colombia and Venezuela*. Final report. CSIRO, Townsville, Australia.
Edye, L.A. and Maass, B.L. (1997) Recent advances in studies of anthracnose of *Stylosanthes*. I. The biogeography of *Stylosanthes hamata*, *S. scabra* and '*Stylosanthes seabrana*'. *Tropical Grasslands* 31, 417–423.
Edye, L.A., Burt, R.L., Williams, W.T., Williams, R.J. and Grof, B. (1973) A preliminary agronomic evaluation of *Stylosanthes* species. *Australian Journal of Agricultural Research* 24, 511–525.
Edye, L.A., Burt, R.L., Nicholson, C.H.L., Williams, R.J. and Williams, W.T. (1974) Classification of the *Stylosanthes* collection 1928–69. *Division of Tropical Agronomy Technical Paper No. 15.* CSIRO, Clayton, Australia, pp. 1–28.
Edye, L.A., Hall, T.J., Clem, R.L., Graham, T.W.G., Messer, W.B. and Rebgetz, R.H. (1998) Sward evaluation of eleven '*Stylosanthes seabrana*' accessions and *S. scabra* cv. Seca at five subtropical sites. *Tropical Grasslands* 32, 243–251.
Estrada-Loera, E. (1991) Phytogeographic relationships of the Yucatan Peninsula. *Journal of Biogeography* 18, 687–697.
Ewel, J.J. and Putz, F.E. (2004) A place for alien species in ecosystem restoration. *Frontiers in Ecology and Environment* 2, 354–360.
Eyles, A.G. and Cameron, D.G. (1985) Pasture research in Northern Australia – its history, achievements and future emphasis. In: Hacker, J.B. (ed.) *Research Report No. 4*. CSIRO, Brisbane, Australia.
FAO/UNESCO (1973) *Soil Map of the World. Vol. III. Mexico and Central America*. FAO/UNESCO, Paris.
Flores, M. (1995) Why do farmers go for fertilizer bean? *ILEIA Newsletter*, March.
Fornasero, L.V., Del Papa, M.F., Lopez, J.L., Albicoro, F.A., Zabala, J.M., Toniutti, M.A., Pensiero, J.F. and Lagares, A. (2014) Phenotypic, molecular and symbiotic characterisation of rhizobial symbionts of *Desmanthus paspalaceus* (Lindm.) Burkart that grow in the Province of Santa Fe, Argentina. *PLoS ONE* 9(8), e104636. doi:10.1371/journal.pone.0104636.
Friedel, M.H., Foran, B.D. and Stafford Smith, D.M. (1990) Where the creeks run dry or ten feet high: pastoral management in arid Australia. *Proceedings of the Ecological Society of Australia* 16, 185–194.
Furley, P.A. (ed.) (1968) *General Report of the Expedition to British Honduras – Yucatan 1966.* University of Edinburgh, Edinburgh, UK.
Gama Castro, J.E. (2006) Rendzinas: a soil group with primitive original attributes? *18th World Congress of Soil Science* 9–15 July 2006, Philadelphia, Pennsylvania.
Gardener, C.J., McIvor, J.G. and Williams, J. (1990) Dry tropical rangelands: solving one problem and creating another. *Proceedings of the Ecological Society of Australia* 16, 279–286.
Gardiner, C.P. (1992) An evaluation of herbaceous tropical pasture legumes with particular reference to *Desmanthus virgatus*. MSc thesis. JCU, Townsville, Australia.
Gardiner, C.P. (2003) An evaluation of tropical pasture legumes on gidgee soils in the semi-arid tropics. *Proceedings of the 11th Australian Agronomy Conference*, 2–6 February 2003, Geelong, Australia.
Gardiner, C.P. (2014a) Unlocking Norths seed secrets. DAF Future Beef, *Northern Muster* (https://futurebeefnew-daff.netdna-ssl.com/wp-content/uploads/Northern-Muster-Issue-34-April-2014.pdf) Accessed November 2015.
Gardiner, C.P. (2014b) Stock do it for seeds. DAF Future Beef, *Northern Muster* (https://futurebeefnew-daff.netdna-ssl.com/wp-content/uploads/NorthernMuster-Issue36.pdf) Accessed November 2015.
Gardiner, C.P. and Burt, R.L. (1995) Performance characteristics of *Desmanthus virgatus* in contrasting tropical environments. *Tropical Grasslands* 29, 183–188.
Gardiner, C.P. and Parker, A. (2012) Steer liveweight gains on Progardes™/buffel pastures in Qld. *Proceedings of the 29th Biennial Conference of the Australian Society of Animal Production*, Christchurch, New Zealand, July 2012.
Gardiner, C.P. and Rangel, J. (1996) Agronomic and nutritional aspects of *Desmanthus*. *The Future of Tropical Savannas Symposium*, Townsville, 19–22 July 1994.

Gardiner, C.P. and Swan, S. (2008) Abandoned pasture legumes offer potential economic and environmental benefits in semiarid clay soil rangelands. *Proceedings of the Australian Rangeland Conference*, Charters Towers, 28 September–2 October 2008.

Gardiner, C.P., Bielig, L., Schlink, A., Coventry, R. and Waycott, M. (2004) *Desmanthus* – a new pasture legume for the dry tropics. *Proceedings of the 12th Australian Agronomy Conference*, 2–6 February 2003, Brisbane, Australia.

Gardiner, C.P., Wright, C. and Coventry, M. (2012) The germination, passage and viability of *Desmanthus virgatus* (L.) Willenow seed through sheep and its implication for dispersal in tropical rangelands. *Proceedings of the 16th Australian Society of Agronomy Conference*, UNE, October 2012.

Gardiner, C.P., Kempe, N., Hannah, I. and McDonald, J. (2013) Progardes: a legume for tropical/subtropical clay soils. *Tropical Grasslands – Forrajes Tropicales* 1, 78–80.

Gillard, P. and Fisher, M.J. (1978) The ecology of Townsville stylo-based pastures in northern Australia. In: Wilson, J.R. (ed.) *Plant Relations in Pastures*. CSIRO, Australia, pp. 340–352.

Glover, J.D. and Wilhelm, W. (eds) (2005) Perennial grain crops: an agricultural revolution. Special issue of *Renewable Agriculture and Food Systems* 20, no. 1.

Glover, J.D., Cox, C.M. and Reganold, J.P. (2007) Future farming: a return to roots? *Scientific American* (August), 66–73.

Gokbulak, F. and Call, C.A. (2004) Grass seedling recruitment in cattle dungpats. *Journal of Range Management* 57, 649–655.

Good, R. (1947) *The Geography of the Flowering Plants*, 4th edn. Longman, London.

Gottlieb, L.D. (1977) Electrophoretic evidence and plant systematics. *Annals of the Missouri Botanic Gardens* 64, 161–180.

Government of British Honduras (1934) *Blue Book 1933*. Government Printer, Belize.

Government of British Honduras (1957) *British Honduras Development Plan Part III 1955–1960*. Printing Department, Belize.

Government of British Honduras (1960) *Report of the Committee on Agriculture*. Government Printer, Belize.

Government of British Honduras (1969) *Annual Abstract of Statistics 1969, No. 7.* Government Printer, Belize.

Government of British Honduras (n.d.a) *British Honduras Development Plan 1964–1970*. Office of the Premier, Government Printer, Belize.

Government of British Honduras (n.d.b) *British Honduras Development Plan Part II: Immediate Plan (1952–1955).* Printing Department, Belize.

Graham, T.W.G., Clem, R.L., Hall, T.J., Porter, F.J., Quirk, M.F. and Warrian, C.M. (1991a) *Desmanthus virgatus* – a summer legume for clay soils. *Tropical Grasslands* 25, 239.

Graham, T.W.G., Hall, T.J. and Clem, R.L. (1991b) *Release of Three Cultivars of Desmanthus virgatus*. Report of the Queensland Department of Primary Industries, Brisbane, Australia.

Gramshaw, D. and McKeon, G.M. (1985) Establishment of surface sown shrubby Stylos. *Proceedings of the 3rd Australia Conference on Tropical Pastures* 8–12 July 1985, Tropical Grassland Society.

Gramshaw, D., Pengelly, B.C. Muller, F.W., Harding, W.A.T. and Williams, R.J. (1987) Classification of a collection of the legume *Alysicarpus* using morphological and preliminary agronomic attributes. *Australian Journal of Agricultural Research* 38, 355–372.

Guarino, L., Maxted, N. and Sawkins, M. (1999) Analysis of georeferenced data and the conservation and use of plant resources. In: Greene, S.L. and Guarino, L. (eds) *Linking Genetic Resources and Geography: Emerging Strategies for Conserving and Using Crop Biodiversity.* Crop Science Society of America Special Publication 27, 1–24.

Hacker, J.B. (1990) *A Guide to Herbaceous and Shrub Legumes of Queensland*. University of Queensland Press, St. Lucia, Australia.

Hafenrichter, A.L., Mullen, L.A. and Brown, L. (1949) *Grasses and Legumes for Soil Conservation in the Pacific North-west*. USDA Miscellaneous Publications No. 678.

Hall, T.J. (1985) Adaption and agronomy of *Clitoria ternatea* L. in northern Australia. *Tropical Grasslands* 19, 156–163.

Hall, T.J. (2014) Rehabilitating degraded frontage soils in tropical north Queensland. *Tropical Grasslands – Forrajes Tropicales* 2, 66–67.

Harlan, J.R. (1971) Agricultural origins: centers and noncenters. *Science* 174, 468–474.

Harlan, J.R. (1983) The scope for collection and improvement of forage plants. In: McIvor, J.G. and Bray, R.A. (eds) *The Genetic Resources of Forage Plants.* CSIRO, Melbourne, Australia, pp. 4–14.

Harper, J.L. (1977) *Population Biology of Plants.* Academic Press, London.

Hartley, E. and Gemell, G. (2015) Assessment of rhizobial isolates from *Neptunia* and *Desmanthus*. Laboratory report. Australian Inoculants Research Group. NSW DPI, Gosford, Australia.
Hartley, W. (1954) The agrostological index: a phytogeographic approach to the problems of pasture plant introduction. *Australian Journal of Botany* 2, 1–21.
Hartley, W. (1963) The phytogeographical basis of pasture plant introduction. *Genetica Agraria* 17, 135–160.
Hegarty, P. (2012) Merino ewe nutrition search comes up trumps with 100% plus lambing. *Australian Farm Journal* May, 33–35.
Heywood, V.H. (1973) Ecological data in practical taxonomy. In: Heywood, V.H (ed.) *Taxonomy and Ecology*. Academic Press, London, pp. 329–347.
Hill, I.D. (1966) *Soil and Land Use Surveys No. 19 A and B. Antigua and Barbuda*. Regional Research Centre, University of West Indies, St. Augustine, Trinidad.
Hill, J.O., Clem, R.L., Robertson, M.J., Pengelly, B.C. and Whitbread, A.M. (2004) Beef production from tropical pasture legumes on cropping soils. *Animal Production in Australia* 25, 256.
Holdridge, L.R. (1947) Determination of world plant formations from simple climatic data. *Science* 105, 367–368.
Holdridge, L.R. and Hooper, E.D.M. (1966) The life zone system. *Andansonia* 6, 199–203.
Hooker, J.D. (1879) *The Flora of British India. II*. L. Reeve & Co. Ltd., Ashford, UK.
Hopkinson, J.M. and English, B.H. (2004) Germination and hardseededness in *Desmanthus*. *Tropical Grasslands* 38, 1–16.
Howard, R.A. (1973) The vegetation of the Antilles. In: Graham, A. (ed.) *Vegetation and Vegetational History of Northern Latin America*. Elsevier, Amsterdam, pp. 1–38.
Humphreys, L.R. and Partridge, I.J. (1995) *A Guide to Better Pastures for the Tropics and Subtropics, rev*. 5th edn. New South Wales Agriculture, Newcastle, Australia.
Hunt, D.R. (1962) Some notes on the pines of British Honduras. *Empire Forestry Review* 41, 134–145.
Hunt, D.R. (1970) Some observations on the pine savannas of British Honduras. Internal report, Royal Botanical Gardens, Kew, UK.
Hymowitz, T. (1971) Collection and evaluation of tropical and subtropical Brazilian forage legumes. *Tropical Agriculture (Trinidad)* 48, 309–315.
Illing, T. (2009) The changing face of the Australian tropical pasture seed industry. Presidential address. *Tropical Grasslands* 43, 1–3.
Jain, H.K. and Mehra, K.L. (1980) Evolution, adaptation relationships and uses of the species of *Vigna* cultivated in India. In: Summerfield, R.J. and Bunting, A.H. (eds) *Advances in Legume Science*. Royal Botanic Gardens, Kew, UK, pp. 459–468.
Jansen, P.I. and Edye, L.A. (1996) Variation within *Stylosanthes* sp. aff. *scabra*, and comparisons with its closest allies, *S. scabra* and *S. hamata*. *Australian Journal of Agricultural Research* 47, 985–996.
Jenkin, R.N., Rose-Innis, R., Dunsmore, J.R., Walker, S.H., Birchall, C.J. and Briggs J.S. (1976) *The Agricultural Development Potential of the Belize Valley, Belize*. Ministry of Overseas Development, Land Resources Division, Land Resource Study 24.
Jimoh, A.O., Chika, A., Umar, M.T., Adebisi, I. and Abdullahi, N. (2011) Analgesic effects and anti-inflammatory properties of the crude methanolic extract of *Schwenckia americana* Linn (Solanaceae). *Journal of Ethnopharmacology* 137, 543–546.
Johnson, M.S. and Chaffey, D.R. (1973) A forest inventory of part of the Mountain Pine Ridge, Belize. *Land Resource Study No. 13*. Land Resources Division, Surbiton, UK.
Jones, P.G., Beebee, S.E., Tohme, J. and Galway, N.W. (1997) The use of geographical information systems in biodiversity exploration and conservation. *Biology and Conservation* 6, 947–958.
Jones, R.M. (2014) The rise and fall of Siratro (*Macroptilium atropurpureum*) – what went wrong and some implications for legume breeding, evaluation and management. *Tropical Grasslands– Forrajes Tropicales* 2, 154–164.
Jones, R.M. and Brandon, N.J. (1998) Persistence and productivity of eight accessions of *Desmanthus virgatus* under a range of grazing pressures in subtropical Queensland. *Tropical Grasslands* 32, 145–152.
Jones, R.M. and Carter, E.D. (1989) Demography of pasture legumes. In: Marten, G.C., Matches, A.G., Barnes, R.F., Brougham, R.W., Clements, R.J. and Sheath, G.W. (eds) *Persistence of Forage Legumes. Proceedings of a Trilateral Workshop on Persistence of Forage Legumes*, Honolulu, Hawaii, 18–22 July 1988.
Jones, R.M. and Clem, R.L. (1997) The role of genetic resources in developing improved pastures in semi-arid and subhumid northern Australia. *Tropical Grasslands* 31, 315–319.
Jones, R.M. and Rees, M.C. (1997) Evaluation of tropical legumes on clay soils at four sites in southern inland Queensland. *Tropical Grasslands* 31, 95–106.

Jones, R.M., Bishop, H.G., Clem, R.L., Conway, M.J., Cook, B.G., Moore, K. and Pengelly, B.C. (2000) Measurements of nutritive value of a range of tropical legumes and their use in legume evaluation. *Tropical Grasslands* 34, 78–90.

Kanani, J., Lukefahr, S.D. and Stanko, R.L. (2006) Evaluation of tropical forage legumes (*Medicago sativa*, *Dolichos lablab*, *Leucaena leucocephala* and *Desmanthus bicornutus*) for growing goats. *Small Ruminant Research* 65, 1–7.

Kellman, M. (1975) Evidence for late glacial age fire in a tropical montane savanna. *Journal of Biogeography* 2, 57–63.

Kemp, E.D.S., McKenzie, R.M. and Romney, D.H. (1961) Productivity of pasture in British Honduras III. Jaragua grass. *Tropical Agriculture (Trinidad)* 38, 161–171.

Keoghan, J.M. (1980) Adaptable and productive forage legumes and grasses for more intensive small ruminant livestock systems in the Caribbean. *Tropical Animal Production* 5, 8–14.

Kerridge, P.C. and Hardy, W. (1994) *The Biology and Agronomy of Arachis*. CIAT Publication 240. CIAT, Cali, Colombia.

Kerridge, P.C. and Skerman, P.J. (1968) The distribution and growth characteristics of the native legume, *Psoralea eriantha*. *Tropical Grasslands* 2, 41–50.

Kleinschmidt, H.E. and Johnson, R.W. (1977) *Weeds of Queensland*. Government Printer, Brisbane, Australia.

Kretschmer, A.E. and Brolmann, J.B. (1984) Global ventures in *Stylosanthes* II. USA and Caribbean. In: Stace, H.M. and Edye, L.A. (eds) *The Biology and Agronomy of Stylosanthes*. Academic Press, Sydney, Australia, pp. 467–485.

Kretschmer, A.E. Jr and Pitman, W.D. (2001) Germplasm resources of tropical forage legumes. In: Sotomayor-Rios, A. and Pitman, W.D. (eds) *Tropical Forage Plants*. CRC Press, Boca Raton, Florida.

Krull, C.F. and Borlaug, N.E. (1970) The utilization of collections in plant breeding. In: Frankel, O.H. and Bennett, E. (eds) *Genetic Resources in Plants, their Exploration and Collection*. I.B.P. Handbook 11, Blackwood, Oxford, pp. 427–440.

Kulakow, P.A. (1999) Variation in Illinois bundleflower (*Desmanthus illinoensis* (Michaux) MacMillan): a potential perennial grain legume. *Euphytica* 110, 7–20.

Kulakow, P.A., Benson, L.L. and Vail, J.G. (1990) Prospects for domesticating Illinois bundleflower. In: Janick, J. and Simon, J.E. (eds) *Advances in New Crops*. Timber Press, Portland, Oregon, pp. 168–171.

Lawrence, J., Bell, L., Breton, M. (2012) Hard seed breakdown rates of six tropical ley pasture legumes. *Proceedings of the 16th Australian Society of Agronomy Conference*, UNE, October 2012 (http://www.regional.org.au/au/asa/2012/pastures/7965_lawrencejrh.htm) Accessed November 2015.

Lazier, J.R. (1978) Forage potential of selected forage species and legumes under Belizean conditions. PhD thesis, University of the West Indies, St Augustine, Trinidad.

Lazier, J.R. (1980a) Productivity of four commercial forage legumes and two native lines under clipping in Belize, C.A. *Tropical Agriculture (Trinidad)* 57, 343–351.

Lazier, J.R. (1980b) Performance of four persistent forage legumes with *Brachiaria mutica* in Belize under regular grazing using observational techniques and clipping. *Tropical Agriculture (Trinidad)* 57, 353–361.

Lazier, J.R. (1981a) Dry matter productivity of eighteen native Belizean legumes and *Codariocalyx gyroides* with Para grass (*Brachiaria mutica*) under clipping. *Tropical Agriculture (Trinidad)* 58, 221–234.

Lazier, J.R. (1981b) Performance of three persistent native Belizean legumes and *Codariocalyx gyroides* with *Brachiaria mutica* under grazing. *Tropical Agriculture (Trinidad)* 58, 235–244.

Lazier, J.R. (1981c) Effect of cutting height and frequency on dry matter production of *Codariocalyx gyroides* syn. *Desmodium gyroides*) in Belize C.A. *Tropical Grasslands* 15, 10–16.

Leng, R.A. (1986) Determining the nutritive value of forage. In: Blair, G., Ivory, D. and Evans, T. (eds) *Forages in SE Asian and South Pacific Agriculture*. ACIAR Proceedings Series No. 12. ACIAR, Canberra, pp. 111–123.

Leslie, J.K., Montgomery, V.E., Cameron, G. and Hogarth, D.H. (1987) Plant genetic constraints in tropical agriculture. *57th ANZAS Congress*, James Cook University, Townsville, Australia, August 1987, pp. 24–28.

Liu, C.J. (1997) Geographical distribution of genetic variation in *Stylosanthes scabra* revealed by RAPD analysis. *Euphytica* 98, 21–27.

Liu, C.J. and Musial, J.M. (1997) *Stylosanthes* sp. aff. *S. scabra*: a putative diploid progenitor of *Stylosanthes scabra*. *Plant Systematics and Evolution* 208, 99–105.

Loch, D.S. (2015) *Desmanthus* varieties JCU1-5 descriptors. *Plant Varieties Journal* 28, 59–63 and 159–192.

Loch, D.S. and Harvey, G.L. (est. 1997) Screening of selective post-emergence herbicides (Legumes for Clay Soils). Internal Queensland Government Report, Gympie, Australia.

Loveless, A.R. (1960) The vegetation of Antigua, West Indies. *Journal of Ecology* 48, 495–527.

Lubanga, S. (1993) The provision of pasture legumes for the Lake Victoria Crescent in Uganda. The use of climatic homologues to aid their introduction and the potential utility of ruminants in pasture establishment. Thesis, James Cook University, Townsville, Australia.

Luckow, M. (1993) Monograph of *Desmanthus* (Leguminosae-Mimosoideae) *Systematic Botany Monographs* 38, 1–166.

Ludemann, C.I., Jacobs, J.L. and Smith, K.F. (2015) The economic significance of maintaining pasture production at its peak value. *Crop and Pasture Science* 66, 205–213.

Lundell, C.L. (1940) Botany of the Maya area: miscellaneous papers, XIV. The 1936 Michigan–Carnegie botanical expedition to British Honduras. *Publications of the Carnegie Institution* 522, 1–57.

Maass, B.L. (1989) Die tropische Weideleguminose *Stylosanthes scabra* Vog. – Variability, leistungastand und Moglichkeiten zuchterischer verbesserung. PhD dissertation. Landbauforschung Volkenrode, Braunschweig, Sonderheft 97.

Maass, B.L. and Sawkins, M. (2004) History, relationships and diversity among *Stylosanthes* species of commercial significance. In: Chakraborty, S. (ed.) *High-Yielding Anthracnose-Resistant Stylosanthes for Agricultural Systems*. ACIAR, Canberra, pp. 9–26.

Maass, B.L. and t'Mannetje, L. (2002) *Stylosanthes seabrana* (*Leguminosae: Papilionoideae*): a new species from Bahia, Brazil. *Novon* 12, 497–500.

Mace, G., O'Regain, J. and Buck, S. (2015) Perennial pasture legumes: plan ahead. *Beeftalk, Queensland Country Life* 30 July, 36.

Maxted, N. and Guarino, L. (1997) Ecographic surveys. In: Maxted, N., Ford-Lloyd, B.U. and Hawkes, J.G. (eds) *Plant Genetic Conservation: The In-Situ Approach*. Chapman and Hall, London, pp. 69–87.

Maxted, N. and Guarino, L. (2006) Genetic erosion and genetic pollution of crop wild relatives. In Ford-Lloyd, B.V., Dias, S.R. and Bettencourt, E. (eds) *Genetic Erosion and Pollution Assessment Methodologies.* IPGRI, Rome, pp. 35–46.

Maxted, N., Scholten, M., Codd, R. and Ford-Lloyd, B. (1997) Creation and use of a national inventory of crop wild relatives. *Biological Conservation* 140, 142–159.

Maxted, M., Mabuza, P. and Kell, S. (2000) Ecogeographic techniques and conservation: a case study for the legume genus *Vigna* in Africa. *7th MADD International Workshop on Genetic Resources Part 1. Wild Legumes*, AFFRC and NIAR, Japan, pp. 63–90.

McCosker, T., McLean, D. and Holmes, P. (2010) *Northern Beef Situation Analysis 2009.* B.NBP.0518 Final Report. Meat and Livestock Australia, North Sydney, Australia.

McIvor, J.G. (2002) Pastures. In: McIntyre, S., McIvor, J.G. and Heard, K.M. (eds) *Managing and Conserving Grassy Woodlands*. CSIRO Publishing, Melbourne, Australia, pp. 61–78.

McKeon, G. (2006) *Living in a Variable Climate*. SoE 2006. Department of the Environment, Canberra.

McKinzie, W. (1977) *Classification of Belize Valley Soils*. Dept. Agric. Belize. Mimeo. Agricultural Library, Belize.

McLennan, S.R., Poppi, D.P. and Hendricksen, R.E. (1998) Sown pastures – limitations to increased nutrient supply. *Proceedings of a Workshop on Sown Pasture R&D*, Queensland, Brisbane, Australia, 5–7 October 1998.

McMeniman, N., Beale, I. and Murphy, G. (1986) Nutritional evaluation of south-west Queensland pastures. 1. The botanical and nutrient content of diets selected by sheep grazing on Mitchell grass and mulga/grassland associations. *Australian Journal of Agricultural Research* 37, 289–302.

McSweeney, C. and Burt, R.L. (1987) Meeting mineral requirements of grazing ruminants: variation within tropical legumes with reference to Caribbean stylo. *Proceedings of the Conference on Herbivore Nutrition Research*, 2nd International Symposium on the Nutrition of Herbivores, University of Queensland, Brisbane, Australia, pp. 211–212.

Mengistu, A. (1985) Feed resources in Ethiopia. In: Kategile, J.A., Said, A.N. and Dzowela, B.H. (eds) *Animal Feed Resources for Small-Scale Livestock Producers.* Proceedings of the Second PANESA Workshop, Nairobi, Kenya, 11–15 November (http://www.fao.org/wairdocs/ILRI/x5548E/x5548e03.htm#feed%20resources%20and%20status) Accessed November 2015.

Miller, V.J. (1960) Temperature effect on the rate of apparent photosynthesis of Seaside Bent and Bermuda qrass. *Proceedings of the American Society for Horticultural Science* 75, 700–703.

Minson, D.J. (1977) The chemical composition and nutritive value of tropical legumes. In: Skerman, P. (ed.) *Tropical Forage Legumes*. FAO, Rome, pp. 186–218.

MLA (Meat and Livestock Australia) (2010) *National Beef Production RD&E Strategy*. Meat and Livestock Occasional Report. Meat and Livestock Australia, Canberra.

MLA (Meat and Livestock Australia) (2012) Antimethanogenic bioactivity of Australian plants for grazing systems. Final report Project code: B.CCH.1012. Meat and Livestock Australia, Canberra.

MLA (Meat and Livestock Australia) (2014) Northern Australian beef fertility project: Cash cow project B. NBP.0382 Final report. Meat and Livestock Australia, Canberra.
Mohlenbrock, R.H. (1957) A revision of the genus *Stylosanthes*. *Annals of the Missouri Botanic Gardens* 44, 299–355.
Molina, A. (1975) Enumeración de las plantas de Honduras. *Ceiba* 19, 1–118.
Moore, R.M. (1967) The naturalization of alien plants in Australia. *Proceedings and Papers of the IUCN 10th Technical Meeting.* IUCN Publ. New Series No. 9, 82–97.
Muir, J.P. and Pitman, W.D. (2004) Establishment of *Desmanthus* species in existing grass stands. *Native Plants Journal* 5, 5–13.
Muir, J.P., Butler, T.J. and Ocumpaugh, W.R. (2005) The long road to developing native herbaceous summer forage legume ecotypes. In: Milne, J.A. (ed.) *Pastoral Systems in Marginal Environments*. Wageningen Academic Publications, Wageningen, The Netherlands, p. 107.
Mullen, B.F., Frank, V.E. and Date, R.A. (1998) Specificity of rhizobial strains for effective N2 fixation in the genus Leucaena. *Tropical Grasslands* 33, 110–117.
Murata, Y. and Iyama, J. (1963) Studies on the photosynthesis of forage crop 2. Influence of air temperature upon photosynthesis of some forage and grain crops. *Proceedings of the Crop Society of Japan* 31, 315–322.
Myers, L.F. (1968) New procedures for plant introduction through CSIRO, Canberra. *Plant Introduction Review* 5, 26–27.
Ngo, T. (2012) The effects of diet preference on feed intake, digestibility and nitrogen balance of sheep given Flinders grass (*Iseilema* spp.) hay and or *Desmanthus leptophyllus ad libitum.* MSc thesis. James Cook University, Townsville, Australia.
NSW Ag (2004) Desmanthus, *Desmanthus* spp. *Agnote DPI-392*. New South Wales Agriculture. DPI, Orange, Australia.
O'Keeffe, P.A., Orchard, D.J., Orchard, B.A., Piltz, J.W. and Clayton, E.H. (2015) Native Australian shrub legume species may provide an alternative feed resource for livestock. *Animal Production Science* 55, 1090–1096.
Ockerby, S. (2013) Progardes™ as a break crop in sugarcane farming systems. AgSight Opinions on Agriculture (http://www.agsight.com.au/) Accessed November 2015.
Ocumpaugh, W.R., Archer, S. and Stuth, J.W. (1996) Switchgrass recruitment from broadcast seed versus seed fed to cattle. *Journal of Range Management* 49, 368–371.
Ocumpaugh, W.R., Grichar, W.J. Jr, Hussey, M.A., Abrameit, A.H., Owens, M.K., Reed, R.L, Muir, J.P., Bade, D. and Reilley, J.L. (2004) Registration of 'BeeTAM-06' Bundleflower. *Crop Science* 44, 1860–1861.
Onim, J.F.N. and Dzowela, B.H. (n.d.) The distribution of *Sesbania* species in the PANESA region. FAO Corporate Document Repository (http://www.fao.org/wairdocs/ilri/x5491e/x5491e0a.htm) Accessed November 2015.
Orr, D. (1975) A review of *Astrebla* (Mitchell grass) pastures in Australia. *Tropical Grasslands* 9, 21–36.
Orr, D.M. and Phelps, D.G. (2013) Impacts of level of utilisation by grazing on an *Astrebla* (Mitchell grass) grassland in north-western Queensland between 1984 and 2010. 2. Plant species richness and abundance. *The Rangeland Journal* 35, 17–28.
Otsyina, R.M., Norton, B.W. and Djimde, M. (1999) Fodder trees and shrubs in arid and semi-arid livestock production systems. *Proceedings of the XVIII International Grasslands Congress, New Zealand and Australia*. Section 23 – Grasslands in Arid and Semi-Arid Regions, pp. 429–438.
Papadakis, J. (1966) *Climates of the World and their Agricultural Potentialities*. Published by the author, Buenos Aires Argentina.
Parsons, D. (1996) Protropico. University of the Yucatan.
Partridge, I. (1997) Stylos – the good, the bad and the cure. *Tropical Grassland Society of Australia, News and Views* 13, 5.
Pearce, S.C. (1979) Experimental design: R.A. Fisher and some modern rivals. *The Statistician* 28, 153–161.
Peck, G., Buck, S., Hofmann, A., Holloway, C., Johnson, B., Lawrence, D. and Paton, C. (2011) *Review of Productivity Decline in Sown Pastures.* Final report of MRC project B.NBP.0624. Meat and Livestock Australia, North Sydney, Australia.
Peck, G., Hall, T., Silcock, R., Clem, B., Buck, S. and Kedzlie, G. (2012) Persistence of pasture legumes in southern and central Queensland. *Proceedings of the 16th Australian Agronomy Conference*, 14–18 October 2012, UNE, Armidale, Australia.
Pengelly, B.C. and Conway, M.J. (1998) The evaluation of tropical legumes for use in ley pastures in central and southern Queensland. *Proceedings of the 9th Australian Agronomy Conference*, Wagga Wagga, Australia, pp. 163–166.

Pengelly, B.C. and Conway, M.J. (2000) Pastures on cropping soils: which tropical pasture legumes to use? *Tropical Grasslands* 34, 162–168.

Pengelly, B.C. and Eagles, D.A. (1995) Geographical distribution and diversity in a collection of the tropical legume *Macroptilium gracile* (Poeppiga ex Bentham) urban. *Australian Journal of Agricultural Research* 46, 569–580.

Pengelly, B.C. and Liu, C.J. (2001) Genetic relationships and variation in the tropical mimisoid legume *Desmanthus* assessed by random amplified polymorphic DNA. *Genetic Resources and Crop Evolution* 48, 91–99.

Pengelly, B.C. and Staples, I.B. (1996) *Development of New Legumes and Grasses for the Cattle Industry of Northern Australia.* Final report of MRC projects CS.054/185 and DAQ.053/081, Meat and Livestock Australia, North Sydney, Australia.

Polhill, R.M. and Raven, P.H. (eds) (1971) *Flora of Tropical East Africa. Leguinosae sub-family Papilioniodeae*. Royal Botanic Gardens, Kew, UK.

Poppi, D.P. and McLennan, S.R. (2010) Nutritional research to meet future challenges. *Animal Production Science* 50, 329–338.

Prideaux, G.J., Roberts, R.G., Megirian, D., Westaway, K.E., Hellstrom, J.C. and Olley, J.M. (2007) Mammalian responses to Pleistocene climate change in southeastern Australia. *Geology* 35, 33–36.

Queensland Government (2011) *Prospects for Queensland's Primary Industries*. Prospects Report, September 2011. Queensland Government, Brisbane, Australia.

Ramesh, C.R., Melkania, M.P., Desale, J.S., Sonona, A.H., Manjara, M.R., Patel, J.R., Yadavendra, D.P., Patel, C.C., Nair, M.A., Lakshmi, S. *et al.* (2004) Regional evaluation of *Stylosanthes* germplasm in India. In: Chakraborty, S. (ed.) *High Yielding Anthracnose-Resistant Stylosanthes for Agricultural Systems*. ACIAR, Canberra, pp. 135–142.

Ramírez-Restrepo, C.A., Xixi, L., Durmic, Z., Vercoe, P., Gardiner, C., O'Neill, C., Coates, D. and Charmley, E. (2011) Assessment of tropical legumes to reduce methane emissions from pastoral systems using in vitro fermentation and near infrared reflectance spectroscopy methodologies. *Greenhouse 2011 Conference*, Cairns, Australia, 4–8 April 2011.

Rangel, J.H. de A. (2005) Agroecology of tropical legumes with especial reference to the *Desmanthus virgatus* genepool. PhD thesis. James Cook University, Townsville, Australia.

Rangel, J.H. de A. and Gardiner, C.P. (1996) Effect of fire and heat of seed germination of *Desmanthus virgatus* accessions. *Proceedings of the International Symposium on the Future of Tropical Savannas: An Australian Perspective*. Poster papers. Townsville, 1994.

Rangel, J.H. de A. and Gardiner, C.P. (2009) Stimulation of wool growth by *Desmanthus* spp. as a supplement to a diet of Mitchell grass hay. *Tropical Grasslands* 43, 106–111.

Rangel, J.H. de A., Schlink, A.C. and Gardiner, C.P. (1996) Stimulation of wool growth by *Desmanthus virgatus*: a legume for tropical savannahs. *Proceedings of the International Symposium on the Future of Tropical Savannahs: An Australian Perspective*. Poster paper. Townsville, 1994.

Rangel, J.H. de A., Gardiner, C.P. and Burt, R.L. (2015) Dormancy releasing mechanisms in soil seed banks of *Desmanthus* genotypes. *Revista Caatinga* 28, 90–99.

Rao, I., Peters, M., Castro, A., Schultze-Kraft, R., White, D., Fisher, M. *et al.* (2015) LivestockPlus: the sustainable intensification of forage-based agricultural systems to improve livelihoods and ecosystem services in the tropics. *Tropical Grasslands – Forrajes Tropicales* 3, 59–82.

Reichard, S.H. (2004) Conflicting values and common goals: codes of conduct to reduce the threat of invasive species. *Weed Technology* 18, 1503–1507.

Reichard, S.H. and White, P. (2007) Horticulture as a pathway of invasive plant introductions in the United States. *BioScience* 51, 103–113.

Reid, R. (1983) Pasture plant collecting in Mexico with emphasis on legumes for dry regions. *Australian Plant Introduction Review* 15, 1–11.

Reid, R., Ryan, D.M. and Burt, R.L. (1976) Exploration for, and utilization of, collections of tropical pasture legumes II. The Papadakis system of climatic classification as applied to testing areas in northern Australia. *Agro-Ecosystems* 2, 309–318.

Reid, R. and Sinclair, D.F. (1980) An evaluation of a collection of *Clitoria ternatea* for forage and grain production. *Genetic Resources Communication No. 1*. CSIRO Division of Tropical Crops and Pastures, Brisbane, Australia.

Rejmanek, M., Richardson, D.M., Higgins, S.I., Pitcairn, M. and Gotkepp, E. (2005) Ecology of invasive plants: state of the art. In: Mooney, H.A., McNeely, J.A., Neville, I., Schei, P.J. and Waage, J.K. (eds) *Invasive Alien Species: A New Synthesis*. Island Press, Washington, DC, pp. 104–161.

Robertson, F.A., Myers, R.J.K. and Saffigna, P.G. (1997) Nitrogen cycling in brigalow clay soils under pasture and cropping. *Australian Journal of Soil Research* 35, 1323–1339.
Robinson, P.J. and Megarrity, R.G. (1975) Characterization of *Stylosanthes* introductions by using seed protein patterns. *Australian Journal of Agricultural Research* 26, 467–479.
Robinson, P.J., Date, R.A. and Megarrity, R.G. (1976) Nodulation of *Stylosanthes guyanensis*: prediction of Rhizobium effectiveness response with seed enzyme patterns. *Australian Journal of Agricultural Research* 27, 381–389.
Robinson, P.J., Burt, R.L. and Williams, W.T. (1980) Network analysis of genetic resource data: II The use of isozyme data in elucidating geographical relationships. *Agro-Ecosystems* 6, 111–118.
Romney, D.H. (1960) Productivity of pasture in British Honduras I. Natural pasture. *Tropical Agriculture (Trinidad)* 37, 135–142.
Romney, D.H. (1961) Productivity of pasture in British Honduras II. Pangola pastures. *Tropical Agriculture (Trinidad)* 38, 39–47.
Rose-Innis, R. (1973) Grass productivity trials in Belize 1968–70. *Foreign and Commonwealth Office Overseas Development Administration Misc. Rept. 172.*
Rotar, P.P. (1983) Legumes in action. In: Burt, R.L., Rotar, P.P., Walker, J.L. and Silvey, M.W. (eds) *The Role of Centrosema, Desmodium and Stylosanthes in Improving Tropical Pastures.* Westview Press, Boulder, Colorado, pp. 21–22.
Russi, L., Cocks, P.S. and Roberts, E.H. (1992) The fate of legume seeds eaten by sheep from a Mediterranean grassland. *Journal of Applied Ecology* 29, 772–778.
Sanchez, P.A., Cheryl, A.P. and Buol, S.W. (2003) Fertility capability soil classification: a tool to help assess soil quality in the tropics. *Geoderma* 114, 157–185.
Sanchez, P.A., Couto, W. and Buol, S.W. (1982) The soil fertility classification system: interpretation, applicability and modification. *Geoderma* 27, 283–309.
Sarwatt, S.V., Mkiwa, F.E.J., Lwoga, A.B. and Dzowela, B.H. (n.d.) Nutritive value of *Crotalaria ochroleuca:* II the effect of supplementation on feed utilization and performance of growing sheep.
Sawkins, M.C., Maxted, N. Jones, P.G., Smith, R. and Guarino, L. (1999) Predicting species distributions using environmental data: case studies using *Stylosanthes* Sw. In: Greene, S.L. and Guarino, L. (eds) *Linking Genetic Resources and Geography: Emerging Strategies for Conserving and Using Crop Biodiversity*. Crop Science Society of America Special Publication 27, 87–99.
Scattini, W.J. and Bartholomew, R.B. (1998 unpub.) Review of improved pasture, ley pasture and forage crop research and development in the Darling Downs and Maranoa. Internal Report of the Department of Primary Industries Extensive Livestock Subprogram, Brisbane, Australia.
Schlink, A.C. and Burt, R.L. (1993) Assessment of the chemical composition of selected tropical legume seeds as animal feed. *Tropical Agriculture* 70, 169–173.
Schultze-Kraft, R., Reid, R., Williams, R.J. and Corodin, L. (1984) The existing *Stylosanthes* collection. In: Stace, H.M. and Edye, L.A. (eds) *The Biology and Agronomy of Stylosanthes.* Academic Press, Sydney, Australia, pp. 125–145.
Shelton, M. (2014) *Breeding a Psyllid Resistant Leucaena Hybrid for Northern Australia – phase 2*. Final report of MLA project B.NBP.0610. Meat and Livestock Australia, North Sydney, Australia.
Shelton, M. and Dalzell, S. (2007) Production, economic and environmental benefits of leucaena pastures. *Tropical Grasslands* 41, 174–190.
Simao Neto, M. and Jones, R.M. (1987) Recovery of pasture seed ingested by ruminants 2. Digestion of seed in sacco and in vitro. *Australian Journal of Experimental Agriculture* 27, 247–251.
Simon, B.K. (1980) *A Check-list of Queensland Grasses*. Technical Bulletin No. 5. Department of Primary Industries, Brisbane, Australia.
Skerman, P.J. (1957) 'Bullaman lucerne' – *Psoralea eriantha* Benth. – a plant worth watching. *Journal of the Australian Institute of Agricultural Science* 23, 373.
Smartt, J. (1976) Comparative evolution of pulse crops. *Euphytica* 25, 139–143.
Smartt, J. (1990) *Grain Legumes: Evolution and Genetic Resources*. Cambridge University Press, Cambridge.
Snaydon, R.W. (1978) Genetic changes in pasture populations. In: Wilson, J.R. (ed.) *Plant Relations in Pastures*. CSIRO, Melbourne, Australia, pp. 253–269.
Snook, L. (1968) Comments on the potential value of the legumes native to British Honduras. Dept. Agric. Belize. Memo to CAO 14/4/68. Agricultural Library, Belize.
Spurgeon, D. (1977) Grow your own grass. *Nature* 266, 207.
Stace, H.C.T., Hubble, G.D., Brewer, R., Northcote, K.H., Sleeman, J.R., Mulcahy, M.J. and Hallsworth, E.G. (1968) *A Handbook of Australian Soils*. Rellim, Glenside, Australia.

Staples, I.B., Reid, R. and Wilson, G.P.M. (1985) Plant introduction for specific needs in northern Australia. *Proceedings of the 3rd Australia Conference on Tropical Pastures*, 8–12 July 1985, Tropical Grassland Society.

Steele, W.M. and Mehra, K.L. (1980) Structure, evolution and adaptation to farming systems and environments in *Vigna*. In: Summerfield, R.J. and Bunting, A.H. (eds) *Advances in Legume Science.* Royal Botanic Gardens, Kew, UK, pp. 393–404.

Stehle, A. (1956) *Survey of Forage Crops in the Caribbean*. Caribbean Commission Central Secretariat, Port of Spain, Trinidad.

Sukkasame, P. and Phaikaew, C. (1998) Utilisation of *Desmanthus virgatus* as protein supplement for fattening cattle in Southern Thailand. In: de la Viña, A.C. and Moog, F.A. (eds) *Integrated Crop–Livestock Production Systems and Fodder Trees*. 6th Meeting of the Regional Working Group on Grazing and Feed Resources for Southeast Asia, Legaspi City, Philippines. FAO, Rome, pp. 157–159.

Summerfield, R.J. and Bunting, A.H. (1980) *Advances in Legume Science*. Volume 1 of the Proceedings of the International Legume Conference, Kew, pp. 1–668.

Sykes, J. (1997) Priorities for new ley legumes for the tropical and subtropical grains industries. *Tropical Grasslands* 31, 273–275.

t'Mannetje, L. and Jones, R.M. (1992) *Plant Resources of South-East Asia: 4. Forages*. Backhuys Scientific Publishers, Wageningen, The Netherlands.

Takahashi, M. and Ripperton, J.C. (1949) Koa haole (*Leucaena glauca*); its establishment, culture, and utilization as a forage crop. *Hawaii Agricultural Experimental Station Bulletin* 100, 56.

Teitzel, J.K. and Middleton, C.H. (1983) The wet tropics: the South Johnstone story. In: Burt, R.L., Rotar, P.P., Walker, J.L. and Silvey, M.W. (eds) *The Role of Centrosema, Desmodium and Stylosanthes in Improving Tropical Pastures*. Westview Press, Boulder, Colorado, pp. 29–42.

Thomas, D. and Grof, B. (1986) Some pasture species for the tropical savannas of South America. 1. Species of *Stylosanthes*. *Herbage Abstracts* 10, 445.

Tothill, J.C. and Gillies, C. (1992) *The Pasture Lands of Northern Australia*. Tropical Grassland Society of Australia Occasional Publication no. 5. Cranbrook Press, Toowoomba, Australia.

Turner, B.L. (1950) Mexican species of *Desmanthus* (Leguminosae). *American Midland Naturalist* 46, 82–92.

Turner, B.L. (1959) *The Legumes of Texas*. University of Texas Press, Austin, Texas.

UNAM (1970) *Cartas de Climas*. Secreataria de la Presidencia, Direccion de Planeacion, Talles Graficos de la Nacion, Mexico.

USDA (1975) *Soil Taxonomy*. Soil Conservation Service, USDA Agriculture Handbook no. 43.

Van Klinken, R.D., Flack, L.K. and Pettit, W. (2006) Wet season dormancy release in seedbanks of a tropical leguminous shrub is determined by wet heat. *Annals of Botany* 98, 875–883.

Vandermeulen, S., Ramírez-Restrepo, C.A., Singh, S., Kinley, R., Gardiner, C.P., Holtum, J., Bindelle, J. and Hannah, I. (2015) Intra-cultivar potential of *Desmanthus* spp. as a greenhouse gas mitigation strategy for tropical livestock pastoral systems. *TropAg 2015 Tropical Agriculture Conference,* Brisbane, Australia, 16–18 November 2015.

Viets, F.G. Jr and Lindsay, W.L. (1973) Testing soils for zinc, copper, manganese and iron. In: Walsh, L.M. and Beaton, J.D. (eds) *Soil Testing and Plant Analysis.* Science Society of America, Madison, Wisconsin, pp. 153–172.

Waghorn, G.C., Tavendale, M.H. and Woodfield, D.R. (2002) Methanogenesis from forages fed to sheep. *Proceedings of the New Zealand Grassland Association* 64, 167–171.

Walker, B. and Weston, E.J. (1990) Pasture development in Queensland – a success story. *Tropical Grasslands* 24, 257–268.

Walker, B., Baker, J., Becker, M., Brunkhorst, R., Heatley, D., Simms, J., Skerman, D.S. and Walsh, S. (1997) Sown pasture priorities for the subtropical and tropical beef industry. *Tropical Grasslands* 31, 266–272.

Walker, J.W., Johnson, J.L. and Taylor, C.A. Jr (2005) Challenges and opportunities for sustainable rangeland pastoral systems in the Edwards Plateau of Texas. In: Milne, J.A. (ed.) *Pastoral Systems in Marginal Environments. Proceedings of a Satellite Workshop of the XXth International Grasslands Congress*, Glasgow, Scotland, Wageningen Academic Publications, Wageningen, pp. 71–80.

Walker, S.H. (1973) *Summary of Climatic Records for Belize*. Foreign Commonwealth Office, Overseas Development Administration, Land Resources Division. Supplemental report no. 3.

Wang, C.M. and Chen, C.H. (2012) Two newly naturalised plants of the Leguminosae in Taiwan: *Macroptilium bracteatum* (Nees & Mart.) Maréchal & Baudet and *Cyamopsis tetragonoloba* (L.) Toubert. *Collection and Research* 25, 17–23.

Weston, E.J. and Turner, E.J. (1987) Our grazing lands are a precious resource. *Queensland Agricultural Journal* January–February 1987. QAJ reprint QA87006.
Weston, E.J., Doughton, J.A., Dalal, R.C., Strong, W.M., Thomas, G.A., Lehane, K.J., Cooper, J.C., King, A.J. and Holmes, C.J. (2000) Managing long-term fertility of cropping lands with ley pastures in southern Queensland. *Tropical Grasslands* 34, 169–176.
Whitbread, A.M. and Pengelly, B.C. (eds) (2004) Tropical legumes for sustainable farming systems in southern Africa and Australia. *ACIAR Proceedings No. 115.*
Whitbread, A.M., Pengelly, B.C. and Smith, B.R. (2005) An evaluation of three tropical ley legumes for use in mixed farming systems on clay soils in southern inland Queensland, Australia. *Tropical Grasslands* 39, 9–21.
White, M.E. (1986) *The Greening of Gondwana*. Reed Books Pty Ltd, Sydney, Australia.
White, M.E. (1994) *After the Greening: The Browning of Australia*. Kangaroo Press, Sydney, Australia.
Whitehead, D.C. (1966) Nutrient minerals in grassland herbage. CAB mimeo Publication 1/1966.
Whiteman, P.C. (1980) *Tropical Pasture Science*. Oxford University Press, Oxford.
Williams, R.J., Burt, R.L. and Strickland, R.W. (1976) Plant introduction. In: Shaw, N.H. and Bryan, W.W. (eds) *Tropical Pasture Research. Principles and Methods*. CAB Bulletin No. 51, pp. 77–100.
Williams, R.J., Reid, R., Schultze-Kraft, R., Sousa Costa, N.M. and Thomas, B.D. (1984) Natural distribution of *Stylosanthes.* In: Stace, H.M. and Edye, L.A. (eds) *The Biology and Agronomy of Stylosanthes*. Academic Press, Sydney, Australia, pp. 73–101.
Williams, W.T. and Burt, R.L. (1982) A re-appraisal of Hartley's agrostological index. *Journal of Applied Ecology* 19, 159–166.
Williams, W.T. and Burt, R.L. (1983) A multidisciplinary approach to tropical pasture improvement. In: Burt, R.L., Rotar, B.P., Walker, J.L. and Silvey, M.W. (eds) *The Role of Centrosema, Desmodium and Stylosanthes in Improving Tropical Pastures*. Westview Press, Boulder, Colorado, pp. 257–287.
Williams, W.T., Burt, R.L., Pengelly, B.C. and Robinson, P.J. (1980) Network analysis of genetic resource data: I Geographical relationships. *Agro-Ecosystems* 6, 99–109.
Wright, A.C.S. (1967) *The Soil Reconnaissance of the Yucatan Peninsula, Mexico*. Final report. Plan Chiapas. FAO Rome.
Wright, A.C.S., Romney, D.H., Arbuckle, R.H. and Vial, V.E. (1959) *Land in British Honduras*. Report of the British Honduras Land Use Survey Team. HMSO, London.
Wynia, R. (2008) Illinois Bundleflower *Desmanthus illinoensis* (Mich.) Macmill. ex B.L. Rob. & Fernald. USDA NRCS Plant Guide (http://plants.usda.gov/core/profile?symbol=DEIL) Accessed November 2015.

Index

Page numbers in **bold** type refer to figures and tables.